62.00 80E

UNSTABLE CURRENT SYSTEMS
AND PLASMA INSTABILITIES IN ASTROPHYSICS

INTERNATIONAL ASTRONOMICAL UNION
UNION ASTRONOMIQUE INTERNATIONALE

UNSTABLE CURRENT SYSTEMS AND PLASMA INSTABILITIES IN ASTROPHYSICS

PROCEEDINGS OF THE 107th SYMPOSIUM
OF THE INTERNATIONAL ASTRONOMICAL UNION
HELD IN COLLEGE PARK, MARYLAND, U.S.A.,
AUGUST 8–11, 1983

EDITED BY

MUKUL R. KUNDU

and

GORDON D. HOLMAN

*Astronomy Program, University of Maryland,
College Park, Maryland, U.S.A.*

D. REIDEL PUBLISHING COMPANY

A MEMBER OF THE KLUWER ACADEMIC PUBLISHERS GROUP

DORDRECHT / BOSTON / LANCASTER

Library of Congress Cataloging in Publication Data

Main entry under title:

Unstable current systems and plasma instabilities in astrophysics.

"Symposium no. 107 held in College Park, Maryland, U.S.A., August 8–11, 1983."
At head of title: International Astronomical Union.
Includes index.
1. Astrophysics–Congresses. 2. Space plasmas–Congresses.
3. Plasma instabilities–Congresses. 4. Syrovatskiı̆, S. I. (Sergei Ivanovich), 1925–1979. I. Kundu, Mukul Ranjan, 1930–
II. Holman, Gordon D. III. International Astronomical Union.
IV. Title: Unstable current systems and plasma instabilities in astrophysics.
QB460.U57   1985   523.01   84-24899
ISBN 90-277-1886-5
ISBN 90-277-1887-3 (pbk.)

*Published on behalf of
the International Astronomical Union
by
D. Reidel Publishing Company, P.O. Box 17, 3300 AA Dordrecht, Holland*

*All Rights Reserved
© 1985 by the International Astronomical Union*

*Sold and distributed in the U.S.A. and Canada
by Kluwer Academic Publishers,
190 Old Derby Street, Hingham, MA 02043, U.S.A.*

*In all other countries, sold and distributed
by Kluwer Academic Publishers Group,
P.O. Box 322, 3300 AH Dordrecht, Holland*

No part of the material protected by this copyright notice may be reproduced or utilized in any form or by any means, electronic or mechanical, including photocopying, recording or by any information storage and retrieval system, without written permission from the publisher.

Printed in The Netherlands

TABLE OF CONTENTS

FOREWORD xi

LIST OF PARTICIPANTS xix

M. R. KUNDU: Introductory Remarks 1

SESSION I

B. U. Ö. SONNERUP: Magnetic Field Reconnection in Cosmic Plasmas (Invited Review). 5

C. T. RUSSELL: Patchy Reconnection and Magnetic Ropes in Astrophysical Plasmas (Invited Review) 25

R. C. ELPHIC: Magnetic Flux Ropes of Venus: A Paradigm for Helical Magnetic Structures in Astrophysical Systems (Poster Paper) 43

R. L. STENZEL and W. GEKELMAN: Laboratory Experiments On Current Sheet Disruptions, Double Layers, Turbulence and Reconnection (Invited Review) 47

J. F. DRAKE: Reconnection in Sheared Magnetic Fields in Space and Astrophysics (Invited Review) 61

SESSION II

C. A. NORMAN: Formation, Equilibrium and Stability of Jets (Invited Review) 85

J. HEYVAERTS: Energy Dissipation Mechanisms in the Solar Corona (Invited Review) 95

R. A. SMITH: On the Role of Double Layers in Astrophysical Plasmas (Invited Contribution) 113

W. LOTKO: Particle Energization in Stochastic Double Layers 125

G. BENFORD: Links Between Jet Instabilities, Radiation and Propagation in Astrophysics (Invited Contribution) 131

J. A. IONSON: A Unified Theory of Coronal Heating 139

SESSION III

A. BRATENAHL and P. J. BAUM: Laboratory Experiments on Reconnection in Current Sheets (Invited Review) 147

J. BIRN: Computer Simulation of Reconnection in Planetary Magnetospheres (Invited Review) 167

M. R. KUNDU: Observational Evidence for Magnetic Reconnection in Microwave Solar Bursts 185

G. D. HOLMAN: Acceleration of Runaway Electrons and Joule Heating in Solar Flares 191

T. TAJIMA, F. BRUNEL, J.-I. SAKAI, L. VLAHOS, and M. R. KUNDU: The Coalescence Instability and Solar Flares (Invited Contribution) 197

T. SATO: 3D Simulation of Externally Driven Reconnection 211

J. J. ALY: Opening of the Magnetic Field Lines in a Fast Rotating Magnetosphere, with an Application to Jupiter 217

J. J. ALY: Quasi-Static Evolution of Force-free Magnetic Fields and a Model for Two-Ribbon Solar Flares (Poster Paper) 221

B. LOKANADHAM, P. K. SUBRAMANIAN, M. S. REDDY, B. M. REDDY, and D. R. LAKSHMI: Solar Microwave Emission in Active Regions (Poster Paper) 225

SESSION IV

E. R. PRIEST: Current Sheets in Solar Flares (Invited Contribution) 233

D. J. MULLAN: Energy Dissipation Mechanisms in Flare Stars (Invited Review) 245

G. VAN HOVEN: A Unified Treatment of the Filament and Flare Instabilities (Invited Contribution) 263

R. S. STEINOLFSON and G. VAN HOVEN: Nonlinear Evolution of the Resistive Tearing Mode (Poster Paper) 273

P. BATISTONI, G. EINAUDI, F. RUBINI, C. CHIUDERI, and G. TORRICELLI: Resistive Instabilities in Astrophysical Conditions: A Critical Discussion (Poster Paper) 277

Y. UCHIDA and T. SAKURAI: Magnetodynamical Processes in Interacting Magnetospheres in RS CVn Binaries 281

Y. UCHIDA and K. SHIBATA: 'Sweeping Pinch' Mechanism and the Acceleration of Jets in Astrophysics (Poster Paper) 287

P. A. STURROCK, P. KAUFMANN and D. F. SMITH: Energy Release in Solar Flares 293

V. KRISHAN and M. R. KUNDU: An Interpretation of Millisecond Time Variation in Hard X-ray Solar Flares ... 299

A. T. Y. LUI: Observations of the Earth's Cross-Tail Current Sheet and Their Implications ... 303

M. ANDRÉ: Plasma Instabilities Generated by Streaming Particles ... 309

SESSION V

J. D. HUBA: Anomalous Transport in Current Sheets (Invited Review) ... 315

C. T. DUM: Anomalous Transport Induced by Field Aligned Currents and its Relation to Electromagnetic Coupling (Invited Review) ... 329

D. G. WENTZEL: Self-Confined Cosmic Rays (Invited Review) ... 341

S. R. SPANGLER and J. P. SHEERIN: Nonlinear Effects and the Limitation of Electron Streaming Instabilities in Astrophysics ... 355

A. H. NELSON: Cosmic Rays and the Parker Instability (Poster Paper) ... 361

L. NOCERA, B. LEROY and E. R. PRIEST: Phase Mixing of Propagating Alfvén Waves ... 365

S. MIGLIUOLO: Velocity Shear Instabilities in the Anisotropic Solar Wind (Poster Paper) ... 371

T. P. RAY and A. I. ERSHKOVICH: Kelvin-Helmholtz Instabilities in a Magnetised Compressible Plasma (Sheared Flow) ... 375

SESSION VI

A. HASEGAWA: Plasma Heating by Alfvén Waves - Kinetic Properties of Magnetohydrodynamic Disturbances (Invited Review) ... 381

S. M. MAHAJAN: Kinetic Theory of Alfven Wave Heating (Poster Paper) ... 391

A. FERRARI: MHD Equilibrium and Instabilities in Extragalactic Jets (Invited Review) ... 393

R. N. HENRIKSEN: Bursting Particle Acceleration in Radio Jets (Invited Contribution) ... 413

D. EICHLER: Instabilities in Astrophysical Jets: Disease and Cure (Invited Contribution) ... 425

J. A. EILEK: Current Systems in Radio Jets ... 433

P. E. HARDEE: Is the Jet in M87 Magnetically Confined? ... 439

SESSION VII

C. S. LIU: Laboratory Plasma Processes of Astrophysical Interest (Invited Review) 447

F. V. CORONITI: Accretion Disk Electrodynamics (Invited Review) .. 453

G. LAKE and R. E. PUDRITZ: Ultra-High Energy Cosmic Ray Production by Current Disruption in Active Galactic Nuclei (Poster Paper) 471

M. J. KESKINEN: Theory of Strongly Turbulent Two-Dimensional Cross Field Convection of Current Carrying Space Plasmas (Abstract) 475

D. GILDEN and T. TAJIMA: Magnetic Field Reconnection in Differentially Rotating Accretion Disks 477

A. RAY: Coronal Arcades in the Sun and Their Hydromagnetic Stability 483

S. T. WU, J. F. WANG, and E. TANDBERG-HANSSEN: MHD Analysis of the Evolution of Solar Magnetic Fields and Currents in an Active Region (Poster Paper) 487

G. BERTIN and B. COPPI: Bending Waves on Current Sheets 491

K. TSINGANOS, A. FERRARI, and R. ROSNER: Quasi-Two-Dimensional Cosmic Jets 497

SESSION VIII

M. TANAKA and K. PAPADOPOULOS: Creation of High Energy Electron Tails by Lower-Hybrid Waves and its Relevance to Type II and III Bursts 505

D. F. SMITH: Current Status of the Dissipative Thermal Model for Solar Hard X-ray Bursts (Poster Paper) 509

J. SAKAI and R. SUGIHARA: Non-Stochastic Acceleration of Protons in the Magnetic Neutral Sheet 513

D. S. SPICER and R. N. SUDAN: Beam Return Current Systems in Solar Flares 519

L. VLAHOS and H. L. ROWLAND: Collisionless Effects on Beam-Return Current Systems in Solar Flares 521

SESSION IX

V. M. VASYLIUNAS: Summary of Conference (Invited Summary) 529

TABLE OF CONTENTS

C. F. KENNEL, J. ARONS, R. BLANDFORD, F. CORONITI, M. ISRAEL, L. LANZEROTTI, A. LIGHTMAN, K. PAPADOPOULOS, R. ROSNER, and F. SCARF: Perspectives on Space and Astrophysical Plasma Physics 537

CONTRIBUTIONS FROM THE U.S.S.R.

G. Ya. SMOLKOV: Spatio-Temporal Features of the Development of Microwave Emission of Active Regions and Flares 555

V. A. MAZUR and A. V. STEPANOV: Concerning the Dynamics of Energetic Protons in Coronal Magnetic Loops: Dispersion Effects of Alfvén Waves 559

SUBJECT INDEX 561

FOREWORD

In the past decade rapid development has occurred in the fields of astrophysics, space science, and plasma physics. The new generation of space observations has led to an increasing requirement for a thorough understanding of processes that occur in magnetized plasmas. The realization that essentially the same plasma processes must be understood for many problems related to astrophysical, space, and man-made plasmas has led to a greater need for interdisciplinary meetings involving experts from these diverse fields. This Symposium, "Unstable Current Systems and Plasma Instabilities in Astrophysics", represents the first meeting within the International Astronomical Union to bring together scientists from these disciplines. It was jointly sponsored by IAU Commissions 40, Radio Astronomy, 12, Solar Radiation and 10, Solar Activity. It was co-sponsored by the Scientific Committee on Solar-Terrestrial Physics (SCOSTEP) and by the Committee on Space Research (COSPAR). The Symposium, No. 107, was held at the University of Maryland in College Park, Maryland, August 8-11, 1983.

The Scientific Organizing Committee of the Symposium consisted of M. R. Kundu (Chairman), A. Bridle, A. A. Galeev, J. Heyvaerts, D. B. Melrose, K. Papadopoulos, E. R. Priest, B. V. Somov, D. S. Spicer, S. K. Trehan, Y. Uchida, and V. Vasyliunas. The topics and speakers were chosen in order to emphasize the common physics underlying a diversity of astrophysical topics, and to present the most recent work on these topics and the relevant physics. Physical processes such as magnetic reconnection, the development of plasma microturbulence, plasma wave acceleration and heating of particles, and the onset of MHD instability were common themes throughout most of the symposiumm papers and discussion. Papers on topics as diverse as jets from the nuclei of active galaxies, solar flares, and planetary magnetospheres were presented and discussed by the symposium participants. These papers and most of the subsequent discussion are reproduced in this volume.

The meeting was hosted by the University of Maryland Astronomy Program. The Local Organizing Committee consisted of D. G. Wentzel (Chairman), G. D. Holman, J. A. Ionson, M. R. Kundu, J. D. Trasco, and L. Vlahos. The committee was assisted by Betty Stevenson. The International Astronomical Union, the National Aeroauntics and Space Administration and the National Science Foundation provided grants for the support of the meeting.

We have been helped considerably by several people during the editing of these proceedings for publication. The discussion remarks were recorded by S. F. Fung and T. N. LaRosa. Betty Stevenson typed the discussions as well as several of the contributed papers. We are grateful to each of these people for their help.

This symposium was dedicated to the memory of the late Professor Sergei Ivanovitch Syrovatskii who passed away on September 26, 1979. He was 54 years old at the time of his death. His great contributions to the physics of solar and other astrophysical plasmas will be remembered by the Scientific community for a long time. A short bibliographical sketch written by one of Syrovatskii's colleagues, Dr. Volodya Dogiel of the P.N. Lebedev Physical Institute, appears elsewhere in this volume.

Mukul R. Kundu
Gordon D. Holman

SERGEI I. SYROVATSKII
1925 - 1979

DR. S. I. SYROVATSKII

Sergei Ivanovitch Syrovatskii was born on March 2, 1925 in the city of Bereznegovatoye. In 1941 when the second World War broke out, sixteen years old Syrovatskii went to war. He was awarded two orders and many medals for his valiant conduct during the war.

After graduation from the Physics faculty of the Moscow State University in 1951, Syrovatskii started post graduate studies in the P.N. Lebedev Physical Institute, and studied under the direction of Prof. S. Z. Belenky. Since that time he has worked in the Theoretical Department of the Institute. The first works of Syrovatskii were devoted to magnetohydrodynamics, in particular to different types of surface discontinuities and shock waves. "The Theory of Discontinuities in Magnetohydrodynamics" was the subject of his dissertation in 1954. In 1957 his results on this work were published and were among the first fundamental researches on magnetohydrodynamics carried out in the USSR.

In 1959 Syrovatskii published in the Soviet Astronomical Journal an analysis of relativisitic electron distribution in the Galaxy and of the determination of the Galactic background radio emission. This paper led to the beginning of a long collaboration between Syrovatskii and Ginzburg in radioastronomy.

It was a great achievement in cosmic ray physics when a phenomenological diffusion model of cosmic ray propagation was developed by Ginzburg and Syrovatskii. In the 1960's Syrovatskii along with Ginzburg and Sazonov solved many problems associated with magnetobremsstrahlung radiation, and elaborated many of its characteristic features. Using the first measurements of relativistic galactic cosmic ray electron flux and of the Galactic background radio intensity, Ginzburg and Syrovatskii investigated in 1961 the question of origin of relativistic electrons in the Galaxy. They showed that direct acceleration of electrons in the Galactic sources was important for radio emission rather than their origin as secondary products due to nuclear interactions in cosmic rays. In 1963, Ginzburg and Syrovatskii published the famous monograph "The Origin of Cosmic Rays".

In mid 1960's, Syrovatskii started studies of plasma behavior in strong magnetic fields. He showed that in the neighborhood of the zero point of magnetic fields in a plasma, there can develop a current sheet with a width much larger than the thickness. These studies were used for modelling plasma behaviour in the solar corona, where the magnetic field energy density is much higher than the plasma energy density. The processes of formation, stability and disruption of current sheets has been studied for many years theoretically, numerically and under laboratory conditions by Syrovatskii and his group. A very important property of current sheets is their stability. In the typical solar corona their lifetime is several hours. This property enables a very large energy to be accumulated in the solar atmosphere in the form of magnetic energy of a current sheet. Current sheet disruption causes release of the accumulated energy in different forms, which is a solar flare in the Syrovatskii's model.

Over the years, the problems studied by Syrovatskii's group involved many new branches of solar physics. These problems include hydrodynamics of the convective zone of the Sun, solar dynamo, particle acceleration in solar flares and secondary processes on the Sun. Syrovatskii had more than 200 scientific publications to his credit.

Syrovatskii became a Professor at the Lebedev Phyisical Institute in 1979. On September 26, 1979 he died of a heart attack.

In 1982, Sergei Syrovatskii (posthumously) and his collaborators were awarded the State Prize of the USSR for a series of papers on "Dynamics of Current Sheets and Solar Activity".

- Volodya Dogiel
P.N. Lebedev Physical Institute
Moscow, U.S.S.R.

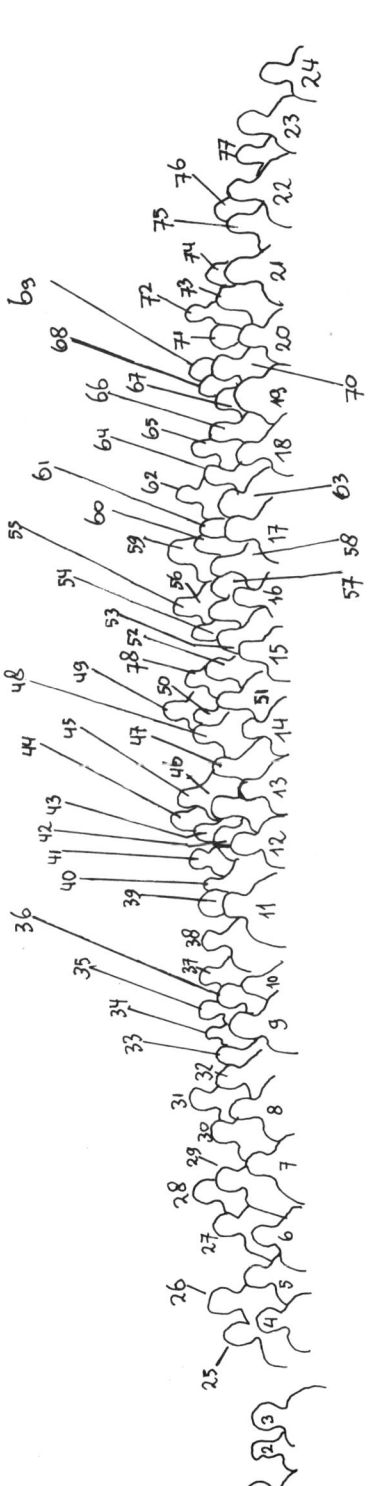

1. Tanaka   2. Lui   3. Elphic   4. Sato   5. Goossen   6. Trehan   7. Mahajan   8. Lotko   9. Smith
10. Krishan   11. Waldron   12. Underhill   13. Degaonkar   14. Uchida   15. Priest   16. Kundu
17. Stenzel   18. Hardee   19. Eilek   20. Benford   21. Davila   22. Ionson   23. Smith   24. Wu
25. Gilden   26. Kaastra   27. Wentzel   28. Van Hoven   29. Shevgaonkar   30. Ray   31. Steinolfson
32. Birn   33. Cheng   34. Coroniti   35. Nocera   36. Papadopoulos   37. Eichler   38. Vasliunas
39. Karpen   40. Sonnerup   41. Ray   42. Krautter   43. Goodrich   44. Huba   45. Bratenahl
46. Henriksen   47. Mullan   48. Ghosh   49. Dum   50. Chiuderi   51. Sturrock   52. Nelson   53. Aly
54. Witta   55. Norman   56. Sakai   57. Nishikawa   58. Goldstein   59. Spicer   60. Lokanadham
61. Tsinganos   62. Rust   63. Spangler   64. Berger   65. Gergely   66. Bridle   67. Heyvaerts
68. Schmahl   69. Holman   70. Ferrari   71. Einaudi   72. Peacock   73. Nordlund   74. Migliuolo
75. Bell   76. Drake   77. Hasegawa   78. Vlahos.

LIST OF PARTICIPANTS

Dr. M. L. Aizenman, Div. of Astronomy Science, National Science Foundation, Washington, D. C. 20550.
Dr. J. J. Aly, Service d´Astrophysique, Cen-Scalay, DPhG/SAP, 91191 GIF-SUR-YVETTE, CEDEX, France.
Dr. M. André, Kiruna Geophysical Institute, University of Umea, S-901 87 Umea, Sweden.
Dr. R. A. Bell, Astronomy Program, University of Maryland, College Park, MD 20742
Dr. G. Benford, Department of Physics, University of California, Irvine, CA 92717.
Dr. M. A. Berger, Center for Astrophysics, Harvard University, Cambridge, MA 02138
Dr. J. Birn, Los Alamos Nat. Lab., M.S. D 438, Los Alamos, NM 87545.
Dr. A. Bratenahl, Institute of Geophysics and Planetary Physics, University of California, Riverside, CA 92521.
Dr. A. Bridle, VLA Program, NRAO, P.O. Box 0, Socorro, NM 87801.
Dr. C. C. Cheng, Code 4170 CC, NRL, Washington, DC 20375.
Dr. C. Chiuderi, Instituto di Astronomia, Universita di Firenze, Largo E. Fermi, 5 50125 Firenze, Italy.
Dr. B. Coppi, MIT, Cambridge, MA 02139.
Dr. F. Coroniti, Department of Physics, University of California, Los Angeles, CA 90024.
Dr. J. M. Davila, NASA/Goddard Space Flight Center, Code 682, Greenbelt, MD 20771.
Dr. S. Degaonkar, Astronomy Program, University of Maryland, College Park, MD 20742
Dr. J. Drake, Lab. for Plasma & Fusion, Energy Studies, University of Maryland, College Park, MD.
Dr. Ch. Dum, Max-Planck-Institut fur Plasmaphysik, D-8-46 Garching Munchen, Germany.
Dr. D. Eichler, Astronomy Program, University of Maryland, College Park, MD 20742.
Dr. J. Eilek, Physics Department, New Mexico Tech, Socorro, NM 87801.
Dr. R. C. Elphic, NASA/Goddard Space Flight Center, Code 961, Greenbelt, MD 20771.
Dr. G. Einaude, Scuola Normale Superiore, Piazza Cacalieri 7, 5600 Pisa, Italy.
Dr. A. Ferrari, Instituto di Fisica Generale, Universita Di Torino, Gorso Massimo d´Azeglio, 46, I-10125 Torino, Italy.
Dr. W. Gekelman, Dept. of Physics, University of California, 405 Hilgard Avenue, Los Angeles, CA 90024.
Dr. T. E. Gergely, Astronomy Program, University of Maryland, College Park, MD 20742.
Dr. D. Gilden, Institute for Advance Study, School of Natural Sciences, Olden Lane, Bldg. E, Princeton, N.J. 08540.
Dr. S. J. Goldstein, Dept. of Astronomy, University of Virginia, Charlottesville, VA 22903
Dr. C. C. Goodrich, Astronomy Program, University of Maryland, College Park, MD 20742

LIST OF PARTICIPANTS

Dr. M. Goossens, Katholieke Universiteit Leuven, Astronomisch Institute, Naamsestraat 61, B-3000 Leuven, Belgium.
Mr. S. Ghosh, Astronomy Program, University of Maryland, College Park, MD 20742
Dr. P. E. Hardee, Dept. of Physics & Astronomy, University of Alabama, Box 1921, University, AL 35486
Dr. A. Hasegawa, Bell Laboratories, Murray Hill, N.J. 07974.
Dr. A. Hassam, Dept. Physics & Astronomy, University of Maryland, College Park, MD 20742.
Dr. T. Heckman, Astronomy Program, University of Maryland, College Park, MD 20742.
Dr. R. N. Henriksen, Dept. of Physics, Astronomy Group, Queens University at Kingston, Stirling Hall, Kingston, ON K7L 3N6, Canada.
Dr. J. F. Heyvaerts, Observatorie de Paris, Section d'Astrophpysique de Meudon, 92190 Meudon, France.
Dr. G. D. Holman, Astronomy Program, University of Maryland, College Park, MD 20742.
Dr. J. D. Huba, Naval Research Lab., Washington, DC 20375.
Dr. J. Ionson, NASA-Goddard Space Flight Center, Code 682, Greenbelt, MD 20771.
Dr. J. S. Kaastra, Sterrekundig Instituut der Rijksuniversiteit Utrecht, Servaas Bolwerk 13, 3512 NL Utrecht, The Netherlands.
Dr. J. Karpen, Code 4040, NRL, Washington, DC 20375.
Dr. C. F. Kennel, Dept. of Physics, University of California, Los Angeles, CA 90024.
Dr. M. J. Keskinen, Naval Research Lab., Washington, DC 20375.
Dr. V. Krishan, Indian Institute of Astrophysics, Bangalore 34, India.
Dr. A. Krautter, Dept. of Physics, Queen's University, Kingston, ON K7L 3N6, Canada.
Dr. M. R. Kundu, Astronomy Program, University of Maryland, College Park, MD 20742.
Dr. G. Lake, Astronomy Department, University of California, Berkeley, CA 94720.
Dr. C. S. Liu, Dept. of Physics & Astronomy, University of Maryland, College Park, MD 20742.
Dr. B. Lokanadham, Dept. of Astronomy, Osmania University, Hyderabad 500 007, India.
Dr. W. Lotko, Space Sciences Laboratory, University of California, Berkeley, CA 94720.
Dr. A.T.Y. Lui, Applied Physics Lab.. The Johns Hopkins University, Laurel, MD 20707.
Dr. S. M. Mahajan, Fusion Res. Center, Dept. of Physics, University of Texas at Austin, Austin, TX 78712.
Mr. R. McDowell, GSFC/NASA, Greenbelt, MD 20771
Dr. S. Migliuolo, High Altitude Observatory, P.O. Box 3000, Boulder, CO 80307.
Dr. D. J. Mullan, Bartol Research Foundation of the Franklin Institute, University of Delaware, Newark, Delaware 19711.
Dr. A. H. Nelson, Dept. of Applied Math & Astronomy, University College, Cardiff, United Kingdom
Dr. K. I. Nishikawa, Plasma Physics Lab., Princeton, NJ. 08544

LIST OF PARTICIPANTS

Dr. L. Nocera, Applied Maths. Dept., University of St. Andrews, North Haugh, St. Andrews KY 16 9SS Great Britian.
Dr. A. Nordlund, HARO/NCAR, P.O. Box 3000, Boulder, CO 80307.
Dr. C. A. Norman, Institute of Astronomy, Madingley Road, Cambridge, CB3 OHA, England.
Dr. K. Papadopoulos, Astronomy Program, University of Maryland, College Park, MD 20742.
Dr. D. Peacock, National Science Foundation, Washington, D. C. 20550.
Dr. E. R. Priest, Dept. of Applied Mathematics, North Haugh, St. Andrews KY16 9SS, United Kingdom.
Dr. R. E. Pudritz, Astronomy Department, University of California, Berkeley, CA 94720.
Dr. A. Ray, Theoretical Astrophysics Group, Tata Institute of Fundamental Res., Homi Bhabha Road, Bombay 400 005, India.
Dr. T. P. Ray, Physics Department, University College, Stillorgan Road, Dublin 4, Ireland.
Dr. W. K. Rose, Astronomy Program, University of Maryland, College Park, MD 20742
Dr. H. L. Rowland, Astronomy Program, University of Maryland, College Park, MD 20742
Dr. C. T. Russell, Inst. of Geophysics and Planetary Physics, UCLA, Loa Angeles, CA 90024.
Dr. D. Rust, Applied Phys. Lab., John Hopkins Univ., John Hopkins Road, Laurel, MD 20707
Dr. J. Sakai, Dept. of Appl. Math. & Physics, Faculty Engineering, Toyama University, Takaoka, Toyama 933, Japan.
Dr. T. Sato, Institute for Fusion Theory, Hiroshima University, Higashisenda-machi, Hiroshima 730, Japan.
Dr. E. J. Schmahl, Astronomy Program, University of Maryland, College Park, MD 20742.
Dr. R. Sharma, Astronomy Program, University of Maryland, College Park, MD 20742
Dr. R. K. Shevgaonkar, Astronomy Program, University of Maryland, College Park, MD 20742
Dr. D. F. Smith, Berkeley Research Associates, P.O. Box 241, Berkeley, CA 94701.
Dr. R. A. Smith, Science Applications Inc., Plasma Physics Div., T-4, 1710 Goodridge Drive, McLean, VA 22102.
Dr. B. Sonnerup, Thayer School of Engineering, Dartmouth College, Hanover, NH 03755.
Dr. S. R. Spangler, Dept. of Physics and Astronomy, University of Iowa, Iowa City, IA 52242.
Dr. D. S. Spicer, U.S. Naval Research Lab., Code 4780, Plasma Physics Division, Washington, D.C. 20375.
Dr. R. S. Steinolfson, Dept. of Physics, University of California, Irvine, CA 92717.
Dr. R. L. Stenzel, Physics Department, University of California, 405 Hilgard Avenue, Los Angeles, CA 90024.
Dr. P. A. Sturrock, Institute for Plasma Research, ERL 306, Stanford University, Stanford, CA 94305.

Dr. T. Tajima, Physics Department, University of Texas, Austin, TX 78712.

Dr. M. Tanaka, Dept. of Physics & Astronomy, University of Maryland, College Park, MD 20742.

Dr. J. D. Trasco, Astronomy Program, University of Maryland, College Park, MD 20742

Dr. S. K. Trehan, Dept. of Mathematics, Simov Fraser University, Burnaby V5A1S6, B.C. Canada

Dr. K. Tsinganos, Harvard University, 60 Garden Street, Cambridge, MA 02139

Dr. Y. Uchida, Tokyo Astronomical Obs., University of Tokyo, Mitaka, Tokyo, Japan.

Dr. A. Underhill, NASA/Goddard Space Flight Center, Greenbelt, MD 20771

Dr. G. Van Hoven, Department of Physics, University of California, Irvine, CA 92717.

Dr. V. Vasyliunas, Max-Planck Institut für Aeronomie, Postfach 20, D-3411 Katlenbrug-Lindau 3, West Germany.

Dr. L. Vlahos, Astronomy Program, University of Maryland, College Park, MD 20742.

Dr. D. G. Wentzel, Astronomy Program, University of Maryland, College Park, MD 20742.

Dr. A. S. Wilson, Astronomy Program, University of Maryland, College Park, MD 20742.

Dr. P. J. Wiita, Dept. of Astronomy & Astrophysics, University of Pennsylvania, Philadelphia, PA 19104.

Dr. S. T. Wu, Dept. Mechanical Engineering, University of Alabama in Huntsville, Huntsville, AL 35899.

INTRODUCTORY REMARKS

M. R. Kundu
Astronomy Program
University of Maryland
College Park, MD 20742

I believe this is the first IAU Symposium with the objective of bringing together researchers in different fields, plasma physicists, space physicists and astrophysicists, to permit cross-fertilization of knowledge and ideas on related problems. These problems refer to understanding collective effects and their signatures in unstable magnetized plasmas. Similar processes play an important role in problems as diverse as the stability of jets in extragalactic radio sources and the release of energy by interacting loops in solar flares, for example. In this symposium we cover five different areas. In order of increasing distance from us, they are laboratory plasma experiments, magnetospheric current systems, solar flares, stellar flares and extragalactic sources. The active ejecta of relativistic particles and hot plasma seen in solar flares provide the only examples of a violent phenomenon subject to detailed diagnostic examination. Many active stars and galaxies display ejecta of similar appearance. The application of plasma physics ideas to astrophysics has been, until now, inversely proportional to the distance of the object from us. The laboratory plasma physicists and magnetospheric plasma physicists are the only ones who can conduct experiments and in-situ measurements of plasmas and compare the results with their theoretical ideas. They are beginning to be interested in solar plasmas, less so in stellar plasmas, and even less in extragalactic systems. The objective of the symposium was to get plasma physicists interested in astronomical problems, and the astronomers in plasma physics. In that sense it was a unique symposium. And the interactions that took place between the plasma physicists and the astronomers in this symposium lead us to believe that the object of the symposium was accomplished.

The history of the symposium goes back to April 1981, when I was visiting, as a guest of the U.S.S.R. Academy of Sciences, the P.N. Lebedev Physical Institute, in fact the laboratory and theoretical group established by Syrovatskii. Several things happened at that time: Syrovatskii had died a year or so before then; the Soviets had organized a conference somewhere in Belourussia in memory of Syrovatskii on "Current Sheets in Astrophysics", a subject which Syrovatskii had pioneered and promoted so vigorously for a good fraction of his life; and Syrovatskii's Annual Reviews of Astronomy and Astrophysics article "Pinch Sheets and Reconnection in Astrophysics" was being finished posthumously by Bulanov, Frank and Sasorov, three loyal and devoted colleagues of Syrovatskii. It is there the idea of this symposium was really conceived. What better way to pay tribute to a man like Syrovatskii! In honor of the man who has done so much to promote the field, we have dedicated this symposium to the memory of Sergei Ivanovitch Syrovatskii!

This symposium was sponsored by the IAU and three of its commissions 10, 12 and 40. It was cosponsored by COSPAR and SCOSTEP. Financial support was received from the IAU and the U.S. National Science Foundation and the National Aeronautics and Space Administration.

SESSION I

MAGNETIC FIELD RECONNECTION IN COSMIC PLASMAS

B. U. Ö. Sonnerup
Dartmouth College, Hanover, NH 03755 U.S.A.

ABSTRACT

A brief review is presented of the concept of magnetic field reconnection or merging. This process occurs whenever an electric field is present along a separator line in the magnetic field. The basic properties of reconnection are discussed in the context of the classical MHD models by Sweet and Parker and by Petschek. Attention is then focussed on reconnection in collision-free plasmas. The energization of charged particles during their interaction with the current layers associated with the reconnection geometry is discussed and the nature of the processes occurring in the so-called diffusion region which surrounds the separator is considered. Finally, comments are made on the nonsteady aspects of reconnection at the earth's magnetopause.

1. INTRODUCTION

Magnetic field reconnection, or merging, is a universal process for the conversion of magnetic energy into plasma kinetic and thermal energy. The process, which may occur either impulsively or in a steady state, taps the magnetic free energy associated with electric current sheets and other sheared field configurations. It is believed to be important in a variety of cosmic situations: solar flares, solar magnetic-field evolution, and perhaps coronal heating; planetary magnetopauses and tails; cometary tails; accretion disks, etc. The process has also been studied extensively in a variety of laboratory devices and simulations as well as in computer simulations.

This paper has two purposes: to provide a brief review of the two classical reconnection models by Sweet-Parker and by Petschek; and to discuss some of the current concerns and findings about reconnection in its magnetospheric setting. For more detailed discussion, the reader is referred to the review papers by Vasyliunas (1975) and Sonnerup (1979).

In the magnetosphere we have a unique opportunity to learn about reconnection in a cosmic plasma from direct in situ measurements and it is expected that the information thus obtained can be translated to other cosmic situations, at least to some extent. However, since its introduction into magnetospheric physics by Dungey (1961), the reconnection concept has been a source of continual controversy. Initially, the evidence for reconnection was mostly indirect but the process nevertheless proved to be a powerful organizing concept for a great variety of observations. Recently, more direct evidence for the occurrence of reconnection has become available. But the controversy has not disappeared, for the reconnection process has proved to be far more complicated than originally envisaged. Some of the complexity of the process will become apparent in the following pages.

2. DEFINITION

In order to deal with reconnection in an organized fashion, it is desirable first to provide a simple and unambiguous definition of the process. This definition contains four parts:

(i) A "separatrix" is a magnetic field line surface which separates different magnetic cells as illustrated in Figure 1.

(ii) A "separator" is the line of intersection of a separatrix with itself or perhaps with another separatrix.

(iii) "Reconnection" or "merging" occurs when an electric field $E_\parallel$ is present along a separator.

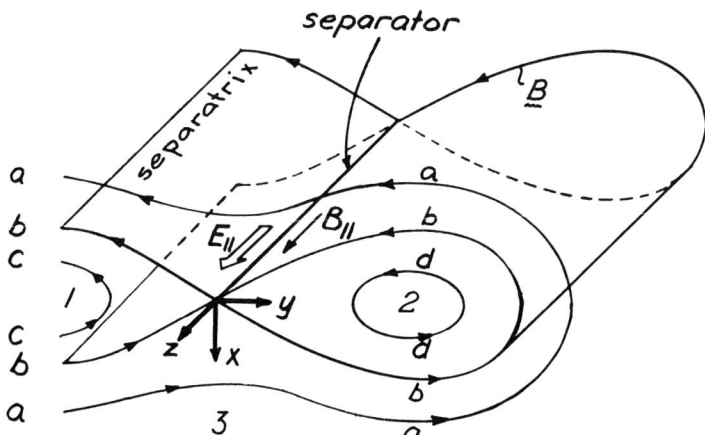

Figure 1. Basic reconnection geometry. The field line "a", originally located in cell 3 moves toward the separatrix surface and lies in that surface at location "b". Reconnection occurs at the separator and the field line is broken into two parts, "c" and "d", located in cells 1 and 2.

(iv) The "reconnection rate" is given by $E_\parallel$ and is equal to the amount of magnetic flux transported per unit time across a unit length of the separator.

Several notes on these definitions should be made:

(a) In general, a separatrix surface is associated with two hyperbolic null points in the magnetic field (see Dungey, 1963). The case shown in Figure 1 is degenerate in the sense that the magnetic field is zero along the entire separator. Usually, this is not the case but a field component $B_\parallel$ is present along most of that line.

(b) The term "separator" is synonymous with "reconnection line," "merging line," or "X line." The phrases "neutral line" and "null line" are sometimes used as well but they are somewhat misleading because the field is in general not equal to zero along the separator.

(c) The definition draws only upon the universally accepted concepts of electric and magnetic fields. A number of magnetospheric scientists are uncomfortable with the ideas of moving field lines and frozen magnetic fields so that a definition based on fundamentals only is desirable, even though the term "reconnection" itself invokes the idea of moving field lines. Note that the definition is such that reconnection can occur freely in a vacuum. However, as a process, it is only of interest in the presence of a highly conducting plasma in which the electromagnetic fields and currents are generated self-consistently by the differential motion of ions and electrons.

(d) The definition emphasizes flux transport rather than plasma transport between different topological cells. A well-known mathematical theorem states that $\underline{E} \cdot \hat{B} = 0$ is a sufficient condition for the flux transport velocity $\underline{v}_\phi = \underline{E} \times \underline{B}/B^2$ to move points that were once joined by a field line in such a manner that they remain joined by a field line at all later times. It is on this fact that the concept of moving field lines is based. Figure 1 illustrates how a field line in cell 3 moves towards the separator. Along this line $\underline{E} \cdot \hat{B} \neq 0$ so that the theorem is violated. Reconnection occurs, the result being a transport of magnetic flux from cell 3 to cells 1 and 2. No plasma physics has been introduced into the above discussion but it is the presence of a highly conducting plasma that assures that the condition $\underline{E} \cdot \hat{B} = 0$ is satisfied everywhere except at the separator. It is also known that $\underline{E} \times \underline{B}/B^2$ is the electric drift velocity of charged particles. Other drifts such as inertia and gradient drifts are unimportant in most of the external flow but become significant near the separator. It follows from these facts that in a highly conducting plasma, the definition presented here is in all practical respects indistinguishable from the nonlocal definition in terms of "plasma flow across a separatrix" adopted by Vasyliunas (1975).

(e) The amount of magnetic flux transported per unit time across a length element $\underline{d\ell}$ of the separator is

$$d\Phi_m = \underline{B} \cdot \{\underline{v}_\phi \times d\underline{\ell}\} \tag{1}$$

which, upon use of the expression for $\underline{v}_\phi$ simplifies to

$$\frac{d\Phi_m}{d\ell} = E_\parallel \tag{2}$$

It is for this reason that $E_\parallel$ is used as a measure of the reconnection rate. It is however often useful to define a nondimensional rate as

$$M_{A1} \equiv E_\parallel / v_{A1} B_1 \tag{3}$$

where v_{A1} and B_1 are the Alfvén speed and magnetic field at a chosen reference point and reference time. This usage is particularly common in steady-state two-dimensional reconnection models in which $E_\parallel$ can be shown from Faraday's law to be constant throughout the plane perpendicular to the separator. In that case $E_\parallel / B_1$ represents the flux transport velocity (or the electric drift velocity) v_1 at the reference point so that $M_{A1} = v_1/v_{A1}$ is the Alfvén Mach number at that location. The definition remains arbitrary in the sense that the location of the reference point relative to the separator may be chosen differently. This point will be taken up later on.

(f) It is noted that $E_\parallel$ and thus the reconnection rate is invariant under Galileo transformations. In other words, it does not matter whether $E_\parallel$ is measured in a frame of reference in which the separator (or a segment thereof) is at rest or whether it is measured at the instant a moving separator passes the observation point.

(g) Finally, the definition is local in nature and thus cannot and does not distinguish between electrostatic and inductive contributions to $E_\parallel$.

3. SWEET-PARKER MODEL

The Sweet-Parker model of reconnection (Sweet, 1958; Parker, 1963) is shown in Figure 2a. It describes the slow steady-state inflow of two oppositely and strongly magnetized plasmas towards a current sheet and the subsequent rapid outflow of weakly magnetized plasma along the sheet. The basic features of this model can be understood in three simple steps.

First, the pressure at the separator (which is perpendicular to the plane of the figure) is equal to $P_\infty + B_\infty^2/2\mu_o$ where the subscript ∞ denotes upstream conditions. The excess pressure, $B_\infty^2/2\mu_o$, is used to accelerate the outflowing plasma along the sheet so that, in accordance with Bernoulli's law, we have $P_\infty + B_\infty^2/2\mu_o = P_\infty + \frac{1}{2}\rho v_{out}^2$. The pressure at the exit is assumed to be the same as in the inflow and, for simplicity, the density ρ is taken to be a constant. It follows that the

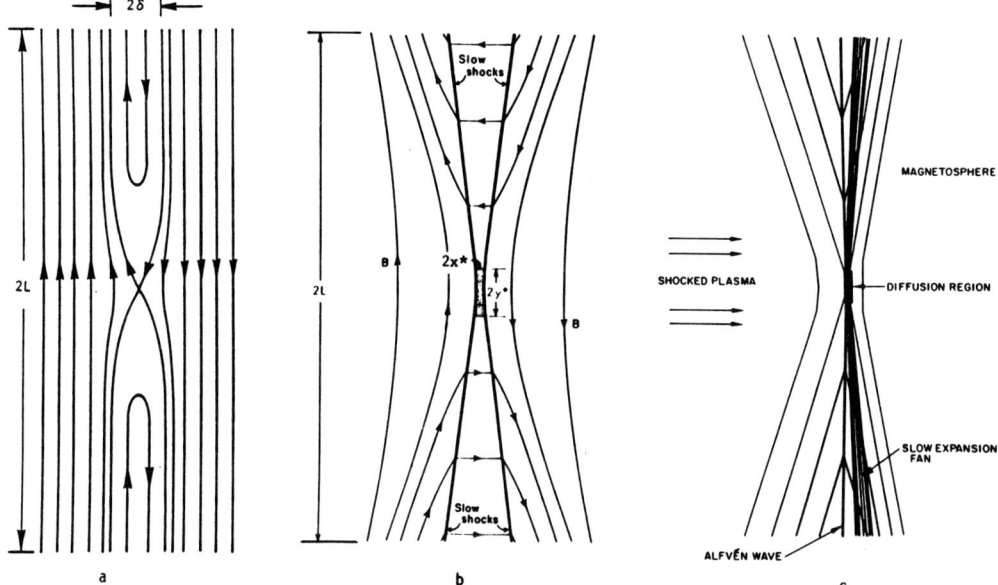

Figure 2. (a) Sweet-Parker model. Slow plasma inflow from the left and right with rapid outflow at top and bottom. (b) Petschek model. Slow plasma inflow from both sides. Acceleration into the wedge-shaped outflow regions by slow shocks. (c) Levy et al. model. Plasma inflow from the left only, with vacuum conditions on the right. Plasma acceleration into the outflow at top and bottom by large-amplitude Alfvén waves (rotational discontinuities). (After Petschek, 1964.)

outflow velocity is

$$v_{out} = B_\infty/\sqrt{\mu_o \rho} \equiv v_{A\infty} \qquad (4)$$

In other words, the outflow velocity is equal to the Alfvén speed $v_{A\infty}$ based on the inflow conditions.

Second, the mass conservation law with constant density yields $v_{in} L = v_{out} \delta$ where δ and L are the half width and half length of the layer, as shown in the figure. Thus the ratio of these two lengths is

$$\delta/L = v_{in}/v_{out} = v_{in}/v_{A\infty} \equiv M_{A\infty} \qquad (5)$$

where $M_{A\infty}$ is the nondimensional reconnection rate. Equation (5) indicates that the larger the reconnection rate, the thicker the layer must be (for given L).

Third, Ohm's law has the form $\underline{j} = \sigma \underline{E}$ near the separator where the plasma is semistagnant and has electrical conductivity σ. According to Ampère's law, we have $j \simeq 2B_\infty/2\mu_o \delta$ and since $E = v_{in} B_\infty$ Ohm's law yields

$$\mu_o \sigma v_{in} \delta = 1 \tag{6}$$

In other words, the magnetic Reynolds number based on the current layer half width δ and the inflow speed v_{in} is equal to unity. This statement expresses the fact that in a steady state the width δ must be such that the inflow speed is equal and opposite to the resistive diffusion speed.

If the width δ is eliminated between (5) and (6), one obtains the well-known Parker formula for the reconnection rate

$$M_{A\infty} = R_{mL}^{-\frac{1}{2}} \tag{7}$$

where R_{mL} is the magnetic Reynolds number based on $v_{A\infty}$ and L, i.e., $R_{mL} = \mu_o \sigma v_{A\infty} L$. Since the value of R_{mL} is extremely large in most cosmic applications, mostly because L is large, one is tempted to conclude from this analysis that reconnection is an insignificant process in cosmic physics. However, as discussed in the next section, cosmic plasmas may have the ability to manufacture small-scale dissipative structures, i.e., L values that are sufficiently small to permit $M_{A\infty}$ values of order unity.

The formula (7) also appears to be relevant to reconnection in tokamaks. Park et al. (1983) have generalized (7) to include the effects of viscosity, their result being

$$M_{A\infty} \sim R_{mL}^{-\frac{1}{2}} (1+\mu_o \sigma \nu)^{-\frac{1}{4}} \tag{8}$$

where ν and σ are the kinematic viscosity and the electrical conductivity, respectively. According to Spitzer (1962), the product $\mu_o \sigma \nu_\perp$, evaluated in transport perpendicular to a strong magnetic field, is

$$\mu_o \sigma \nu_\perp = \frac{3\sqrt{2}}{40\pi} (T_e/T_i)^{\frac{1}{2}} (m_i/m_e)^{\frac{1}{2}} \beta_e \tag{9}$$

where β_e is the ratio of electron pressure to magnetic pressure. This formula indicates that the modified Parker scaling (8) should be used whenever the β_e value of the plasma is of order unity or greater, as is frequently the case in cosmic plasmas.

In two-dimensional tokamak computer simulations performed by Park et al. (1983), it appears that in the m = 1 flip (internal kink) the reconnection rate is governed by (8).

A family of exact solutions of the MHD equations with constant density has been described by Sonnerup and Priest (1975) for the case of resistive stagnation point flow at a current sheet which is essentially the Sweet-Parker geometry. These authors also formulated the problem for the case where viscosity is important.

4. PETSCHEK MODEL

In order to overcome the difficulty with the small reconnection rate in the Sweet-Parker geometry, Petschek (1964) devised his now famous model in which resistive diffusion is important, not over the entire length, 2L, of the current layer, but only over a short distance, 2y*, around the separator, as illustrated in Figure 2b. In the remainder of the flow field, electromagnetic energy is converted to plasma kinetic energy and heat in a set of standing slow-mode shocks originating at the separator.

The region surrounding the separator in which resistive diffusion leads to a violation of the frozen magnetic field condition is called the "diffusion region." It has cross section 2x* by 2y* as shown in the figure. The Parker formula (7) applies to this region but with L replaced by the small length y*. The nature of the processes in the diffusion region in collision-free plasmas will be discussed in Section 6.

Away from the immediate vicinity of the separator, the frozen magnetic field condition holds except in the slow shocks. The constancy of the electric field component tangential to a shock surface guarantees that in the xy plane the magnetic field appears to be frozen across the shocks. However, the charged particles undergo a displacement in the z direction as they cross the shock whereas there is no corresponding displacement of the magnetic field lines. Thus the frozen-field condition is in fact violated in the shocks. The z displacement of the charged particles is in the direction of the reconnection electric field, $E_{\parallel}$, and it is therefore the means by which particles are energized in the shocks. Further discussion of this effect is given in Section 5. Here we simply note that the outflow speed of the plasma in the narrow wedges between the shocks can be obtained directly from the conservation laws for mass, magnetic flux, and tangential (y) momentum:

$$\left. \begin{array}{c} v_{in} L = v_{out} \delta \\ B_{in} \delta = B_{out} L \\ \rho v_{in} v_{out} = B_{in} B_{out}/\mu_o \end{array} \right\}$$

By elimination of v_{in} and B_{out} between these equation one obtains

$$v_{out} = B_{in}/\sqrt{\mu_o \rho} \tag{10}$$

As in the Sweet-Parker model, the outflow velocity is equal to the Alfvén speed based on the magnetic field in the inflow. This result is independent of the reconnection rate.

In qualitative terms, the geometrical behavior of Petschek's reconnection model for different reconnection rates is as follows. For very small rates the diffusion region length y* is equal to L in which case no shocks develop and the geometry is that of the Sweet-Parker model. As the rate increases, y* and x* decrease and shock pairs appear on the upper and lower sides of the diffusion region. The wedge angle between these shocks is initially very small. As the rate increases further, y* and x* continue to decrease and the wedge angle between the shocks increases. At the maximum reconnection rate, given by Petschek as

$$M_{A\infty} = \pi[8\ln(2M_{A\infty}^2 R_{mL})]^{-1} \tag{11}$$

(for the incompressible case; a correction by Vasyliunas (1975) has been included), the diffusion region (y*) may be extremely small and the wedge angle substantial. The logarithmic dependence of $M_{A\infty}$ on R_{mL} permits reconnection rates $M_{A\infty} \simeq 0.1$-0.2 in typical cosmic applications.

Note that the maximum reconnection rate given by (11) is the Alfvén Mach number far upstream of the reconnection region where Petschek assumed the flow and magnetic field to be uniform. If a reference point on the x axis immediately outside the diffusion region is used instead, one finds the magnetic field to be substantially weaker there and the inflow speed substantially larger so that the inflow Alfvén Mach number is of order unity. Thus the logarithmic factor in (11) is the result of the specific upstream boundary conditions used by Petschek. Vasyliunas (1975) has pointed out that these conditions correspond to fast-mode expansion in the two inflow regions. Other boundary conditions (Sonnerup, 1970; Yeh and Axford, 1970) may lead to different upstream values $M_{A\infty}$ and different behavior (slow-mode expansion) in the inflow regions. But the result that the maximum reconnection rate corresponds to an Alfvén Mach number of order unity immediately adjacent to the reconnection region is likely to be valid regardless of the external boundary conditions on the inflow side, as long as the outflow remains unimpeded. For this reason the Petschek reconnection rate is often quoted simply as $M_A \simeq 1$.

Park et al (1983) have generalized this rate to include viscous effects, the result being

$$M_A \sim (1+\mu_0 \sigma \nu)^{-\frac{1}{2}}. \tag{12}$$

This rate is not observed in computer simulations of tokamak reconnection. The reason is that the Petschek model assumes free and unimpeded outflow from the reconnection region, a condition not satisfied in tokamaks but likely to be valid in a number of cosmic situations, e.g., reconnection at the earth's magnetopause.

A precise mathematical analysis of the Petschek model may be found in Soward and Priest (1977;1982) and Soward (1982).

Petschek's model can also be developed for the case of magnetopause reconnection where the plasma and magnetic field conditions are dissimilar on the two sides of the current sheet. The special case with a vacuum on one side of the layer was discussed by Levy et al. (1964) and is shown in Figure 2c. In general, asymmetric models contain a rotational discontinuity across which most of the requisite field direction change occurs. In addition, slow shocks or slow expansion fans may be present.

Note that even a small asymmetry in the plasma density on the two inflow sides will lead to the occurrence of a rotational discontinuity. The reason for this is as follows. The slow shocks on the low density side of the configuration will have to be somewhat weaker and those on the high density side somewhat stronger than in the symmetric model in order to match flow velocities and field directions in the outflow wedges. But the strongest permissible slow shock is a switch-off shock in which the field on the downstream side of the shock is along the shock normal. If an even stronger tangential momentum change is needed on the high density side, it has to be provided by a rotational discontinuity initially followed by a slow shock which reduces the excessive momentum change provided by the rotational discontinuity, as illustrated in Figure 3.

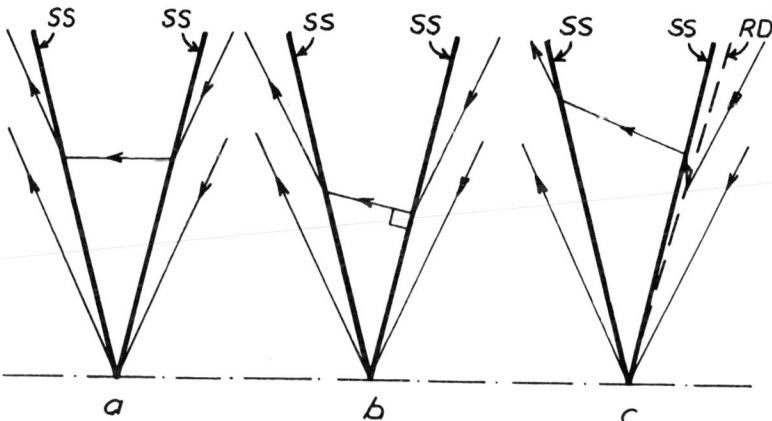

Figure 3. Asymmetric reconnection. Upper half of configuration is shown. (a) Symmetric case. (b) Density increased on the right and decreased on the left until right-hand slow shock (SS) is a switch-off shock. Left-hand shock gets weaker. (c) Density asymmetry increased further. A rotational discontinuity (RD) appears on the right, followed by a slow shock. Left-hand slow shock weakens further.

5. PARTICLE ACCELERATION IN CURRENT SHEETS

The classical reconnection models discussed in the two previous sections are magnetohydrodynamic in nature. However, the magnetospheric plasma is collision-free and it is not entirely clear how this influences the geometry or how slow shocks manifest themselves in such a medium. This point is illustrated in Figure 4 which shows symmetric

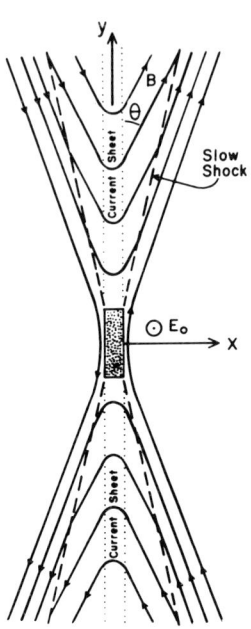

Figure 4. Hill's collision-free reconnection model. The outflow wedges are located between the central current layer and the slow shocks.

collision-free reconnection in the geomagnetic tail, as envisaged by Hill (1975). Individual particles interact with, and are energized at, a central current sheet (which in MHD terminology is a contact discontinuity and can exist only when the pressure is suitably nonisotropic), being either reflected or transmitted there. The outflowing particles form wedge-shaped regions on either side of the current sheet, regions that are somewhat similar in nature to the so-called foreshock region upstream of the earth's bow shock. In these wedges, the plasma consists of a mixture of particles that flow slowly toward the central current sheet and particles that have already interacted with it and have gained kinetic energy during the interaction. This latter population is streaming with a large field-aligned velocity component away from the diffusion region at the center of Figure 4. The outer edge of each outflow wedge is marked by particles that have passed through the diffusion region. At this location the magnetic field changes direction and magnitude in order to accommodate the higher plasma pressure in the wedge. This discontinuity is presumably a slow shock, although the usual jump conditions probably need to be modified to allow for heat flow behind it.

Since only a very small portion of the inflowing particles passes through the diffusion region, while the overwhelming majority interacts with the central current sheet, it is logical first to examine the interaction of particles with a one-dimensional laminar current sheet which may be either a shock, a rotational discontinuity or a contact discontinuity.

The first important point is that such current sheets have a non-vanishing normal magnetic field component. This feature allows one to transform away the reconnection electric field, $E_{\parallel}$, by examining the particle orbits in a frame of reference that slides along the current sheet, away from the diffusion region, with speed $v_t = E_{\parallel}/(B\sin\theta)$ where θ is defined in Figures 4 and 5. This frame is often referred to as the de Hoffmann-Teller (dHT) frame. The principal advantage of this procedure is that, in the moving frame, a particle either conserves its energy or changes it in a known manner in response to the electric potential structure, $\Phi(x)$, of the current sheet. In particular, reflected particles must have the same energy before and after reflection.

As illustrated in Figure 5 for a symmetric field-reversing current sheet with $\Phi = 0$, a particle moving with guiding-center velocity v_{1g} toward the sheet in the stationary frame appears to be moving along the magnetic field with guiding-center velocity v'_{1g} and pitch angle α_1 in the dHT frame (Fig. 5a). Assume that it traverses the sheet and leaves on the other side, moving along $\underline{B}$, with guiding-center velocity v'_{2g} and pitch angle α_2 (Fig. 5b). Conservation of energy in the dHT frame implies that $v'_{1g}/\cos\alpha_1 = v'_{2g}/\cos\alpha_2 \equiv v'$. In the stationary frame (Fig. 5c), we then see that v_{2g} can be much larger than v_{1g}. The energy change in this frame is $\Delta\varepsilon = m(v_2^2 - v_1^2)/2$ where (for both subscripts 1 and 2) $v^2 = v'^2 + (v't\tan\alpha)^2$. It is easy to show from the triangles in Figure 5 that the energy increase may be written

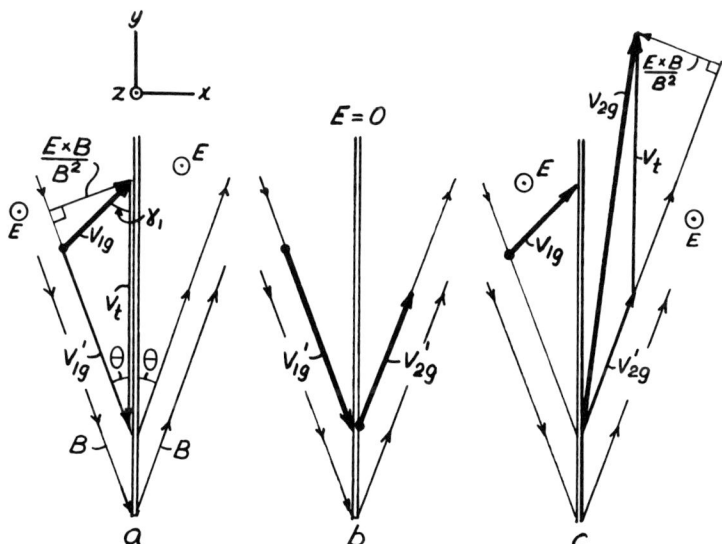

Figure 5. Particle energization in a current sheet. (a) Particle moves towards sheet with guiding center speed v_{1g}. Transformation velocity v_t along the sheet is added so that in the moving (dHT) frame the velocity is v'_{1g} along $\underline{B}$. (b) In the moving frame the electric field $\underline{E}=0$. The particle exits with velocity v'_{2g} along $\underline{B}$. (c) Return to stationary frame in which the exit velocity v_{2g} is large.

$$\Delta\varepsilon = m \frac{E_\parallel}{B\sin\theta} v' \cos\theta (\cos\alpha_1 - \cos\alpha_2) \quad (13)$$

or

$$\Delta\varepsilon = |q| E_\parallel R_{Lx} \cos\theta (\cos\alpha_1 - \cos\alpha_2) \quad (14)$$

where $R_{Lx} \equiv mv'/|q|B_x$, q being the particle charge and B_x the normal magnetic field component. Thus R_{Lx} is the particle gyroradius in the normal field.

Several general comments should be made:

(i) The particle energization is proportional to the particle mass. Thus electrons pick up very little energy and different species of ions pick up energy in proportion to their mass. At the earth's magnetopause and magnetotail where trace amounts of a variety of ion species may be present, this feature provides an opportunity for a persuasive test of the reconnection hypothesis.

(ii) The energization formulas (13,14) work for reflected as well as transmitted particles. Generalization to cases where the magnetic field and plasma flow on the two sides of the layer do not lie in a single plane is straightforward.

(iii) The energization formulas suggest large energization when θ is small. This is indeed the case for particle reflection in the earth's bow shock (e.g., Sonnerup, 1969) where the electric field is fixed. However, in the reconnection case, self consistency of magnetic fields and plasma motion implies that for a typical plasma ion $v'_{1g} = v' \cos\theta \cos\alpha_1 \simeq v_A$. For small angles θ and with the angle γ_1 in Figure 5 sufficiently different from 0 or π, we then have approximately $v_t = E_\parallel/(B\sin\theta) \simeq v_A$. In other words, $E_\parallel$ is proportional to $\sin\theta$ and the energization of a typical plasma particle is independent of θ and of the reconnection rate $E_\parallel$. Since $v_t \simeq v_A$ it is clear that for such a particle $v_2 \simeq 2v_A$. As noted below, these results agree with the predictions of MHD theory in which the geometrical effects discussed here are embodied in the law of conservation of tangential momentum (see e.g., Hudson, 1970; Sonnerup et al., 1981).

(iv) For the symmetric contact discontinuity in the geomagnetic tail (Fig. 4), the flow in the exit wedges contains a mixture of incoming particles with essentially zero flow speed ($v_{1g} \sim 0$) and outflowing particles with speed $v_{2g} \simeq 2v_A$ directed nearly along $\underline{B}$. This feature cannot be described by the MHD model, but the average velocity in the outflow regions is v_A in agreement with that model. For the dayside magnetopause, the magnetosheath plasma flows across a rotational discontinuity and, in the MHD model, acquires a tangential velocity of $2v_A$ in agreement with the single particle results given above. As long as only average plasma properties are considered, the MHD model gives the correct results but detailed distribution functions such as are now measured in the geomagnetic tail and at the magnetopause can of course not be obtained from that model.

(v) The single particle considerations discussed here, along with actual orbit calculations in a model tail current sheet, have been used by Lyons and Speiser (1982) to predict distribution functions in the exit wedges and good agreement with observations is obtained. To date, the corresponding calculations have not been performed at the magnetopause which usually has a much more complicated structure. However, on the fluid level, agreement between theory and observations is reasonably good (Paschmann et al., 1979; Sonnerup et al., 1981). Observations include energization of transmitted as well as reflected magnetosheath ions. Recently, the full fluid energy balance has been checked (Paschmann, private communication, 1983). It has been found that, in addition to the electromechanical energy conversion described above, the plasma enthalpy and entropy increase substantially in a rotational discontinuity. Furthermore, reflected ions may occasionally provide an important heat flow away from the discontinuity.

(vi) The energization of a particle during a single encounter with a current sheet is relatively small. In order to increase the energy to large values, multiple encounters are needed. In the geomagnetic tail, this may occur as a result of reflection near the earth in the magnetic mirrors provided by the dipole field. In other circumstances, scattering due to electromagnetic irregularities may bring some particles back to the current sheet.

(vii) One may conclude from (14) that, during its interaction with the current sheet, a particle must experience a displacement Δz in the z direction, i.e., along $E_\parallel$ given by

$$\Delta z = R_{Lx} \cos\theta (\cos\alpha_1 - \cos\alpha_2) \tag{15}$$

This displacement can indeed be derived directly from the conservation of energy in the dHT frame and the conservation of the generalized particle momenta associated with the two cyclic coordinates tangential to the current sheet (Cowley, 1978). The result (15) is represented graphically in Figure 6 for the case of reflected particles. Looking along the magnetic field, one finds that the particle orbit before and after reflection must be located inside an ellipse of major and minor axes $2R_{Lx}$ and $2R_{Lx}\sin\theta$. Inside the ellipse are circles, each labeled with a specific pitch angle α, which represent the projections of the helical

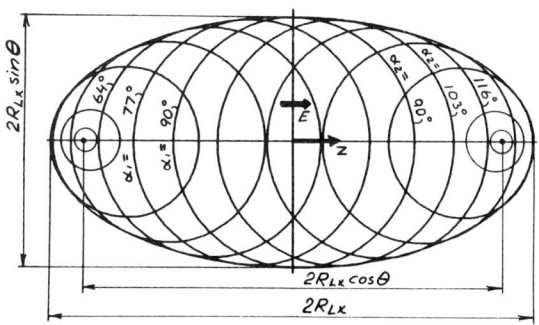

Figure 6. View along **B** of particle reflection in a current sheet. Incident helical orbit is a circle on the left, labeled by α_1; exit helix is circle on the right, labeled by α_2. Guiding-center displacement is purely along z (see Fig. 5).

particle orbits. The centers of these circles, i.e., the guiding centers, are located at $z = \pm R_{Lx}\cos\theta\cos\alpha$ and, except for small pitch angles, the circles touch the ellipse. The two foci of the ellipse occur at $z = \pm R_{Lx}\cos\theta$ and correspond to pitch angles $0°$ and $180°$. As a particle approaches the current layer, it moves on that circle, centered at $z<0$, which corresponds to its pitch angle $\alpha_1 (<90°)$. As it leaves the current sheet after reflection, it moves on a circle, centered at $z>0$, which corresponds to its pitch angle $\alpha_2 (>90°)$. Maximum displacement of the guiding center occurs for $\alpha_1 = 0°$, $\alpha_2 = 1 \cdot 0°$ in which case the orbit moves from the left-hand to the right-hand focus of the ellipse. Similar diagrams may be constructed for transmitted particles both in the planar case (e.g., Fig. 5) and in the case where the rotation angle of the tangential magnetic field is arbitrary. Except for thick current layers where $\alpha_1 = \alpha_2$, the actual relationship between the entry and exit pitch angles must be obtained from detailed orbit calculations.

6. DIFFUSION REGION

In discussing various physical processes in the diffusion region, it is desirable to examine the generalized Ohm's law:

$$\underline{E} + \underline{v} \times \underline{B} = \underline{j}/\sigma + (\underline{j} \times \underline{B})/ne - (\underline{\nabla} \cdot \underline{\underline{P}}_e)/ne$$
$$+ \{\partial \underline{j}/\partial t + \underline{\nabla} \cdot (\underline{j}\underline{v} + \underline{v}\underline{j})\} m_e / ne^2 \qquad (16)$$

Away from the separator and the slow shocks, the electric field $\underline{E}$ is balanced by $\underline{v} \times \underline{B}$, i.e., the frozen magnetic field condition holds. In the diffusion region, $\underline{E}$ remains nonzero but, since the plasma is semi-stagnant, the $\underline{v} \times \underline{B}$ term is small. Additionally, for $B_{\|} = 0$ the magnetic field vanishes at the separator. It is then necessary that one or more of the terms on the right-hand side of (16) becomes sufficiently large to balance $\underline{E}$. In a collisional plasma the term $\underline{j}/\sigma$ provides this balance. To do so, the current density $\underline{j}$ must be large when the conductivity σ is large; this leads to a very narrow diffusion region (small x^*), as discussed in Section 4.

In a collision-free plasma an effective resistivity $1/\sigma$ may be provided by plasma microinstabilities driven by the strong currents or gradients in the diffusion region (J.D. Huba, this Symposium). Another possibility is that electron inertial or pressure effects contained in the last two terms on the right in (16) are dominant, as discussed by Vasyliunas (1975) and Sonnerup (1979). There is no doubt that these terms must become important at the separator itself for without them, and with $\sigma = \infty$, the generalized Ohm's law implies that the magnetic field is frozen into the electron fluid which would make reconnection impossible. These terms produce the electron inertial length, λ_e, and gyroradius, R_{Le}, as important scale sizes in the diffusion region.

However, the question remains whether <u>all</u>, or even most, of the diffusion-region current is carried by electrons so that $x^* \sim \lambda_e$ (or R_{Le}).

Sonnerup (1979) has argued that the Hall term $(\mathbf{j}\times\mathbf{B})/ne$ plays an important role and leads to a much larger diffusion region width, $x^* \sim \lambda_i$ (or R_{Li}), with a substructure of size $\sim \lambda_e$ at the separator. In this situation, most of the current in the diffusion region is carried by ion drift in the z direction rather than by electron drift. Another effect is the formation of Hall-current loops in the xy plane with associated regions of positive and negative B_z values in the diffusion region: this may have a bearing on the observed magnetic field orientations in the plasmoid formed in the geomagnetic tail during reconnection (J. Birn, this Symposium). A recent analysis of the Hall effect in collisional tearing (Terasawa, 1983) is also relevant.

The inertial effects of electrons and ions discussed above can be dealt with in a qualitative manner in terms of an "inertial" conductivity $\sigma = ne^2\tau/m$ where τ is the residence time of a particle in the diffusion region, estimated as $\tau = x^*/v_{in}$. Thus the condition $\mu_o \sigma x^* v_{in} = 1$, which for constant σ leads to $x^* \sim v_{in}^{-1}$, in this case produces a value of x^* that is independent of v_{in}, i.e., of the reconnection rate. This value is

$$x^* = \sqrt{m/\mu_o ne^2} \qquad (17)$$

which is the definition of the inertial length of electrons or ions depending on whether the mass m_e or m_i is used.

Microinstabilities may well be present in the diffusion region and may provide an important signature of this region and the separatrix surfaces attached to it (Scudder et al., 1983). But it is not clear that such instabilities play a significant role in allowing reconnection to occur in a collision-free plasma. The argument that in their absence electrons would be able to move quickly along the separator to cancel out $E_\parallel$ (in particular when $B_\parallel \neq 0$) fails to take account of the short residence time τ of most particles in the diffusion region. For the same reason, the diffusion region may not be a prodigious source of high energy particles, accelerated by a single large displacement along the separator. A detailed study of particle orbits near the separator would be of great interest, in particular for $B_\parallel \neq 0$, but a realistic model of the electric and magnetic fields should be used ($E_x, E_y \neq 0$; $B_z = B_z(x,y)$).

7. NONSTEADY EFFECTS

A basic discussion of nonsteady reconnection must start with an examination of the tearing mode, a topic dealt with elsewhere (J.F. Drake, this Symposium). Thus my presentation will be limited to several remarks about observed, nonsteady aspects of magnetopause reconnection.

Although the situation at the magnetopause would seem ideal for the occurrence of reconnection in a quasisteady state, whenever the interplanetary magnetic field has a substantial southward component, $B_z < 0$, observations indicate that the process is mostly patchy, or at least limited to a narrow longitude segment, and that it is highly time-dependent. It is not entirely clear whether the recently discovered "flux transfer events" correspond to localized holes in the magnetopause moving away from the equatorial region, as visualized by Russell and Elphic (1978) (see also C.T. Russell, this Symposium), or to field lines connected across an open strip of the magnetopause, with the observation site located away from that strip. In either case, the observations suggest the existence of a threshold, other than $B_z<0$, for the onset and switch off of magnetopause reconnection. The nature of this threshold is not understood but the following scenario may account for the observations.

Assume that a bundle of interplanetary magnetic field lines gets hung up, perhaps in a small indentation, over the subsolar magnetopause. The plasma will escape from the bundle by streaming along $\underline{B}$, the result being a lowering of the plasma density n, of the β value (in particular $β_\parallel$), and of the Alfvén Mach number, M_A. It may be argued that each of these factors is conducive to the onset of reconnection over the narrow longitude segment occupied by the bundle. As soon as reconnection has started, two effects occur: a deepening indentation in the magnetopause develops; and the region originally occupied by the bundle gets replenished with fresh magnetosheath flux and plasma in which n, β, and M_A return to their normal values. The latter effect may lead to the switch-off of reconnection, while the former creates a suitable site at which a new magnetosheath field bundle may get hung up. It seems possible that for varying plasma conditions, this kind of model may lead either to a succession of flux transfer events, to the occurrence of quasisteady reconnection in a narrow longitude segment, or to reconnection that ultimately spreads over a substantial longitude segment. Detailed theoretical and observational studies guided by this scenario should provide important insights into reconnection in its magnetopause version.

ACKNOWLEDGEMENT

The research was supported by the Division of Atmospheric Sciences, National Science Foundation, under Grant ATM-8201974 to Dartmouth College.

REFERENCES

Cowley, S.W.H., 1978, Planet. Space Sci., 26, p. 539.
Dungey, J.W., 1961, Phys. Rev. Lett., 6, p. 47.
Dungey, J.W., 1963, "Geophysics of the Earth's Environment," (C. deWitt, J. Hieblot, and L. leBeau, eds.), Gordon and Breach, p. 503.
Hill, T.W., 1975, J. Geophys. Res., 80, p. 4689.
Hudson, P.D., 1970, Planet. Space Sci., 18, p. 1611.
Levy, R.H., H.E. Petschek, and G.L. Siscoe, 1964, AIAA J., 2, p. 2065.
Lyons, L.R., and T.W. Speiser, 1982, J. Geophys. Res., 87, p. 2276.

Park, W., D.A. Monticello, and R.B. White, 1983, "Reconnection Rates of Magnetic Fields," Princeton Plasma Phys. Lab., PPL 2014 (submitted to Phys. Fluids).
Parker, E.N., 1963, Astrophys. J. Suppl. Ser., 8, p. 177.
Paschmann, G., B.U.Ö. Sonnerup, I. Papamastorakis, N. Sckopke, G. Haerendel, S.J. Bame, J.R. Asbridge, J.T. Gosling, C.T. Russell, and R.C. Elphic, 1979, Nature, 282, p. 243.
Petschek, H.E., 1964, "The Physics of Solar Flares," (W.N. Hess, ed.), NASA SP-50, p. 425.
Russell, C.T., and R.C. Elphic, 1978, Space Sci. Rev., 22, p. 681.
Scudder, J.D., K.W. Ogilvie, and C.T. Russell, 1983, J. Geophys. Res. (submitted).
Sonnerup, B.U.Ö., 1969, J. Geophys. Res., 74, p. 1301.
Sonnerup, B.U.Ö., 1970, J. Plasma Phys., 4, p. 161.
Sonnerup, B.U.Ö., 1979, "Solar System Plasma Physics," Vol. III, (L.T. Lanzerotti, C.F. Kennel, and E.N. Parker, eds.), North Holland, p. 45.
Sonnerup, B.U.Ö., and E.R. Priest, 1975, J. Plasma Phys., 14, p. 283.
Sonnerup, B.U.Ö., G. Paschmann, I. Papamastorakis, N. Sckopke, G. Haerendel, S.J. Bame, J.R. Asbridge, J.T. Gosling, and C.T. Russell, 1981, J. Geophys. Res., 86, p. 10049.
Spitzer, L., Jr., 1962, "Physics of Fully Ionized Gases," Interscience.
Soward, A., 1982, J. Plasma Phys., 28, p. 415.
Soward, A.M., and E.R. Priest, 1977, Phil. Trans. Royal. Soc. London, 284, p. 369.
Soward, A., and E.R. Priest, 1982, J. Plasma Phys., 28, p. 335.
Sweet, P.A., 1958, Proc. IAU Symposium No. 6, "Electromagnetic Phenomena in Cosmic Physics," (B. Lehnert, ed.), Cambridge U. Press, p. 123.
Terasawa, T., 1983, Geophys. Res. Lett., 10, p. 475.
Vasyliunas, V.M., 1975, Revs. Geophys. Space Phys., 13, p. 303.
Yeh, T., and W.I. Axford, 1970, J. Plasma Phys., 4, p. 207.

DISCUSSION

Sturrock: In order to explain a solar flare, we need a system which exhibits an "explosive" or "hard" instability. What do we know about the conditions which determine whether or not reconnection is an explosive process?

Sonnerup: In my view, not much. A few suggestions for explosive instability behavior have been made. For example, explosive behavior of the ion tearing mode has been predicted by Galeev and coworkers. However, this result has been contested by Pellat.

On the whole, I believe that the type of behavior observed in solar flares is not determined solely by local conditions at the reconnection line, but by the global configuration and the dynamic accessibility of the free energy stored in it.

Mahajan: What is the definition of a "classical picture"? Does it have anything to do with the form of the Ohm's law?

Sonnerup: No, it does not refer to classical versus anomalous resistivity. I used the phrase "classical reconnection models" for the Sweet-Parker and the Petschek models.

Vasyliunas: On the definition of reconnection: 1) As you pointed out, definitions in terms of plasma flow or in terms of electric field are practically equivalent; the advantage of the former is that the trivial vacuum case is excluded. 2) It may be preferable to define reconnection globally, as palsma flow across an electric field E in the separatrix surface; the existence of E along the separator line can then be deduced as a consequence. 3) Various terms: reconnection, field line merging (preferable in my opinion, for linguistic reasons), field annihilation, etc. are all synonymous.

Sonnerup: (1) I prefer to think of flux transfer as the principal characteristic of reconnection. Since such transfer occurs unimpeded in a vacuum, that case must of course be included in the definition; I do not see that as a disadvantage. The definition in terms of plasma flow across a separatrix does, on the other hand, have the disadvantage that the case of annihilation of exactly antiparallel fields (e.g., the stagnation point flows studied by Priest and myself a few years ago) does not qualify as reconnection under such a definition. (2) I actually prefer the local definition because the reconnection actually takes place at the separator. From this definition one can then deduce (with minimal additional assumptions) that an electric field is also present elsewhere on the separatrix surface. (3) It is true that the terms reconnection, merging, and annihilation are used more or less interchangeably. However, in my view it would be desirable to make a slight distinction: reconnection describes the case of a distinct separator line; annihilation describes the case where the separator has degenerated to a surface (this occurs in a current sheet without a normal magnetic field component). Finally, merging could be used to incorporate both of the preceding situations.

Bratenahl: I am fascinated with Nancy Crooker's idea that the maximum reconnection rate occurs when the fields are antiparallel. With the new calculational machinery, are we on the track of being able to settle this appealing idea of Crooker?

Sonnerup: Ordinary reconnection theory predicts a simple formula for the reconnection electric field as a function of the angle between the two reconnecting magnetic fields. This formula gives a maximum when the fields are anti-parallel. Nancy Crooker's proposal is more radical. She argues that reconnection may occur only when the two reconnecting magnetic fields are exactly, or very nearly, antiparallel. I do not know of any strong theoretical or observational support for this idea. In a tokamak the reconnecting fields form only a very small angle. On the other hand, the recent collision free electron tearing mode analysis by Coroniti and Quest does indicate a fairly strong dependence of the growth rate on the angle between the fields.

A definite answer to the question must await a better understanding of the role played by B in the diffusion region.

Van Hoven: Can the width of diffusion region be uniquely specified for the case of steady reconnection?

Sonnerup: In the case where classical resistivity η multiplied by the current density is used to balance the electric field in the diffusion region, the width x^* is such that the magnetic Reynolds number $M_o v_1 x^*/\eta \simeq 1$. In that case x^* simply becomes as small as is required in order to maintain this approximate equality. If, on the other hand, the effective resitivity is a function of current density or of the reconnection rate, then the situation may be different. For example, in my paper I have shown that the width of a diffusion region in which ion inertial effects dominate is always of the order of the ion inertial length λ_i; (or perhaps the ion gyroradius). Similarly, a diffusion region dominated by electron inertial effects would have width $x^* \sim \lambda_e$.

PATCHY RECONNECTION AND MAGNETIC ROPES IN ASTROPHYSICAL PLASMAS

C.T. Russell
Department of Earth and Space Sciences
and
Institute of Geophysics and Planetary Physics
University of California, Los Angeles
California, USA, 90024

ABSTRACT

Reconnection is clearly observed at the terrestrial magnetopause but seldom in the simple geometry originally proposed. Most often reconnection is patchy, forming tubes of twisted flux. The passage of one of these twisted tubes has been called a flux transfer event. Similar twisted tubes, or flux ropes, are formed at Venus by velocity shear. These tubes become so highly twisted that they become kink unstable. The presence of the kink instability suggests a way of creating compound flux ropes as have been postulated to be necessary to explain photospheric magnetic structure.

1. INTRODUCTION

Reconnection in astrophysical plasmas is difficult to treat theoretically and difficult to study observationally. Most places where we believe reconnection is occurring such as on the sun, in pulsar magnetospheres and in radio galaxies the site of reconnection is inaccessible to direct probing. Thus, we must infer what is occurring from indirect evidence. In planetary magnetospheres, however, we do have in-situ data and can directly probe the reconnection site. Nevertheless, we still must infer, rather than observe, many of the properties of the reconnection process. Reconnection, perhaps because it is a time-varying phenomenon, or perhaps because of the intrinsic three-dimensionality of planetary magnetospheres, produces three-dimensional structures. These three-dimensional structures are difficult to investigate with our spacecraft because they are carried over the point of observation with usually unknown velocities. Further, these structures may not be time stationary. If these complexities were not enough to slow down progress, there is yet one more complication, velocity shear. Reconnection, occurring on the forward hemisphere of planetary magnetospheres, does so at the interface between two flowing plasmas whose direction of flow might be quite different. The existence of velocity shear in addition to magnetic shear is often not appreciated in treating reconnection at the magnetopause.

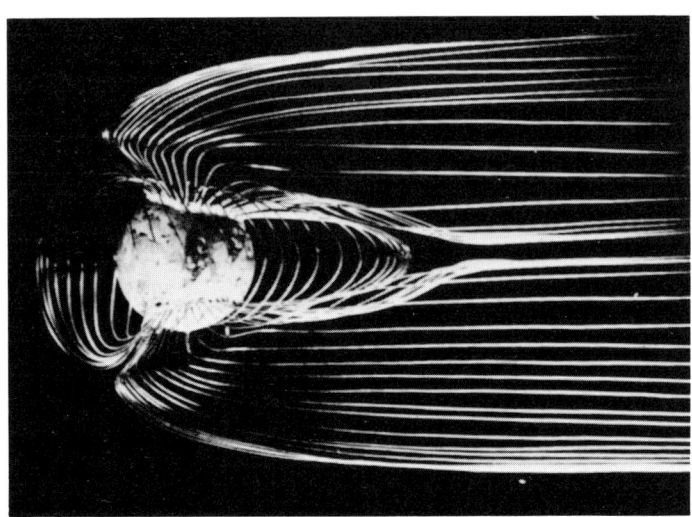

Figure 1. Wire model of magnetosphere.

Velocity shear produces flux ropes, twisted bundles of magnetic field lines. At a planetary magnetopause such flux ropes connect the magnetospheric plasma and field to that of the shocked solar wind. Flux ropes are also found in the Venus ionosphere and on the sun. In order to understand some of the behavior of such flux ropes, we will examine the observed properties of Venus flux ropes and draw some inferences regarding the possible behavior of solar flux ropes.

2. RECONNECTION AT THE TERRESTRIAL MAGNETOPAUSE

The geometry of the reconnection process at the terrestrial magnetopause is intrinsically three-dimensional. Figure 1 shows a three-dimensional wire model of the terrestrial magnetic field (Podgorny, 1976). The polar field lines are swept back by the tangential stresses of the solar wind plasma, including the Maxwell stress due to reconnection. The solar wind plasma is super-magnetosonic and thus a standing, detached shock wave forms in front of the terrestrial magnetosphere. The interplanetary magnetic field is carried by the solar wind through this bow shock and it becomes distorted as the shocked solar wind is deflected around the magnetospheric cavity. Not only is this distortion difficult to treat analytically but also the magnetic stresses affect the flow field. Thus, global treatments of this problem generally involve computer models and simulations.

If we assume that the reconnection process is steady-state and that one need not consider the global geometry, then the configuration sketched in Figure 2 applies. Magnetized plasma flows from the left (shocked solar wind) and the right (magnetosphere), and exits top and bottom, being accelerated by the magnetic sling shot configuration. In this way magnetic tension produces directed flow. A spacecraft such as the

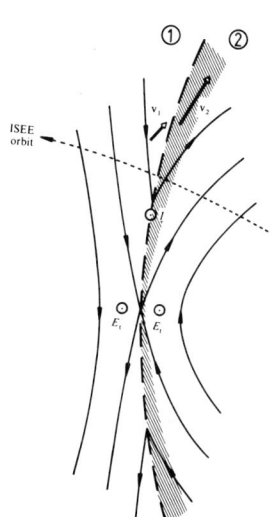
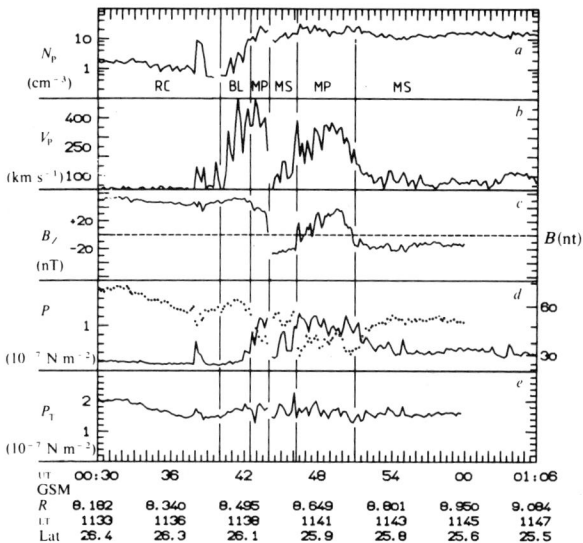

Figure 2. Reconnection at magnetopause.

Figure 3. ISEE observations of reconnection.

International Earth-Sun Explorer (ISEE) would see these various flows sequentially as the magnetopause crossed the spacecraft. ISEE, in fact, does see such flows on occasion as shown in Figure 3 (Paschmann et al., 1979). As indicated in the top panel (plasma number density, N_p), the spacecraft passes out of the ring current plasma (RC), into the boundary layer plasma (BL) which is denser and cooler than the magnetospheric ring current plasma. Then the spacecraft moves through the magnetopause (MP) and into the magnetosheath (MS). The magnetopause is oscillating at this time and moves back over the spacecraft again but the spacecraft does not return entirely to the magnetosphere. Rather the magnetopause moves back towards the earth and the spacecraft enters the magnetosheath plasma once and for all. While in the magnetopause, and to a certain extent on either side of it, vertical (northward directed) flows are observed quantitatively, not just qualitatively, as expected for reconnection. The middle panel, B_z, shows the vertical component of the magnetic field. The sudden changes in this quantity marks the current layer which is the magnetopause. The bottom two panels give respectively the two components of the plasma pressure, kinetic and magnetic, and their sums. The magnetic pressure (dots) dominates here, both in the magnetosphere and the magnetosheath, but not in the magnetopause.

This early observation with the ISEE mission was an important one because it showed for the first time that plasma was accelerated at the magnetopause as predicted by simple reconnection theory. However, such behavior was not always observed. Nevertheless, a sufficient number of similar events occurred so that it was clear that such reconnection was an important contributor to magnetospheric energetics (Sonnerup et al., 1981).

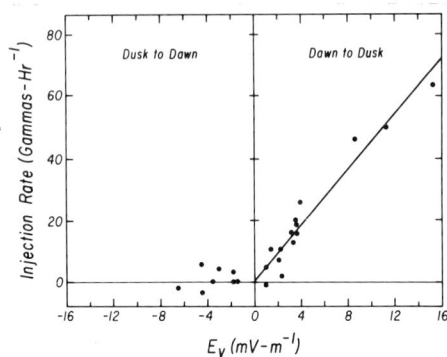

Figure 4. Ring current injection rate as a function of the interplanetary electric field.

These in-situ observations prove that reconnection occurs on the dayside magnetopause, but there are too few of these observations to establish clearly the factors that control the rate of reconnection. It has been postulated that the magnetosphere behaves as a half-wave rectifier, reconnecting with the magnetosheath magnetic field whenever the angle between the magnetospheric and magnetosheath fields exceeds 90°. The support for this postulate comes from indirect evidence, i.e., from the response of geomagnetic activity to varying directions of the interplanetary magnetic field. Figure 4 shows that the energization rate of the ring current, which is the main energy storage reservoir during a magnetic storm, is proportional to the east-west interplanetary electric field, i.e., the product of the solar wind velocity and the north-south component of the interplanetary electric field, whenever the magnetic field has a southward component, and is zero otherwise.

The site of reconnection is also under study. It has been proposed that reconnection occurs most rapidly where the magnetosheath magnetic field is exactly antiparallel to the magnetospheric magnetic field (Crooker, 1979). To calculate where these sites are located one must calculate how the magnetosheath magnetic field is distorted by the magnetosheath flow field as well as take into account the distortions in the magnetosheath field. This has been done (Luhmann et al., 1983) and the predicted sites of reconnection are found to be quite sensitive to the direction of the interplanetary magnetic field. Attractive as the Crooker conjecture appears, it does not appear that the magnetopause behaves as predicted. Sonnerup et al. (1981) find that a near equatorial merging site could explain all their data, whereas the Crooker reconnection sites only occur near the equator for a small range of orientations of the interplanetary field. More recently Daly et al. (1983) have demonstrated that patchy reconnection in the ISEE data appears always to start in the equatorial region.

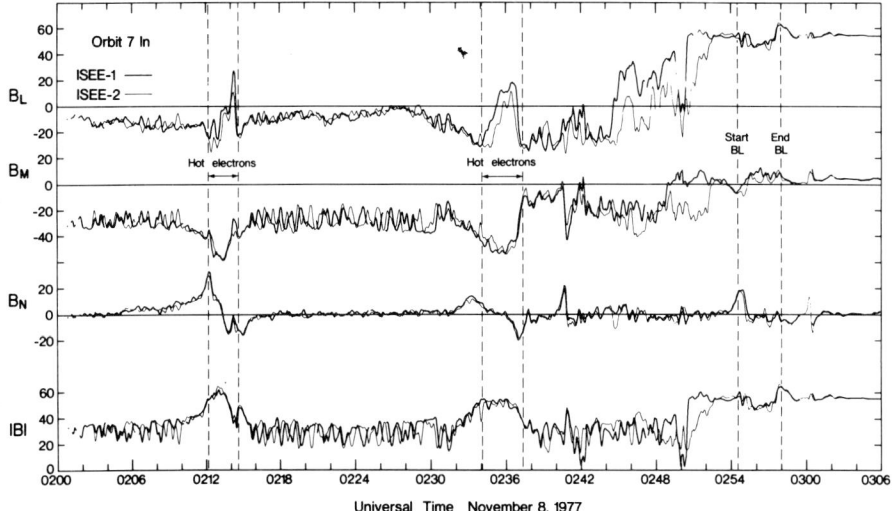

Figure 5. ISEE-1 and -2 magnetometer observations of flux transfer events in the magnetosheath near the magnetopause.

3. PATCHY RECONNECTION

The ISEE-1 and -2 spacecraft were launched into the same orbit with the capability of varying their separation to allow the measurement of boundary velocities and hence allow their thicknesses to be determined. Figure 5 shows magnetic field measurements from the two ISEE spacecraft surrounding a magnetopause traversal that occurred at about 0250 UT (Russell and Elphic, 1978; 1979). The magnetic field is displayed in boundary normal coordinates. The BN component is along the expected boundary normal and the BM component is in the plane of the boundary roughly along the direction of the magnetospheric field. This figure illustrates the irregularity of the magnetosheath field even under moderately quiet conditions as we have here. The magnetic field profile across the magnetopause looks quite different at the two spacecraft. These differences are in part due simply to the variability of the magnetopause velocity. Not only is this velocity large but it is quite variable and irregular.

Two very strange features appear in the data at 0214 and 0236 UT. The field first points outward from the magnetopause and then points inward. At the same time the BM component strengthens in the downward direction and the BL component varies in a manner suggestive of a partial magnetopause traversal. The plasma data at this time indicates the spacecraft did not leave the magnetosheath. However, the energetic electron and ion data resemble magnetospheric data.

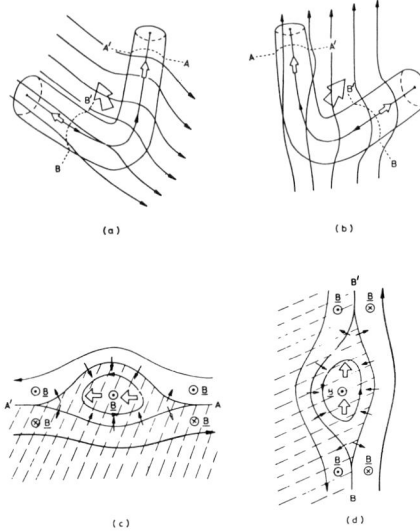

Figure 6. Schematic of a flux transfer event.

Our explanation of this phenomenon is shown in Figure 6 (Cowley, 1982). A flux tube of magnetosheath field has become connected to the magnetospheric field and energetic particles are leaking out. Panel (a) shows the tube as seen from the magnetosheath. Magnetosheath field lines drape up and over it. Panel (b) shows the tube as seen from the magnetosphere. The magnetospheric field lines are also distorted by the presence of the tube. As the tube moves across the magnetopause, both carried along by the flow and because of the straightening up of the flux tube, a characteristic +/- signature is created in the normal component of the magnetic field. In panels (c) and (d) are shown cross sections of the flux tube. This interpretation is supported by various studies of the particle signatures (Scholer et al., 1982; Daly and Kepler, 1982, 1983a,b; Daly et al., 1981, 1983).

Such structures occur in pairs, one connected to the southern hemisphere and one to the north. If they are formed close to the north-south dividing line of the flow field, they will be pulled to the south and north, respectively. However, it is possible that southern connected events get pulled northward and vice-versa. Energetic particle data allow this to be sorted out (Daly et al., 1983). ISEE measurements were initially in the northern hemisphere so that only northward moving FTE's were seen at the start of the mission. However, later southward moving FTE's were detected (Rijnbeek et al., 1982).

The motion of FTE's and the observed particle anisotropy points to the near equatorial region as the site of the initiation of reconnection (Daly et al., 1983). Steady-state reconnection is thought to occur there too (Sonnerup et al., 1981). In fact, it is not clear whether

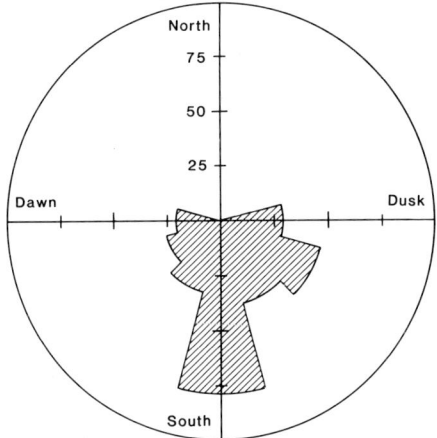

Figure 7. Flux transfer event occurrence rate as a function of the direction of the interplanetary magnetic field in Y-Z plane.

the patchy reconnection characteristic of FTE's is independent of steady-state reconnection. On several occasions both phenomena are observed on the same pass (Rijnbeek et al., 1983). It is possible that flux transfer events are formed at the edges of the "steady-state" reconnection site.

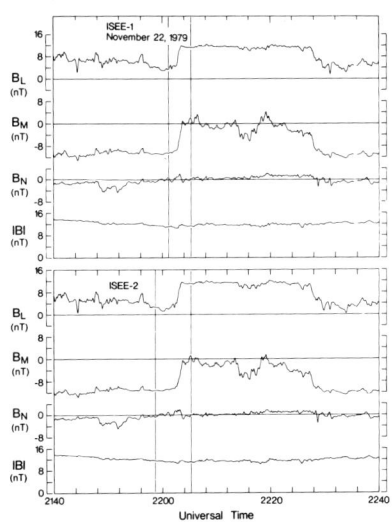

Figure 8. ISEE-1 and -2 magnetometer observations of quiet magnetopause.

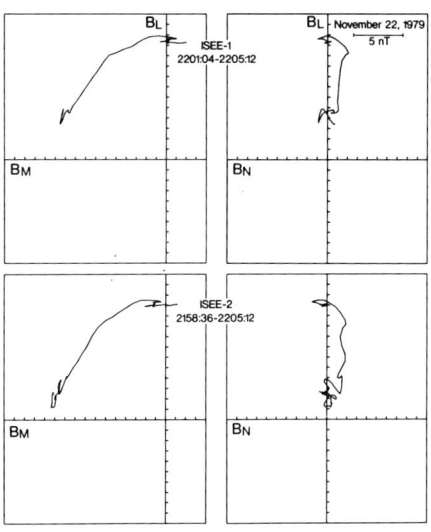

Figure 9. Hodograms of quiet magnetopause crossing.

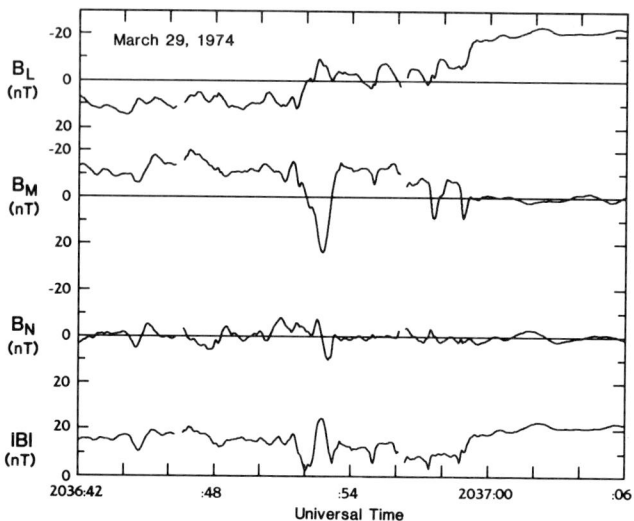

Figure 10. Flux transfer event at Mercury.

The occurrence of flux transfer events is controlled intimately by the direction of the interplanetary magnetic field (IMF). When the IMF is northward in solar magnetospheric coordinates, flux transfer events do not occur. When the IMF is southward, FTE's occur from 50 to 100% of the time depending on the strictness of the definition used. This is shown in Figure 7.

It should be emphasized that the "classical" flux transfer event is the end member of a continuum of structure. Figure 8 shows the magnetic field observed by ISEE-1 and -2 through the magnetopause during a very quiet period when the solar wind was sub-Alfvenic (Gosling et al., 1982). The important point to note is that there is still structure in the normal component. Another representation of the same data is shown in Figure 9. Here hodograms are plotted of the tip of the magnetic field vector in the plane of the boundary on the left and in the orthogonal plane on the right. There is still much structure in the normal component B_N. This structure is probably evidence for tearing of the boundary.

We note that patchy reconnection is not restricted to the terrestrial magnetosphere. Figure 10 shows a flux transfer event occurring at the magnetopause of Mercury and Figure 11 shows a flux transfer event at the Jovian magnetopause (Walker and Russell, 1983).

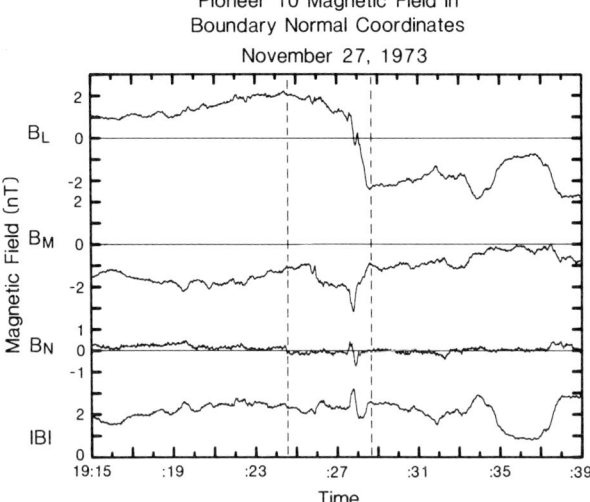

Figure 11. Flux transfer event at Jupiter.

The +/- signature in the normal component as seen in these Figures as well as Figure 5 is only partially due to motion and the draping of field lines. The signature in BN internal to the flux rope suggests this (Paschmann et al., 1982; Cowley, 1982) as does the BM component, especially the BM component seen simultaneously on two opposite sides of an FTE (Saunders and Russell, 1983). This twist probably arises from velocity shear as the flux rope is rolled across the magnetopause while one of its ends is "frozen" into the ionosphere and cannot rotate. The fact that the flux tube is twisted is very important for it helps maintain its shape and hence the distortion of the surrounding field. Thus, patchy reconnection is controlled by two shears, that of the magnetic field and that of the velocity field. This is a very complex situation. Fortunately, there is a simpler situation in which flux ropes are formed in velocity shears and we can study these ropes to learn more about the formation and evolution of magnetic flux ropes without having to worry about the simultaneous effects of reconnection.

4. VENUS FLUX ROPES

The Pioneer Venus Orbiter was placed into orbit about Venus on December 4, 1978 and periapsis gradually lowered so that the spacecraft could study Venus' upper atmosphere and ionosphere. The first deep penetration of the ionosphere was a complete surprise to the magnetometer investigators. These measurements in spacecraft coordinates are shown in Figure 12. The high field regions on the extreme right and left are the magnetosheath. The region of low fields in the middle is the ionosphere and the dashed line marks periapsis. The surprise was the turbulent-appearing impulsive high fields seen within the ionosphere. While

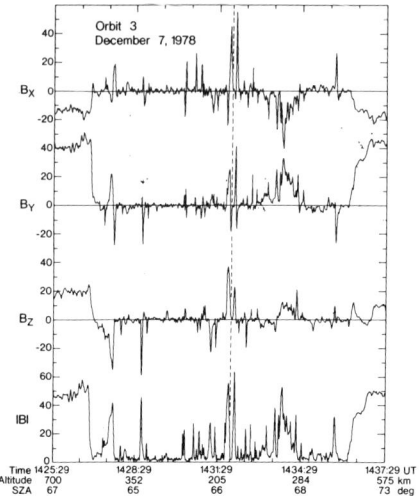

Figure 12. Flux ropes in Venus ionosphere.

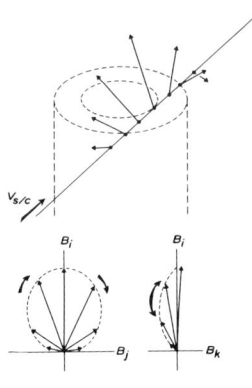

Figure 13. Schematic flux rope traversal.

at first glance these fields seem to be chaotic, there is much order to them at least internally. Figure 13 shows a schematic of the results of minimum variance analysis (bottom panel) and the spatial variation along the trajectory through one of these features (top panel). The hodogram shows that the field variation is intrinsically three-dimensional rather than two-dimensional. A simple three-dimensional structure which can explain the variation is a flux rope in which concentric shells have varying field strength and pitch relative to the rope axis (Russell and Elphic, 1979). Figure 14 (top) shows the structure we deduce for a Venus flux rope. In the center the field lines are straight. As the distance from the center increases, the field weakens and the field twists to a larger and larger pitch angle with respect to the rope axis. The bottom panel presents a picture of the distribution of flux ropes within the ionosphere.

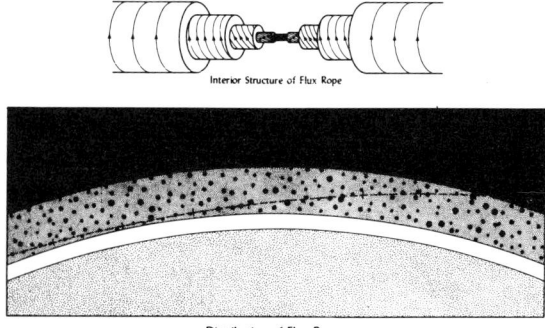

Figure 14. Flux rope schematic and distribution.

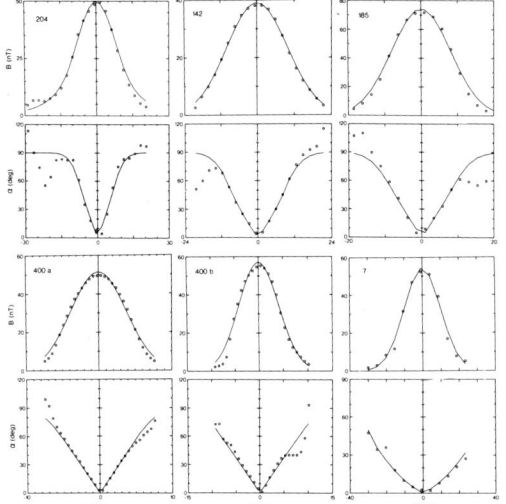
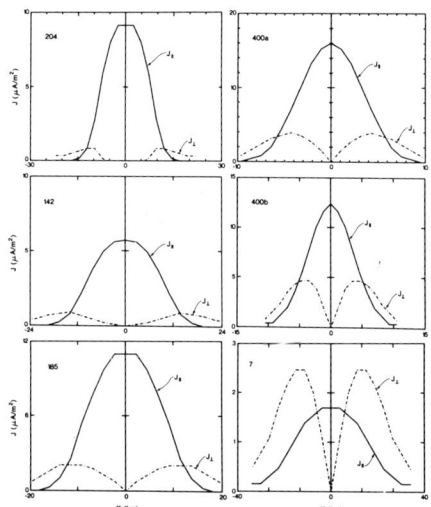

Figure 15. Observed field strength and pitch angle in Venus flux ropes.

Figure 16. Modeled flux rope currents.

We can model the observed flux ropes in an attempt to deduce their physical properties. Figure 15 shows analytic fits to six flux ropes (Elphic and Russell, 1983). It is possible to model these structures quite well with only a few parameters. Although the field variations in these six ropes look qualitatively very similar, the currents we deduce from our model fit shows quite a variation. Figure 16 presents the parallel and perpendicular currents in each rope. On orbit 204, (top left panel), the rope consists almost entirely of currents flowing parallel to the magnetic field. Whereas on orbit 7 the current is predominantly perpendicular to the field with parallel current only in the center of the rope. In other words, the flux rope on orbit 204 is highly twisted and on orbit 7 is weakly twisted. The rope on orbit 204 is essentially what is called a force-free structure. The magnetic tension in the twist balances the magnetic pressure gradient. The rope on orbit 7 is far from self-balancing and must be contained in part by plasma pressure gradients. Such inferences from the magnetic structure are, in fact, confirmed by comparing with the plasma data on board. Weakly twisted ropes have associated plasma pressure changes; highly twisted ropes do not (Elphic et al., 1980).

It seems quite reasonable to assume that Venus flux ropes arise in much the same way as flux transfer events except that reconnection is not involved here. That mechanism is through velocity shear, perhaps coupled to the Kelvin-Helmholtz instability. The region of velocity shear is principally at the ionopause where there is a steep gradient in magnetic field and large ionospheric flows ∿5 km/sec (Knudsen et al., 1980) which decrease with decreasing altitude. Another possible source is the upper edge of the low altitude magnetic belt that often appears during periods of high solar wind dynamic pressure (Luhmann et al., 1980).

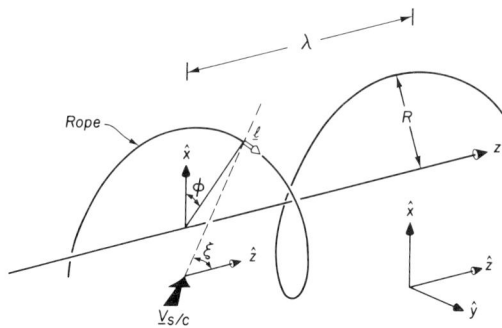

Figure 17. Kink unstable flux rope.

Flux ropes are most greatly twisted at low altitudes on the dayside of Venus. However, their axes seem to become randomized here. This puzzled us at first until we realized that the ropes were becoming so tightly twisted that they were becoming kink unstable (Elphic and Russell, 1983b). Figure 17 shows the resulting corkscrew pattern of the flux rope axis. The fact that Venus flux ropes become kink unstable when they are sufficiently twisted is interesting enough but there are important implications to the fact that the flux rope axis is now three-dimensional. The flux rope can now interact with other flux ropes and get intertwined with them due to the motions of the surrounding plasma. It appears possible to start building more complicated structures such as that shown in Figure 18 in which a compound flux rope is braided out

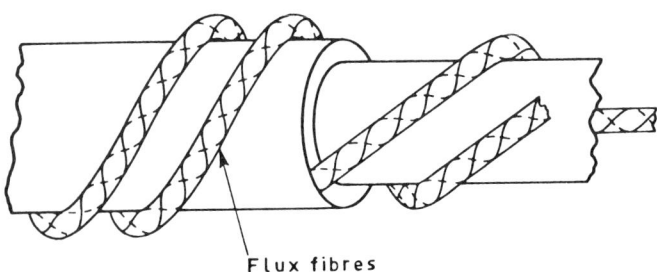

Figure 18. Compound flux ropes.

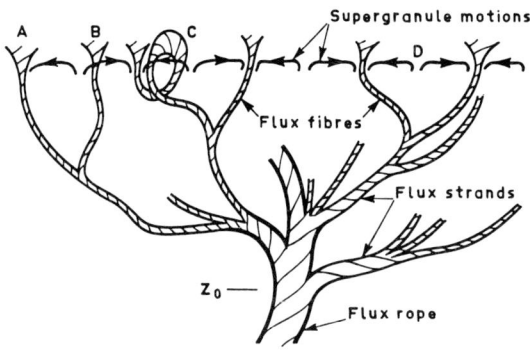

Figure 19. Solar flux rope.

of simpler flux fibers (which correspond to our original flux ropes). We would be hard pressed to see this in the Venus ionosphere where we have to seek out the simplest structures in order to make any sense out of the data. We do not anticipate ever having the luxury of a multiple spacecraft, ISEE-like, mission to Venus. However, there is another region in the solar system in which compound flux ropes have been proposed to exist, that is the sun. Figure 19 shows a compound flux rope, coming up out of the convection zone and splaying out in the photosphere (Piddington, 1981).

5. SOLAR FLUX ROPES

The physics of the photosphere demands the existence of some magnetic structure such as the flux rope depicted here but our ignorance of the convection zone makes it difficult to deduce what goes on there. Thus, one can only conjecture. However, if one were to conjecture based on our Venus experience, we would postulate that deep in the convection zone there is a source of steady magnetic field, probably driven by a solar dynamo. Velocity shear above this magnetized zone wraps up the field into ropes and continued twisting leads to corkscrew flux ropes through the kink instability. Turbulent motions lead to compound intertwined rope structures perhaps through several cycles of kink instability and turbulent intertwining. Finally, perhaps through another kink instability cycle a compound flux rope emerges through the solar surface. At this point the ropes can start to interact through reconnection producing solar flares, etc.

6. SUMMARY AND CONCLUSIONS

Magnetic shear is certainly important for the reconnection process, but velocity shear is also a crucial element in shaping the magnetic structures that we observe in astrophysical plasmas. The simple geometries that we would like to study rarely ever occur. At the terrestrial magnetopause two-dimensional merging occurs either rarely or over a limited region of space. Over most of the magnetopause we find twisted tubes of connected flux. At Venus we also find twisted tubes of flux. Here reconnection seems not to be a critical process in the evolution of these flux ropes. Studying their behavior leads us to the conclusion that they become kink unstable when they become too twisted. The resulting kink instability twists the axis of the flux rope into a corkscrew and enables it to become intertwined with its neighbors. The two key elements in this process are a steep gradient in magnetic field between a magnetized and unmagnetized plasma and velocity shear near the boundary. While we are ignorant of the magnetic properties of the solar convection zone, we conjecture that such a field gradient and velocity shear can be found deep in the convection zone. If so, then they would lead naturally to the formation of compound flux ropes through repeatedly going kink unstable and mixing with other ropes. In conclusion, in studying astrophysical current instabilities, it is often just as important to understand the velocity field as it is to understand the magnetic field.

ACKNOWLEDGMENTS

The research described herein has resulted from the interaction with many individuals. R.C. Elphic deserves special mention in this regard as he did much of both the FTE and the magnetic flux rope analysis. This research was supported by the National Aeronautics and Space Administration under contracts NAS5-25772 and NAS2-9491.

REFERENCES

Cowley, S.W.H.: 1982, Rev. Geophys. Space Phys. 20, pp. 531-565.

Crooker, N.U.: 1979, J. Geophys. Res. 84, pp. 951-959.

Daly, P.W., and Keppler, E.: 1982, Planet. Space Sci. 30, pp. 331-337.

Daly, P.W., and Keppler, E.: 1983a, J. Geophys. Res. 88, pp. 3971-3980.

Daly, P.W., and Keppler, E.: 1983b, J. Geophys. Res. in press.

Daly, P.W., Williams, D.J., Russell, C.T., and Keppler, E.: 1981, J. Geophys. Res. 86, pp. 1628-1632.

Daly, P.W., Saunders, M.A., Rijnbeek, and Sckopke, N.: 1983, Planet. Space Sci. submitted.

Elphic, R.C., and Russell, C.T.: 1983a, J. Geophys. Res. 88, pp. 58-72.

Elphic, R.C., and Russell, C.T.: 1983b, Geophys. Res. Lett. 10, pp. 459-462.

Elphic, R.C., Russell, C.T., Slavin, J.A., and Brace, L.H.: 1980, J. Geophys. Res. 85, pp. 7679-7696.

Gosling, J.T., Asbridge, J.R., Bame, S.J., Feldman, W.C., Paschmann, G., Sckopke, N., and Russell, C.T.: 1982, J. Geophys. Res. 87, pp. 2147-2158.

Knudsen, W.C., Spenner, K., Miller, K.L., and Novak, V.: 1980, J. Geophys. Res. 85, 7803-7810, 1980.

Luhmann, J.G., Elphic, R.C., Russell, C.T., Mihalov, J.D., and Wolfe, J.H.: 1980, Geophys. Res. Lett. 7, pp. 917-920.

Luhmann, J.G., Walker, R.J., Russell, C.T., Crooker, N.U., Spreiter, J.R., and Stahara, S.S.: 1983, J. Geophys. Res. submitted.

Paschmann, G., Sonnerup, B.U.O., Papamastorakis, I., Sckopke, N., Haerendel, G., Bame, S.J., Asbridge, J.R., Gosling, J.T., Russell, C.T., and Elphic, R.C.: 1979, Nature, pp. 243-246.

Paschmann, G., Haerendel, G., Papamastorakis, I., Sckopke, N., Bame, S.J., Gosling, J.T., and Russell, C.T.: 1982, J. Geophys. Res. 87, pp. 2159-2168.

Piddington, J.H.: 1981, Cosmic Electrodynamics, 361 pp., R.E. Krieger Publ. Co., Malabar, Florida.

Podgorny, I.M.: 1976, Physics of Solar Planetary Environments, pp. 241-254, American Geophysical Union, Washington, D.C.

Rijnbeek, R.P., Cowley, S.W.H., Southwood, D.J., and Russell, C.T.: 1982, Nature 300, pp. 23-26.

Rijnbeek, R.P., Cowley, S.W.H., Southwood, D.J., and Russell, C.T.: 1983, J. Geophys. Res. submitted.

Russell, C.T., and Elphic, R.C.: 1978, Space Sci. Rev. 22, pp. 681-715.

Russell, C.T., and Elphic, R.C.: 1979, Nature 279, pp. 614-616.

Russell, C.T., and Elphic, R.C.: 1979, Geophys. Res. Lett. 6, pp. 33-36.

Saunders, M.A., and Russell, C.T.: 1983, Geophys. Res. Lett. submitted.

Scholer, M., Hovestadt, D., Ipavich, F.M., and Gloeckler, G.: 1982, J. Geophys. Res. 87, pp. 2169-2175.

Sonnerup, B.U.O., Paschmann, G., Papamastorakis, I., Sckopke, N., Haerendel, G., Bame, S.J., Ashbridge, J.R., Gosling, J.T., and Russell, C.T.: 1981, J. Geophys. Res. 86, pp. 10049-10067.

Walker, R.J., and Russell, C.T.: 1983, Paper PA.20, presented at IAGA Symposium, Hamburg, August, 1983.

DISCUSSION

Priest: A lot of work has been done on the kink instability of flux tubes in the solar context, especially by Van Hoven, Chiuderi, Einaudi, Hood and Priest. Such tubes go unstable very easily, so what is keeping them stable for most of their time? Is it shear or the anchoring of the tubes in denser layers of the ionosphere?

Russell: It is difficult for us to determine why the Venus flux ropes remain stable as long as they do. They are anchored in the magnetosheath and they are immersed in a shear layer. We believe the important points in our observations are that the ropes are formed and transported some distance before they become unstable. In the subsolar region at high altitudes, presumably close to where the ropes are forming, the rope axes appear to be mainly horizontal; whereas at low altitudes the rope axes assume a variety of orientations.

Sturrock: What are the observed dimensions of the FTE flux tubes? More important, we would like to know what determines the dimensions in order to infer what their dimension would be in the sun's atmosphere, if they occur there.

Russell: Flux ropes vary in diameter from about 10 km at 500 km in the subsolar Venus ionosphere to about 6 km below 200 km altitude. They are nearly twice that width near the terminator (Elphic and Russell, J. Geophys. Res. 88, 1993-3003, 1983). These diameters are from 2 to 4 times either the ion gyroradius using the axial field strength in the rope or the ion inertial length. Since the beta of the plasma in the rope is close to unity, we would expect these scale lengths to be similar

We do not know whether it is coincidental that the ropes have this scale size, or not. My intuition would tell me that the scale size of the velocity shear would play a role.

D. Smith: Although reconnection may not play the dominant role in the sense that flux ropes evolve to the point of becoming kinks unstable, doesn't it play an important role in their evolution? What is the difference between the cases you showed for the terrestrial magnetosphere and the ionosphere of Venus; i.e. why is there no reconnection in the latter?

Russell: Flux ropes and what we have termed flux transfer events both occur in the presence of velocity shear and both evolve into highly twisted structures. The Venus flux ropes occur at the interface between a magnetized and unmagnetized palsma. The terrestrial flux transfer events occur at the interface between two magnetized plasmas. Spatially and temporally limited reconnection forms a flux tube which then rolls along the interface. The absence of an intrinsic planetary magnetic field at Venus keeps the ionosphere field-free and limits reconnection there. Flux ropes could reconnect with themselves when they went kink unstable. This would produce a straighter rope plus a field torus. I do not believe we would recongize such a structure with a single spacecraft if we saw it.

Goldstein: If your spacecraft passes a small current loop, you would see a triple event instead of the double you call an FTE. Have you seen any triples?

Russell: In the well-developed, well-isolated events we have studied closely we have not seen any triples. However, if one examines finer and finer detail in the observations in situations where many events are occurring, one could find the occurrence of triples. I would hesitate to attribute such structure to a current loop without corroborating data from several spaced satellites.

Tsinganos: I wonder if you had some idea as to how these complicated "flux tree" ropes are kept in a steady equilibrium; because from the theoretical point of view, there are indications that these asymmetric and complex structures are unlikely to be found in equilibrium.

Russell: It is not obvious to me that solar magnetic structures are in equilibrium. They are constantly evolving. The way such structures might exist is that they are formed stably in one region and are transported to another in which we observe them out of equilibrium.

Krishan: How do you separate the spatial and temporal fluctuations in the magnetic field?

Russell: At the earth we are probing these structures with dual satellites separated by a controllable separation of from 100 to 1000 km. To the extent that the structure is moderately simple and moving past our two spacecraft rapidly, we can separate temporal from spatial changes. At Venus our one spacecraft is passing through the ionosphere at a super-Alfvenic velocity so that the structure does not have time to change during the measurement interval.

Vlahos: In one of your figures the interacting fields were homogeneous. What localized the ropes?

Russell: The figure to which I believe you refer concerned the formation of Flux Transfer Events at the earth's magnetopause. These events appear to result from temporally and spatially localized reconnection. I presume that whatever modulates the reconnection rate controls the size of an event. Perhaps fluctuations in the direction of the interplanetary magnetic field do this. Perhaps the scale size of natural variations in the boundary structure such as caused by the Kelvin-Helmholtz instability are responsible.

MAGNETIC FLUX ROPES OF VENUS: A PARADIGM FOR HELICAL MAGNETIC STRUCTURES IN ASTROPHYSICAL SYSTEMS

R. C. Elphic, NAS/NRC Research Associate, Laboratory for
Planetary Atmospheres, NASA/Goddard Space Flight Center,
Greenbelt, MD 20771

The magnetic flux ropes of Venus are small scale (ion gyroradius) cylindrically symmetric structures observed in situ by the Pioneer Venus orbiter in the largely magnetic field-free ionosphere of the planet. They are so named because of their helical magnetic structure, which in turn is due to primarily field-aligned currents within the rope. Empirical models can be used to examine the current structure in detail, and these models indicate that flux ropes may be unstable to the helical kink mode. Statistics of rope distribution and orientation also support this instability picture. The results of investigations into the direct measurements of Venus flux ropes may be relevant to certain astrophysical phenomena that must be observed remotely.

Flux ropes are basically tubes of magnetic flux; the field is strongest at the center where it is also purely axial in orientation. At greater distances from the axis, the field becomes weaker and more helical until at large distances it is very weak and almost entirely azimuthal. The structures envisioned here are cylindrically symmetric, in that if z is taken along the rope axis the field is

$$\underline{B}(\rho) = B_\phi(\rho) \hat{\phi} + B_z(\rho) \hat{z}$$

where ρ and ϕ correspond to the usual cylindrical coordinates. Such a structure automatically satisfies $\nabla \cdot \underline{B} = 0$. The field components could be further described as $B_\phi(\rho) = \overline{B}(\rho) \sin\alpha(\rho)$ and $B_z(\rho) = B(\rho) \cos\alpha(\rho)$ where $\alpha(\rho)$ is the helical pitch of the field, which satisfies the boundary conditions $\alpha(\rho) \to 0$ as $\rho \to 0$, and $\alpha(\rho) \to \pi/2$ as $\rho \to \infty$.

The magnetic field structure of flux ropes can be explicitly modeled for cases in which the spacecraft passed very close to the center of the rope. Elphic and Russell (1983a) have fit a model to magnetic field data, and took the curl of the model to obtain the current densities flowing locally parallel and perpendicular to the magnetic field, shown in Figure 1. In most cases, as here, field-aligned currents dominate the rope structure, suggesting that $\underline{J} \times \underline{B}$ forces are small and that ropes tend toward a force-free configuration.

Modeling the detailed magnetic structure of ropes allows us to evaluate their stability to pinch-related perturbations. For example, we may use the 'sausage' pinch (m = 0) instability or 'helical kink' (m = 1 or -1) instability criteria to evaluate the susceptibility of

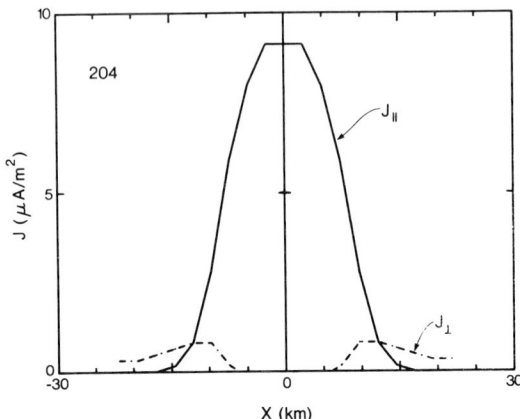

Figure 1. Current densities parallel and perpendicular to local magnetic field.

ropes to these modes. None of the flux ropes heretofore modeled even approach instability to the sausage pinch. The 'helical kink' however has a lower threshold for instability than the sausage pinch. If we approximate the flux rope magnetic field as $B_z = B_{z0}$, $B_\phi = 0$ for $0 < \rho \leqslant a$, and $B_z = 0$, $B_\phi = B_{\phi 0} a/\rho$ for $\rho > a$ then the ropes are unstable to the spiral kink for wavenumber k (in the limit ka << 1) if $B_{\phi 0} > ka\ B_{z0}$ (Hasegawa, 1975). Taking characteristic values of $B_{\phi 0}$, B_{z0} and rope scale radius a (5-10 km), the wavelength of marginal stability is about 90 km, longer than rope scale size a, but shorter than characteristic ionospheric horizontal length scales (thousands of kilometers).

One piece of evidence supporting the helical kink instability is the relative occurrence or fractional volume occupied by flux ropes as a function of altitude. As shown in Figure 2 of Elphic and Russell (1983b), the fractional volume occupied by ropes increases with decreasing altitude, particularly in the region between 0° and 45° solar zenith angle. This is to be expected if ropes convect from the ionopause (the upper boundary of the ionosphere) down to lower altitudes due to magnetic tension. As they move into denser ionosphere at lower altitudes, the ropes must push through more mass and, consequently, move more slowly. As they slow, the vertical separation between ropes decreases; more ropes are observed per unit volume. The convection speed V_c varies inversely as the square root of the height-dependent ionospheric mass density, and the flux rope volume density N can be determined from observations of occurrence and scale size. The product of N and V_c should be a constant over all altitudes. Using the actual values, however, the supposedly constant product above in fact varies with altitude, growing systematically from a normalized value of 1 at 300 km and above to 2.8 at 160 km. This suggests that there are more ropes at low altitudes than would be predicted by such a mechanism. In fact, however, the rope volume density N is valid only for downwardly convecting horizontal, straight flux tubes. If the helical kink

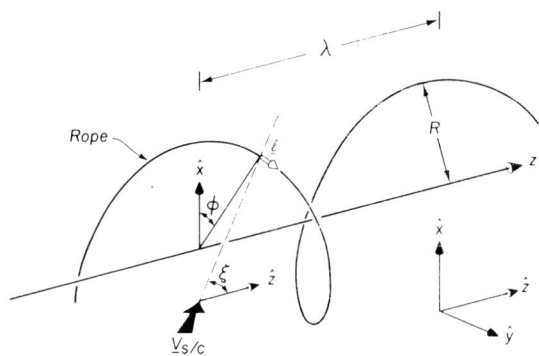

Figure 2. Arbitrary traversal of a flux rope that was helical kink unstable.

instability is producing "corkscrew shaped" flux ropes, the actual volume occupied by the rope increases over that of a simple straight flux rope, since the volume increases as the length of the tube increases due to the helical kink. To match the values above, a helix of pitch $\sim 70°$ must form by the time ropes convect down to 160 km.

We can independently check the foregoing explanation for the increase in fractional volume occupied by flux ropes by investigating the observed distribution of rope orientations at low altitudes. The results of Figure 6 of Elphic and Russell (1983b) suggest that, for the subsolar regions, the low altitude cases tend to favor vertical orientations, while the cases above 200 km tend to be found in more nearly horizontal orientations. This is qualitatively to be expected of a rope that has gone helical kink unstable.

We can compute the expected distribution of occurrence frequency with inclination i by evaluating the overall probability of observing a flux rope with inclination i, given that it is part of a large scale helical structure whose axis is horizontal. Because the spacecraft trajectory is mainly parallel to the planet's surface at these altitudes, the problem simplifies to the situation shown in Figure 2. Here, the spacecraft trajectory forms the approach angle ξ with the kink spiral axis, and the spacecraft passes through the rope at a helical aximuth ϕ. The probability of observing any particular segment of the helix depends on the effective cross-section of that segment, which in turn depends on whether the flux tube element ℓ is perpendicular or parallel to the spacecraft trajectory. The probability of observing the rope is maximum when ℓ is at right angles to the line of sight, but vanishingly small when ℓ is along the line of sight. The probability of observing a rope at inclination i can be calculated based on helical azimuth ϕ, spiral pitch Θ and approach angle ξ. This quantity must be averaged over all approach angles, in effect summing over all the spacecraft trajectories, to yield an expectation distribution of occurrence of ropes with inclination i.

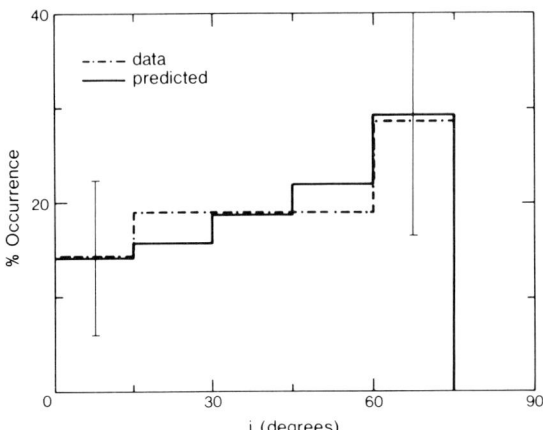

Figure 3. Distribution of the expected and observed occurrence frequency of ropes of inclination i relative to local horizontal. Helical kink pitch is 70°.

Figure 3 shows the resulting distribution of probable occurrence frequency over inclination i given the helical kink pitch angle (solid line) together with the actual observed distribution for the low solar zenith angle cases below 200 km altitude (dashed line). The inclination used here is actually the value defined by $\sin i = |\sin \theta \sin \phi|$ which folds both downward and upward pointing ropes into the same bin. The pitch ϕ chosen for this distribution is the value producing the best fit, in this case 70°, as compared to a value of 70° inferred from the occurrence vs. altitude results discussed earlier. A pitch of either 60° or 80° gives a poorer fit, suggesting that an average pitch of roughly 70° indeed develops by the time ropes have gotten to altitudes below 200 km.

We may conclude that the magnetic structure of flux ropes indicates they are helical kink unstable for $\lambda > 100$ km. The altitude distribution of the fractional volume occupied by ropes can be explained by invoking the helical kink instability as ropes convect to lower altitudes. Furthermore, the low altitude orientations of ropes can also be explained by helical kink instability. It may be that this instability is at work in other astrophysical phenomena. For example, it may cause filaments of sub-photospheric magnetic flux to protrude into the solar atmosphere.

References

Elphic, R. C., and Russell, C. T.: 1983a, J. Geophys. Res. 88, 58.
Elphic, R. C., and Russell, C. T.: 1983b, J. Geophys. Res. 88, 2993.
Hasegawa, A.: 1975, Plasma Instabilities and Nonlinear Effects: Physics and Chemistry in Space, Springer-Verlag, New York.

LABORATORY EXPERIMENTS ON CURRENT SHEET DISRUPTIONS, DOUBLE LAYERS TURBULENCE AND RECONNECTION

R. L. Stenzel and W. Gekelman
Department of Physics, University of California, Los Angeles,
CA 90024

ABSTRACT

The role of laboratory experiments to the understanding of current systems in space plasmas is reviewed. It is shown that laboratory plasmas are uniquely suited to make detailed investigations of basic physical processes in current-carrying plasmas. Examples are given for double layers, current-driven instabilities, and the plasma dynamics at magnetic neutral points during reconnection. Observations of current sheet disruptions show the coupling between local plasma phenomena (double layers) and global circuit properties (magnetic energy storage).

INTRODUCTION

Although laboratory plasmas and astrophysical plasmas consider a vastly different parameter regime, unstable current systems are encountered in both cases. Although the observational methods are opposite, local versus remote, there is evidence for common physical processes: instabilities, plasma energization, reconnection. The investigation of these processes from various viewpoints contributes to the general understanding even though a one-to-one correspondence is often not established.

Carefully designed laboratory experiments establish stable plasmas with simple, known boundary conditions, and a variety of diagnostic tools for in-situ time and space resolved measurements of fields, waves, particle distributions. These features are especially suited for observing localized processes such as double layers which have been well established in the laboratory (Torvén, 1979; Sato, 1982). Similarly, electrostatic and short-scale length electromagnetic turbulence has been successfully studied in many laboratory plasmas carrying currents (Gekelman and Stenzel, 1978; Hollenstein and Guyot, 1983; Stenzel, 1977; Gekelman and Stenzel, 1983). The investigation of magnetic reconnection requires plasmas of larger dimensions, on the order of the ion collisionless skin depth c/ω_{pi}. In earlier investigations (Baum and Bratenahl, 1980; Frank, 1976; Ohyabu et al., 1974) this has been established by high densities ($n_e \simeq 10^{15}$ cm^{-3}) and small physical sizes (1 ... 10 cm) which makes the diagnostics difficult. Recent reconnection experiments at UCLA (Stenzel and Gekelman, 1979; Gekelman et al., 1982; Stenzel et al., 1983a) have the opposite approach; large scales (1 ... 2m) and medium densities ($n_e = 10^{12}$ cm^{-3}) so as to optimize the diagnostic aspect. In these experiments a classical magnetic neutral sheet has been established; reconnection and particle energization are observed facts. The current sheet associated with the magnetic topology of a flattened X-point exhibits microscopic instabilities (magnetic turbulence) and macroscopic instabilities (current disruptions). The latter case is most interesting since reconnection, current sheet disruptions and double layers occur simultaneously. It is an example of the complicated but probably realistic nonlinear plasma dynamics at magnetic null points which also requires the knowledge of the global circuit properties.

EXPERIMENTAL CONFIGURATION

The UCLA reconnection experiments are performed in a device shown schematically in Fig. 1. A linear discharge plasma column is generated

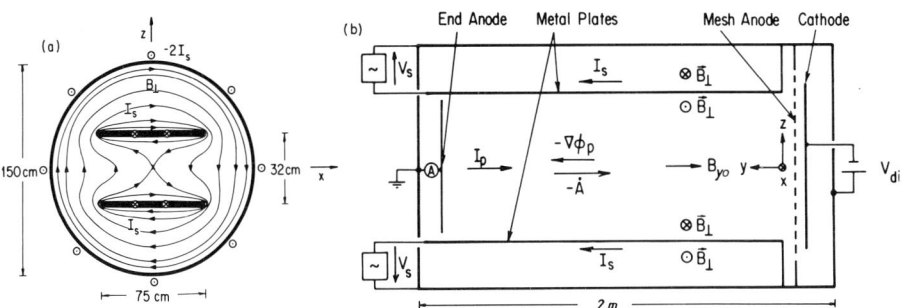

Fig. 1. Schematic picture of the experimental arrangement. (a) Cross-sectional view showing parallel plate electrodes with pulsed currents I_s and magnetic field lines $\vec{B}_\perp$ without plasma. (b) Side view of the device with main electrodes, currents (I_p, I_s) electric fields ($\vec{E} = \vec{A} - \nabla\phi_p$), and magnetic fields ($\vec{B} = \vec{B}_\perp + \vec{B}_{yo}$). The coordinate system common in magnetospheric physics has been adopted where y is along the neutral line (device axis), x is along the horizontal neutral sheet, and z is normal to the sheet.

with a 1 m diameter cathode. Detailed plasma diagnostic tools are employed in conjunction with a state-of-the-art digital data processing system. Time and space resolved probe measurements of fields ($\vec{E}$, $\vec{B}$), plasma properties (n_e, T, ϕ, $\vec{v}$) and distributions [$f(\vec{v}, \vec{r}, t)$] are performed. The analog data are digitized with 100 MHz, 32K A-D converters, evaluated on-line with an array processor linked to a Cray computer.

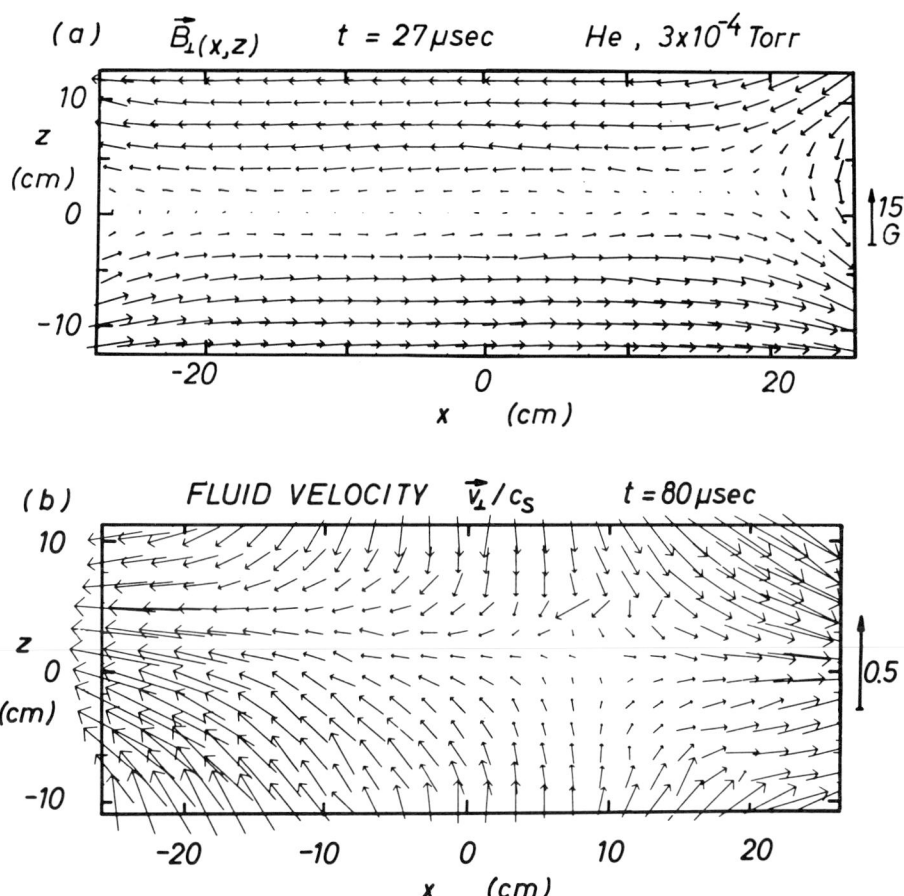

Fig. 2. Measured magnetic fields and flow field during reconnection. (a) Transverse field $\vec{B}_\perp$ (x, z) at a fixed time t, position y = 137 cm, showing the classical neutral sheet topology. (b) Transverse ion flow velocity, $\vec{v}_\perp$ (x, z), normalized to the local sound or Alfvén speed, $c_s = (kT_e/m_i)^{1/2} \simeq c_A = (B^2/2\mu_0 nm_i)^{1/2}$. Vertical compression and horizontal jetting are in qualitative agreement with reconnection models.

The plasma of typical density $n_e \simeq 10^{12}$ cm^{-3} and temperature $T_e \simeq 10$ eV is uniform, quiescent, essentially collisionless, and highly reproducible in pulses of duration $t_p \simeq 5$ msec, repeated every $t_r = 2$ sec.

After preparing the plasma in a uniform axial magnetic field ($B_y \simeq 10G$) a pulsed transverse magnetic field ($<B_\perp> \simeq 10G$) is applied. Its topology vacuum contains an X-type neutral point along the axis of the device. In the presence of plasma, electron currents are induced which flow preferentially in regions of $B_\perp \simeq 0$. They are so large ($I \simeq 1000A$) as to modify the field topology self-consistently, forming a Dungey (1958) type neutral sheet during the interval of rising applied fields ($0 < t \lesssim 100\mu sec$) shown in Fig. 2a. The corresponding current sheet ($\vec{J} = \nabla \times \vec{B}/\mu_0$) has a thickness ($\Delta z \simeq 5cm$) between the

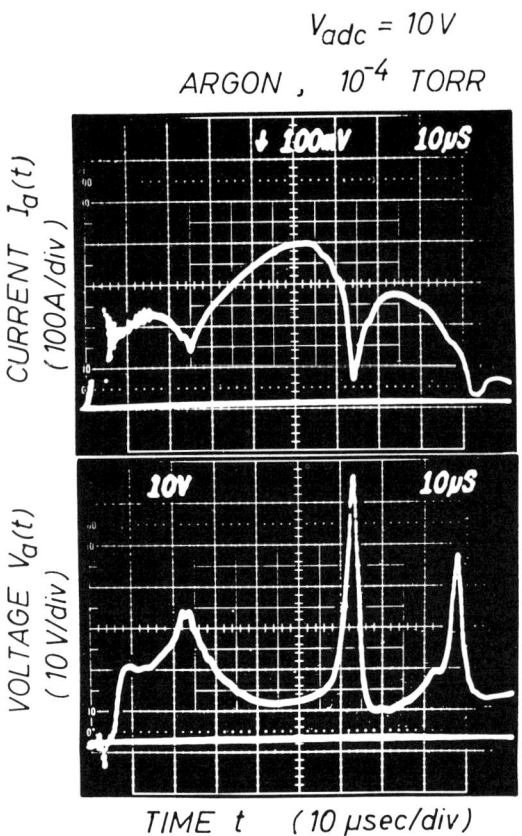

Fig. 3. Disrupted center plate current $I_a(t)$ (top trace) and instantaneous plate voltage $V_a(t)$ (bottom trace) measured in situ. At each disruption a voltage spike is generated owing to the distributed circuit inductance L. Note that $LdI_a/dt \gg V_{adc} = 10V$.

collisionless electron skin depth ($c/\omega_{pe} \simeq 0.6$ cm) and the ion Larmor radius ($r_{ci} \simeq 30$ cm). Thus, we are investigating the fine structure of the diffusion region which is beyond the resolution of present solar observations and has neither been observed in the magnetosphere. Reconnection occurs in our experiment at a rate indicated by the axial electric field along the separation ($E_y \simeq 0.5$ V/cm) (Sonnerup, 1979) or the normalized inflow velocity of the fluid into the field reversal region ($v/v_A \simeq 0.3$) shown in Fig. 2b (Vasyliunas, 1975). For uniform plasmas and modest current densities (normalized electron drifts $v_d/v_e \simeq 0.1$) the current sheet is macroscopically stable on a time scale long compared with the Alfvén transit time across the sheet.

CURRENT DISRUPTIONS AND DOUBLE LAYERS

The stability of the current sheet with respect to increasing current densities at the separator has been investigated. This is accomplished by raising the potential of the central portion of the end anode on which the plasma terminates axially. When monitoring collected current I_a to this center electrode, we find that at increasing current densities ($v_d/v_e \gtrsim 0.3$) spontaneous current disruptions develop (see Fig. 3a). The cause for this current switch-off has been inferred from detailed diagnostics of the local plasma properties. During a disruption the plasma potential rises in the perturbed current channel to a value much larger than the dc potential ($V_{dc} \simeq 10$V) applied to the end plate (see Fig. 3b). Simultaneously, the plasma density decreases. These processes have a finite axial extent. In particular, the plasma potential exhibits an abrupt axial drop ($\Delta\phi \simeq 30$V in $\Delta y \simeq 5$mm $\simeq 100\lambda_D$), i.e., a double layer is formed in the region where the current is disrupted.

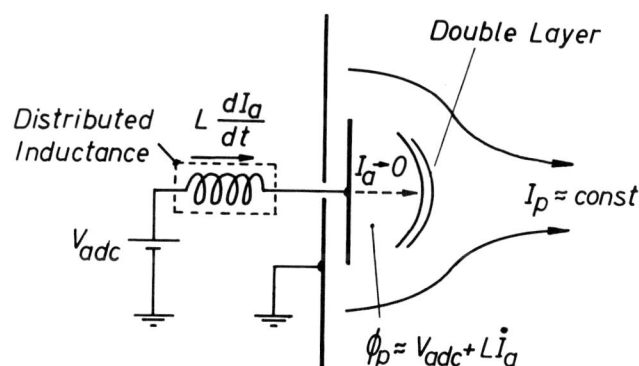

Fig. 4. Schematic diagram of the important elements in the physical model for the disruptive instability

The large positive plasma potential can only be explained in terms of the circuit properties. Fig. 4 shows schematically that the extended current path has a distributed inductance L which, in the presence of a current drop results in an inductive voltage rise LdI/dt.

This voltage drives the plasma potential positive and causes an expulsion of ions, hence a density and current drop. The current decrease in turn reinforces the inductive voltage so that an explosive disruption develops. The current lost in the center of the original sheet is redirected to the sides.

At the double layer particles are accelerated on expense of magnetic field energy stored at different locations in the current system. The kinetic energy first resides in particle beams. Subsequent beam-plasma instabilities transfer the directed energy into waves and heat. For example, microwave emissions at the electron plasma frequency are caused by electron beam-plasma interactions (Whelan and Stenzel, 1981).

Many of the observed phenomena of current sheet disruptions may apply to space plasmas as well. The dynamic current modifications during substorms and solar flares accelerate particles to high energies. The formation of field aligned potential structures is well known from auroral physics but it is possible that nonstationary double layers arise from inductive voltages where the energy storage is remote from the region of dissipation. Type III solar radio bursts involve beam-plasma instabilities possibly analogous to the present emission process.

TURBULENCE IN CURRENT SHEETS

Even in the absence of current disruptions the macroscopically stable neutral sheet exhibits a significant level of turbulence. Currents in collisionless plasmas create various fluid and kinetic instabilities (Das, 1981). The first step in the observation of turbulence is the mode identification from the frequency and wavenumber spectrum. Then, the instability mechanism has to be isolated which involves particle distribution measurements. Finally, the effect of the instability on the macroscopic transport properties is of interest. Some of these questions have been addressed in the present experiment.

Three characteristic modes have been identified. An enhanced level of microwave emissions in the range of plasma frequencies ($6 \leq f \leq 12$ GHz) is observed during reconnection (see Fig. 5 top). These nonthermal emissions are due to electron plasma waves excited by energetic electrons. A second spectrum of waves below the ion plasma frequency has been investigated (Fig. 5 bottom). With two-probe cross-correlation measurements the dispersion $\omega(k)$ is obtained which identifies the noise to consist of ion acoustic modes ($\omega \simeq kc_s$). Finally, low frequency magnetic fluctuations above the lower hybrid frequency ($f_{\ell h} \simeq 200$ kHz) are observed (Fig. 6a) and studied in depth.

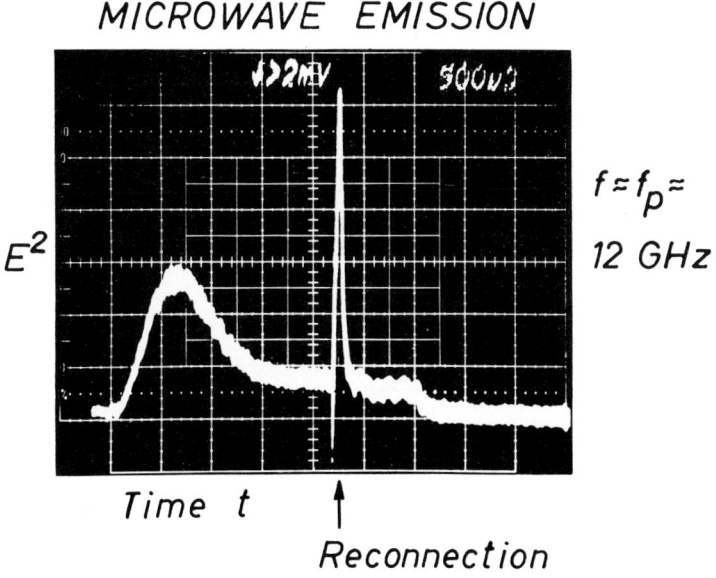
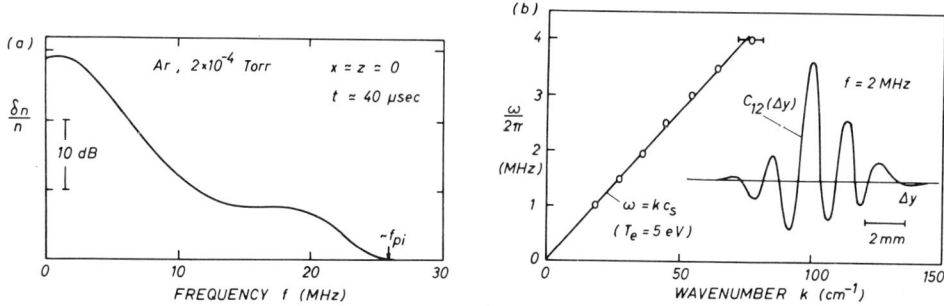

Fig. 5. Turbulence spectra in a current sheet. Top: Microwave emission near the electron plasma frequency is greatly enhanced during the reconnection pulse. Bottom: (a) Electrostatic turbulence spectrum below the ion plasma frequency (~25 MHz) sampled during reconnection. (b) Dispersion relation $\omega(k)$ obtained from measured cross-correlation functions (see insert) identifies spectrum to consist of ion acoustic modes.

The nonstationary magnetic turbulence is analyzed by ensemble averaging two-probe vector cross-correlations. Digital Fourier analysis in time and space reveals a multitude of modes which fall near the average dispersion surface for whistler modes (Fig. 6b). By selecting individual modes the polarization properties $\vec{B}(\Delta \vec{r}, \Delta t)$ are found to be right hand circular, confirming that the modes are indeed whistlers.

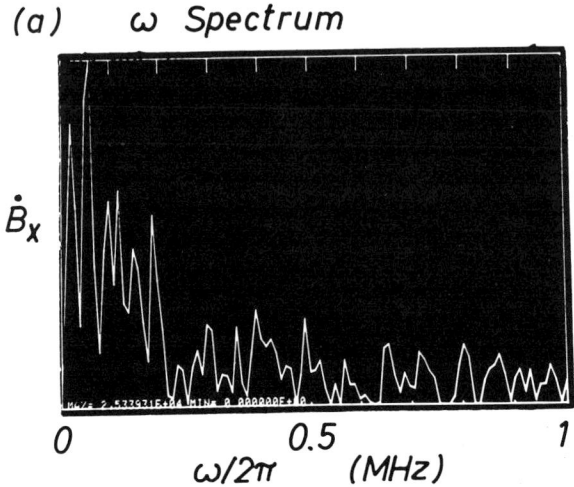

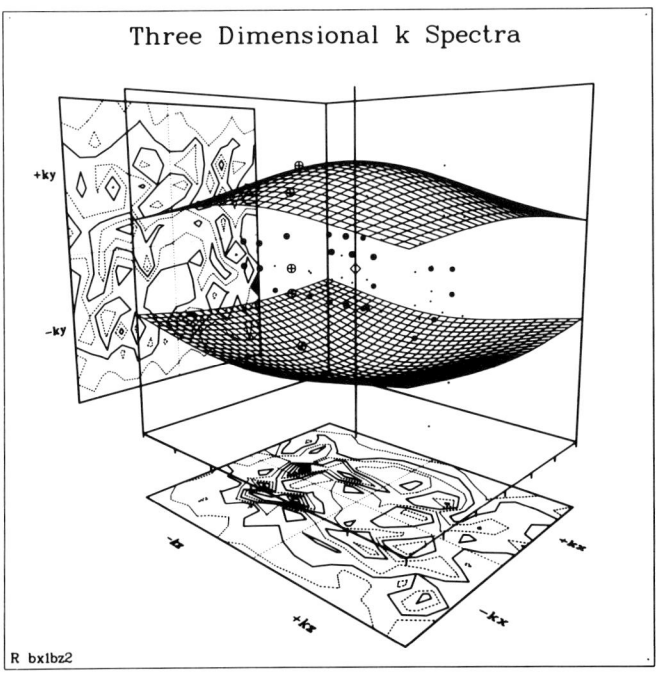

Fig. 6. Magnetic turbulence in a current sheet. (a) Frequency spectrum (lower hybrid $\omega_{\ell h}/2\pi \simeq 0.1$ MHz; electron cyclotron $\omega_{ce}/2\pi \simeq 30$ MHz). (b) Three-dimensional wave vector $\vec{k}$ space with unstable modes of the magnetic turbulence. The cross spectral function $<B_{x1} B_{z2}>$ at $f = 1$ MHz has been measured in two orthogonal planes and spatially Fourier transformed. The data points are scattered around the theoretical dispersion surfaces of whistler waves.

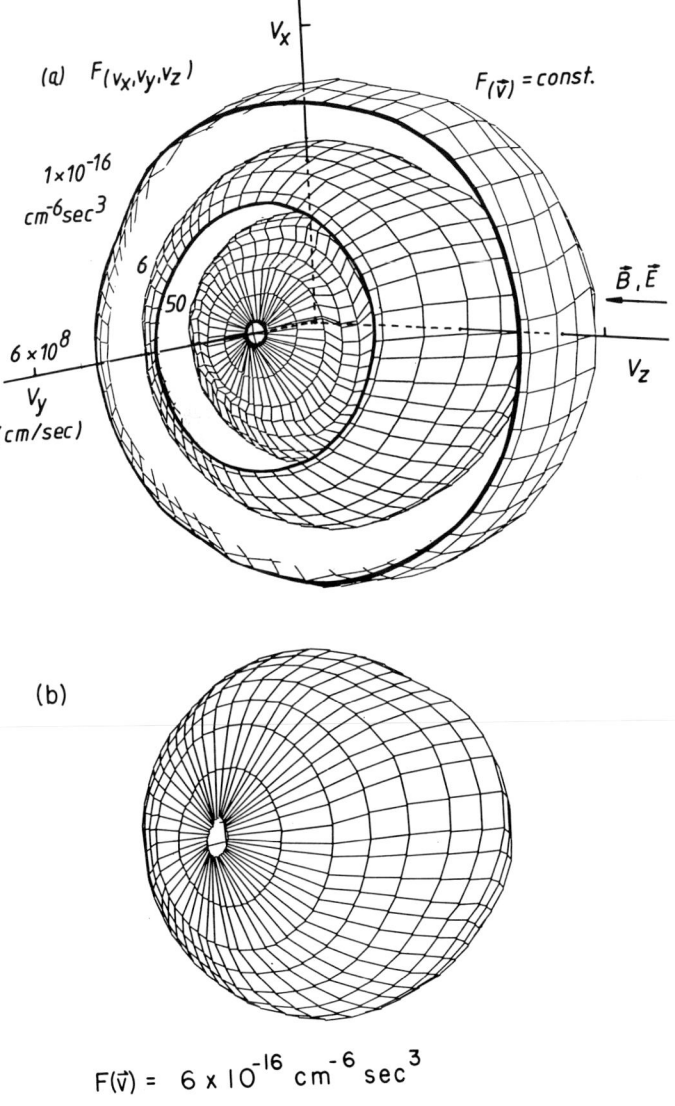

Fig. 7. Display of the distribution function in three-dimensional velocity space as surfaces of constant $f(v_x, v_y, v_z)$. (a) Three nested surfaces cut for purpose of display. Radius expands with decreasing value of $f(\vec{v})$. (b) Complete display of the middle surface showing velocity space anisotropy (nonspherical shape) due to runaway electrons in a current sheet.

Since whistler instabilities can arise from velocity space anisotropies we have investigated the electron velocity distribution function. These measurements are performed with a novel velocity analyzer with good angular resolution ($\Delta\Omega \simeq 10^{-2}$ sterad) (Stenzel et al., 1983b). The three-dimensional distribution $f(v_x, v_y, v_z)$ is displayed in Fig. 7a as surfaces of constant value $\vec{f}$, which are nested with the maximum ($f_{max} \simeq 10^{-14}$ cm^{-6} sec^3) near the origin (v = 0). Anisotropies cause deviations from spherical surfaces, clearly visible for the contour $f = 6 \times 10^{-16}$ cm^{-6} sec^3 displayed separately in Fig. 7b.

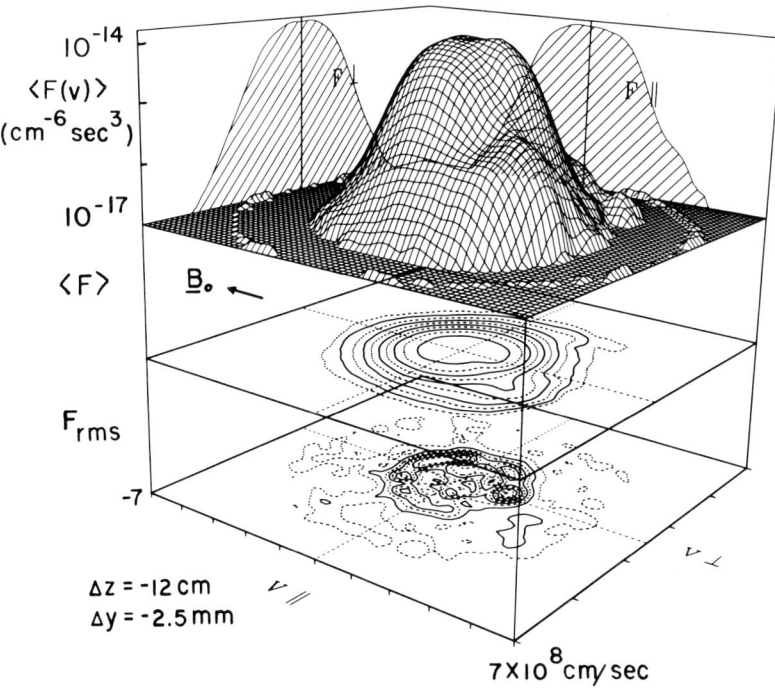

Fig. 8. Comparison of ensemble average distribution $\langle f(\vec{v})\rangle$ and root-mean-square fluctuations $f(\vec{v})_{rms}$. Upper panel shows the mean $\log_{10} \langle f(v_\parallel, v_\perp)\rangle$ and two one-dimensional cuts, $\log_{10} \langle f(v_\parallel, 0)\rangle$ and $\log_{10} \langle f(0, v_\perp)\rangle$. Middle plane shows contours of constant $\langle f(v_\parallel, v_\perp)\rangle$. 3 dB per contour. Lower plane shows fluctuations $f_{rms}(v_\parallel, v_\perp)$, ~$4 \times 10^{-16}$cm^{-6} sec^{+3} per contour. Note anisotropic distribution of fluctuations with approx. maxima where $f(\vec{v})$ exhibits strong gradients.

The elongation along the magnetic field is produced by energetic electrons which runaway in the electric field along the separator. Runaway electrons are observed only inside the neutral sheet. At the sheet edge the normal field component B_z prevents the free acceleration of electrons.

Kinetic instabilities can be predicted both from the shape of the average distribution function (e.g. $\partial f/\partial \vec{v} > 0$) and by fluctuations in the distribution involving wave-particle resonances. From a large ensemble of repeated measurements we have determined both the mean value and the rms fluctuations which are displayed in Fig. 8 in a two-dimensional velocity space. Comparing the contour plots of f_{mean} ($v_\parallel$, $v_\perp$) and f_{rms} ($v_\parallel$, $v_\perp$) we find enhanced fluctuations near the anisotropic runaway electrons. The particle velocities correspond to the Doppler shifted phase velocities of unstable whistlers, and their fluctuation spectra overlap. Thus, the origin of the whistler wave turbulence appears to lie in the observed nonequilibrium distributions in the current sheet.

The complete knowledge of the distribution function $f(\vec{v}, \vec{r}, t)$ provides us with the space-time resolved fluid properties of which we have calculated the particle density, current density, mean energy and heat flow. Transport parameters have been studied with the help of an array probe which provides the bulk electron temperature with high time resolution (Wild, et al., 1983). Correlation analysis of temperature fluctuations in the current sheet indicate that heating occurs in bursts of short durations ($\Delta t \simeq 3\mu sec$) generated in narrow regions (≤ 3 cm diam.) propagating and diffusing across the field at approximately the Alfvén speed. Heat bursts correlate with current pulses. The cross-field heat conductivity is larger than the classical value. The microscopic heating and transport processes in collisionless plasmas undoubtedly involve plasma turbulence effects, but further investigations are required to understand the processes in detail. Laboratory plasmas are highly suited for this task.

ACKNOWLEDGMENTS

The authors acknowledge helpful assistance in this work from N. Wild and J. M. Urrutia. This research was sponsored in part by the National Science Foundation (ATM 81-19717, PHY 82-09524), the National Space and Aeronautics Administration (NAGW-180) and the Air Force Office of Scientific Research (F19628-82-K0019).

REFERENCES

Baum, P. J. and Bratenahl, A.: 1980, in C. Marton (ed.), *Advances in Electrons and Electron Physics*, Academic, New York, Vol. 54, p. 1.
Das, A. C.: 1981, in H. Kikuchi (ed.), *Relation Between Laboratory and Space Plasmas*, D. Reidel Publ. Co., Dordrecht, Holland, p. 241.
Dungey, J. W.: 1958, *Cosmic Electrodynamics*, Cambridge University Press, New York.
Frank, A. G.: 1976, *Proc. P. N. Lebedev Phys. Inst. Acad. Sci. USSR Engl. Transl.*, 74, 107.

Gekelman, W. and Stenzel, R. L.: 1978, *Phys. Fluids* 21, 2014.
Gekelman, W., Stenzel, R. L., and Wild, N.: 1982, *Physica Scripta* T2/2, 277.
Gekelman, W. and Stenzel, R. L.: 1983, J. Geophys. Res. (submitted).
Hollenstein, Ch. and Guyot, M.: 1983, *Phys. Fluids* 26, 1606.
Ohyabu, N., Okamura, S., and Kawashima, N.: 1974, J. Geophys. Res. 79, 1977.
Sato, N.: 1982, in P. Michelsen and J. J. Rasmussen (eds.), *Symposium on Plasma Double Layers, Risø National Laboratory, June 16-18, 1982*, Risø-R-472, p. 116.
Sonnerup, B. U. O.: 1979, in *Space Plasma Physics: The Study of the Solar System*, National Academy of Sciences, Washington D. C., Vol. 2, p. 879.
Stenzel, R. L.: 1977, J. Geophys. Res. 82, 4805.
Stenzel, R. L. and Gekelman, W.: 1979, *Phys. Rev. Lett.* 42, 1055.
Stenzel, R. L., Gekelman, W., and Wild, N.: 1983a, J. Geophys. Res. 88, 4393.
Stenzel, R. L., Gekelman, W., Wild, N., Urrutia, J. M., and Whelan, D.: 1983b, *Rev. Sci. Instrum.* 54 (in press).
Torvén, S.: 1979, in P. J. Palmadesso and K. Papadopoulos (eds.), *Wave Instabilities in Space Plasmas*, D. Reidel, Hingham, Mass., p. 109.
Vasyliunas, V. M.: 1975. Rev. Geophys. Space Phys. 13, 303.
Whelan, D. A. and Stenzel, R. L.: 1981, *Phys. Rev. Lett.* 47, 95.
Wild, N., Stenzel, R. L., and Gekelman, W.: 1983, *Rev. Sci. Instrum.* 54, (in press).

DISCUSSION

Sonnerup: What is the ion gyroradius? What is the reconnection rate?

Stenzel: The calculated ion Larmor radius is $r_{ci} \sim 30$ cm for Argon, $r_{ci} \sim 10$ cm for Helium at $\langle B \rangle \sim 20$ G. The reconnection rate, as determined by the observed upstream inflow velocity, is given by

$$M = \frac{v}{v_A} \sim 0.2.$$

Van Hoven: What is the driving electric field strength ($E/E_{Dreicer}$) when you show disruptions? Is there any E-field threshold for this behavior?

Stenzel: We are in the regime where the electron drift velocity approaches the thermal velocity or $E/E_{Dreicer} \sim 1$. There are no disruptions observed for $v_d/v_e \sim 0.1$; they build up in the range $0.2 \lesssim v_d/v_e \lesssim 1$.

Benford: Your 6 GHz radiation seems to be evidence for strong electromagnetic emission near ω_p. What is the efficiency of this radiation, and how does it compare, roughly, to microwave emission from flaring loops and type III bursts?

Stenzel: Microwave signals at $f \sim f_p \sim 6$ GHz are observed both as electrostatic signals in the beam-plasma region and as electromagnetic waves outside the plasma. The mode conversion on density gradients and scattering off ion acoustic modes has a rather low efficiency, estimated to be $P_{em}/P_{es} \sim 10^{-4}$. I have pointed out the qualitative analogy between type III solar radio bursts and our ω_p emission which occur during impulsive reconnection events.

Sato: Can you distinguish between double-layer acceleration and the acceleration by reconnection (slow shocks, for example)?

Stenzel: Yes, the acceleration by the double layer is mainly in the direction of the separator (y-axis) while the acceleration due to the reconnection is in the direction of the neutral sheet (x-axis).

Kundu: With regard to 6 GHz plasma emission, I believe that the time structure of these microwave bursts is less than 10 µs. Does it mean that in an astrophysical object such as the solar flare plasma, one should be able to produce such fine time structure bursts intrinsically, except for the fact that other effects may modify such time structures?

Stenzel: The time scale for the current distruption and associated microwave bursts is roughly given by the ion transit time through the perturbed current channel, here $\tau \sim 5$ cm/(5×10^5 cm/sec) ~ 10 µsec. If a solar flare of similar plasma parameters would exhibit a multitude of unstable current filaments of comparable scale size one may indeed expect the emission to be randomly modulated at the above mentioned time scale.

Birn: In magnetotail simulations the asymmetry of the configuration with gradients parallel to the current sheet and a magnetic field component perpendicular to it seems to have an important effect. Could you study such effects with your device?

Stenzel: Yes, such studies could be done. We could establish such asymmetries by inclining the current-carrying plates at an angle with respect to one another.

Dum: You pointed out the importance of the electrodynamics of the plasma and external circuit for Double Layer (DL) formation. I think this is a very significant point for our going beyond the usual electrostatic laminar solution of the one dimensional Vlasov equation that is offered as an explanation even for such highly dynamic processes as auroras. Nevertheless, particles are also important. What do you think are the particle boundary conditions essential to DL formation such as injection or reflection from electrodes, etc. In the very recent experiment of Guyot and Hollenstein (Phys. Fluids 1983) DL formation seems to depend on whether one electrode is heated or not. How do you trigger DL in a controlled manner, as you mentioned?

Stenzel: The anode, which is the only relevant boundary, absorbs all electrons; thus we have no control over the injected particle distributions as in Hollensteins work.

I mentioned that we now trigger current disruptions (not necessarily double layers) in a controlled manner. This is done by pulsing a thin slab of magnetic field (B_z) transverse to the current sheet (J_y) which causes a total current disruption. Large inductive voltages drop off across the slab, possibly forming again double layers.

Bratenahl: Did I understand that you are now looking for spontaneous development of double layers; that is, double layer development without artificial stimulation?

Stenzel: The work reported did describe spontaneous development of double layers without artificial stimulations. We are now working on controlled disruptions of the entire current sheet.

D. Smith: You mention that when a double layer forms, a 3-D potential results with radial ion acceleration. Could you explain in more detail how this occurs?

Stenzel: The potential structure is three-dimensional, since only the perturbed central part of the current sheet rises to a large positive potential over a finite axial length. Since the ions are unmagnetized, they are accelerated in both radial and axial directions. This leads to a large ion, hence density loss from the perturbed current channel. The pump-out of plasma ultimately causes the current loss.

Henriksen: Does the electron distribution have a characteristic shape in velocity space?

Stenzel: Yes, we have made extensive measurements of the electron distribution function and found asymmetric distributions even without current dusruptions. The current sheet contains a large number ($\sim 10\%$) of runaway electrons which are accelerated by the inductive electric field along the separator. These particles strongly influence the transport processes and provide a source of free energy for driving various microinstabilities.

RECONNECTION IN SHEARED MAGNETIC FIELDS IN SPACE AND ASTROPHYSICS

James F. Drake
Laboratory for Plasma and Fusion Energy Studies
University of Maryland
College Park, Maryland 20742

ABSTRACT

The current theoretical understanding of the linear and nonlinear evolution of resistive tearing instabilities in sheared magnetic fields is reviewed. The physical mechanisms underlying this instability are emphasized. Some of the problems which are encountered in developing a model of magnetic energy dissipation in coronal loops are discussed and possible solutions are suggested.

I. INTRODUCTION

The conversion of magnetic to particle energy is a fundamental process which underlies such diverse phenomena in nature as disruptions in laboratory plasmas,[1] magnetic substorms in the magnetosphere of the earth,[2] and solar flares.[3] The development of theoretical models of this energy conversion process is therefore of central importance to our understanding of these basic phenomena.

The most simple models of magnetic energy conversion are based on a slab equilibrium such as that shown in Fig. 1. A plasma carrying a current $J_{z0}(x)$ generates the magnetic field $B_{y0}(x)$. When a uniform guide field $B_{z0} \gg B_{y0}$ is included in the configuration the equilibrium is unchanged but the magnetic field rotates through a finite angle as the region of current carrying plasma is crossed. For $B_{z0} = 0$ the magnetic field simply reverses. Only the former case, that of a sheared magnetic field, is considered in this paper.

When collisions are included in this 1-D equilibrium, the resistive diffusion of the magnetic field B_{y0} dissipates both magnetic energy and magnetic flux and the flux annihilation velocity V_B is given by a/τ_r, where $\tau_r = 4\pi a^2/\eta c^2$ is the resistive diffusion time based on the resistivity η and equilibrium scale size "a". In most applications the resistive diffusion time is far too long to explain the observed energy release time scales.

Faster rates of magnetic energy dissipation can be achieved by considering perturbations of the equilibrium in Fig. 1 which change the topology of the magnetic field by forming one or more x-points as shown in Figs. 2 and 3a. In Fig. 2 the plasma and magnetic field at $x = \pm\infty$ flow toward a single x-line at $x = 0$ as indicated by the short arrows. The inflowing magnetic field reconnects in the vicinity of the x-line and flows out at $y = \pm\infty$. In Fig. 3a the magnetic field reconnects to form a periodic set of x and 0-points with a flow pattern indicated by the thin arrows. In both configurations the flow is driven by the release in the tension of the reconnecting magnetic field, which will be discussed in more detail later.

While in the case of magnetic dissipation in the one-dimensional model, only one time scale, τ_r, entered the problem, the bending of the magnetic field in the x-y plane in Figs. 2 and 3 brings in a second fundamental time scale, the shear Alfvén time, $\tau_{Ay} = a/c_{Ay}$, where $c_{Ay} = (B_{y0}^2/4\pi m_i n)^{1/2}$ is the Alfvén velocity associated with the magnetic field B_{y0}. The ratio of these two times, the Lundquist number $S \equiv \tau_r/\tau_{Ay}$, is a large parameter in high temperature plasma. The existence of this second time scale implies that the characteristic time for magnetic energy dissipation can be a hybrid of τ_r and τ_{Ay} which can be much shorter than τ_r. Note that the compressional Alfvén time $\tau_{Az} = a/c_{Az} \ll \tau_{Ay}$ does not enter the problem because the large B_{z0} field contains no free energy and is simply convected with the fluid.

A number of steady state models of reconnection have been proposed which are based on the single x-line configuration shown in Fig. 2. Vasyliunas[4] has discussed these models in detail so we only briefly mention some essential features of these theories. In the Sweet-Parker[5] model magnetic reconnection takes place in an elongated layer (the region where resistivity is important) of width $\Delta x \sim a/S^{1/2}$ in the x direction and the system size in the y direction. The outflow velocity is the Alfvén speed c_{Ay} and the merging velocity V is given by

$$V/c_{Ay} \sim S^{-1/2},$$

so that the time scale for dissipation of magnetic energy is of order $(\tau_r/\tau_{Ay})^{1/2}$. In Petschek's[6] model the reconnection region is much smaller, scaling as $\Delta x \sim aS^{-1}\ln S$ and $\Delta y \sim S^{-1}\ln^2 S$ in the x and y directions, respectively. The outflow velocity is again the Alfvén speed with the merging velocity

$$V/c_{Ay} \sim (\ln S)^{-1},$$

much larger than the Sweet-Parker rate. In Petschek's model a set of slow shocks form outside the resistive region along the separatrices in Fig. 2. These slow shocks link the magnetic fields and flows within the outflow region to those in the inflow region. The time development of these slow shocks has been observed in numerical simulations[7] in which the inflow velocity V in Fig. 2 is imposed as a boundary condition.

As emphasized very strongly by Vasyliunas, the Petschek solution actually represents an upper bound on the reconnection rate and therefore neither the steady reconnection models nor simulations of forced reconnection where the inflow velocity is imposed externally can answer the important question: what is the characteristic magnetic energy dissipation rate or time scale which is generated by the hydromagnetic forces within the current carrying plasma? In order to address this question it is necessary to include all the forces within the plasma which are driving the reconnection and follow the time development of the system from some appropriate initial conditions. Such a calculation essentially reduces to solving for the nonlinear development of the tearing instability (or reconnection driven by an ideal mode such as the kink) in the geometry of interest. In the remainder of this paper, the basic properties of the linear resistive tearing instability and our current understanding of the nonlinear behavior of these and related instabilities is discussed. Because of space limitations, the tearing instability in the collisionless and long mean-free path regimes is not discussed. The interested reader is referred to Refs. 8 and 9 for a discussion of the linear and nonlinear behavior of the instability in these limits.

II. LINEAR TEARING INSTABILITY

The tearing instability is the mechanism by which a current carrying plasma spontaneously evolves to a lower magnetic energy state. The magnetic energy released goes into bulk plasma motion as shown in Fig. 3a and into the internal plasma energy by Joule heating. The resistive tearing instability is described by the one-fluid magnetohydrodynamic equations,

$$\rho d\underline{v}/dt = \underline{J} \times \underline{B}/c - \nabla p \, , \qquad (1)$$

$$dp/dt + (\gamma_s - 1) p \nabla \cdot \underline{v} = (\gamma_s - 1) \eta J^2 \, , \qquad (2)$$

$$\nabla \times \underline{B} = 4\pi \underline{J}/c \, , \qquad (3)$$

$$\nabla \times \underline{E} + c^{-1} \partial \underline{B}/\partial t = 0 \, , \qquad (4)$$

$$\eta \underline{J} = \underline{E} + \underline{v} \times \underline{B}/c \, , \qquad (5)$$

$$\nabla \cdot \underline{B} = 0 \, ; \quad \nabla \cdot \underline{J} = 0 \, ; \quad d/dt = \partial/\partial t + \underline{v} \cdot \nabla \, , \qquad (6)$$

where the mass density ρ is taken to be a constant, γ_s is the ratio of specific heats and other notation is standard. The forces driving the tearing instability and the fundamental structure of the mode are the same in slab and cylindrical geometry. For simplicity, we therefore first discuss the instability in the slab geometry of Fig. 1 in detail and then extend the results to a cylinder.

The tearing mode is driven by the relaxation in the tension of the reconnecting magnetic field lines. This can be easily shown by using Eqs. (1) and (3) to write the force density $\underline{F}$ on the local magnetofluid as

$$\underline{F} = -\nabla(p+B^2/8\pi) + \underline{B}\cdot\nabla\underline{B}/4\pi .$$

Since the time scale associated with the growth of magnetic islands is much longer than compressional Alfvén time τ_{Az}, the plasma flow shown in Fig. 3a is incompressible ($\nabla\cdot\underline{v} = 0$). In this limit the gradient term in the force density does not contribute to the energetics of the tearing mode. The instability is driven only by the magnetic tension. The forces driving the instability are illustrated in Fig. 3a. The force density, given by the local magnetic tension, is shown by the thick arrows and the flow velocity by the thin arrows. In the region outside of the separatrix of the magnetic island the plasma flow opposes the magnetic force. In this region the flow stretches the magnetic field lines, locally increasing the magnetic energy. Within the magnetic island, however, the flow and magnetic forces are aligned so that it is the relaxation of the magnetic stress of the reconnecting field lines which drives the flow and the magnetic energy is reduced in this region. It is quite clear from Fig. 3a that in the absence of resistivity, so that reconnection of the magnetic field can not take place, all perturbations increase the magnetic energy so there is no instability. Moreover, to the extent that the resistivity in a high temperature plasma is small (S is large), the tearing instability is a slowly growing mode.

We now simplify the magnetohydrodynamic equations given in Eqs. (1)-(6) to study the tearing mode in more detail. For simplicity we treat the case where $\partial/\partial z = 0$. In the low frequency limit $\partial/\partial t \ll \tau_{Az}^{-1}$, the plasma motion is nearly incompressible and the velocity can be represented by the stream function ϕ as

$$\underline{v} = \hat{z} \times \nabla\phi . \tag{7}$$

An equation for the stream function is obtained by taking the curl of the momentum equation [Eq. (1)] and then taking the dot product with the unit vector $\hat{z}$,

$$\rho d(\nabla^2\phi)/dt = c^{-1}\underline{B}\cdot\nabla J_z . \tag{8}$$

The magnetic field can similarly be expressed in terms of the $\hat{z}$ component of the vector potential as

$$\underline{B} = B_{z0}\hat{z} - \hat{z} \times \nabla A_z , \tag{9}$$

and an equation for A_z follows from the $\hat{z}$ component of Eq. (3),

$$\nabla^2 A_z = -4\pi J_z/c . \tag{10}$$

Since $\partial/\partial z = 0$, the vector potential A_z is also a flux function,

$\underset{\sim}{B} \cdot \nabla A_z = 0$, so that the flux surfaces are surfaces of constant A_z. From the z component of Ohm's law in Eq. (5), we obtain an equation for J_z,

$$\eta J_z = E_z - \underset{\sim}{B} \cdot \nabla \phi / c . \tag{11}$$

The electric field $E_z = - c^{-1} \partial A_z / \partial t$ follows from Eqs. (9) and (4), so that (11) becomes

$$\eta J_z = - c^{-1} (\partial A_z / \partial t + \underset{\sim}{B} \cdot \nabla \phi) . \tag{12}$$

Equations (6)-(10) and (12) provide a complete set of equations for both the linear and nonlinear tearing mode. In the incompressible limit considered here neither the pressure p nor B_{z0} enter the equations for the tearing mode. Of course, when $B_{z0} = 0$, the electrons and ions are demagnetized near x = 0 and the fluid equations can not be used to describe the plasma dynamics. The resistivity has been taken to be constant in time in Eq. (12) so that neither the rippling[10,11] nor the thermal instabilities[12,13] are described by these equations.

In the limit of $B_{z0} \gg B_{y0}$, the $\hat{z}$ direction is nearly aligned along the magnetic field, $A_z \simeq A_\parallel$, $J_z \simeq J_\parallel$, and Eq. (12) can be interpreted as the parallel Ohm's law,

$$\eta J_\parallel = E_\parallel = - c^{-1} (\partial A_\parallel / \partial t + \underset{\sim}{B} \cdot \nabla \phi) , \tag{13}$$

where the subscript $\parallel$ denotes the component along $\underset{\sim}{B}$. The stream function ϕ is now the electrostatic potential and the flow in Eq. (7) is simply given by the $\underset{\sim}{E} \times \underset{\sim}{B}$ drift associated with this potential. In the two fluid description of the tearing mode in this limit, Eq. (8) is the charge neutrality condition $n_i = n_e$ where n_i on the left side of Eq. (8) is the divergence of the ion polarization drift and the right side of (8) results from the parallel bunching of electrons.[8]

To study the linear tearing instability, we linearize the equations assuming all perturbed quantities vary as $\exp(\gamma t + i k_y y)$. The coupled second order equations for $\tilde{A}_z$ and $\tilde{\phi}$ are

$$\nabla^2 \tilde{A}_z = - 4\pi \tilde{J}_z / c , \tag{14}$$

$$\gamma \nabla^2 \tilde{\phi} = i \underset{\sim}{k} \cdot \underset{\sim}{B} \tilde{J}_z / c + i k_y \tilde{A}_z J'_{z0} / c , \tag{15}$$

$$\eta \tilde{J}_z = \tilde{E}_\parallel = - (\gamma \tilde{A}_z + i \underset{\sim}{k} \cdot \underset{\sim}{B} \tilde{\phi}) / c , \tag{16}$$

where the prime denotes d/dx. Because of the x dependence of $\underset{\sim}{k} \cdot \underset{\sim}{B} = k_y B_{y0}(x)$ and $J_{z0}(x)$, these equations must be solved as an eigenvalue problem subject to the boundary conditions $\tilde{\phi}, \tilde{A}_z \to 0$ as $|x| \to \infty$. Further insight into the nature of the tearing mode can be gained by constructing an energy integral from Eqs. (14)-(16),

$$\frac{\partial}{\partial t} \int d\underline{x} \, [\frac{1}{2} \rho |\nabla \tilde{\phi}|^2 + \frac{1}{8\pi} |\nabla \tilde{A}_z|^2 + \frac{1}{8\pi} \frac{B''_{y0}}{B_{y0}} |\tilde{A}_z|^2]$$

$$= - \int d\underline{x} \, \eta \tilde{J}_z (\tilde{J}_z + \frac{J'_z}{B_{y0}} \tilde{A}_z) \, . \tag{17}$$

The first term on the left side of this equation is the flow kinetic energy, the second two terms combine to give the total change in magnetic energy, while the resistive term is the rate of dissipation of energy by Joule heating. To have instability we require that the magnetic energy decrease or[14]

$$\delta W_B = \frac{1}{8\pi} \int d\underline{x} (|\nabla \tilde{A}_z|^2 + \frac{B''_{y0}}{B_{y0}} |\tilde{A}_z|^2) < 0 \, . \tag{18}$$

The first term in the integral, which is positive and therefore stabilizing, is the square of the first order fields, $\tilde{B}^{(1)} \cdot \tilde{B}^{(1)}$, while the second comes from the second order field, $\tilde{B}_y^{(2)} B_{y0}$. For the equilibrium of Fig. 1, the second term is destabilizing (negative) and scales as $a^{-2} |\tilde{A}_z|^2$, where "a" is the characteristic length of the equilibrium magnetic field, while the first is stabilizing and scales as $k_y^{-2} |\tilde{A}|^2$. For instability we therefore require $k_y a < 1$ so that the tearing instability is inherently a long wavelength phenomena. Note also that $B''_{y0} = J'_{z0}$ so that the instability is driven by the gradient of the equilibrium current.

We now return to discuss the solutions to Eqs. (14)-(16). In high temperature plasma where $S \gg 1$ the resistivity η in Eq. (16) can be neglected throughout most of the plasma so that $\tilde{E}_\parallel = 0$, i.e., the parallel induction and electrostatic fields balance. The cancellation between these terms occurs because of the high electron mobility along $\underline{B}$. The bunching of electrons parallel to $\underline{B}$ driven by the induction field produces an electrostatic field which effectively shorts out the induction field. The condition $\tilde{E}_\parallel = 0$ can be rewritten as

$$\partial \tilde{A}_z / \partial t + \tilde{v}_x \partial A_z / \partial x = 0 \, , \tag{19}$$

so that the rate of change of flux in the frame of the moving fluid is zero. In the "ideal" region the plasma and flux are frozen together. The inertia term on the left-hand side of Eq. (15) can be neglected in the ideal region so the two terms on the right side of the equation balance and

$$\tilde{J}_z = - (\tilde{A}_z / B_{y0}) J'_{z0} \, . \tag{20}$$

The plasma displacement Δx can be calculated using Eq. (19) as $\Delta x = - \tilde{A}_z / B_{y0}$. Thus, the perturbed current in the ideal region simply results from the displacement of the equilibrium current J_{z0}. Equations

(14) and (20) yield a single second order equation for $\tilde{A}_z$ valid in the ideal region,

$$\nabla^2 \tilde{A}_z - (B_{y0}''/B_{y0})\tilde{A}_z = 0 . \tag{21}$$

The parallel induction and electrostatic fields balance everywhere except where the parallel electrostatic field is zero in the vicinity of $\underline{k} \cdot \underline{B} = 0$ [see Eq. (16)], or for the equilibrium of Fig. 1 at $x = 0$. In this region $\tilde{E}_\parallel \neq 0$ and since η is small a large parallel current $\tilde{J}_z$ is driven. For an ideal mode, where η is taken to be strictly zero, $\tilde{A}_z(x=0) = 0$ and since $\tilde{B}_x = ik_y \tilde{A}_z = 0$, no magnetic island is formed. However, for a resistive mode such as the tearing mode, this constraint does not apply and $\tilde{B}_x(x=0) \neq 0$ allowing a magnetic island to form as in Fig. 3a. Because η is small, the region Δ around $x = 0$ where $\tilde{E}_\parallel \neq 0$ is also small ($\Delta/a \ll 1$) so that Eq. (21) for $\tilde{A}_z$ is valid throughout most of the plasma.

A typical solution of the ideal equation for $\tilde{A}_z$ is shown in Fig. 4a for a mode with $k_y a < 1$. The slope of $\tilde{A}_z$ is discontinuous across $x = 0$ where the ideal equation for $\tilde{A}_z$ breaks down. In the region $x \sim \Delta$ resistivity must be retained in the equation for $\tilde{A}_z$ and the solutions of this equation must be matched to the ideal solutions so that the complete solution for $\tilde{A}_z$ is continuous everywhere. Nevertheless, the ideal solution can be characterized by the discontinuity in the slope of $\tilde{A}_z$ at $x = 0$ as measured by the quantity[10]

$$\Delta' = [\partial \tilde{A}_z(x=0^+)/\partial x - \partial \tilde{A}_z(x=0^-)/\partial x]/\tilde{A}_z(0) . \tag{22}$$

For the Harris equilibrium, $B_{y0} = B_{y00}\tanh(x/a)$, the parameter Δ' can be calculated analytically as

$$\Delta' a = 2(1 - k_y^2 a^2)/|k_y|a . \tag{23}$$

The parameter Δ' is related to the magnetic energy released by the tearing mode. This relation can be derived by multiplying Eq. (21) by $\tilde{A}_z$ and integrating over the region $|x| > \Delta$. After an integration by parts, we find

$$\Delta' = - 8\pi \delta W_B / \int dy |\tilde{A}_z(x=0)|^2 . \tag{24}$$

The tearing mode should be unstable for $\delta W_B < 0$ or $\Delta' > 0$.

The equations governing the resistive region $|x| \sim \Delta \ll a, k_y^{-1}$ are obtained from Eqs. (14)-(16) by neglecting J_{z0}' in (15) and approximating $\nabla^2 \simeq \partial^2/\partial x^2$ and $\underline{k} \cdot \underline{B} = k_y B_{y0}' x$,

$$\tilde{A}_z'' = (4\pi\gamma/\eta c^2)(\tilde{A}_z + x\hat{\phi}/\Delta) , \tag{25}$$

$$\hat{\phi}'' = (x/\Delta^3)(\tilde{A}_z + x\hat{\phi}/\Delta) , \tag{26}$$

$$\Delta/a = (\gamma\tau_{Ay}/Sk_y^2 a^2)^{1/4}, \qquad (27)$$

where Δ is the scale length of the resistive layer and $\hat{\phi} \equiv ik_y B_0' \Delta\tilde{\phi}/\gamma$. The equations describe the transition from $\tilde{E}_\| = -\gamma\tilde{A}_z/c$ at $x = 0$ to $\tilde{E}_\| \approx 0$ for $x \gg \Delta$ as illustrated in Fig. 4b. While the resistive layer equations can be solved exactly, in the limit $4\pi\gamma\Delta^2/\eta c^2 \ll 1$, the flux perturbation $\tilde{A}_z$ can readily diffuse across the resistive layer Δ during the growth time γ^{-1} of the mode so that $\tilde{A}_z(x)$ is nearly constant in the resistive region and can be approximated by $\tilde{A}_z(0)$ on the right side of Eqs. (25) and (26). This is the widely invoked "constant ψ" approximation.[10] Equation (26) can now be readily inverted to obtain ϕ and Eq. (25) can then be integrated to obtain the jump in $\partial\tilde{A}_z/\partial x$ across the resistive layer. Matching this jump with that obtained from the ideal solution in (23), we obtain the growth rate of the resistive tearing mode,[10]

$$\gamma\tau_{Ay} = S^{-3/5}[(1 - k_y^2 a^2)\Gamma(1/4)/\pi\Gamma(3/4)]^{4/5}(k_y a)^{-2/5} \qquad (28)$$

For $k_y a \lesssim 1$, $\Delta/a \sim S^{-2/5} \ll 1$ so the resistive layer is indeed small. The expression for the growth rate in (28) implies that the fastest growing modes are those with $k_y a \ll 1$. The "constant ψ" approximation breaks down for $k_y a$ too small since both Δ and γ increase with decreasing k_y. The fastest growing modes are those with $k_y a \sim S^{-1/4}$, for which $4\pi\gamma\Delta^2/\eta c^2 \sim 1$, i.e., the growth time and diffusion time across the resistive layer are comparable.[15] In practical applications where S is very large, however, the finite geometry of the system often prevents $k_y a$ from being as small as $S^{-1/4}$.

In deriving the dispersion relation in Eq. (28), we assumed $\partial/\partial z = ik_z = 0$ so that $\underline{k}\cdot\underline{B} = 0$ at $x = 0$ and the magnetic island forms symmetrically in this region as shown in Fig. 3a. When $k_z \neq 0$, the linear tearing mode equations in (14)-(16) are still valid as long as $B_{z0} \gg B_{y0}$. Resistivity is again important where

$$\underline{k}\cdot\underline{B} = k_y B_{y0}(x) + k_z B_{z0} = 0,$$

which is now at $x = x_0 \neq 0$. The magnetic islands form around x_0 and are driven by the reconnection of the magnetic flux corresponding the component of $\underline{B}$ along $\underline{k}$, which reverses sign at $\underline{k}\cdot\underline{B} = 0$. The calculation of the growth rate of the nonsymmetric islands is basically unchanged from that presented for the symmetric case. The growth rates are generally smaller than when $k_z = 0$ because Δ' is reduced.

The location and size of the magnetic islands formed can be calculated by constructing the flux function defined by

$$\underline{B}\cdot\nabla\psi = B_{z0}\partial\psi/\partial z - \hat{z}\times\nabla A_z \cdot \nabla\psi = 0. \qquad (29)$$

Such a function ψ can generally only be constructed for 2-D problems in which $\psi = \psi(x,y+pz)$, where p is a constant. For this case $\partial/\partial z = p\partial/\partial y$ and

$$\psi = pB_{z0}x - A_z(x,y+pz) . \tag{30}$$

The islands form where the equilibrium flux function, ψ_0, has an extremum, or

$$\partial\psi_0/\partial x = [pB_{z0}x - A_{z0}(x)]' = (k_z B_{z0} + k_y B_{y0})/k_y = 0 , \tag{31}$$

i.e., where $\underline{k}\cdot\underline{B} = 0$. Note that ψ_0 is the flux associated with the component of $\underline{B}$ along $\underline{k}$. The island size is calculated by including the perturbation $\tilde{A}_z = \tilde{A}_z(x)\cos k_y(y+pz)$ in Eq. (30) and expanding x around x_0,

$$\psi = B'_{y0}(x-x_0)^2/2 + \tilde{A}_z(x_0)\cos[k_y(y+pz)] . \tag{32}$$

Since ψ is a constant along a given magnetic field line, the structure of the magnetic island can be mapped out by calculating the constant ψ surfaces. The island width w is simply

$$w = [2\tilde{A}_z(x_0)/|B''_{y0}|]^{1/2} . \tag{33}$$

Before discussing the nonlinear behavior of the tearing instability, we will complete our description of the linear mode by discussing the case of a cylindrical equilibrium. Consider a plasma carrying a current $J_{z0}(r)$ which produces a poloidal field $B_{\theta 0}(r)$ and is periodic in z over the length L. For simplicity, we again consider the limit of a large axial magnetic field $B_{z0} \gg B_{\theta 0}$. The pitch of the magnetic field line is measured by the quantity

$$q = 2\pi r B_{z0}/L B_{\theta 0}(r) . \tag{34}$$

The parameter q increases monotonically with increasing r for current profiles peaked around r = 0. As in the slab geometry, a flux function can be defined for the 2-D case where $\psi = \psi(r,\theta + 2\pi pz/L)$. The flux function can be easily calculated as

$$\psi = pB_{z0}\pi r^2/2 - A_z(r,\theta + 2\pi pz/L) ,$$

and again reconnection can occur where $\underline{k}\cdot\underline{B} = 0$. In a cylinder periodicity in θ and z requires $k_\theta = m/r$ and $k_z = -2\pi n/L$, where m and n are integers, so that

$$\underline{k}\cdot\underline{B} = [m-nq(r)]B_{\theta 0}(r)/r , \tag{35}$$

and $p = -m/n$. Tearing modes can therefore only form at discrete rational surfaces r_0 where $q(r_0) = m/n$.

The essential physics of tearing in a cylinder is unchanged from that in slab geometry and in particular the linearized Eqs. (14)-(16) are also valid for the cylinder if $k_y J'_{z0} \to k_\theta \partial J_{z0}/\partial r$ and k_z^2 is neglected compared with k_θ^2 in the ∇^2 operators. Because of geometrical effects, the magnetohydrodynamic driving energy Δ' is approximately an order of magnitude larger in the cylinder than in the slab.[16]

The most significant new feature in a cylinder is the existence of the m=1 kink mode.[17-19] For the slab tearing mode it was previously shown (Fig. 3) that the stretching of the magnetic field in the region outside of the magnetic island was stabilizing. This argument does not apply to the kink mode as can be seen in Fig. 5. The inner region of the plasma, where $q < n^{-1}$ is displaced uniformly while the outer region, where $q > n^{-1}$, remains stationary. Large poloidal flows emanating from the region of the x-line enable the flow to remain nearly incompressible. A uniform displacement of the plasma does not cause any distortion of the magnetic field lines so there is no restoring force from magnetic tension as in the slab or as with cylindrical modes with m≠1. Indeed, the magnetic forces in the ideal region are destabilizing where $q(r) < n^{-1}$ and the kink mode is unstable when $q(0) < n^{-1}$ even in the ideal limit.[17,18]

A detailed calculation of the stability of the kink tearing mode has been carried out previously[19] and we do not repeat the details here. The procedure is basically the same as that used to calculate the growth rate of tearing modes in slab geometry. The ideal equations are solved away from the rational surface and the discontinuity in the ideal solutions across the rational surface is bridged with the solutions of the resistive equations similar to those in Eqs. (25) and (26). The result is that the growth rate of the kink-tearing mode is the larger of the ideal,[18]

$$\gamma \tau_{A\theta} = \frac{\pi n}{|B_\theta^2 q'|_{r_0}} \left(\frac{2\pi}{L}\right)^2 \int_0^{r_0} dr B_\theta^2 (1 - nq)(3nq + 1)$$

$$\sim (2\pi r_0/L)^2 , \tag{36}$$

and resistive,

$$\gamma \tau_{A\theta} = (\tau_{A\theta}/\tau_r)^{1/3} (q' r_0 n)^{-4/3} , \tag{37}$$

growth rates where $\tau_{A\theta}$ is Alfvén time based on $B_\theta(r_0)$ and the length r_0 and τ_r is the resistive time based on r_0. It should be emphasized that the "constant ψ" approximation does not apply to the m=1 mode. In fact, the dispersion relation of the kink tearing mode in Eq. (37) can be expressed as

$$4\pi\gamma\Delta^2/\eta c^2 \sim 1 , \tag{38}$$

where

$$(\Delta/r_0) = (\gamma \tau_{A\theta}^2 / \tau_r q'^2 r_0^2)^{1/4} \tag{39}$$

is the resistive layer thickness [from Eqs. (15) and (16)]. The growth time of the resistive kink mode is therefore given by the flux diffusion time across the resistive layer.

When the kink mode is discussed in the literature, the ideal external kink, in which $q(r) < n^{-1}$ throughout the entire plasma and the rational surface falls in the vacuum region,[12] and the ideal internal kink in which $q(0) < n^{-1}$ but the rational surface lies within the plasma,[17,18] are often treated as entirely separate instabilities. The growth rate of the external kink is of order $\tau_{A\theta}^{-1}$, which is much larger than that of the internal mode given in Eq. (36). A vacuum can be considered as the limit of a very resistive plasma, so when resistivity is included there is a continuous transition between these two modes. When the rational surface falls in the region of high conductivity, the growth rate is given in Eq. (36). As the rational surface moves towards larger r, where η is larger, the growth rate increases until $\tau_{A\theta} \sim \tau_r$ when $\gamma \sim \tau_{A\theta}^{-1}$. At this point, of course, Δ is no longer small so that the matching procedure which leads to the growth rates in Eqs. (36) and (37) is invalid and inertia must be retained through the entire plasma.

An important point which is often overlooked in invoking the tearing mode to explain physical phenomena is that the linear treatment of this instability mode is valid only when the width of the magnetic island w is smaller than the resistive layer thickness Δ. When $w \sim \Delta$, the island structure strongly modifies the magnetic geometry of the dissipation region. Since $\Delta \ll a$, for a high temperature plasma the linear theory only applies when w is extremely small. As a consequence, the linear growth time of the mode has no relation with the time required to dissipate a significant fraction of the magnetic free energy in a current carrying plasma. The nonlinear development of the tearing mode must be studied to address this problem.

III. NONLINEAR TEARING INSTABILITIES

When the width w of the magnetic island of the tearing mode becomes comparable to the width of dissipation region Δ, the magnetic structure of the dissipation region is strongly modified. Since the growth rate of the tearing mode is sensitive to the plasma dynamics in this resistive region, the evolution of the mode should be strongly affected once $w > \Delta$.

For simplicity, we consider the slab geometry tearing mode of Fig. 3 with $\partial/\partial z = 0$. Equations (7)-(10) and (12) describe the nonlinear behavior of the 2-D structure of the mode in the x-y plane. When $\Delta < w \ll a$, the dominant nonlinear behavior of the tearing mode is produced by the $\underset{\sim}{B}\cdot\nabla$ operators in Eqs. (8) and (12) in the resistive layer. We therefore first focus on this region where $\nabla^2 \approx \partial^2/\partial x^2$ and equilibrium currents can be neglected. In the linear phase the left and right sides of Eq. (8) are comparable. When the island width w exceeds the tearing width the scale size of the resistive region increases so that the inertia term scales as $\rho\gamma\tilde{\phi}/w^2$ and is therefore reduced. By contrast, the right side of (8) should scale as $c^{-1}B'_{y0}w\tilde{J}_z$, which increases with w. To lowest order, the inertia in Eq. (8) can therefore be neglected, leaving[20]

$$\underset{\sim}{B}\cdot\nabla\tilde{J}_z = 0 , \tag{40}$$

where $\underset{\sim}{B} = \underset{\sim}{B}_0 + \underset{\sim}{\tilde{B}}$ is the total field. Nonlinearly, $\tilde{J}_z$ is a constant along the magnetic field line and Eq. (12) can therefore be averaged over a field line,

$$\eta\tilde{J}_z = \eta\langle\tilde{J}_z\rangle = -c^{-1}\langle\partial\tilde{A}_z/\partial t\rangle , \tag{41}$$

where $\langle\ \rangle$ denotes the average over one period of the tearing mode and $\langle\underset{\sim}{B}\cdot\nabla\tilde{\phi}\rangle = 0$ since $\tilde{\phi}$ is periodic. Combining Eq. (41) with Ampere's law, we obtain a nonlinear equation for $\tilde{A}_z$,

$$\tilde{A}''_z = -(4\pi/\eta c^2)\langle\partial\tilde{A}_z/\partial t\rangle . \tag{42}$$

Note that although $\tilde{\phi}$ does not appear in Eq. (42), $\tilde{\phi}$ is certainly not negligible. In fact, the parallel bunching of electrons along $\underset{\sim}{B}$ by $\partial\tilde{A}_z/\partial t$ produces the potential $\tilde{\phi}$ which forces $\tilde{E}_\parallel$ to be constant along $\underset{\sim}{B}$. If we represent $\tilde{A}_z$ by a single harmonic $\cos k_y y$ and invoke the "constant ψ" approximation, $\langle\partial\tilde{A}_z/\partial t\rangle \simeq (\partial\tilde{A}_z(0)/\partial t)\langle\cos k_y y\rangle$. It is easily seen by comparing Figs. 3a and 3b that $\langle\cos k_y y\rangle \sim 1$ within the magnetic island and is zero outside. Nonlinearly, the resistive layer width is therefore given by the width of the magnetic island. Integrating Eq. (42) across the island and matching to the outer ideal solution as in the linear theory, we obtain the Rutherford expression for the rate of increase of the magnetic island,[20]

$$d(w/a)/dt \sim \Delta'a/\tau_r . \tag{43}$$

The island width w grows linearly in time. Most significantly, w becomes of order "a" on the resistive time scale τ_r so that the standard tearing mode does not lead to a significant enhancement in the rate of dissipation of magnetic flux. It is clear from Eq. (43) that inertia does not play any role in the nonlinear growth of the mode. Consequently, the tearing mode simply represents the diffusive evolution of the initial state to a more complex equilibrium. The expression for dw/dt in (43) has been extended to include the quasilinear modification of the equilibrium current profile, the result being that $\Delta'(0)$ in (43) is replaced by $\Delta'(w)$, the discontinuity in the slope of the ideal solution across the width of the island.[21] As can be seen in Fig. 4, $\Delta'(w) = 0$ for large enough w so the magnetic island eventually saturates. The nonlinear evolution of slab modes with $k_z \neq 0$ and cylindrical modes is basically unchanged from that of the symmetric slab mode.

The previous calculation of the nonlinear evolution of tearing modes does not apply to modes which violate the "constant ψ" approximation since the induction field $\partial \tilde{A}_z/\partial t$ may be very nonuniform within the magnetic island. Such modes include the very long wavelength tearing modes, the kink-tearing mode, the double tearing mode, and the coalescence instability. The double tearing mode occurs when $\underline{k} \cdot \underline{B} = 0$ at two locations which are close enough so that the flows associated with the reconnection at the two reversal layers drive each other.[22] This instability is stable when $\eta = 0$ but has a structure and growth rate which is similar to the kink-tearing mode for η not too small. The coalescence instability is driven by the attraction of the current filaments of adjacent magnetic islands in an existing island chain.[23] This is an ideal instability which, like the kink tearing mode, also drives reconnection when $\eta \neq 0$. The nonlinear behavior of the long wavelength and double tearing modes is currently not well understood.

The coalescence instability and the kink-tearing mode are the most likely candidates for producing fast reconnection since they are both ideally unstable and the flows are therefore strongly internally driven. Computer simulations of both of these instabilities have been performed and the qualitative features of the reconnection process are the same for both instabilities. During the initial ideal phase of the coalescence instability as shown in Fig. 6, two adjacent magnetic islands accelerate together and form a quasineutral layer at the location of the original x-point between the two.[24] Similarly in the cylinder the central region of current carrying plasma kinks and forms a quasineutral around the x-line shown in Fig. 5.[25,26] At this point the flow towards the quasineutral layer in both instabilities begins to force the reversed magnetic flux to reconnect. In both cases the reconnection velocity is roughly given by $v \sim a/(\tau_A \tau_r)^{1/2}$ and the vertical scale size of the quasineutral layer is independent of η, in agreement with the Sweet-Parker model. However, the dynamics of the process are much more complex than in the Sweet-Parker model. The neutral layer tends to break up into multiple magnetic islands and reconnection of the reversed flux occurs in bursts. It should also be emphasized that the boundaries in both simulations are far from the region of plasma flow and therefore do not inhibit the rate of reconnection. The clear conclusion from these simulations is that reconnection of magnetic flux does not typically occur at Petschek's rate but much slower. Of course, one can not eliminate the possibility that under some very special conditions faster reconnection may be possible.

The discussion of nonlinear tearing instabilities up to this point is valid for 2-D systems where the magnetohydrodynamic variables depend only on x and y+pz in the slab and r and $\theta - 2\pi mz/mL$ in a cylinder. While in the 2-D case the tearing mode grows rather slowly and then saturates in a rather benign manner, the evolution of the instability in 3-D can be much more violent. In 3-D simulations of the tearing mode in a cylinder, the magnetic island of the m/n = 2/1 tearing mode can reach large amplitude, w/a ~ .4 - .5, when the initial current profile is fairly flat. When the magnetic island of the 2/1 mode approaches the q = 3/2 and 5/3 rational surfaces, the m/n = 3/2 and 5/3 modes are strongly destabilized. These modes subsequently destabilize even higher mode numbers, the entire process culminating in a broad spectrum of magnetic turbulence.[27] The destruction of the magnetic surfaces associated with the formation of the turbulent bath and the associated loss of particle and energy confinement has been correlated with the major disruptions in tokamak discharges. In detailed studies of the properties of this magnetic turbulence it has been shown that the dissipation of magnetic energy is dominated by the short wavelength component of the spectrum and that the effective resistivity, given by

$$\eta_{eff} = \eta_0 \int d\underline{x}\, \eta \tilde{J}^2 / \int d\underline{x}\, \eta J_0^2 ,$$

where η_0 is the spatial averaged value of $\eta(r)$, is nearly independent of η_0.[28]

The physical mechanism behind the destabilization of the short wavelength tearing modes has recently been investigated in some detail by examining the linear stability of a cylindrical equilibrium containing a large amplitude m/n = 2/1 magnetic island.[29] The 2/1 tearing mode grows by feeding off the current gradient in region $r < r_0$, where r_0 is the location of the 2/1 rational surface. As a consequence, the growth of the 2/1 mode actually steepens the current profile as shown in Fig. 7. The solid (dashed) line is the current profile after (before) the 2/1 mode reaches large amplitude. The arrows mark the position of inside and outside separatrices of the 2/1 magnetic island. In a 2-D simulation all modes have rational surfaces at the q=2 surface so no modes can be driven by the steep gradient in Fig. 7. In 3-D simulations other modes, such as the 3/2 and 5/3 modes, have rational surfaces in the region of large current gradient produced by the 2/1 mode. These modes are strongly destabilized with growth rates which are nearly independent of the resistivity.[29] It is also interesting to observe that this calculation of the onset of 3-D magnetic turbulence in the cylinder is quite similar to that of Poiseuille flow in fluids.[30]

IV. PROBLEMS IN UNDERSTANDING RECONNECTION IN SPACE AND ASTROPHYSICAL PLASMAS

Until recently, much of the linear and especially nonlinear work on tearing modes has been carried out for laboratory applications. Applications to space and astrophysical plasmas are complicated by generally more complex geometries and extremely large values of the Lundquist number $S = \tau_r/\tau_A$. As a specific example, we discuss efforts to understand magnetic energy dissipation in coronal loops. The structure and nonlinear evolution of collisionless and collisional tearing modes in the Earth's magnetosphere are also being studied.[31-33]

The dissipation of magnetic energy in coronal loops was proposed as a flare model a number of years ago.[34] Loops such as that shown in Fig. 8 are observed to be stable for many Alfvén times and then to release large amounts of energy ranging from $10^{28}-10^{30}$ ergs on a time scale of order 10^3 sec. The major radius of the loop is of order 10^9 cm and minor radius 10^8 cm. At the top of the loop in the corona (T ~ 100 ev, n_e ~ $10^9/cm^3$, B ~ 200 G) $\tau_{A\theta}$ ~ 0.7 sec and τ_r ~ 1.3×10^{12} sec while at the base in the photosphere (T ~ 1 ev, n_e ~ $10^{13}/cm^3$, n_n ~ $10^{16}/cm^3$ and B ~ 10^3G) $\tau_{A\theta}$ ~ 14 sec with τ_r ~ 10^7 sec. The flare time scale is therefore intermediate between the Alfvén and resistive time scales.

The equilibrium of a loop such as shown in Fig. 8 is fully 3-D and therefore the equilibrium and ideal stability of the loop must be studied concurrently, i.e., no equilibrium can be found if the loop is unstable to an ideal MHD mode. The computation of an equilibrium is further complicated by the strong variation of the plasma parameters with altitude. It is not obvious that flux surfaces even exist for such a configuration. Sakurai has attempted to numerically solve for a 3-D force free equilibrium for a loop by specifying the normal magnetic flux at the photosphere.[35] He shows that the loop twists and expands upwards with increasing field aligned current. Greater resolution would improve these computations. Xue and Chen show than an axisymmetric equilibrium requires the pressure outside the loop to be greater than that inside to prevent the hoop force from causing the loop to expand.[36] A much more concerted effort will be required before the equilibrium of these loops is understood.

In the absence of a proper 3-D equilibrium the stability of coronal loops has been investigated by approximating the loop as a cylinder and applying boundary conditions at the ends to mock-up the effect of the photosphere. For ideal modes since $\tau_{A\theta}$(corona) << $\tau_{A\theta}$(photosphere), the appropriate boundary condition at the ends of the cylinder are $\underset{\sim}{v} = 0$. This constraint has a stabilizing influence on the ideal kink mode.[37] For resistive modes this boundary condition does not seem appropriate. The growth time of the tearing mode based on photospheric parameters is ~10^5 sec, which is shorter than the ~10^7 sec. time scale for the corona.

Because of the very large Lundquist number, S ~ 10^{12}, for the corona, numerical calculations of the evolution of resistive modes with realistic parameters are not possible and the magnetic energy dissipation time scale for the corona must be extrapolated from the scaling of the dissipation rate in simulations at smaller values of S. The dissipation times of order τ_r for the standard tearing mode are far too long to explain the flare time scale. The Sweet-Parker scaling of reconnection rate for the kink-tearing mode or coalescence mode yields time scales of order 10^4 sec, which are much closer to the observed energy release times. Moreover, since the current density J near the neutral layer of the Sweet-Parker model scales as $\eta^{-1/2}$, which is very large for small η, nonclassical processes may limit J, thereby producing an anomalous resistivity and faster reconnection. In the standard

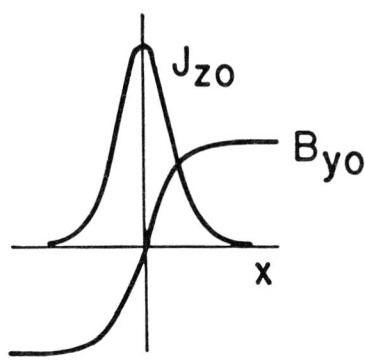

Fig. 1 Slab equilibrium

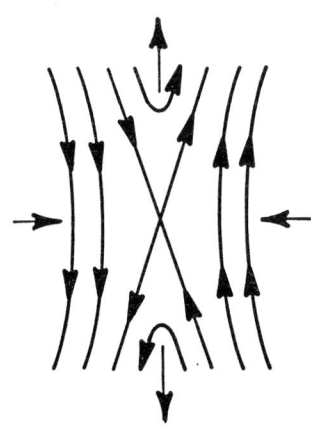

Fig. 2 Reconnection at a single x-line

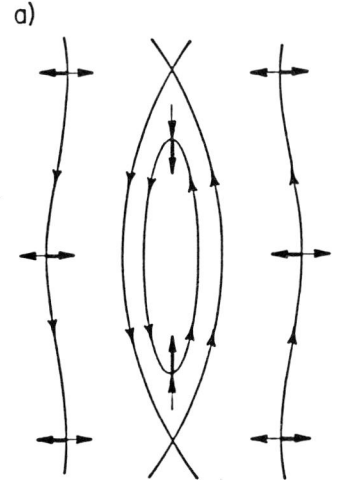
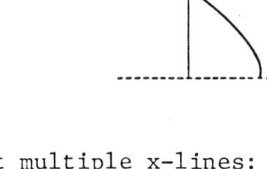

Fig. 3 Reconnection at multiple x-lines: the tearing instability

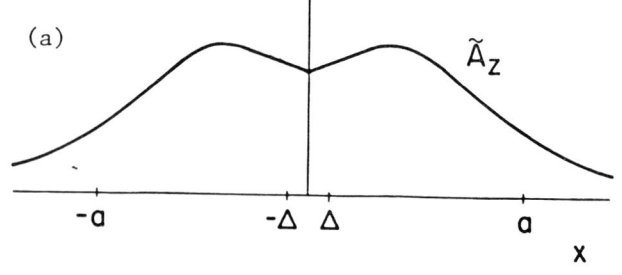
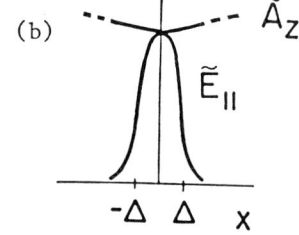

Fig. 4 Mode structure of tearing instability from ideal equation (a) and resistive equations (b)

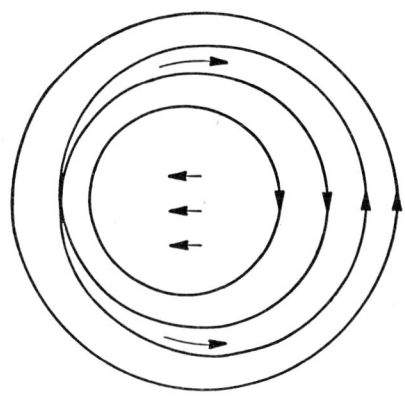

Fig. 5   Kink-tearing instability

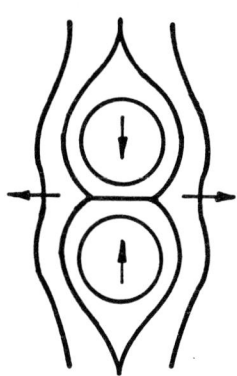

Fig. 6   Coalescence instability

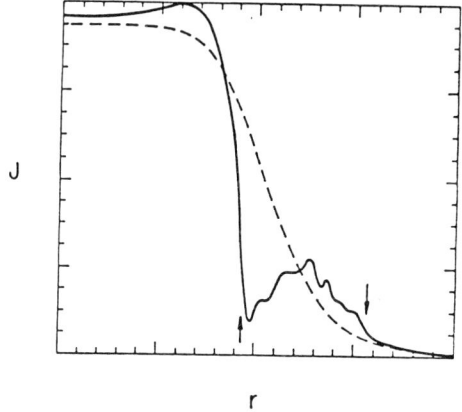

Fig. 7   Current profile at t = 0 (dashed) and after saturation (solid) of m/n = 2/1 tearing instability

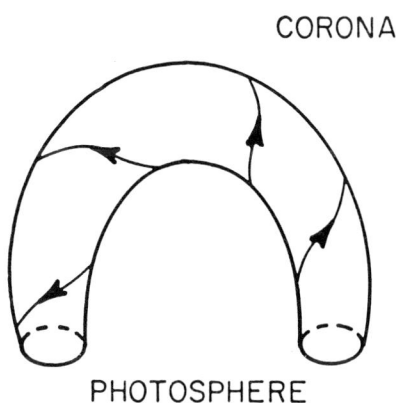

Fig. 8   Magnetic loop in the sun's corona

tearing mode where $J \propto \eta^0$ this would not occur.

The onset of the 3-D tearing mode turbulence is also fast enough to explain the observed energy release time scales. In such a scenario a small change in equilibrium parameters of the loop could cause an existing large amplitude tearing mode to exceed the rather sharp threshold for the onset of the secondary instabilities. The rapid growth of 3-D magnetic modes would rapidly dissipate the magnetic energy.

ACKNOWLEDGEMENTS

I would especially like to thank Dr. A. B. Hassam for many illuminating discussions on the subject of reconnection in solar plasmas and Dr. R. G. Kleva for Fig. 7 (to be published in Ref. 29). Discussions with Dr. J. Chen, Dr. J. Finn, Dr. G. Van Hoven, and Dr. D. Spicer are also acknowledged. This work was supported in part by the U.S. Department of Energy and Office of Naval Research.

REFERENCES

1. N. R. Sauthoff, S. Von Goeler, and W. Stodiek, Nucl. Fusion 18, 1445 (1978).
2. S.-I. Akasofu, The Physics of Magnetospheric Substorms (D. Reidel, Dordrecht, Holland, 1977).
3. Solar Flares, edited by P. A. Sturrock (Colorado Assoc. Univ. Press, Denver, Co., 1980).
4. V. M. Vasyliunas, Rev. Geophys. Space Phys. 13, 303 (1975).
5. P. A. Sweet, in Proceedings of the Astronomical Union Symposium on Electromagnetic Phenomena in Cosmic Physics, No. 6, Stockholm, 1956 (unpublished) p. 123; E. N. Parker, Astrophys. J. Suppl. Ser. No. 77, 8, (1963).
6. H. E. Petschek, in AAS/NASA Symposium on the Physics of Solar Flares, edited by W. N. Hess (National Aeronautics and Space Administration, Washington, D.C., 1964) p. 425.
7. T. Sato and T. Hayashi, Phys. Fluids 22, 1189 (1979).
8. J. F. Drake and Y. C. Lee, Phys. Fluids 20, 1341 (1977).
9. J. F. Drake and Y. C. Lee, Phys. Rev. Lett. 39, 453 (1977).
10. H. P. Furth, J. Killeen, and M. N. Rosenbluth, Phys. Fluids 6, 459 (1963).
11. B. Coppi, Phys. Fluids 8, 2273 (1965); A. B. Hassam and J. F. Drake, Phys. Fluids 26, 133 (1983).
12. B. B. Kadomtsev, Reviews of Plasma Phys. (M. A. Leontovich, Ed., Consultants Bureau, N.Y., 1966) Vol. 2, p. 153.
13. G. Van Hoven, R. S. Steinolfson, and T. Tachi, Astrophys. J. 268, 860 (1983).
14. H. P. Furth, Nucl. Fusion Suppl. Pt. 1, 169 (1961).
15. R. S. Steinolfson and G. Van Hoven, Phys. Fluids 26, 117 (1983).
16. H. P. Furth, P. Rutherford, and H. Selberg, Phys. Fluids 16, 1054 (1973).

17. W. A. Newcomb, Ann. Phys. 10, 232 (1960).
18. M. N. Rosenbluth, R. Y. Dagazian, and P. H. Rutherford, Phys. Fluids 16, 1894 (1973).
19. B. Coppi, R. Galvao, R. Pellat, M. Rosenbluth, and P. Rutherford, Fiz. Plazmy 2, 961 (1976) [Sov. J. Plasma Phys. 2, 533 (1976)].
20. P. Rutherford, Phys. Fluids 16, 1903 (1973).
21. R. B. White, D. A. Monticello, M. N. Rosenbluth, and B. V. Waddell, Phys. Fluids 20, 800 (1977); R. B. White, D. A. Monticello, and M. N. Rosenbluth, Phys. Rev. Lett. 39, 1678 (1977).
22. P. L. Pritchett, Y. C. Lee, and J. F. Drake, Phys. Fluids 23, 1368 (1980).
23. J. M. Finn and P. K. Kaw, Phys. Fluids 20, 72 (1977); P. L. Pritchett and C. C. Wu, Phys. Fluids 22, 2140 (1979); A. Bondeson, Phys. Fluids 26, 1275 (1983).
24. D. Biskamp and H. Welter, Phys. Rev. Lett. 44, 1069 (1980).
25. W. Park, D. A. Monticello, R. B. White, and S. C. Jardin, Nucl. Fusion 20, 1181 (1980).
26. W. Park, D. A. Monticello, and R. B. White, Phys. Fluids 26, XXX (1983).
27. B. V. Waddell, B. Carreras, H. R. Hicks, and J. A. Holmes, Phys. Fluids 22, 896 (1979); Phys. Fluids 23, 1811 (1980).
28. D. Biskamp and H. Welter, in *Plasma Physics and Controlled Nuclear Fusion Research* (IAEA, Vienna, 1982) CN-41-TI.
29. R. G. Kleva, J. F. Drake, and A. Bondeson (submitted to Phys. Fluids); A. Bondeson (submitted to Phys. Fluids).
30. S. A. Orszag and A. T. Patera, Phys. Rev. Lett. 45, 989 (1980).
31. K. B. Quest and F. V. Coroniti, J. Geophys. Res. 86, 3289 (1982) and references therein.
32. J. G. Lyon, S. H. Brecht, J. D. Huba, J. A. Fedder, and P. J. Palmadesso, Phys. Rev. Lett. 46, 1038 (1981).
33. J. Birn (this meeting and references therein).
34. D. S. Spicer, Solar Phys. 53, 308 (1976); 71, 149 (1981).
35. T. Sakurai, Solar Phys. 69, 343 (1981).
36. M. L. Xue and J. Chen, Solar Phys. 84, 119 (1983).
37. A. W. Hood and E. R. Priest, Geophys. Astrophys. Fluids Dynamics 17, 297 (1981); G. Einaudi and G. Van Hoven, Solar Phys. 88, 163 (1983).

DISCUSSION

Wu: Did you consider gravitational effects in these studies? If not, would you care to comment on the effects of gravity on those coronal loops?

Drake: I have not considered the effect of gravity. Gravity is probably important. Otherwise how can one find an equilibrium in which the density at the bottom of the loop is 10^6 times larger than at the top?

Migliuolo: The density scale height becomes small near the photosphere. What are the effects of this inhomogeneity on the tearing mode?

Drake: The effects of density and temperature gradients along $\underset{\sim}{B}$ are completely unknown.

Birn: I would like to point out again that an asymmetry of the current sheet in the direction parallel to it can also change the possible saturation of a tearing mode. Whereas in a plane current sheet with periodic boundary conditions the location of the center of magnetic islands is fixed in space, an asymmetric situation could produce moving islands being pushed out from the diffusion region as shown in magneto-tail simulations. This can lead to larger energization than possible in a slab geometry with periodic boundary conditions.

Drake: I agree with your observation. In the magneto-tail, the magnetic fields can act like a slingshot in ejecting plasma and flux away from the earth.

I would also note, more generally, that the magnetic configuration may strongly affect the nonlinear behavior of the reconnection process. It is therefore important to carry out careful simulations of the particular configuation of interest.

Kundu: I just wanted to alert you to the fact that there exist at present values of temperature, density and magnetic field as a function of height in a 3-D coronal loop. This is possible due to measurements made in the radio (microwave) domain with the Very Large Array (VLA) at several frequencies, which really explore different layers of the loop. Also, there are soft x-ray measurements of coronal loops.

Drake: More detailed measurements will certainly help in building models of energy dissipation in coronal loops.

Priest: What is the effect of the growth of the $m = 2$ mode on the $m = 1$ mode and on the basic current profile?

Drake: The $m = 2$ mode does not strongly affect the $m = 1$ mode because the rational surface of the $m = 1$ mode is rather far away from the magnetic island of the $m = 2$ mode.

The current profile before and after the $m = 2$ mode has reached large amplitude is shown in Fig. 7 (more details can be found in Ref. 29). The $m = 2$ mode basically dissipates most of the current in the vicinity of its magnetic island, leaving an elliptically shaped current pedestal.

Sturrock: What conditions are required to lead to the mode interaction process investigated by Carreras and others? Are these conditions likely to be met in coronal loops?

Drake: All that is required is a current profile which is somewhat flat near the center with the $q = 2$ surface just outside of the steep gradient (see dashed curve in Fig. 7). This current profile is not in any sense unusual and therefore could very possibly be present in coronal loops.

Van Hoven: i) I disagree with your statement that the tearing rate is not anomalous. Tearing self-consistently generates a narrow enough layer to be diffusive. However, it is anomalous on the global scale.
ii) I believe that very long wavelength modes can be added to your list of $m = 1$ and double tearing modes as good nonlinear-release candidates. We have done nonlinear calculations which show this.

Drake: The standard "constant ψ" tearing mode evolves to large amplitude on the resistive time, and, is therefore not in any sense anomalous.

The nonlinear behavior of very long wavelength modes is currently not well understood. I think it is an important area of research.

Mahajan: If the minimum of current is formed in the plasma, then there is a possibility of exciting large growth rate double tearing modes discussed by Mahajan and Hazeltine, typically 10^3 times larger. Have those been considered?

Drake: The standard double tearing mode has been previously considered by D. Spicer. The mode which you refer to has not been considered. I would emphasize again that the linear growth rate does not really give the time scale for the release of magnetic energy. It is essential to carry out a nonlinear calculation to find the energy dissipation rate.

Henriksen: Would there be any difference in the enhanced growth rates if the magnetic islands were formed by collision and coalescence of previously existing flux tubes?

Drake: The coalescence instability is an ideal mode and the reconnection time scales as $(\tau_r \tau_A)^{1/2}$, which is much faster than the standard tearing mode.

D. Smith: Van Hoven and one of his students showed that when you consider line tying, only the m = 1 mode is linearly unstable. How would this affect the possibility of 3-D magnetic turbulence?

Drake: As I have mentioned, the growth time of the tearing mode based on the parameters of the photosphere is actually shorter than the growth time based on the parameters of the corona. So it is not obvious to me that line tying boundary conditions at the photosphere are appropriate for resistive modes. Line tying conditions are appropriate for ideal modes.

SESSION II

FORMATION, EQUILIBRIUM AND STABILITY OF JETS

Colin A. Norman
Institute of Astronomy, Cambridge, U.K.,
Sterrewacht, Leiden, Netherlands, and
European Southern Observatory, Garching, F.R.G.

ABSTRACT
   Consideration of the many observed types of jets on scales ranging from parsecs to megaparsecs seen in radio, optical, infrared and X-ray wavebands with a variety of morphologies both in galactic and extragalactic systems leads to some constraints on their fundamental nature. Jet formation is introduced with the concept of the Laval nozzle and related points include the problem of maintaining the nozzle, Mach disk effects due to under and over-expansion and the potential importance of magnetic confinement and focussing. Current ideas on jet formation at the black hole and accretion disk are given with emphasis on the plasma physics associated with black-hole electrodynamics, thermal and magnetically driven winds and thick disks. Stability of jet propagation is reviewed with emphasis on magnetised and unmagnetised Kelvin-Helmholtz instabilities and the various dominant modes. The particle acceleration physics of shocks, wave-particle interactions and turbulence is summarised while noting some outstanding plasma physics problems. Jet equilibrium associated with the non-linear saturation of instabilities, the formation of cocoons, shock stabilisation and magnetic fields is discussed. Detailed plasma physics studies that could significantly clarify jet physics are indicated.

I. INTRODUCTION

   The range of scientific phenomena associated with the formation, stability and equilibrium of jets is extremely broad. Originally proposed to explain powerful double radio sources (Rees, 1971), the jet hypothesis has found widespread application to observations in all wavebands on both galactic and extragalactic scales. In this review I shall concentrate on exhibiting the important plasma astrophysics problems in the study of jets. I shall be posing more problems to this talented interdisciplinary group from many different fields of plasma physics, space physics and astronomy than I can answer, and I hope to stimulate a productive cross discipline interaction.

It is very difficult to outline the well motivated plasma problems that can be directly inferred from the observations. We have nothing like a laboratory set-up such as Dr. Stenzel's or even satellite experiments such as Dr. Kennel's where the relevant microscopic processes can be isolated experimentally. It is a situation perhaps akin to early solar physics where our <u>models</u> are what motivates the interesting plasma problems. With this in mind the plasma astrophysicist is then challenged to solve some rather interesting physics problems concerning, for example, relativistic collisonless shock structure, energy flow in shearing turbulent fluids, particle acceleration, magnetic confinement mechanisms for jets and magnetic reconnection studies in turbulent jet flow. Specifically in the area of jet formation models one is faced with modelling, for example, black hole and accretion disk coronae and magnetospheres complete with general relativistic effects on electric circuit models of magnetospheres!

The definition of a jet is best obtained by inspecting some really excellent examples of various jet classes. This can be found in the proceedings of the recent IAU meeting No. 98 and the Turin Astrophysical Jets workshop. Here I shall merely state the examples of various classes of jet that I selected to illustrate the talk. Jets range from being quite straight over hundreds of kiloparsecs (HB 13) and highly collimated on all scales (NGC 236, NGC 6251), with occasionally an extreme widening or flaring at larger distances (IC 4296), to exhibiting considerable bending (NGC 449, NGC 326) with the classic head tail radio source NGC 1265 as an extreme case. VLBI jets such as 3C 236, 3C 303.1 and 3C 138 often show clear alignment with large scales. Severe bending is often seen in intermediate scale maps from the MERLIN array (3C 371, 3C 454.3) and small scale inner jets are seen in Seyfert and Markarian galaxies. Jets are also seen in the X-ray (Centaurus-A) and in the optical (3C 273, 3C 305 and Coma A). In our own Galaxy Coma A is a miniature, classic extragalactic double radio source, SS 433 has a well defined jet outflow, bipolar or collimated flows around protostars in molecular clouds are now often found (S 140, S 68, Mon R2) and there is even a potential jet candidate in the Galactic centre. The very high resolution numerical simulations produced by M. Norman and Winkler using the Garching Cray machine are now sufficiently detailed to provide us with excellent observational material to be assimilated with the real world. So much for the observations. Their direct interpretation is still rather controversial and the physical parameters are not particularly well constrained.

Extragalactic jets could still be either relativistic or non-relativistic. Superluminal motion observed in some radio cores seems compelling evidence for relativistic bulk motion (Blandford et al., 1977) although we should await the results of the new VLBI surveys of faint radio cores before being too definite. The large bending angles observed in extended radio jets seem most consistent with non-relativistic speeds (van Groningen et al., 1980). To slow a jet from relativistic in the galaxy core to non-relativistic further out requires the dissipation of most of the initial jet luminosity which would in general violate the observational limits, but energisation of the broad line region may be a real effect (Norman and Miley, 1983). Many jets look confined rather than free, for example HB 13. In a number of cases, for example 4C 32.69 (Potash and Wardle, 1980), extreme pressures are demanded for thermal confinement and magnetic confinement is most plausible. Whether jets could be generally magnetically, as opposed to thermally, confined is unclear. A related question is whether jets are heavy or light in terms of density contrast with their environment. Light jets will be buoyant and dense jets may develop knots and thermal instabilities. Knot structure in jets could well be due to cooling instabilities or alternatively to shock structures in a time variable jet flow. Convolved in with our basic jet model is the brightness function of jets associated with particle acceleration processes for the synchrotron radiating relativistic electrons. One-sided jets are seen with symmetrical hot spots and lobes on both sides of the parent galaxy. Therefore jets are either intrinsically invisible, glowing when triggered by a propagation instability, or always glowing, but rendered invisible by a physical process such as Doppler fading.

II. JET FORMATION

Why is the jet phenomena so ubiquitous? The most obvious concept of focussing a point explosion in an inhomogeneous atmosphere does not work. As shown by Sanders (1976), the focussing angle is only 25°-40° from the vertical direction for an exponential atmosphere where the explosion breaks free within a few scale heights. At break out, the cone angle is typically the inverse of the Mach number $\sim M^{-1}$ where M is the ratio of shock velocity to post shock sound speed. Only in the case where the density profile of a rifle barrel was used could any significant focussing be found and in no case would it account for jet collimation over a scale range of at least 10^6!

Continuous injection and outflow models can avoid at least some of these problems. The simplest and first version was the de Laval nozzle concept introduced by Blandford and Rees (1974) who noted that the transition from subsonic to supersonic in a preexisting one-dimensional flow involves a pinching or nozzling of the jet cross-section. A detailed numerical study of this process in an inhomogeneous atmosphere has been presented by Smith et al. (1981, 1983) who made several important clarifications. The nozzle point (at r_n) and the spherical shock between the injected fluid and the pre-nozzle cavity (at r_c) are rather close to each other; r_n/r_c = 1.18 and 1.05 for γ = 5/3 and 4/3 respectively for strong shocks, so that the one-dimensional nozzle approximation may not be valid or alternatively the shocks may be weak. Interesting instability regions were isolated as a function of jet power. For a specified cavity sound speed, at low luminosity the nozzle becomes Kelvin-Helmholtz unstable resulting in "bubbling" at the nozzle, at higher luminosities the nozzle stabilises as the jet flow convection beats the instability and thus results in a stable flow regime. However, this regime has an upper luminosity limit where the cavity itself becomes Rayleigh-Taylor unstable creating clouds to be injected into the flow.

The pressures required to confine the nozzle for high luminosity jets are rather extreme (Blandford and Rees, 1978) and in the case of thermal confinement the gas would have to be dense X-ray emitting material which is not observed. Magnetic nozzling is a good candidate here but may be viewed as part of the class of direct jet formation models associated with black hole and accretion disk physics. Thick accretion disks develop very deep vortex funnels at their centres (Lynden-Bell, 1978). These are rifle barrels of the type referred to earlier. Abramowicz and Piran (1980) showed that a highly collimated radiation flux will flow up the funnel acting on matter injected from the funnel-disk surface. Very effective collimation occurs but the particles are so locked to the radiation flow that bulk Lorentz factors of at most only a few are produced which could lead to problems in certain models of superluminal sources.

While analysing accretion disk electrodynamics, Blandford (1976) found a similarity solution for the free-free magnetosphere above an accretion disk that could transport a sufficient flow of energy and angular momentum necessary for a steady flow in a Keplerian accretion disk. The poloidal field lines are paraboloidal with foci on the rotation axis and at larger distances the field becomes predominantly toroidal leading to focussed magnetically confined jets. Analogous self-similar solutions are given in Blandford and Payne (1982) for the magneto-hydrodynamic field structure above a Keplerian disk. Similar flow patterns result. It is interesting to note here that both these models give quite a good qualitative understanding of the recent magnetic jet formation work of Uchida and Shibata (1983) who applied their numerical results mainly to the high beta to low beta transition for flux loops emerging into the solar corona and found that the emerging field evolves into a predominantly toroidal configuration focussed along the vertical axis, as in the above examples.

Black hole electrodynamics incorporates one of the most promising jet formation models (Blandford and Znajek, 1977, MacDonald and Thorne, 1982). A black hole is threaded by field lines that are due to currents flowing in a surrounding accretion disk or torus. The black hole has a surface resistance of 377 ohms and, as for any magnetospheric problem, an equivalent electric circuit can be drawn (Ionson, these proceedings). Incidentally, this is also true for the accretion disk electrodynamics problem. The maximum energy output from the hole occurs when the external impedance matches that of the hole. This seems to be related to the dissipative microphysical processes occurring in the far magnetosphere but there may be a more fundamental basis (Phinney, 1983). This question and the detailed examination of black holes magnetospheres seem a rich field of research for the current audience.

An interesting plasma physics problem arises when considering the ion-pressure supported torus model of galactic nuclei (Rees et al., 1982, Phinney, 1982). For very low accretion rates, thick disks or tori of hot ions will circulate a hole, their electrons having cooled via synchrotron or Compton losses, the ions remaining hot since they spiral into the hole before temperature equilibration due to Coulomb collisions occurs. Fields generated by currents in this torus thread the hole and extract energy from it. A two year old challenge is to prove this is instable. It is not obviously so. Characteristic scales are so large that current driven instabilities are unlikely to occur. Coupling between electrons and ions is unlikely due to the large frequency and wave number mismatch. As yet uncalculated anisotropic effects may do it, Dr. Coroniti may elaborate on these topics. Finally, it is important to mention the nearby jet-like systems in our own Galaxy and their formation mechanism. Particularly exciting is the possibility of observing jet flows form in real time over the next two decades using the Space Telescope on protostellar jets in the nearby Taurus-Aurigae association - a mere $10^{2.3}$ pc away!

III. JET INSTABILITY

Jet stability begins classically with the Klevin-Helmholtz instability (cf. Ferrari, these proceedings). For a cylindrical jet with shear flow, streaming through an ambient medium one applies a perturbation $\sim g(r)e^{i(kz+n\theta-wt)}$ for jet radius r, θ a polar coordinate on its circumference, the jet axis as the z-axis, t being time, and w, k and n are perturbation frequency, wave-number of the z coordinate and polar coordinate respectively. The standard ordinary mode classification calls n = 0 the Pinching mode, n = 1 the Helical or Kink mode and n $\geq$ 2 the Flute mode. Effects due to the finite geometry are very important via the reflection models which can often dominate the physics of the jet instability.

Briefly summarising the instability behaviour (Hardee, 1982a,b; Ferrari et al., 1980, 1981, 1982; Cohn, 1983) of a jet with initial radius a, long wavelengths (with ka << 1) are always unstable if the propagation angle is large enough and here the helical and flute modes dominate. Short wavelength purely longitudinal modes (ka >> 1) are stable for Mach numbers greater than $2\sqrt{2}$ but, for finite propagation angles ($\sim 90°$), they are always unstable. The reflection mode is always unstable for ka $\sim$ 1 and Mach numbers $\gtrsim 2\frac{1}{2}$. Pinching dominates for large Mach numbers and in general the growth rates are faster than the ordinary mode.

Growth rates can be reduced by a density contrast for either heavy or light jets. For relativistic flows the ordinary modes have zero growth as the bulk Lorentz factor γ_b goes to infinity but for the reflection mode there is no real suppression except that ka tends to zero as γ_b goes to infinity. For strongly magnetised jets where the Alfven velocity v_A greatly exceeds the sound speed c_s all modes stabilise. The pinch can be stabilised for Mach numbers $\leq 2 \frac{v_A}{c_s}$ and all modes are stable for M < 1. Velocity profiles with a shear scale $\sim$ h stabilize the short wavelength modes for k h $\gtrsim$ 1 and smoother profiles show growth for ka $\lesssim$ a/n which can be very important for reflection modes. Hardee (1983) incorporated jet expansion effects and showed that exponential growth rates become secular but remember how straight HB 13 is over most of its length.

Observable consequences of the instability calculation are that reflection modes may produce quasi periodic knot structures (Norman and Winkler, 1983) and that helical modes may produce twists and bends in, for example, M87 (Hardee, 1982) and flaring in NGC 236. The non-linear development of the Kelvin-Helmholtz instability is an important factor in considering jet equilibrium.

IV. NON-LINEAR EFFECTS AND EQUILIBRIUM

The actual jet equilibrium is related to the observations by a convolution with the relativistic electron and magnetic field dependent brightness function. What we see indeed looks turbulent and highly non-linear: an equilibrium only with a considerable coarse graining of the observations. The numerical work (Norman et al., 1981, 1983) shows the Kelvin-Helmholtz instability saturates in shocks, Mach disks, knot structures and back flowing cocoon around the jet.

The development of shocks during the Kelvin-Helmholtz instability occurs when the transverse expansion velocity of the perturbation exceeds the sound speed. Shocks then develop, making an angle $\sim 1/M$ with the jet axis. Further growth of the instability is slowed from exponential to secular (Benford, 1981). Of particular interest for knots are the intersecting shock structures that develop from the reflection mode. Cocoons drive shock structures in the jets as the back flowing vortices pinch the jet in the axially symmetric case. Clearly three-dimensional work will be important here and the slender jet approximation may prove a useful first estimate (Smith and Norman, 1981a,b).

Thermally confined jets break free if the external pressure drops as $r^{-\alpha}$ where α exceeds 2 and consequently the transverse velocity becomes supersonic (Smith, 1982). Jets do not in general look free with a characteristic opening angle $\sim 1/M$. A solution is to reconfine the flow which, as shown by Sanders (1983) results in a series of Mach disks spaced roughly at the Prandtl wave length λ_p = Ma. Internal shocks could cause reheating and a quasi periodic free expansion and reconfinement process. Of particular observational relevance are the gaps in jets observed between VLBI scales and, say, the Holmberg radius of the parent galaxy. The jet may be switched on by the reconfinement shocks due to particle acceleration processes in shocks. Quasi-periodic knot structures may be found in the train of Mach disks created by overexpansion or reconfinement.

Magnetic confinement mechanisms are very plausible (Benford, 1978, 1981, 1983) and solar wind techniques have proved most useful (Chan and Henriksen, 1980; Achterberg et al., 1983). Making a self-similar hypothesis for the MHD flow variables an effective potential can be found for the behaviour of the jet radius where the effects of toroidal field, internal pressure, longitudinal field and toroidal flow are included. Motion in this effective potential is seen to be quasi-periodic and can be modelled to fit 3C 31 jet structure if the energy density in the toroidal field is roughly 1% (Bridle et al., 1980) of the bulk kinetic energy density.

In principle the Kelvin-Helmholtz instability can be saturated by particle acceleration (Lacombe, 1977; Benford et al., 1980; Henriksen et al., 1982; Bicknell and Melrose, 1983; Krautter et al., 1983). The reasoning is that the Kelvin-Helmholtz instability drives a turbulent cascade from wave numbers ka ~ 1 that is subsequently dominated either by hydrodynamic effects (Henriksen, these proceedings) or collisionless wave-wave interactions (Benford et al., 1980). At a particular wave number the cascade dissipates either by Fermi acceleration, resonant wave-particle interactions, weak shocks or heating. In practice one would expect that only some of the available free energy could be saturated in the particle acceleration process. The turbulence and particle acceleration regions must diffuse inward from the jet cocoon shear layer since limb brightening is not, in general, found (Eilek, 1982). No significant magentic field generation occurs as the Kelvin-Helmholtz does not generate helicity (de Young, 1980). A real weakness of both theory and data is that most of the above mentioned models can fit the observations!

Jet brightness is inconsistent with flux freezing and adiabatic expansion of the relativistic electrons. Magnetic field reconnection could provide the energy source for the particle acceleration but the reconnection process would have to be driven on very small scales since for typical jet parameters the Reynolds number is 10^{15}! Numerical simulations (cf. de Young, 1980) are one approach but possibly a minimum energy type variational principle could be constructed allowing an estimate of the available free energy in analogy to the solar corona (Norman and Heyvaerts, 1983, Heyvaerts and Priest, 1983).

A final word about shocks. That they develop in jet flow and can accelerate particles is unquestionable (cf. Blandford and Eichler, 1983). However, there is no real consensus yet on what limits the particle acceleration efficiency: shock precursor pressure of the accelerated particles, heating of the thermal component, escape of the highest energy particles or the limitations of the Alfven wave scattering centre amplitude to a value $\delta B/B \sim 1$ (McKenzie and Völk, 1981). Since it is electrons that radiate we require not only that the protons, but also electrons, are injected into the shock acceleration regime; possible mechanisms are preacceleration due to electrostatic processes in perpendicular shocks or Fermi acceleration. Note that whistlers as well as Alfven waves may become important scattering centers. Many shocks will not have time to develop into a stationary state, the upper energy cut-off to the accelerated particle distributions being given by equating the acceleration time to the shock age as is observed in the interplanetary medium.

These questions of non-linear self-consistent, self-regulating shock structures and particle injection seem one of the most precisely formulated challenges confronting the plasma astrophysicist. Considerable help is on the way from the study of quasi parallel shocks in the interplanetary medium (Kennel, these proceedings).

V. SUMMARY

Jet physics embraces many major plasma astrophysics problems including the magnetospheres and coronae of black holes and accretion disks, fully developed turbulence in magnetised, shearing jet flows, and the understanding of detailed shock structures and particle acceleration processes.

Emission line structures are now seen associated with jets and lobes and in these cases densities, temperatures and velocities can be constrained. This is the observational area that can allow much more detailed modelling to be done (cf. Norman and Miley, 1983). Much help can be gained from interdisciplinary collaboration in even just helping pose the correct question (cf. Kennel, these proceedings)! The problems are hard but involve extremely challenging and profound concepts at the frontiers of current astrophysics.

It is a pleasure to thank many colleagues for stimulating discussions and particularly P. Barthel, R. Blandford, J. Hayvaerts, G. Miley, M. Norman, S. Phinney, M. Rees, R. Schilizzi, M. Smith, R. Strom, and K.-H. Winkler.

REFERENCES

Abramowicz, M.A., and Piran, T.: 1980, Ap.J. (Letters) 241, L7.
Achterberg, A., Blandford, R.D., and Goldreich, P.: 1983, submitted to Nature.
Benford, G.: 1978, MNRAS 183, p. 29.
Benford, G.: 1981, Ap.J. 247, p. 792.
Benford, G.: 1983, preprint.
Benford, G., Ferrari, A., and Trussoni, E.: 1980, Ap.J. 241, p. 98.
Bicknell, G.V., and Melrose, D.B.: 1982, Ap.J. 262, p. 511.
Blandford, R.D.: 1976, MNRAS 176, p. 465.
Blandford, R.D., and Eichler, D.: 1983, Ann. Rev. Astron. Astrophys., in preparation.
Blandford, R.D., and Znajek, R.L.: 1977, MNRAS 179, p. 433.
Blandford, R.D., McKee, C.F., and Rees, M.J.: 1977, Nature 267, p. 211.
Blandford, R.D., and Rees, M.J.: 1974, MNRAS 169, p. 395.
Blandford, R.D., and Rees, M.J.: 1978, Physica Scripta 17, p. 265.
Blandford, R.D., and Payne, D.G.: 1982, MNRAS 199, p. 883.
Bridle, A.H., Henriksen, R.M, Chan, K.L., Fomalont, E.B., Willis, A.G., and Perley, R.A.: 1980, Ap.J. (Letters) 241, L145.
Chan, K.L., and Henriksen, R.N.: 1980, Ap.J. 241, p. 534.
Cohn, H.: 1983, Ap.J. 269, p. 500.
de Young, D.S.: 1980, Ap.J. 241, p. 81.

Eilek, J.A.: 1982, Ap.J. 254, p. 472.
Ferrari, A., Trussoni, E., and Zaninetti, L.: 1980, MNRAS 193, p. 469.
Ferrari, A., Trussoni, E., and Zaninetti, L.: 1981, MNRAS 196, p. 1051.
Ferrari, A., Massaglia, S., and Trussoni, E.: 1982, MNRAS 198, p. 1065.
Hardee, P.E.: 1982a, Ap.J. 257, p. 509.
Hardee, P.E.: 1982b, Ap.J. 261, p. 457.
Hardee, P.E.: 1983, Ap.J. 269, p. 94.
Henriksen, R.N., Bridle, A.H., and Chan, K.L.: 1982, Ap.J. 257, p. 63.
Heyvaerts, J., and Priest, E.: 1983, Astron. Astrophys., in press.
Krautter, A., Henriksen, R.N., and Lake, K.: 1983, Ap.J. 269, p. 81.
Lacombe, C.: 1977, Astron. Astrophys. 54, p. 1.
Lynden-Bell, D.: 1978, Physica Scripta 17, p. 185.
MacDonald, D., and Thorne, K.S.: 1982, MNRAS 198, p. 345.
McKenzie, J.F., and Völk, H.J.: 1981, Proc. 17th Int. Cosmic Ray Conference (Paris) OG4-27.
Norman, C.A., and Heyvaerts, J.: 1983, Astron. Astrophs. 124, L1.
Norman, C.A., and Miley, G.K.: 1983, Astron. Astrophys., in press.
Norman, M., and Winkler, K.H.: 1983, in preparation.
Norman, M.L., Winkler, K.-H., and Smarr, L.: 1983, Astrophysical Jets, A. Ferrari and A. G. Pacholczyk (eds.), p. 227.
Norman, M.L., Smarr, L., Wilson, J.R., and Smith, M.D.: 1981, Ap.J. 247, p. 52.
Phinney, E.S.: 1982, in Plasma Astrophysics, Varenna ESA SP-161, p. 337.
Phinney, S.: 1983, Ph.D. Thesis, University of Cambridge.
Potash, R.J., and Wardle, J.F.C.: 1980, Ap.J. 239, p. 42.
Rees, M.J.: 1971, Nature 229, p. 312.
Rees, M.J., Begelman, M.C., Blandford, R.D., and Phinney, E.S.: 1982, Nature 295, p. 17.
Sanders, R.H.: 1976, Ap.J. 205, p. 335.
Sanders, R.H.: 1983, Ap.J. 266, p. 73.
Smith, M.D.: 1982, Ap.J. 259, p. 522.
Smith, M.D., and Norman, C.A.: 1981a, MNRAS 194, p. 771.
Smith, M.D., and Norman, C.A.: 1981b, MNRAS 194, p. 785.
Smith, M.D., Smarr, L., Norman, M.L., and Wilson, J.R.: 1981, Nature 293, p. 277.
Smith, M.D., Smarr, L., Norman, M.L., and Wilson, J.R.: 1983, Ap.J. 264, p. 432.
Uchida, Y., and Shibata, K.: 1983, preprint.
van Groningen, E., Miley, G.K., and Norman, C.A.: 1980, Astron. Astrophys. 90, L7.

DISCUSSION

Sturrock: What is the origin of the magnetic field in jets?

Norman: Probably entrainment, since the Kelvin-Helmholtz instability does not generate helicity and the expansion factors $\gg 10^6$ in some cases made primordial fields generated at the jet origin unlikely as the dominant field in the jet at large distances.

Eichler: My impression of the axisymmetric simulations of Norman et al. is that they yield poor collimation. Is that the case?

Norman: Collimation is far better than for a point explosion in an inhomogeneous atmosphere, but for highly collimated flows from 10^{15}–10^{25} cm, some reconfinement and recollimation is necessary.

Vasyliunas: The term "jet" carries the connotation of flow along the observed structure. How strong is the evidence for such a flow, or is it simply that no one can think of another possibility?

Norman: Many other possibilities have been thought of in this context, including low frequency electromagnetic waves, compact objects, γ-rays and electron-position beams.

Chiuderi: Is it true that flow velocities deduced from optical emission lines are much smaller than those required by fluid models?

Norman: The emission line studies give velocities in the range 300–500 km s^{-1}, which is indeed at the lower end of the expected jet velocities. The interpretation of these observed emission line velocities as directly reflecting the jet's velocity is not yet certain.

Krishan: Has anyone tried to study the correlation between the velocity flows (from simulation) and magnetic fields from polarization data?

Norman: Not yet, but I hope this will be done soon.

Uchida: The importance of a large scale magnetic field in guiding the jet or in determining the initial direction of jets hasn't been mentioned. Shibata and myself recently worked on the formation of jets by a toroidal field relaxing into a low-β region along the external magnetic field (in these proceedings).

Norman: As noted in the text, your work can be quite nicely explained in terms of the physics of a magnetically confined jet shown by the simularity solutions.

Bratenahl: Regarding those beautiful numerical jets at the beginning, what were the boundary conditions? I saw no spreading. How was it confined?

Norman: The pictures I showed (due to Norman and Winkler) were taken in the comoving frame of the jet head in the simulation. The confinement is due to the external medium, as is probably the case in the real world.

Priest: What are the values of basic parameters such as the plasma beta and the ratio of the know separation to the mean free path or ion gyroradius?

Norman: The plasma's magnetic field is inferred using equipartition arguments and, therefore, beta is always of order unity. For typical parameters the proton gyroradius is many orders of magnitude less than the characteristic flow scales and, therefore, the fluid approximation is a good one.

Kundu: I have a question related to the physics of jets. I know from observational evidence that galaxies in the process of merging are more violently radio emissive than the interacting galaxies. Can you explain this in physical terms?

Norman: Merging can severely dynamically distort the surroundings of the central engines, repopulating stellar and gaseous orbits that can be used as fuel for the central black hole, say. Some of this has been ruled out in Norman and Silk (Ap.J. 266, 502, 1983) and Lake and Norman (Ap.J. 270, 51, 1983).

ENERGY DISSIPATION MECHANISMS IN THE SOLAR CORONA

J. Heyvaerts
Observatoire de Meudon and Université Paris VII
92195 Meudon - France

ABSTRACT

Present ideas concerning the electric heating of the solar corona are reviewed. We consider in some more detail the dissipation of MHD waves in strong horizontal gradients of the Alfvén velocity. Then we consider the evolution of DC currents in the solar corona. Some theories aiming at the evaluation of the net rate of energy dissipation by such mechanisms will be described. A short account will be given of a recent analytical study based on a generalization of Taylor's hypothesis concerning the evolution of magnetic helicity in plasma with a large magnetic Reynolds number.

1. INEFFICIENCY OF ACOUSTIC HEATING

The fact that the solar corona is made of a very hot ($2\ 10^6$ K), tenuous plasma ($n_e = 10^{10}$ cm^{-3}) has been an intriguing fact for a number of years. The heating of this outer atmosphere of the sun, and of other stars as well, poses a theoretical question that has not yet received an answer. This heating results from the dissipation of mechanical motions, as for example MHD waves generated by the subphotospheric convection zone of the star, or from the continuous release of magnetic energy pumped into the solar corona by the stresses exerted in the photosphere and convection zone on the low lying parts of coronal magnetic field lines. Up to some five years ago, the standard belief was that chromospheric and coronal heating were the result of the dissipation of sound waves emanating from the sun's convection zone, steepening into shocks when reaching the tenuous top of the atmosphere. This classical scenario (Schatzman, 1949) has been elaborated in great detail, and is still a very viable candidate for understanding the heating of the sun's chromosphere. For a review, see Kuperus et al. (1981) and Ulmschneider (1981). However, this view has been strongly modified, concerning the corona, when the Skylab X-ray pictures have shown a solar corona heavily structured by its magnetic field in the form of loops. Also, when reasonably reliable measures of the flux of accoustic energy emanating from the dense atmosphere towards the corona were made, they revealed that this energy flux is insufficient by a factor approximately 10 to feed the energy lost by the X and UV radiating corona Athay and White, 1978 ; Mein et al. 1980 ; Schmieder and Mein, 1980). Results by Mein's and Schmieder include phase information and give a measured flux of $4 \cdot 10^3$ ergs/cm^2/s at 1500 km, to be compared to the $3 \cdot 10^5$ ergs cm^{-2} s^{-1} needed to feed the radiative losses above this level.

2. DISSIPATION OF CURRENTS REQUIRES ENHANCED DAMPING

Currents are driven in the solar corona as a consequence of the motions of the heavy atmosphere ($\beta > 1$) which move the footpoints of coronal field lines more or less at random. If the characteric time of these motions is longer than the Alfvén transit time along closed loops, the coronal magnetic configuration continuously adapts to these changes at the boundary through series of magnetohydrostatic equilibria, at least if such equilibria can be found. These configurations normally carry currents, and the permanent dissipation of these permanently re-created currents could be the cause of coronal heating. If motions of the boundary are faster than an Alfvén transit time, MHD oscillations are set up in the corona. These carry (A.C.) currents too. Of course these oscillations may propagate, be evanescent or form standing waves according to the nature of the coronal environment. The dissipation of these oscillations could also give rise to coronal heating.

However, both of these ideas share a common difficulty: dissipation is impossible by "normal" dissipative processes. This is particularly obvious for DC currents. The time scale for dissipating currents by Joule effect is not less than 10^6 years. Similarly, MHD waves in an homogeneous medium having the properties of the solar corona are damped on very long time scales (Uchida and Kaburaki, 1974). In brief, if the corona is really heated by electric currents, nature must have invented some trick to hasten their dissipation. Much of the recent and past work on solar coronal heating has been concerned with the understanding of that trick.

It is not possible to review here most of the ideas which have been considered, some of which were marginally successful, as, far example the theory based on the non-linear interaction of counterstreaming Alfvén waves, coupling to form a slow mode wave which is efficiently damped (Wentzel, 1974 ; Kabuaki and Uchida, 1971, 1974 ; Chin and Wentzel, 1972). This theory was not pursued after Wentzel (1976), recognized that it could meet the observational requirements only for loops having magnetic fields less than 10 G., the effect being extremely sensitive to the value of the magnetic field.

In a similar way Hollweg et al. (1982) have studied the possibility that Alfvén waves in flux tubes could non-linearly steepen into shock waves in the chromosphere and enter in the corona as so called switch-on shocks. The heating due to trains of such shocks give rise to the following heating rate per unit volume (Hollweg, 1982):

$$Q_{\text{switch on shocks}} = B_o^2/(32\pi\tau) \cdot (\Delta v/v_A)^4$$

where Δv is the transverse velocity jump in the shock and τ is the period of wave trains. The effect scales as B^{-2} and is then found effective only in weak field regions, like coronal holes, if τ is small (100s) and Δv large enough (200 km/s, say) (Hollweg, 1983b). The situation in this respect is similar to that reported by Wentzel for weak turbulence effect on Alfvén waves. The present trend in research is to improve on the geometrical aspects of all these phenomena, because of the strong coronal structuration due to gravitationnal and magnetic fields. WKB, as a rule, is a very poor approximation, and quasi static evolution is heavily dominated by the magnetic configuration.

3. WAVE ACCESS TO THE SOLAR CORONA: REFRACTION AND REFLECTION

Ignoring for conciseness the problem of wave generation (Unno, 1964 ; Stein, 1968 ; Ulmschneider, 1981), it is important to recognize the extreme importance of wave reflection and refraction at the chromosphere-corona interface. In a low β plasma, for example, fast modes obey the dispersion relation $\omega^2 = k^2 v_A^2$. Splitting k into an horizontal (conserved) component k_H and a vertical component k_V yields $k_V^2 = \omega^2/v_V^2 - k_H^2$. This implies wave evanescence for large enough v_A, which is actually the case for the solar corona. Fast waves are trapped in the sun's low atmosphere. Slow mode and Alfvén waves on the other hand are chanelled by the magnetic field.

However, even for field-aligned waves, non WKB effects produce strong reflection. Actually the wave length of oscillations of any given period increases enormously as it propagates into a more tenuous plasma, because c_S and v_A increase. As a rule, the wavelength becomes much larger than the scale height, and a full wave description should be considered. These effects have been considered in a series of papers by B. Leroy, S. Schwarz and N. Bel (Leroy and Bel (1979); Leroy (1980), (1981); Bel and Leroy (1981); Leroy and Schwarz (1982); Schwarz and Leroy (1982) in a model including gravity and a vertical magnetic field. These authors found that the escaping flux does not exceed some 10^6 ergs cm^{-2} s^{-1} in the form of slow modes for large vertical photospheric motions of 100m/s in a field of 3000 G, and is smaller for weaker fields. They similarly found a high reflectivity for Alfvén waves of periods in excess of 100 to 1000 seconds.

The medium has also large horizontal gradients and coronal loops have two feet. Hollweg (1983) has studied wave propagation in loop models consisting of tubes of constant Alfvén velocity (the coronal part) and two ends where the Alfvén velocity increases exponentially to infinity. He considers a transmission-reflection problem, with outgoing waves only on one side. The interesting new result is that the transmission coefficient in the region of incoming waves is very large at resonances of the coronal section, defined by $\omega_{Res} = n \pi v_{Acor}/L$. The transmission resonances have a width which reflects the rate at which energy leaks out of the Alfvén-resonant cavity. The associated quality $Q = \omega_{Res}/\Delta\omega$, is found equal to $L/4\pi h$, where h is the scale height of the atmosphere bordering the coronal section, and is typically of the order of 50. Hollweg estimates that, though the loop picks a small part of the incoming spectrum, due to the resonance, the energy flux entering it may be sufficient for feeding its radiative losses. Actually if P is the incident power, with bandwith B, T_{max} the maximum transmission coefficient, the loop receives a power :

$$F = P \, T_{max} \, \frac{\Delta\omega}{B} \, \frac{\pi}{2}$$

P may be estimated from photospheric fluctuations and flux divergence:

$$P = 2 \, (\rho_{phot} \, v_{phot}^2) \, v_{Aphot} \, \frac{B_{cor}}{B_{phot}} \simeq 1.5 \, 10^8 \, \text{ergs cm}^{-2} \, \text{s}^{-1}$$

With $B = 0.1 \, \text{s}^{-1}$, $Q = 1/50$, $F/P = 0.02$ and $2\pi/\omega_{Res} = 1000$ seconds, Hollweg finds $F \simeq 3 \, 10^6$ ergs cm^{-2} s^{-1} for the first resonance, which is large enough.

An interesting effect of refraction of fast mode waves in the corona has been discussed by Habbal, Leer and Holtzer (1979). These authors postulate the presence of coronal fast modes with very short periods, of the order of several seconds, and show that these waves tend to focus into regions of smaller v_A. So, even if the wave flux were uniform at the base of the corona, it could converge to heat selective regions. This would solve the apparent paradox that an isotropically propagating wave-mode could produce field-aligned heterogeneities. This is beautifully shown by Zweibel (1980). However, as shown before, the presence of a high flux of fast modes in the corona is questionable.

4. WAVES IN MAGNETICALLY STRUCTURED MEDIA

Much of the recent literature has been concerned with wave motions in a magnetically structured corona, with large horizontal gradients of the Alfvén velocity. The new feature, in its simplest expression, is that two media with different Alfvén velocity, in contact at a discontinuity may propagate surface waves along their interface. These modes decay exponentially on both sides of the interface, and propagate along the interface, in the limit $c_S \gg v_A$ at a velocity given by:

$$\frac{\omega^2}{k_\parallel^2} = \frac{B_1^2 + B_2^2}{\mu_0 (\rho_1 + \rho_2)}$$

See Uberoy (1972), Wentzel (1979 a-b), Ionson (1978). More elaborate structures (sheets or tubes) have been examined for wave properties (Spruit (1981), Roberts (1981); Edwin and Roberts (1982); Roberts and Webb (1979); Webb and Roberts (1980a-b)). These waves are dispersive in the presence of gravity, and obey a Klein-Gordon type of equation (Rae and Roberts (1982). They turn non-linearly into solitons (Roberts and Mangeney, 1982), which propagate at velocities in excess of $c_T = v_A c_S / (v_A^2 + c_S^2)^{1/2}$. This phenomenon may delay shock formation. But for effects considered in the next chapter, surface waves dissipate little in the body of the plasma itself (Gordon and Hollweg, 1983).

5 EFFECT OF A FINITE GRADIENT OF ALFVEN VELOCITY

New important effects arise if we consider wave propogation in a region where the Alfvén velocity is inhomogeneous. The standard case considered is an inhomogeneity in the x-direction, with straight field lines in the z-direction. If we consider approximately incompressible motions, the wave equation splits into two parts, according to the polarization of velocity fluctuation (along $\vec{e}_y$ or along $\vec{e}_x$).

(a) $\vec{v} = \vec{e}_y \, \hat{v}(x) \, \exp i \, (k_\perp y + k_\parallel z - \omega t)$

$(\omega^2 - k_\parallel^2 \, v_A^2(x)) \, \hat{v}(x) = 0$

(b) $\vec{v} = \vec{e}_x \, \hat{v}(x) \, \exp i \, (k_\perp y + k_\parallel z - \omega t)$

$\frac{d}{dx} (\omega^2 - k_\parallel^2 \, v^2(x)) \frac{d\hat{v}}{dx} = (k_\perp^2 + k_\parallel^2) \, (\omega^2 - k_\parallel^2 \, v_A^2(x)) \, \hat{v}$

We refer to (a) as shear Alfvén waves and (b) as MHD waves. The solution for case (a) is simply:

$\hat{v}(x) = v_\omega \, \delta \, (x - x_\omega)$

where x_ω is the location of the Alfvén resonance, defined by:

$\omega = k_\parallel \, v_A(x_\omega)$

This means that each magnetic surface oscillates at its own eigenfrequency, independently of the neighbouring surface. For a given $k_\parallel$, there is then a continuum of (singular) modes, with frequencies between $\omega_- = k_\parallel \, v_A(-\infty)$ and $\omega_+ = k_\parallel \, v_A(+\infty)$. The case (b) (MHD waves) is more tricky, because the wave equation is singular in the vicinity of the Alfvén resonant point. It is noteworthy that this equation also arises in the study of plasma oscillations in an inhomogeneous plasma. Simply $k_\parallel v_A(x)$ is replaced by the local plasma frequency, $\omega_p(x)$. Barston (1964) has shown that this equation admits also a continuum of modes for a given $k_\parallel$, with ω between ω_- and ω_+. These modes, however, are singular in the vicinity of their Alfvén resonant point, where they behave like

$\hat{v} = \log | x - x_\omega |$ $\quad (x > x_\omega)$

$\hat{v} = \log | x - x_\omega | + i\pi$ $\quad (x < x_\omega)$

As in other similar problems, these singular modes can be used to construct mode packets which are perfectly regular. Now, how do we recover the surface mode in the limit of an infinitely sharp interface? This point has given rise to some controversy in the literature. It all started with the work of Sedlacek, who solved the initial value problem for the equation:

$$\frac{\partial}{\partial x}\left(\frac{\partial^2}{\partial t^2} + \Omega^2(x)\right)\frac{\partial \hat{v}}{\partial x} = k^2 \left(\frac{\partial^2}{\partial t^2} + \Omega^2(x)\right)\hat{v}$$

by Laplace transform method. Disregarding mathematical details, it suffices here to say that the continuum of modes introduces cuts in the complex Laplace variable plane. On inverting the Laplace transform of the solution, the Bromwich contour is displaced from the original plane into other Riemann sheets connected to it. Sedlacek has found that one of these sheets contains a pole given approximately, for a sharp but finite transition ($k_{\|} a \ll 1$), by:

$$\omega = k_{\|} \sqrt{\frac{B_1^2 + B_2^2}{\mu_0(\rho_1 \rho_2)}} \left(1 - i\pi\left(\frac{\rho_1}{\rho_2}+\frac{\rho_2}{\rho_1}-2\right)^{-1} k_{\|} a \; \frac{|v_{A1} - v_{A2}|}{\sqrt{\frac{B_1^2 + B_2^2}{\mu_0(\rho_1+\rho_2)}}}\right)$$

In the limit $k_{\|} a \to 0$, this pole obviously represents the surface wave. So any surface wave propagating on a finite thickness discontinuity damps in time. There has been some controversy in the literature concerning the question as to whether this damping coefficient really represents "true" damping (Lee, 1980) because the equations from which we start are non dissipative. It is fair to say that this controversy is now quite completely resolved. Perhaps the best account of it is the paper by Rae and Roberts (1981) which gives a detailed description of the evolution of an initial disturbance. It is shown there that the surface wave decays because it couples to ordinary, field aligned, Alfvén modes in the inner part of the interface. There is no dissipation of the wave strictly speaking, but a conversion to the continuum of Alfvén waves. However these waves rapidly phase-mix, and eventually damp, even if the dissipative coefficients like viscosity are very small. This behaviour is well known to laboratory plasma physicists (see for example Tataronis and Grossmann, 1973). We come back to wave damping by phase mixing, for the simpler case of shear Alfvén waves, below. At this point, it is perhaps useful to trace a parallel between these phenomena and the Van Kampen treatment of plasma waves propagation in hot plasmas (see Ecker, 1972, Ch. III). In this case too, eigenmodes form a continuum for a given k. They can be used to solve initial value problems, where the Laplace method discovers a damping (Landau damping) which is not dissipation in the thermodynamical sense of the word, but results from phase mixing of Van Kampen modes. Of course, for most practical purposes, Landau damping behaves like true damping. So does the dissipationless damping found here. Barston modes play the role of Van Kampen modes. Phase mixing is due to a spread of phase velocity with x, instead of v in the Landau problem. Sedlacek's solution is akin to Landau's.

Phase mixing in inhomogeneous structures is an ubiquitous phenomenon which is in no way related exclusively to surface wave damping. The really relevant problem to treat is to find the response of an inhomogeneous structure excited at its base by random given motions. This problem is considered in the next section in more detail, for the simple case of shear Alfvén waves (Heyvaerts and Priest, 1983).

6. HEATING BY PHASE MIXING OF SHEAR ALFVEN WAVES

We consider an equilibrium which is laterally stratified, with an Alfvén velocity $v_A(x)$, and a field $B_o(x)\ \vec{e}z$. Modeling open structures we consider a boundary surface at $z = o$, where the fluid motion, in the y direction, is assumed to be prescribed, or, when we model closed loops, we add a second boundary at $z = \ell$, where the fluid motion is also prescribed. We assume the coronal fluid (region $z > o$ or $o < z < \ell$ for, resp., open and closed configurations) to be weakly dissipative. The equation of motion, for small dissipation is then found to write:

$$\frac{\partial^2 v}{\partial t^2} = v_A^2(x) \frac{\partial^2 v}{\partial z^2} + \nu \left(\frac{\partial^2}{\partial x^2} + \frac{\partial^2}{\partial z^2} \right) \frac{\partial v}{\partial t}$$

where ν is the sum of kinematic viscosity and ohmic diffusity. In the case of an open configuration, it is convenient to consider an harmonic excitation, coherent on the inhomogeneity scale, at $z = o$. Neglecting dissipation, this excitation will propagate at the velocity $v_A(x)$, and as the altitude z grows these oscillations will become more and more out of phase. In closed configurations, it is the parallel wavelength which is fixed, and stationnary oscillations excited in this resonant cavity progressively phase-mix when time elapses. These oscillators are coupled by weak friction. It is possible to approximately solve equation above. The most interesting case is perhaps that of a closed structure with imposed boundary motions. Separating the boundary motion by putting:

$$v(x,z,t) = v(x,o,t) + \frac{z}{\ell}(v(x,\ell,t) - v(x,o,t)) + w(x,z,t)$$

and analyzing w in Fourier series in z on the interval (o,ℓ) (only sine terms contribute), we get for the time dependent coefficients $b_n(x,t)$ the equation ($\Omega_n = 2n\pi v_A(x)/\ell$):

$$\frac{\partial^2 b_n}{\partial t^2} + \Omega_n^2(x) b_n - \nu \frac{\partial^2}{\partial x^2} \frac{\partial}{\partial t} b_n = \frac{(-1)^{n+1}}{n\pi} \frac{\partial^2}{\partial t^2}(v(x,o,t) - v(x,\ell,t))$$

This is the equation of a continuum of forced oscillators (indexed by x), weakly coupled by friction. An approximate solution can be found, because in the limit of small damping and large phase mixing ($\Omega_n(x)t \gg 1$), the Green function is approximately given by:

$$G(t) = \frac{\sin\Omega(x)t}{\Omega(x)t} \exp\left(-\frac{\nu}{6} t^3 \left(\frac{d\Omega}{dx}\right)^2\right)$$

Waves of frequency Ω_n damp in a time

$$\tau_{Damp} = 6^{1/3} \; (\nu(\frac{d\Omega}{dx})^2)^{-1/3}$$

Putting $d\Omega/dx = \Delta\omega/a$, where a is the thickness of the inhomogeneous region, we see that the damping time is a multiple of the phase coherence loss time, $(\Delta\omega)^{-1}$:

$$\tau_{Damp} = (\frac{6\Delta\omega a^2}{\nu})^{1/3} \; \frac{1}{\Delta\omega}$$

The dimensionless quantity in parentheses above is a Reynolds number, $R_e \simeq 6\omega a^2/\nu$, approximately equal to 10^5, typically. The damping time is then around 10 periods, and the quality factor near 10-20. However, as shown below, this laminar calculation gives an upper limit to the damping time. Having obtained the solution for b_n's, and hence for $v(x,z,t)$ in terms of its drive $v(x,0,t)$ and $v(x,\ell,t)$, it is a simple matter to calculate the average rate of energy dissipation:

$$\overline{W} = \frac{1}{2} \rho(x) \; \nu \; \int_0^\ell \frac{dz}{\ell} \; (\frac{\partial v}{\partial x})^2$$

The calculation gives the time Fourier transform of $(\partial v/\partial x)$ in terms of that of its drive in the form of a response-function:

$$(\frac{\partial v}{\partial x})_\omega = Z(x,\omega) \; (v(z=0) - v(z=\ell))_\omega$$

When $(v(z=0)-v(z=\ell))$ is a stationnary random process with a power spectrum $P_v(x,\omega)$ the time average of $\overline{W}$ can be expressed as:

$$\overline{\overline{W}} = 2 \int_0^\infty d\omega \; | \; Z(x,\omega) \; |^2 \; P_v(x,\omega)$$

After some algebra, we get the total heating rate of a slab of width L in the y direction:

$$W = \sum_{n=1}^\infty \; (\ell L \int_{-\infty}^{+\infty} dx) \; \frac{\mu}{\pi} \; \rho(x) \; \Omega_n^2(x) \; P_v(x,\Omega_n(x))$$

μ is a pure number of order unity, and P_v is the power spectrum of velocity at the border. This is an important result. It shows that in this limit of strong phase mixing and weak damping, the average rate of energy dissipation is independent of viscosity, confirming earlier results of Ionson (1982). The heating rate, in that case, depends only on the properties of the photospheric drive. Ionson's view sheds another light on this question. We review it now.

7. THE LRC CIRCUIT ANALOGY

Ionson (1982) took another, simplified, look at the same problem. His starting point is also our basic equation, with two-sided boundary conditions, except that he expressed the equation for the z component of the electric current, instead of velocity. As we do, he also separates that part which describes the corona from that part which describes the photospheric drive. For the coronal part, he obtains the equation:

$$\frac{\partial^2 j}{\partial t^2} = v_A^2(x) \frac{\partial^2 j}{\partial z^2} + \nu \frac{\partial^2}{\partial x^2} \frac{\partial j}{\partial t}$$

which is just the same as Heyvaerts and Priest's (1983). He then proceeds to average this equation over the entire loop volume. Taking into account the photospheric driving terms, he reduced his wave equation into an oscillating LRC equivalent circuit equation.

$$L \frac{d^2 I}{dt^2} + R \frac{dI}{dt} + \frac{I}{C} = \frac{dE}{dt}$$

where $L = \mu_0 \ell$, $R = R_{photosphere} + \mu_0 \ell \nu/a^2$, $C = \ell/\mu_0 \pi^2 v_A^2$, and $E = \pi a\, v_{phot} B_{phot}$. These are respectively the inductance of the loop circuit, its total resistance (photospheric plus coronal), its capacity (mainly storage of mechanical energy), and the applied equivalent driving voltage. The circuit has an eigen frequency $\omega_0 = (LC)^{-1/2}$, obviously the basic period of coronal Alfvén waves, and it has a quality factor due to dissipation:

$$Q = \frac{1}{R} \sqrt{\frac{L}{C}}$$

Note that the effect of the radial structure in the loop (different Alfvén velocity as a function of radius, effects of phase mixing) have been blurred in the averaging procedure. However the formulation has the advantage of simplicity and globality. For example the average rate of energy dissipation in the loop is given by:

$$\overline{W} = \langle R_{loop} I^2 \rangle = R_{loop}\, 2 \int_0^\infty P_I(\omega)\, d\omega$$

where $P_I(\omega)$ is the power spectrum of I, which is related to that of the drive by the resonant circuit equation:

$$P_I(\omega) = |I(\omega)|^2 = \left| \frac{E(\omega)}{1 + i Q (\frac{\omega}{\omega_0} - \frac{\omega_0}{\omega})} \right|^2$$

Putting this is the expression for $\overline{W}$, and taking into account the fact that the resonance function has bandwith which scales as R, we find again the result that underdamped loops dissipate at a rate independent of viscosity. Both these results show that, for underdamped loops, the velocity power spectrum of the photosphere maps into the loop properties. Given such a spectrum and solving the thermal equilibrium problem we obtain satisfactory results (Ionson, 1982), which obey the well known scalling laws (Rosner et al. 1978; Serio et al. 1981). Martens and Kuperus (1982) show that such a heating causes a surprising behaviour of loops thermal equilibrium, with catastrophic changes in the X-ray visibility of loops as they lengthen.

8. PHASE MIXING AND TURBULENCE

The phase-mixed flow in the Alfvén waves may well not remain laminar especially when the mixing is quite complete. This flow has been examined for stability to the Kelvin-Helmholtz and tearing perturbations. The analysis has been restricted to perturbations developing faster than the wave period, and was local. Without going into details, it suffices to say that propagating waves are stable, but that standing waves very easily suffer the K.H. instability at velocity antinodes. The order of magnitude of the growth rate is (Heyvaerts and Priest, 1983):

$$\gamma_{KH} = k_\perp v$$

where v is the velocity in the wave and $k_\perp$ the scale length in the cross-loop direction. This flow ceases then to be laminar when $k_\perp$ has increased due to phase mixing up to the point $\gamma_{KH} > \omega$. This defines naturally a Kelvin-Helmholtz stability time. After that time the flow becomes turbulent. For any reasonable set of parameters this time is less than, or equal to a couple of wave periods. The important consequence is that standing Alfvenic oscillations in inhomogeneous loops must be turbulent. This of course should reduce quite substantially the effective damping time (Heyvaerts and Priest, 1983). Hollweg (1983) has estimated the volumetric heating rate to be expected from a Kolmogoroff cascade. For an observed $<v^2>^{1/2}$ in the corona of 30 km/s, this rate is $8 \cdot 10^{-4}$ ergs cm^{-3} s^{-1}, or an equivalent flux of $8 \cdot 10^6$ ergs cm^{-2} s^{-1}, of the right order of magnitude to heat active region coronal loops.

9. HEATING BY D.C. CURRENTS: GLOBAL ENERGY BALANCE

An alternative class of theories of coronal heating is based on the idea that a part of the energy stored in the corona as a result of slow, quas static evolution be permanently released. It is easy to convince one sel that the energy flux which flows to the corona in that form (it is actually a Poynting flux) may be just of the right order of magnitude, if th corona were able to get rid of that energy into heat at the same pace. Consider for example a loop of length ℓ, radius R and longitudinal field $B_\parallel$. Assume that as a result of boundary motions this loop is twisted by an angle χ. It acquires an azimutal component of B, B_θ say, approximately equal to $\chi B_\parallel (R/\ell)$ and the stored magnetic energy augments by $\Delta W = (B_\parallel^2/8\pi)(R^2/\ell^2) \pi R^2 \ell \chi^2$. This process takes place in a time $t = R\chi/\Delta v$ where Δv is the difference in velocity between the two sides of the loop i.e. that part which induces twist. Then $t = \chi R/(R|\nabla v|) \simeq |\nabla v|^{-1}$, and the rate of energy transfer to that loop is (ϕ is the loop's flux):

$$\dot{W} = \Delta W/t = \frac{\chi}{8\pi} B_\parallel^2 \frac{\pi R^4}{\ell} |\bar{\nabla} v| = \frac{1}{t} \frac{\chi^2}{8\pi^2} \frac{\phi^2}{\ell^2}$$

which corresponds to an average flux:

$$F = \dot{W}/\pi R^2 = (\chi/8\pi) B_\parallel^2 (R/\ell) R |\nabla v|$$

For $B = 100$ G, $R = 5000$ km, $l = 50\,000$ km, $|\nabla v| = (1 \text{km s}^{-1})/1000$ km $= 10^{-3} \text{s}^{-1}$, $\chi = 2\pi$, we get $F = 10^7$ ergs cm^{-3} s^{-1}. The energy is sufficient. The problem is how to dissipate it quickly enough. Some authors simply assume the possibility to exist. Sturrock and Uchida (1981) elaborate on the preceding derivation by recognizing that boundary motions are not persistent but stochastic, so that loops are successively twisted and untwisted. If the correlation time tc of the boundary motion is short as compared to the dissipation time tD, the state of twist of the loop random walks instead of increasing continunously. The stored magnetic energy of a loop twisted by an angle χ, is:

$$W(\chi) = \frac{1}{8\pi^2} \frac{\phi^2}{l} \chi^2$$

as this quantity is quadratic in χ, it increases in time when χ random walks at the rate:

$$\frac{d\langle w \rangle}{dt} = \frac{1}{8\pi^2} \frac{\phi^2}{l} \left\langle \frac{\Delta \chi^2}{\Delta t} \right\rangle$$

where $\langle \Delta \chi^2 / \Delta t \rangle$ is given in terms of the correlation function of the local angular velocity of the fluid at the base of the loop. If the coherence time of these motions is t_c:

$$\left\langle \frac{\Delta \chi^2}{\Delta t} \right\rangle = 2 \langle \omega^2 \rangle t_c$$

Hence the heating rate is:

$$\frac{d\langle w \rangle}{dt} = \frac{\phi^2}{\pi \mu_0 l} \langle \omega^2 \rangle t_c$$

When entered into a thermal balance equation, scaling laws in agreement with data are also obtained.

Parker (1981a-b,1983a-b) makes it convincing that complex boundary motions must tangle the coronal field in such a way that a great many current sheets spontaneously form in the overlying corona. These sheets must suffer rapid reconnection, so relaxing continuously the amount of energy brought to the corona as a result of work done on the foot points of magnetic field lines. The estimation of this work is straightforward. If B is the vertical component of the field and $B_\perp$ its horizontal component which, after stresses have accumulated for a time t, is of order $B_\perp = Bxt/l$, where w is a typical velocity in the photosphere, then the work done per unit surface against the stress $B_\perp/\mu_0$ is:

$$F = w \frac{BB_\perp}{\mu_0} = \frac{B^2}{\mu_0 l} w^2 t$$

This is the vertical component of the Poynting flux as well. Parker estimates, from observation w = 0.4 km/s, using Smithon studies (1973) of the motion of field knots in the photosphere, and he estimates the time by the condition that reconnection has time enough to relax the stress due to field line tangling. He finds, putting $t = h/v_{Rec}$, where h is the tube thickness and v_{Rec} some reconnection velocity, an equivalent heating flux (Parker 1983b):

$$F = w \frac{B^2}{\mu_o} \left(\frac{w}{v_{Rec}}\right) \left(\frac{h}{\ell}\right)$$

Perhaps the most interesting contribution of this series of papers is the detailed demonstration that complex boundary motions must create current singularities and lead to fast dissipation. These results should be put in parallel to those of Syrovatskii (1978) and Bobrova and Syrovatskii (1979) who have obtained similar conclusion for force free fields. In interchange (Uchida and Sakurai, 1977) or the tearing in closed magnetic structures (Galeev et al. 1981) could concur to create fine scale structures and provoque fast magnetic energy release by reconnection.

10. HEATING BY COMPLEX RECONNECTION: THE ROLE OF GLOBAL INVARIANTS

Very turbulent fusion machines were built in the past. May be one of the most intriguing are so called reversed z-pinches, which after a turbulent phase, show a reversal of the sense of the axial magnetic field as compared to the initial situation. Surprisingly enough, the field profile, in the final state is just a constant-α force-free field. Obviously reconnection operates in such discharges during some first phase efficiently enough to change totally the topology of magnetic configuration, but still the field does not relax to a potential field. Only a certain fraction of the magnetic energy can be released quickly to accomodate the topological restructuration of the magnetic field, but not all of it. How much ? Taylor (1974) has suggested that the field will rearrange internally, in such a way as to find itself in the minimum energy state compatible with the conservation of some global invariants, or, more precisely, quasi-invariants, because, as we shall see shortly the quantity which Taylor advocated should be conserved, is in fact not a strict invariant. That quantity is the so called total magnetic helicity:

$$K = \int_{vol} (\vec{A}.\vec{B}) \, d^3r$$

where $\vec{A}$ is a vector potential for $\vec{B}$. In perfect MHD, the magnetic helicity is an invariant for each closed flux tube (Woltjer, 1958), and it has the property that the state of minimum magnetic energy subject to a prescribed distribution of magnetic helicity is some general force-free field. It has not been mathematically proved that the helicity should be more conserved in a dissipative system than other quantities conserved in perfect MHD, though some arguments can be given in favour of this idea (see Heyvaerts and Priest, 1983b, and Norman and Heyvaerts (1983)). Taylor's conjecture has led to a success in explaining the behaviour of the Zeta machine, because the state of lowest magnetic energy having a given total helicity is just a constant-α (linear) force-free field, as observed. Heyvaerts and Priest (1983b) have adapted the same idea to the calculation of the heating of the corona by DC currents. The simplest way is to first assume, that, at each moment of its evolution, complex reconnection phenomena occur very rapidly in the magnetic configuration, so that it is also always observed in the lowest energy state compatible with helicity. The latter evolves according to the equation:

$$\frac{dK}{dt} = \int_{Boundary} (\vec{A}.\vec{v}) (\vec{B}.\vec{dS})$$

Care must be taken of the fact that for an open-ended flux tube K is not a gauge invariant quantity, so that some precise gauge condition has to be imposed. We have treated in detail the specific example of the evolution through the series of force-free configurations given by:

$$B_x = - B_o (\ell/k) \cos kx \exp - \ell z$$

$$B_y = - B_o (1-\ell^2/k^2)^{1/2} \cos kx \exp - \ell z$$

$$B_z = + B_o \sin kx \exp - \ell z$$

This represents a series of linear force-free fields endowed with translational symetry in the y-direction. Some shear motion along y is prescribed at the boundary. So, the flux distribution keeps the same at the base, as it is (for fixed k and different ℓ) for the series of equilibria described above. The evolution of helicity is then calculated and used to find the parameter ℓ as a function of time. Then, comparing the rate of increase of the stored magnetic energy with the integral of the Poynting flux through the boundary gives the heating rate, which is just the difference between the two. It has been found that this difference is systematically zero! This is a consequence of the time it takes for stresses to relax by dissipation being implicitely taken to be zero in this formulation. This result can be understood: Assume that the system performs an ideal MHD displacement during a very small time tD, and then relaxes by reconnection. During the perfect MHD step, the magnetic energy increases due to the work performed by boundary motions by an amount which is $O(\xi)$ where ξ is the fluid displacement ($\xi \simeq v.tD$). In the ensuing relaxation, the displacement will also be $O(\xi^2)$ but, as the final state is an equilibrium, the change in potential energy is only $O(\xi^2)$. In the limit $\xi \to o$, (tD$\to$o), a negligible part of the energy increase goes into heat.

It was then felt necessary to develop a second order calculation based on the scheme sketched above: small perfect MHD displacements, lasting a phenomenological "stress relaxation time" tD are followed by episodes of relaxation to the minimum energy state compatible with the new value of the magnetic helicity. The change in magnetic energy of the configuration in the last step be then calculated. It represents the heating during the time tD. This procedure (see details in the paper) yields an explicit analytical expression for the rate of heating for the particular displacement field considered. This expression is not useful in itself, but in its structure. It shows the following features. a - If the boundary motion is such as to keep the field force-free with constant-α (this may happen if the scale of the magnetic field ℓ_B is much less than that of the velocity, ℓ_v, no heating ensues. b - The rate of heating is proportional to the small parameters: $(tD.v/\ell_B)$, where tD is the phenomenological relaxation time, v the velocity of boundary motions, and ℓ_B the scale of magnetic structures on the boundary. As we noted, heating vanishes when this parameter vanishes. On the other hand boundary motions such that $v \gg \ell_B/tD$ just do not relax, and evolve according to perfect MHD, also with no heating. c - The equivalent heating flux may be written in order of magnitude:

$$F = \frac{B^2}{\mu_0} v \left(\frac{\ell_B}{\ell_B+\ell_v}\right)^2 \left(\frac{t_D.v}{\ell_B}\right)$$

This analysis gives again the order of magnitude estimates of other theories namely $B^2/\mu_0.v$, but this maximum is limited by the rapidity of stress-relaxation (factor $tD.v/\ell_B$), and by the geometrical factor, which measures the importance of relaxable stresses accumulated in each step. The same idea of relaxation with conservation of global invariants can also be used to predict the end result of catastrophic phenomena, like flares (Norman and Heyvaerts, 1983): after a flare, the coronal magnetic configuration should be a constant-α force-free field with the helicity of the configuration prior to flare. Also this analysis traces an interesting border between flares and coronal heating. Both phenomena would result from magnetic stress relaxation by reconnection, but heating achieves this result continuously, because boundary motions are slow and stresses never accumulate very much, while flares occur in configurations which have evolved so quickly that stresses could not relax at the same pace as they were built, and ultimately relax very violently.

11. UNIFICATION: THE LRC CIRCUIT ANALOGUE AGAIN

There is something common to the various processes considered in this review. They all deal with the dissipative response of coronal electric circuits driven at various low or high frequencies by the photospheric driver. So it should be possible to formulate a unique simplified theory incorparating them all. Ionson (1984), also in these proceedings, proposes such an unifying scheme: the LRC circuit analogue. See his communication.

12. REFERENCES

Athay, G., and White, S,: 1979, Astrophys. J. 226, 1135.
Barston, G.: 1964, Ann. Phys. (NY) 29, 282.
Bel, N., and Leroy, B.: 1981, Astron. Astrophys. 104, 203.
Bobrova, N.A., and Syrovatsky, S.I.: 1979, Solar Phys. 61, 379.
Chin, Y., and Wentzel, D.: Astrophys. and Sp. Sc. 16, 465.
Ecker, G.: 1976, Theory of fully ionized plasma, Academic Press.
Edwin, P., and Roberts, B.: 1982, Solar Phys. 76, 239.
Galeev, A.A., Rosner, R., Serio, S., and Vaiana, G.: 1981, Astrophys.J.243,301

Gordon, B.E., and Hollweg, J.V.: 1983, Astronphys. J. 266, 373.
Habbal, S.R., Leer, E., Holtzer, T.: 1979, Solar Phys. 64, 287.
Heyvaerts, J., and Priest, E.R.: 1983, Astron. Astrophys. 117, 220.
Heyvaerts, J., and Priest, E.R.: 1983, submitted to Astron. Astrophys.
Hollweg, J.V.: 1982, Astrophys. J. 254, 806.
Hollweg, J.V.: 1983a, Preprint Univ. of New Hampshire.
Hollweg, J.V.: 1983b, in "Solar Wind V" ed. by M. Neugebauer.
Hollweg, J.V., Jackson, S., and Galloway, D.: 1982, Solar Phys. 75, 35.
Ionson, J.A.: 1978, Astrophys. J. 226, 650.
Ionson, J.A.: 1982, Astrophys. J. 254, 318.
Ionson, J.A.: 1984, to appear in Astrophys. J. (January).
Kaburaki, O., and Uchida, Y.: 1971, Pub. Astr. Soc. Jap. 23, 405.
Kuperus, M., Ionson, J.A., Spicer, D.S.: 1981, Ann. Rev. Astron. Astrophy. 19, 7.
Lee, M.: 1980, Astrophys. J. 240, 693.
Leroy, B.: 1980, Astron. Astrophys. 91, 136.
Leroy, B., and Bel, N.: 1979, Astron. Astrophys. 78, 129.
Leroy, B., and Schwartz, S.: 1982, Astron. Astrophys. 112, 84.
Martens, P., and Kuperus, M.: 1982, Astron. Astrophys. 113, 324.
Mein, N., and Schmieder, B.: 1981, Astron. Astrophys. 97, 310
Mein, P., Mein, N., and Schmieder, B.: 1980, in Proceedings of Japan-France Seminar on Solar Physics, ed. by Moriyama and Hénoux, p.70.
Norman, C., and Heyvaerts, J.: 1983, Astron. Astrophys. Letters to appear.
Parker, E.N.: 1977, Ann. Rev. Astron. Astrophys. 15, 45.
Parker, E.N.: 1981a, Astrophys. J. 244, 631.
Parker, E.N.: 1981b, Astrophys. J. 244, 644.
Parker, E.N.: 1983a, Astrophys. J. 264, 635.
Parker, E.N.: 1983b, Astrophys. J. 264, 642.
Rae, I., and Roberts, B.: 1981, Geophys. and Astrophys. Fluid Dynamics 18, 197.
Rae, I., and Roberts, B.: 1982, Astrophys. J. 256, 761.
Roberts, B.: 1981, Solar Phys. 69, 39.
Roberts, B., and Webb, A.: 1979, Solar Phys. 64, 77.
Rosner, R., Golub, L., Coppi, B., and Vaiana, G.: 1978, Astrophys. J. 222, 317.
Rosner, R., Tucker, W., and Vaiana, G.: 1978, Astrophys. J. 220, 643.
Schatzman, E.: 1949, Ann Astrophys. 12, 203.
Schwarz, S., and Leroy, B.: 1982, Astron. Astrophys. 112, 93.
Sedlacek, S.: 1971, Journal Plasma Phys. 5, 239.
Serio, S., Peres, G., Vaiana, G.S., Golub, L., and Rosner, R.: 1981, Astrophys. J. 243, 288.
Smithon, R.C.: 1973, Solar Phys. 29, 365.
Spruit, H.C.: 1981, Astron. Astrophys. 98, 155.
Stein, R.F.: 1968, Astrophys. J. 154, 297.
Sturrock, P.A., and Uchida, Y.: 1981, Astrophys. J. 246, 331.
Syrovatskii, S.I.: 1978, Solar Phys. 58, 89.
Tataronis, J., and Grossmann, W.: 1973, Zeitschrift für Physik 261, 203.
Taylor, J.B.: 1974, Phys. Rev. Letters 33, 1139.
Uberoi, M.: 1972, Phys. Fluids 15, 1673.
Uchida, Y., and Kaburaki, O.: 1974, Solar Phys. 35, 451.
Uchida, Y., and Sakurai, T.: 1977, Solar Phys. 51, 413.

Ulmschneider, P.: 1981, in "Solar phenomena in stars and stellar systems" R.M. Bonnet and A.K. Dupree , ed., NATO adv. Study Institutes series, (Reidel), pp 239-263.
Unno, W.: 1964, Proc. IAU General Assembly (Hamburg), p. 555.
Wentzel, D.: 1974, Solar Phys. 39, 129.
Wentzel, D.: 1976, Solar Phys. 50, 343.
Wentzel, D.: 1979a, Astrophys. J. 227, 319.
Wentzel, D.: 1979b, Astron. Astrophys. 76, 30.
Woltjer, L.: 1958, Bull. Astron. Netherlands 14, 39.
Zweibel, E.: 1980, Solar Phys. 66, 305.

DISCUSSION

D. Smith: What do you consider the best candidate for coronal heating on the basis of your calculations?

Heyvaerts: Our calculations of DC current heating contain a phenomenological parameter which is unfortunately very ill defined. The maximum conceivable equivalent flux from such mechanisms ($B^2 v/8\pi$) is of a confortable order of magnitude, but the factors which reduce this efficiency, and which appear in all theories developed so far (geometical factors ℓ_B/ℓ_v, loop aspect ratio, R/ℓ, dissipation time scale ($\tau_D v/\ell_B$)) are all not known with enough precision. If we include the most trivial geometrical factors, the predicted flux still keeps larger than the necessary flux by a factor between 5-10 (Heyvaerts and Priest 1983b). So τ_D should really be just a bit shorter than (ℓ_B/v) to meet the requirement, which is not impossible if we judge from linear tearing times. To reach a firmer conclusion we need simultaneous observations of all the physical parameters of some coronal loop, a more detailed understanding of the dissipation time τ_D, and also a more detailed knowledge of the spectrum of the horizontal velocity which enters in the resonant heating theory. Probably both type of heating are active simultaneously (see Jim Ionson's communication, but in different structures).

Sturrock: Most of your talk has been concerned with loop structures. How are we to understand coronal heating in open-field structures such as coronal holes?

Heyvaerts: Any variation imposed at the foot point of an open field line must result in propagating or evanescent oscillations in the overlying corona. Hence wave mechanisms should be responsible for coronal hole heating. It is important to stress that due to solar wind losses, the energy needs of coronal holes are by no means smaller than those of closed field line regions. The subject of coronal hole heating is less developed than that of loop heating. Nevertheless some specific suggestions have been made (Hollweg 1983), namely the conversion of twist Alfvén wave trains in flux tubes into some type of perturbations which dissipate. In a more speculative way, we suggested (Heyvaerts and Priest 1983) that phase mixing of evanescent MHD waves in the corona could drive the Kelvin-Helmholtz instability, and trigger a turbulent cascade. If that idea is correct, holes and loops could be heated by similar mechanism.

Drake: Most of the energy absorbed by Alfvén waves in closed loops will flow down into the photosphere. How is the energy transferred to the open field regions?

Heyvaerts: The open field regions must be subject to special heating mechanisms, evocated in my answer to Dr. Sturrock's question. Loop heating can definitely not be transferred to open regions, at least not deep into them.

Benford: How can observations best distinguish between direct current heating and resonant, or AC heating?

Heyvaerts: The scale of structures involved in subtelescopic; resonant heating by waves should give rise to motions detectable as "microturbulent" line broadening, i.e. to broadening in excess of Doppler. Unfortunately, it is quite difficult to measure lines high enough in the corona, and, on the other hand, the reconnections associated with DC current dissipation also show up as motions. These, however, are expected to be more bursty, and perhaps it would be possible to detect the effect of some modest particle acceleration likely to be associated with this reconnection process (permanent weak X-ray noise, small, but bursty, component on top of the thermal radio emission).

ON THE ROLE OF DOUBLE LAYERS IN ASTROPHYSICAL PLASMAS

Robert A. Smith
Plasma Physics Division
Science Applications, Inc.
1710 Goodridge Drive
McLean, Virginia 22102

    Limitations of current knowledge of plasma double layers create difficulties in extrapolating double-layer concepts for application to astrophysical models. Some problems of this sort are described, and some central issues in structure and dynamics of double layers are identified, which must be addressed in astrophysical contexts. These include the determination of kinetic boundary conditions, and the relations of time and length scales of local dynamics and structure to those of the global circuit in which the double layer is contained.

    There is widespread interest in double layers (DL) as a possible acceleration mechanism in various energetic phenomena in space and astrophysical plasmas. They have been invoked in such diverse contexts as terrestrial auroral discharges, magnetospheric substorms, solar flares, Jovian radio emission, and extragalactic radio sources. In a thought-provoking series of discussions, Alfvén (1979, 1981, 1982) has considered DL to be a central paradigm in plasma astrophysics.

    Our current knowledge of DL physics, however, is insufficient for us to judge with much confidence what roles DL may play in astrophysics. This knowledge is derived from a growing but still limited number (and perhaps more important, a limited class) of experimental, theoretical, and numerical investigations. The application, however tentative, of this knowledge to the scales and conditions of astrophysical phenomena requires conceptual extrapolations which must be quite judicious and which are, in our present state of ignorance, most probably unwarranted. Many basic questions must be addressed before DL can become the fruitful astrophysical paradigm envisioned by Alfvén.

    A first indication of the nature of these questions can be inferred from the current literature. In this paper I shall discuss a few central points concerning both the structure and dynamics of double layers. Consideration of such questions may be essential in assessing the applicability of models invoking DL to various astrophysical phenomena, from the standpoints of both basic physics and applications.

Before the work of Sato and Okuda (1980), the common concept of the DL was that of a "strong", laminar, BGK-type potential structure (Figure 1). Steady-state fluid analyses yield well-known necessary boundary conditions for the existence of such structures. In the asymptotic forms usually cited (Block, 1978), these are the "Bohm criteria"

$$U_{e1}^2 > (\gamma_e T_{e1} + T_{i1})/m_e \quad ; \quad U_{i2}^2 > (\gamma_i T_{i2} + T_{e2})/m_i \qquad (1)$$

and the Langmuir condition

$$F_{e1}/|F_{i2}| = (m_i/m_e)^{1/2} , \qquad (2)$$

where $\gamma_{e,i}$ are the electron and ion adiabatic indices (i.e., $p_{e,i} \sim n_{e,i}^{\gamma_{e,i}}$) and subscripts 1,2 refer to the cathode ($\phi=0$) and anode ($\phi=\phi_{DL}$) sides of Figure 1; U_j and F_j are, respectively, the velocity moment and flux of free particles of species j entering the DL. The Bohm criteria relate to monotonicity of the potential $\phi(x)$; the Langmuir condition relates to overall charge neutrality integrated across the DL.

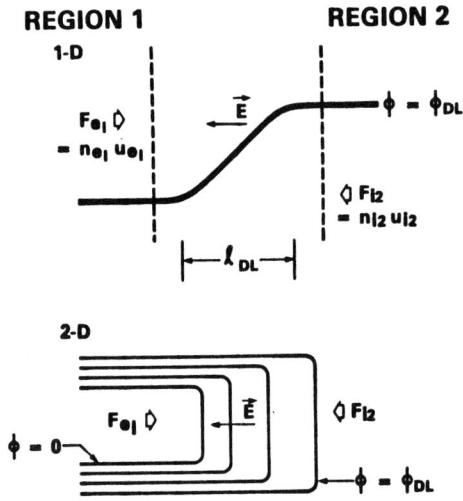

Fig. 1. Schematic of the classical BGK double layer concept, in 1-D (potential profile) and 2-D (potential contours). Plasma in regions 1 and 2 is uniform with respect to the direction of current flow.

In fluid analyses, U_{e1} and U_{i2} are the bulk velocities of electron and ion fluids with temperatures T_{e1} and T_{i2}. As was shown by Levine and Crawford (1980), however, application of adiabatic fluid theory is self-consistent only when the Bohm criteria are in fact satisfied. Moreover, because the DL structure is bounded, its general description requires a kinetic analysis in which the boundary conditions are applied to the free particles on the appropriate half-spaces in velocity; in the geometry of Figure 1, these are $v>0$ at $\phi=0$ and $v<0$ at $\phi = \phi_{DL}$. Thus, for example, in judging whether the boundary conditions for the existence of a BGK DL are met in some situation, it would be erroneous to argue that a relative bulk drift of velocity $(U_{e1} - U_{i2})$ must exist in the plasma. For example, Hubbard and Joyce (1979) found quasi-steady DL in numerical simulations in which the free particles were injected from the half-spaces of nondrifting Maxwellians.

Therefore, although the existence of the BGK double layer must depend on some conditions analogous to (1) and (2), these conditions must be formulated kinetically. Knorr and Goertz (1974) constructed a steady-state waterbag model of DL; to my knowledge, a more general kinetic formulation has not been done. Nor can the Bohm criteria simply be re-interpreted to apply to kinetic moments on a half-space in velocity. Besides the limitations pointed out by Levine and Crawford on the validity of fluid analyses of DL, the indices $\gamma_{e,i}$ in the fluid equation of state really have no analogs in the kinetic formulation: simulations (Smith 1983) reveal that both electrons and ions transport substantial heat flux through the double layer.

In addition to expressing boundary conditions for the existence of BGK DL, the as-yet unknown kinetic Bohm/Langmuir conditions must be either formulated a priori with reference to plasma conditions external to the DL boundaries, or considered a posteriori in this context. The consideration here is the stability of the DL. For example, Hubbard and Joyce (1979) observed disruption of the DL due to trapping of the electron influx in low-frequency waves produced by the accelerated ions. The extent of such trapping will depend on the distributions of the inflowing particles.

Considerations of the Bohm criteria (1) have motivated some useful work in which the question of accessibility of the DL state was addressed, with the Bohm criteria viewed as an initial condition in a current-carrying plasma with no initial dc electric field. (As we shall see below, however, in the evolutionary problem the final boundary conditions are not identical to the initial conditions!) Smith and Goertz (1978) noted that the Bohm criteria (1) were compatible with the threshold condition

$$|U_e - U_i| \gtrsim 1.3 \, V_e \qquad (3)$$

for the Buneman instability, where V_e is the electron thermal velocity. They suggested that DL evolve nonlinearly in an inhomogenous,

Buneman-unstable plasma. When (3) is initially met, this turns out to be the case (Smith 1982a,b; 1983); I shall describe the evolutionary dynamics below. Sato and Okuda (1980) suggested that for $|U_i - U_e| < V_e$, DL-like structures may be driven by particles accelerated in DC electric fields supported by anomalous resistivity provided that the system be "sufficiently long".

In simulations using short systems, Sato and Okuda observed only anomalous resistivity supported by ion-acoustic turbulence. Upon lengthening the system, they found sharp potential spikes embedded in the turbulence (Figure 2); they named these spikes "ion-acoustic double layers" (IA DL). In contrast to the strong, laminar, BGK DL, the IA DL are weak ($e\phi_{DL} \sim T_e$), turbulent, and unstable to emission of ion sound. In very long systems, (Sato and Okuda 1981), they recur on some characteristic length scale which I shall call ℓ_{IA} (in the simulations ℓ_{IA} is of order 1000 λ_e, but this value may be an artifact of the numerical parameters, such as m_i/m_e). On this characteristic scale, IA DL seem to be statistically stationary in the sense that as individual structures decay, others spontaneously appear.

The existence of the scale length ℓ_{IA} is crucial and must be linked to the ion dynamics. Unlike the essentially monotonic BGK DL, the structure of the IA DL includes a sharp negative spike at the leading edge. This negative spike, indicating a local "ion hole" in the ion phase space, dynamically leads to the subsequent potential rise by reflecting current-carrying electrons (Hasegawa and Sato 1982; Schamel 1982; Chanteur et al. 1983). The scale length ℓ_{IA} may be simply a correlation length for the spontaneous formation of an ion hole by constructive interference of random-phased ion acoustic waves (W. Lotko, private communication).

Our current knowledge of DL, then, concerns two strikingly different limiting cases: the classical BGK paradigm and the more

Fig. 2. Recurrent weak double layers embedded in ion-acoustic turbulence. After Sato and Okuda (1981).

recently discovered ion acoustic double layers. Recent spacecraft observations of electric fields in the broad inverted-V auroral region have been interpreted as IA DL (Temerin et al. 1982); BGK DL may exist in the narrower discrete auroral arcs, but observations on this question are not definitive. Nothing that we currently know rules out the possibility that both types of DL exist in space and astrophysical plasmas. There is much, however, that we do not yet know concerning DL structure under different conditions and the dynamic accessibility of the final DL state. Related to these questions is the possibility that the BGK and IA DL are limiting cases of a sequence of intermediate states linked by transitions in structure and underlying dynamics.

Such questions can only be completely addressed by studying the DL as part of a complete electric circuit, a point that has been stressed by Alfvén and others and on which, I believe, there is growing concurrence. The reasons are several; I shall not attempt to compose a formal list, but shall give some examples relating to time and length scales and to the intrinsic nature of DL evolution in a circuit.

Let us first consider some dynamical issues that arise when a DL evolves in a circuit. For visualization, consider the simple model circuit of Figure 3a, which may (or may not!) be of interest for the

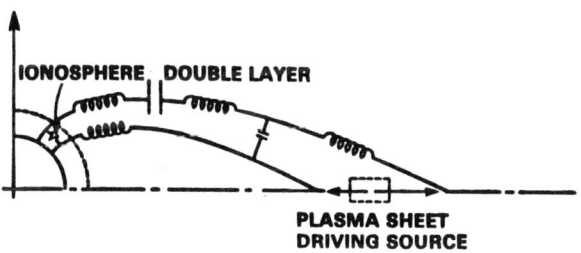

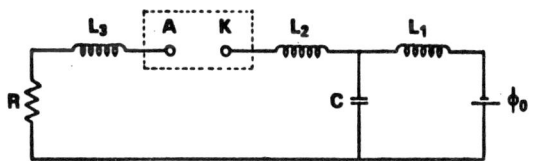

Fig. 3. (a) Schematic of a simplified model circuit for the terrestrial aurora. The circuit parameters are distributed. (b) Lumped network model of the circuit in (a).

terrestrial aurora and, perhaps, topologically similar astrophysical situations. For simplicity, I shall assume here that this circuit can be represented by the lumped-element network of Figure 3b. In Figure 3a the DL is an active nonlinear element, which in Figure 3b is represented by the anode-cathode pair A-K with variable voltage $\phi_{DL}(t)$. Analyzing this circuit, we find three time constants $\tau_m(R, C, L_1, L_2, L_3)$, where m = 1, 2, 3. The current density $J_0(t)$ through the DL is given by

$$J_0(t) = \sum_{m=1}^{3} J_m e^{t/\tau_m} + \int^t dt' \; G \; \lceil \phi_{DL}(t'), \frac{t'}{\tau_1}, \frac{t'}{\tau_2}, \frac{t'}{\tau_3} \rceil, \quad (4)$$

where the J_m are constants and G is a functional which, were it exhibited, would look more complicated than it really is.

We see from (4) that the global currents in the circuit depend on the time history of the DL itself. But $J_0(t)$ provides the boundary condition on the DL evolution, and so there is a mutual feedback effect between the DL and the circuit. Moreover, this feedback involves a nonlocal influence of the DL on the kinetic distributions of the plasma that enters the DL, in order that they provide contributions to $J_0(t)$ that are self-consistent with the contributions of the particles that have traversed the DL. Thus, over the course of the DL evolution the distributions of the incoming particles, which correspond to the boundary conditions in a stationary BGK description, can change considerably from their initial values. To simulate DL evolution in space and astrophysical circuits, boundary conditions which model this nonlocal feedback must be employed. This has not yet been done.

As a second example, the ion transit time

$$\tau_{tr} \simeq \ell_{DL}/U_i \quad (5)$$

is an important time scale for the DL evolution. In (5), ℓ_{DL} is the length of the DL region and U_i the characteristic speed of a free ion entering the DL (ℓ_{DL} and U_i may be time-dependent, in general). The dynamics of the evolution may depend heavily on the value of τ_{tr} relative to the circuit time constants τ_m.

Smith (1982a) simulated DL evolution in a simple LR circuit with a constant voltage source; this circuit has one time constant, τ_{ind} = L/R. In runs with $\tau_{tr} \simeq \tau_{ind}$ and with the initial and injected distributions chosen to satisfy to Bohm criteria (1), Smith observed the following dynamics: (i) In the initial phase, a linear Buneman-like instability develops, with the unstable wave envelope growing spatially in the direction of electron drift; (ii) As electrons become trapped, the potential profile rectifies and its dominant length scale changes from the short wavelength of the unstable waves to a longer scale; (iii) Next, ions become trapped, developing holes in the ion phase space. In this stage, the DL potential is already "strong"; i.e., $e\phi_{DL} \gg T_e$. There are rapid overshoots and undershoots of the potential $\phi_{DL}(t)$, involving complex phase - space dynamics; (iv) The ions became detrapped, damping the potential oscillations and leading to the transition to a strong, laminar BGK state. In the steady state, the current density has diminished by a factor of 15 from its initial value, and ϕ_{DL} is slightly less than the driving potential ϕ_0. The time history of $\phi_{DL}(t)$ in one such case is shown in Figure 4a.

In contrast to Smith, Belova et al. (1980) did not employ a circuit model, but attempted to simulate a constant current source by injecting constant electron and ion distributions, which were also Buneman-unstable. A constant current source corresponds to an LR circuit with $L \to \infty$, and thus $\tau_{ind} \to \infty$. Belova et al. observed recurrent explosive development of the potential, the explosions occuring within a few ion plasma periods and recurring on approximately the τ_{tr} time scale (Figure 4b). The reason for these contrasting results is not yet fully understood, but seems to be related to the differing values of τ_{tr}/τ_{ind}. In the case $\tau_{ind} \simeq \tau_{tr}$ the DL-circuit feedback stabilizes the dynamics, while in the case $\tau_{ind} \to \infty$ the DL continues to grow until the influxes can no longer sustain it (Yu. Sigov, private communication).

Finally, let us consider the influence of length scales on DL structure. A well-known scaling result for strong DL, first found by Goertz and Joyce (1975), is a relation between the potential ϕ_{DL}, scale length ℓ_{DL}, and upstream density n_{e1} (c.f. Figure 1):

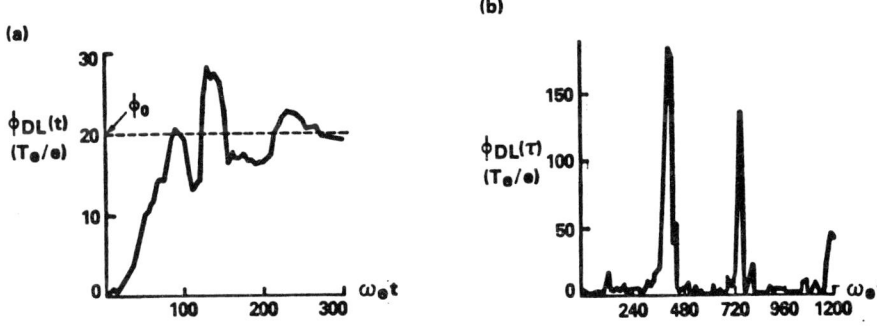

Fig. 4. Time histories of the double layer potential in two numerical simulations. (a) LR circuit with $\tau_{ind} \simeq \tau_{tr}$. After Smith (1982a). (b) Constant injected current, $\tau_{ind} \to \infty$. After Belova et al. (1980).

$$\phi_{DL} = \alpha\, n_{e1}\, \ell_{DL}^2, \tag{6}$$

where α is an empirical constant. Recently, however, Chan and Hershkowitz (1982) showed that there is some length ℓ_* such that, for a given ϕ_{DL} applied across a system of length ℓ_{sys}, a BGK structure of the type described by Eq. (6) develops when $\ell_* > \ell_{sys} > \ell_{DL}(\phi_{DL})$. When the ordering is changed to $\ell_{sys} > \ell_* > \ell_{DL}(\phi_{DL})$, however, the laminar structure splits into a "multiple-DL" structure (Figure 5) in which both electron plasma and ion-acoustic oscillations develop. This phenomenon is not well understood. In particular, we don't know what determines ℓ_*.

We note, through, that the same basic physical phenomena, involving current interruption by particle trapping, seem to be important in both the strong Buneman DL and the weak ion-acoustic DL. The very different phenomenology between the two types indicates important differences in the underlying dynamics, however. The principal reason seems to be that in the Buneman cases of Smith (1982 a, b; 1983), the inertia of the drifting ions causes them to be trapped later than the electrons; the nonlinear phase of the evolution is controlled first by the electrons, while the intermediate and final stages are dominated by the ions. In the ion-acoustic regime, the ion dynamics dominate throughout. On physical grounds, therefore, one expects that in any circuit, the laminar length scale $\ell_{DL}(\phi_{DL})$ is determined by kinematics, ℓ_* is determined by a transition from kinematics to dominance by the ion dynamics, and ℓ_{IA} is determined solely by ion dynamics. The available system length ℓ_{sys}, together with the time constants τ_m, are governed by the circuit topology.

I have chosen to discuss these issues of DL structure and dynamics not only because I believe that they are fundamental ones, but because they may relate to observable consequences for astrophysical models of double layers. For example, the time-dependence of radiation signatures may be related to the dynamical considerations involving the time scales for local and global evolution of circuits containing DL. The

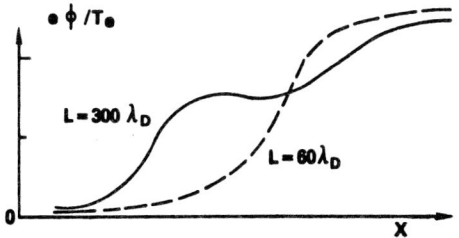

Fig. 5. Transition from single to "multiple" double layers as system length increases (Courtesy of C. Chan and N. Hershkowitz).

spectral characteristics of radiation may be related to the distributions of accelerated particles, which will be nearly monoenergetic for some distance after having transversed a strong BGK DL, but much more thermalized if they have transversed a turbulent region containing ion-acoustic DL.

I shall close this discussion on a frankly speculative note. The transition from BGK to "multiple" DL, with the attendant appearance of IA oscillations, leads one to wonder whether a gross extrapolation from laboratory scales to space or astrophysical scales might lead to the aforementioned sequence of structures, in which the BGK and IA DL are limiting cases. Such a sequence is indicated schematically in Figure 6. This would be appealing, but some caveats should be borne in mind. First, the experimental results on "multiple-DL" are few in number and are limited to showing two such structures, so we don't know whether a "many-DL" structure such as shown in Figure 6c could exist. Second, although it seems plausible to hypothesize that successive fractalizations of the sequence in Figure 6 would involve increasing levels of LF turbulence and might lead to a turbulent region of multiple IA DL as in Figure 6d, it is not yet established that recurring IA DL give a cumulative potential drop.

Such caveats, in fact, are illustrative of the introductory theme; at present, discussion of double layers in space and astrophysical contexts must rely on uncomfortably large extrapolations of our current lore of theory, experiment, and simulation. Such extrapolations require judicious appraisal of the effects of experimental factors (grids, walls, volume ionization, etc.) and boundary conditions on the interpretation of these different types of results. A deeper

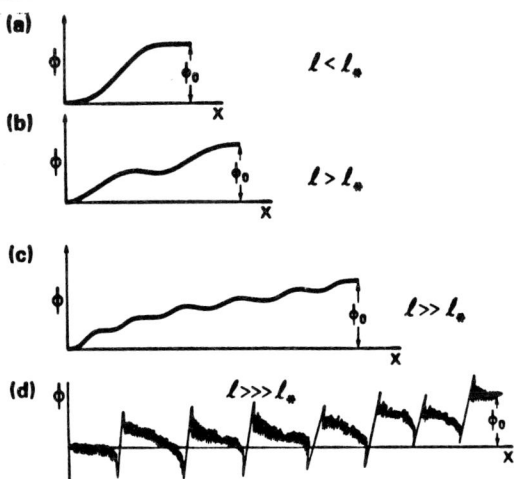

Fig. 6. Hypothetical sequence of double layer structures ordered by system length, with BGK and ion-acoustic double layers as limiting members.

realization of the difficulties in translating our current knowledge of DL into astrophysical contexts, however, can help us direct future work toward considerations of importance for space and astrophysical applications. The experiments of Stenzel, Gekelman, and Wild (1983) already point in this direction. Numerical simulations can put more emphasis on elucidating structure and dynamics under boundary conditions modeling global circuits. Finally, but crucially, all these approaches must ultimately consider observable tests for DL in astrophysical models. Then we can begin to work on Alfvén's paradigm.

I am grateful to C. Chan, N. Hershkowitz, W. Lotko, K. Papadopoulos, D. Spicer, and L. Vlahos for useful and stimulating discussions. In addition, I thank C. Chan and N. Hershkowitz for permission to use the previously unpublished Figure 5. This work was supported by and performed at the Naval Research Laboratory, under Contract No. N00014-83-C-2034.

REFERENCES

Alfvén, H.: 1979, J. Physique 40, pp. C7 1-19.
——————: 1981, "Cosmic Plasma", Reidel, Dordrecht.
——————: 1982, Physica Scripta T2, pp. 10-19.
Belova, N. G., Galeev, A. A., Sagdeev, R.Z., and Sigov, Yu. S.: 1980, JETP Lett. 31, pp. 518-522.
Block, L. P.: 1978, Astrophys. Space Sci. 55, pp. 59-83.
Chan, C., and Hershkowitz, N.: 1982, Phys. Fluids 25, pp. 2135-2137.
Chanteur, G., Adam, J.-C., Pellat, R., and Volokhitin, A. S.: 1983, Phys. Fluids 26, pp. 1584-1586.
Goertz, C. K., and Joyce, G.: 1975, Astrophys. Space Sci. 32, pp. 165-173.
Hasegawa, A., and Sato, T.: 1982, Phys. Fluids 25, pp. 632-635.
Hubbard, R. F., and Joyce, G.: 1979, J. Geophys. Res. 84, pp. 4297-4304.
Knorr, G., and Goertz, C. K.: 1974, Astrophys. Space Sci. 31, pp. 209-223.
Levine, J. S., and Crawford, F. W.: 1980, J. Plasma Phys. 23, pp. 223-247.
Sato, T., and Okuda, H.: 1980, Phys. Rev. Lett. 44, pp. 740-743.
——————————————: 1981, J. Geophys. Res. 86, pp. 3357-3368.
Schamel, H.: 1982, Physica Scripta T2, pp. 228-237.
Smith, R. A.: 1982a, Physica Scripta 25, pp. 413-415.
——————: 1982b, Physica Scripta T2, pp. 238-251.
——————: 1983, prep. SAI-83/1032.
Smith, R. A., and Goertz, C. K.: 1978, J. Geophys. Res. 83, pp. 2617-2627.
Stenzel, R. L., Gekelman, W., and Wild, N.: 1983, J. Geophys. Res. 88, pp. 4793-4804. (c.f. also this volume)
Temerin, M., Cerny K., Lotko, W., and Mozer, F. S.: 1982, Phys. Rev. Lett. 48, pp. 1175-1179.

DISCUSSION

Kundu: Can you give some example of astrophysical situations where the concept of double layer has been applied in detail?

R. Smith: Double layers have been proposed to be important in applications such as auroral arcs, solar flares, the Io-Jupiter circuit, etc., and this morning we saw a beautiful experimental observations of double layers in a reconnecting current sheet. But I think its fair to say that although some authors have tried to approach the evolution analytically, one really can't go very far beyond treating models of possible triggering mechanisms, i.e., one can't take the analysis into the nonlinear stage. As for considering questions of the sort I've tried to raise here, no papers have addressed them.

van Hoven: What sets the width of the spikes you showed for the case in which the external-circuit time constant is very long?

R. Smith: It is between a few ion plasma periods and the ion crossing time. That is because the ion inertia governs the phase-space dynamics both in the later stages of evolution and in the breakup.

Sturrock: What is the nature of the "turbulence" which shows up in systems with multiple double layers?

R. Smith: In the so-called "stairstep" double layers, there is both high frequency (electron plasma frequency) and low frequency turbulence, which I believe the authors identify as ion-acoustic.

Wentzel: If one reads the J. Geophys. Res., one gets the impression that double layers are well established. Why do you disagree?

R. Smith: Generally speaking, one finds two types of articles concerning double layers in the current literature. First, there are various simulation papers with titles referring to double layers in the aurora, for example. They contain some useful results, but one must be very careful because their boundary conditions are not appropriate to treat auroral dynamics, and they are also not related to any circuit concepts. I've published these comments elsewhere. Second, there are reported observations of double layers, from S3-3. These concern the so-called "ion-acoustic" double layers to which I referred at the end. These are very different from the "classical" D.C. concept about which I've been speaking, and seem to be linked with anomalous resistivity.

Bratenahl: Can you quickly tell me what is wrong with the Bohm and Langmuir conditions, i.e., why they are not applicable?

R. Smith: These conditions emerge from fluid-theory analyses, assuming adiabatic acceleration (the Bohm criterion, e.g., is written in terms of the adiabatic indices) and in the limit of infinite potential. But the double layer problem is bounded, and therefore, on the one hand, there is significant heat flux through the double layer, and on the other hand one must consider kinetic descriptions on half spaces in velocity at the boundaries. In general, though there is some condition such as the Bohm condition, it is not straight forward to find, and thinking in terms of fluid drifts is quite misleading.

PARTICLE ENERGIZATION IN STOCHASTIC DOUBLE LAYERS

William Lotko
Space Sciences Laboratory
University of California
Berkeley, California 94720

Electrostatic turbulence develops in current carrying plasmas when the relative electron-ion drift exceeds the critical value for laminar current flow. Recent 2D computer experiments (Barnes, 1982) indicate that many weak ion acoustic double layers form in such turbulence when the plasma is strongly magnetized ($\omega_{ce} \gtrsim \omega_{pe}$), the electron/ion temperature ratio is large ($\gtrsim 10$), and the relative electron-ion drift is comparable to or less than the electron thermal speed. The double layers emerge from the incoherent spectrum of electrostatic ion cyclotron and ion acoustic waves as intense localized electric field structures propagating subsonically relative to the ion bulk flow. The occurrence of weak ion acoustic double layers, excited by field-aligned currents in the Earth's auroral regions, has also been reported from in situ spacecraft measurements (Temerin et al., 1982). An important question concerns the effect of these coherent electric fields on plasma transport properties such as bulk heating and acceleration. For example, one might expect nonlinear diffusion processes, manifested as distinct nonthermal features in the particle spectra, to accompany the quasilinear diffusion of ions as they traverse turbulent regions in space. This idea motivates the work presented here.

A test particle approach is used to determine the spatial evolution of ion velocity distributions through a model turbulent layer containing randomly distributed, relatively propagating electric field pulses of a single polarity. Results of the aforementioned spacecraft observations and numerical simulations are used to construct a reasonable (though non-self-consistent) model for the random double layer electric fields. The basic phenomenology of weak ion acoustic double layers in a magnetized plasma is described briefly, a model turbulent layer containing many weak double layers is then formulated, and the evolution of test ion distributions through the turbulent layer is discussed. Since the calculation does not include quasilinear effects, it should be interpreted as a qualitative model complementing the more traditional approach to turbulent heating and acceleration.

MODEL TURBULENT LAYER

Weak ion acoustic double layers ($e\phi \lesssim T_e$) appear to evolve randomly from current driven turbulence in both 1D simulations (Sato and Okuda, 1981; Hudson and Potter, 1981; Kindel et al., 1981) and 2D simulations (Barnes, 1982). Their mean separation along the magnetic field is about 100-1000 λ_D (Debye lengths), depending on the boundary conditions. Their transverse dimension varies but is generally comparable to or larger than the parallel dimension - typically $10\lambda_D$. The parallel potential profile is nonmonotonic with a negative potential pulse preceding (in space) the double layer potential transition (Hasegawa and Sato, 1982). The transitional part of the waveform evolves from the negative pulse, which resembles a small but finite amplitude wave packet propagating initially with velocity $\frac{1}{2} c_s \lesssim v_p \lesssim c_s$. The localized disturbance intensifies by exchanging momentum with reflected electrons (Chanteur et al., 1983; Lotko, 1983). As $e\phi$ approaches T_e, strong ion trapping in the negative pulse produces an inertial drag, causing a rapid deceleration of the wave to $v_p = 0$. At this time, the ion starved double layer decays within a few ion plasma periods. The net lifetime of the DL is a few ion trapping periods - typically 10 - 100 ω_{pi}^{-1}. Subsequently, another double layer forms at a random location in the system (Okuda and Ashour-Abdalla, 1982; Barnes, 1982).

In order to model this phenomenon, we consider a model system of length 100 $N\lambda_D$ partitioned into N unit cells of length 100 λ_D. The cell length corresponds to the lower limit on the mean double layer spacing and is chosen for computational expediency. At any given instant, each cell contains one DL located at a random position, so the rms spacing differs from the mean spacing. The propagating DL electric field profile is taken to be a delta function whose parameters vary stochastically with time:

$$E_j = \phi_j \, \delta[x - r_j - s_j(t - t_j)] \tag{1}$$

ϕ_j, s_j and r_j are uncorrelated random variables corresponding to the DL potential, velocity, and position in the cell at time $t = t_j = 10j\omega_{pi}^{-1}$, where $j = 0,1,2,...$ To account for the finite DL lifetime, the values of ϕ_j, s_j and r_j are changed every $10\omega_{pi}^{-1}$, whereupon j is incremented by one. The jth values of the random variables are selected from a random number sequence uniformly distributed over the intervals, $0 < e\phi \lesssim \alpha T_e$, $0 \lesssim \bar{v} - s \lesssim c_s/2$, and $0 \lesssim r \lesssim 100\lambda_D$, where T_e, c_s, and $\bar{v}$ are the electron temperature, ion acoustic speed, and ion bulk flow velocity, and $0 < \alpha \lesssim 1$ determines the mean DL amplitude. The parameter α is varied in order to determine the effect of stronger or weaker double layers on the ion energization.

The previously mentioned numerical simulations indicate that the double layer propagates more slowly as its amplitude increases, so at first glance, one might choose these two variables to be correlated. In 2D simulations, however, the nonplanar DL propagates along the magnetic field with the maximum amplitude varying in the transverse direction. Thus, when an ion encounters a double layer, the effective potential depends on the relative position at the time of encounter. For this reason, ϕ and s are taken to be uncorrelated.

The delta function approximation (1) assumes that (i) the double layer spacing is large compared to the double layer width, and (ii) the ion transit time through the double layer is small compared to the double layer lifetime. The first condition is certainly satisfied in view of the simulation results and spacecraft observations. The second is equivalent to assuming that the ion motion is adiabatic during the encounter, i.e.,

$$\tfrac{1}{2} M (v - s)^2 + e\phi = \text{constant} \tag{2}$$

If W and T denote, respectively, the double layer width in λ_D and lifetime in ω_{pi}^{-1}, then (ii) requires that

$$W/T < c_s (e\phi/T_e)^{\tfrac{1}{2}} \tag{3}$$

This condition is at best marginally satisfied, so nonadiabatic effects can be expected to modify somewhat the results presented here. As a final caveat, it should be noted that the waveform (1) does not include the negative potential pulse that precedes the double layer and that locally traps ions. This neglected feature is presumably important for only a small fraction of the ions in any given unit cell.

SPATIAL EVOLUTION OF TEST ION DISTRIBUTIONS

The problem is to determine the velocity increments of test ions traversing the model turbulent layer. As formulated, the process is Markovian and can be characterized by a transition probability, i.e., the probability $P(v_0, \Delta)$ that a particle with initial velocity v entering a unit cell leaves the cell with velocity $v = v_0 + \Delta$. This probability function is constructed numerically from an ensemble of 1000 trials for each initial velocity with each trial characterized by a different random number sequence for the DL parameters. The change in ion velocity after a DL encounter, if one occurs, is determined from (2), taking into account the three distinct types of phase space trajectories corresponding to ions overtaken by, reflected by, or overtaking the double layer.

The transition probability after N unit cells may be determined by iterating the so-called Smoluchowski equation (Wang and Uhlenbeck, 1945). Alternatively, one can iterate directly the integral equation for the particle distribution function:

$$F_{n+1}(v) = \int_0^v du\, F_n(u)\, P(u, v - u) \tag{4}$$

$F_n(v)$ is the velocity distribution of ions entering the nth cell or alternatively, leaving the $(n-1)^{th}$ cell. N iterations of (4) determines the evolution through a turbulent layer N cells long.

A few illustrations would greatly fascilitate the interpretation of the following discussion, but, unfortunately, space limitations preclude their inclusion in this brief report. A complete discussion including detailed diagnostics will be published elsewhere. The main features of the spatial evolution are as follows:

(1) After a few unit cells, a double peaked distribution forms. The low velocity peak is the remnant of the incoming distribution. The higher velocity peak is the accelerated component. Both have comparable densities at this stage.

(2) Since the double layers are all polarized in the same direction ($E > 0$) and propagate in the direction antiparallel to the electric field vector (negative velocities), they always accelerate ions in the positive direction. Thus, the accelerated component appears at positive velocities in the rest frame of the incoming distribution.

(3) The peak energy and thermal spread of the accelerated component scale with the product of the mean double layer amplitude and the number of unit cells transversed, and, by comparison, depend only weakly on the mean double layer velocity and thermal spread of the incoming distribution.

(4) The slope of the remnant distribution at negative velocities increases with the number of cells traversed. This feature would presumably help sustain the instability that causes double layers. On the other hand, incoherent waves may also propagate at these velocities, leading to a quasilinear type diffusion of the velocity distribution. This effect would tend to suppress any instabilities via enhanced ion Landau damping.

(5) After about 10-20 cells, depending on the mean double layer amplitude, the density of the accelerated component exceeds that of the remnant component. At these distances into the turbulent layer, one might expect the nature of the turbulence to differ significantly from that near its leading edge. Double layers may cease to form at this point, and continued iteration of Eq. (4) becomes questionable. This effect may determine in part the spatial extent of the turbulent layer.

In summary, it has been shown that randomly distributed double layers that arise in current driven turbulence accelerate and heat ions. Their effects on the ion distribution function differ substantially from those predicted by quasilinear processes and can be expected to influence significantly the turbulent properties of current carrying plasmas.

ACKNOWLEDGMENT. Work supported by NSF, Grant ATM-8103347 and Cal Space.

REFERENCES

Barnes, C.: 1982, EOS Trans. Am. Geophys. Union 63, p. 1074.
Chanteur, G., J. Adam, R. Pellat, and A. Volokhitin: 1983, Phys. Fluids 26, p. 1584.
Hasegawa, A., and T. Sato: 1982, Phys. Fluids 25, p. 632.
Hudson, M. K., and D. Potter: 1981, in "Physics of Auroral Arc Formation", Geophys. Monogr. Ser. 25, ed. by S.I. Akasofu and J.R. Kan, Am. Geophys. Union, Washington, D.C., p. 260.
Kindel, J., C. Barnes, and D. Forslund: 1981, ibid., p. 296.
Lotko, W.: 1983, Phys. Fluids 26, p. 1771.
Okuda, H., and M. Ashour-Abdalla: 1982, Phys. Fluids 25, p. 1564.
Sato, T., and H. Okuda: 1981, J. Geophys. Res. 86, p. 3357.

Temerin, M., K. Cerny, W. Lotko, and F. Mozer: 1982, Phys. Rev. Lett. 48, p. 1175.
Wang, M., and G.E. Uhlenbeck: 1945, Rev. Mod. Phys. 17, p. 323.

DISCUSSION

R. Smith: I'm worried about the ion-crossing-time problem in Markovian statistics.

Lotko: The ion crossing time problem and the use of Markovian statistics are separate issues. Markovian statistics are applicable here because the successive double layer fields encountered by a particle are uncorrelated after a mean inter-double layer transit time, which, for the bulk, is greater than 10 DL lifetimes. The physical difficulty using Markov's method is to determine the transition probability, wherein the ion crossing time problem enters. As indicated, this effect is important when the ion transit time across a single double layer exceeds the DL lifetime. The δ-function DL model neglects this effect, and so, must be considered as a first approximation. It is, of course, least accurate for particles occupying trajectories near the phase space separatrix, although the stochastic nature of particle/DL encounters ameliorates this problem to some extent. A truly self-consistent calculation (which is currently beyond the memory capability of large scale computers) would presumably yield more bulk heating and less bulk acceleration.

Vlahos: I believe that the test particle orbits by definition cannot be the whole distribution because in this case there is an obvious lack of self consistency. Of course things will be different if you work with 10^{-2} or less particle on the tail, for example.

Lotko: Test particle calculations certainly do not describe the self-consistent evolution of the particle distribution. Nevertheless, they provide useful information about nonlinear interactions between particles and fields and about evolutionary trends in the distribution function, which is the spirit of this calculation.

LINKS BETWEEN JET INSTABILITIES, RADIATION AND PROPAGATION IN ASTROPHYSICS*

Gregory Benford
Department of Physics
University of California
Irvine, California 92717

ABSTRACT

At the head of a jet the confining medium of plasma frequency ν_p is compressed, so that streaming instabilities between relativistic electrons and this plasma produce waves at $\nu_p' > \nu_p$. Considerable power can be lodged in these electrostatic waves, and conversion to electromagnetic waves allows them to propagate far beyond the jet. Emission at $\nu \approx \nu_p'$ or Compton boosted radiation at $\nu \lesssim \gamma^2 \nu_p'$ yields a cone of radiation of angle $\sim 1/\gamma$, which illuminates the region directly in front of the jet. This emission is not absorbed by the surrounding plasma unless a cloud blocks the jet. Absorption in a cloud can lead to tunneling through large clouds, or propelling of smaller clouds out of the jet path. In this fashion jets may clear their way through an inhomogeneous medium, avoiding lateral disturbances and pre-heating their path.

INTRODUCTION

We often assume that the propagation of astrophysical jets is a matter of macroscopic physics -- ram pressure balance, Kelvin-Helmholtz instabilities and the like. Microscopic process are responsible for reacceleration -- and thus the radio emission whereby we see the jets -- but seem to have little impact on large scales.

This need not be so. One of the primary difficulties in understanding jets is how they can pass relatively unperturbed through a medium which must contain irregularities -- clouds of dense gas, random gradients, and perhaps large magnetic fields. Dense clouds particularly seem to be a problem for good jet propagation, since a cloud can deflect a jet abruptly, driving instabilities on a scale of the jet radius. Since it seems likely that clouds are common in intergactic space, propagation along a straight line for megaparsecs, as in Cen A, requires an explanation. Similar difficulties apply to small jets such as SS433 and perhaps the Crab jet, since their environments are certainly highly inhomogeneous.

The primary assumption made in studying such effects is that the beam can deliver energy to the surrounding gas only at the "working surface" -- a region where shocks and rapid energy deposition achieve the tunnel boring, producing synchrotron hot spots. Here I propose a simple mechanism whereby beams can affect the plasma beyond the working surface, perhaps many beam radii ahead. This mechanism <u>selects</u> high density irregularities for energy deposition, and thus acts as an "intelligent" preparer of the downstream environment.

BASIC PICTURE

The jet of density n_j and speed v_j collides with a dense surrounding gas (density n_g) at the working surface, compressing this gas to a higher density n_g'. Ahead of the jet, a dense cloud of density n_c waits. If the jet could not affect the cloud until it struck, deflections could be considerable. (Fig. 1)

However, the beam is a source of electromagnetic radiation. Aside from synchrotron emission, which is of high frequency, there can be copious collective emission at or above the local plasma frequency, $\nu_p(n_g') \gtrsim 100$ Hz. The key features of this mechanism are:

1. The jet compresses the gas at the working surface, so emission is at $\nu_p(n_g') > \nu_p(n_g)$. Since collective processes in the homogeneous gas (n_g) absorb only at $\nu \leq \nu_p(n_g)$, the radiation propagates freely. It will be absorbed <u>only</u> by clouds denser than n_g'.

2. Emission is intense, if our knowledge of Type III solar bursts and of laboratory beam-plasma experiments is any guide.

3. Emission is concentrated into a cone of opening angle $\Omega = 1/\gamma + \phi$, where γ is the Lorentz factor of the beam electrons and ϕ is the average opening angle of the beam electron trajectories in the emitting area. This serves to concentrate the power into the region downstream where the beam must clear a path.

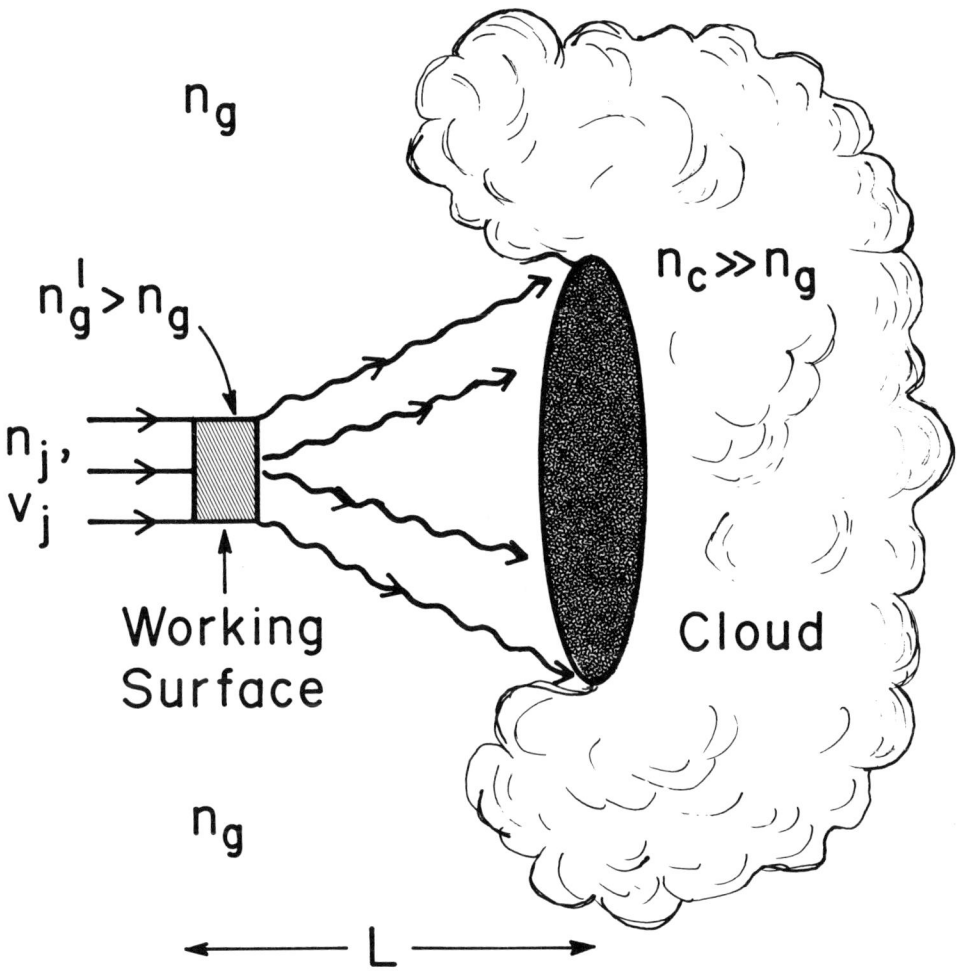

Fig. 1. A jet compresses the surrounding plasma to a higher density, n'_g at the working surface. Electrostatic waves produced by beam-plasma instabilities there lead to electromagnetic emission. A cloud is dispersed by this radiation before the jet arrives.

4. Absorption of the radiation in the denser cloud will heat this cloud, causing it to expand, reducing the density. When n_c drops to n_g', the photons pass through, so they can do further tunneling downstream. Thus energy is delivered to a cloud only so long as it acts as a jarring barrier to the (presumably) smooth tunneling going on.

The plasma waves are produced at the working surface by beam-plasma instability. Presumably these are among the processes which eventually clear the beam path of high density plasma. Nonlinear processes then convert these electrostatic waves to electromagnetic photons of very low frequency, $\nu_p'(n_g') \approx 100 \, n_{-4}^{1/2}$ Hz. These photons are unobservable but carry a high flux of energy.

COLLECTIVE EMISSION PROCESSES

Decades-long study of the Type III solar radio radiation has yielded detailed scenarios for electromagnetic emission (Smith, 1970, and references therein). Recent experiments have verified some aspects of the weak turbulence emission scheme, both for relativistic beams (Benford, et al, 1980) and nonrelativistic cases (Cheung et al, 1982; Whelan & Stenzel, 1981). Such processes can be very efficient because there is spatial amplification of the electromagnetic waves as they propagate through a background of large amplitude electrostatic waves (driven by beam-plasma instability). We shall use the theoretical model of weak turbulence emission at ν_p and apply it to the working surface, where a weak beam penetrates a dense plasma.

The volume emissitivity of the emission (Benford & Smith, 1982) for plasma of temperature T in eV is

$$J = 6 \times 10^{-16} \frac{T k_m^2 \nu_p^2(n_g')}{n_g'} \text{ W}$$

and the power emitted by the beam at $\nu_p(n_g')$ is $P = VJe^{\mu x}$ with V the volume, x the characteristic mean distance an electromagnetic wave travels while being amplified, and

$$\mu = \frac{6.6 \, 10^{-34} \, W \, \nu_p^2(n_g')}{T^{3/2} \, k_p \, n_g'}$$

is the amplification factor. It depends on W, the energy density of the plasma waves, which in turn depends on the nonlinear saturation mechanism. We shall suppose weak turbulence processes saturate the electrostatic waves at a low amplitude given by (Papadopoulos, 1975)

$$W \approx n_j \gamma m c^2 \frac{\delta}{\nu_p(n_g')}$$

where δ is the linear wave growth rate for a warm beam which has a large spread in emission angle,

$$\delta \approx \nu_p(n_g') \eta \frac{1}{\gamma \phi^2}$$

Here ϕ is the angular spread of the beam electrons and $\eta = n_j/n_g$. The typical wavenumber of the turbulence, k_p, is taken as $\sim \nu_p(n_g')/c$. The largest k present, k_m, is taken as the same order.

In this model, emission at $2\nu_p$ is much less powerful, and cannot be amplified by the relatively slow plasma waves. Thus we neglect such radiation.

In terms of astrophysical parameters (cgs units)

$$W \simeq 10^{-9} \eta^2 \phi^{-2} n_{-4}$$

$$J \simeq 2 \times 10^{-32} T n_{-4}^2 (\eta/\phi)^2 \quad (1)$$

$$\mu = 6 \times 10^4 n_{-4}^{1/2} T^{-3/2} (\eta_{-3}/\phi_{-1})^2$$

$$P = 2 \times 10^{29} T n_{-4}^2 (\eta_{-3}/\phi_{-1})^2 R^3 (y/R) e^{\mu x}$$

Here we write $n_{-4} = n_g/10^{-4}$ cm^{-3}, T the plasma temperature in eV., R the beam radius in kpc, η_{-3} the beam-plasma density ratio in units of 10^{-3}, etc. The working surface has a thickness y, and we expect $y/R \sim 0.1$.

Since spatial amplification occurs over an e-folding length $\sim 10^5$ cm., we can expect a great increase in emission above the simple volume emissivity J. This is typical of Type III bursts, and also agrees with laboratory work, where the product $\mu x \sim 25$ is compatible with observed radiated power (Benford & Smith, 1982). Such a short amplification length means that to disrupt amplification requires wave refraction on a scale much less than the typical wavelength of the beam-plasma instability, productive losses -- i.e., those which smooth out the beam path -- occur only to the sides. We shall content ourselves with a naive estimate of the heating time, assuming that simple heating and expansion occur. A tunnel of width R must push cloud matter within a distance R_c, the cloud radius, out of the way to expand, so that heating times are lengthened by the area factor $(R_c/R)^2$. Cases in which the cloud is not much larger than the jet radius clearly will be less "loaded", since the heated matter can escape into lower density regions. When the cloud is smaller than the jet radius, it can even be shoved to the side by a rocket effect.

The beam compresses the surrounding gas to density n_g' over a distance L' and radially over a scale R'. This pressure differential is relieved by sound waves in the cool plasma of speed U. The amplification of density is roughly

$$n_g'/n_g \approx 1 + \frac{U}{v_j} \frac{L'}{R'} \eta^{-1/2}$$

If we estimate $U/v_j \sim 0.1$, $L'/R' \sim 1$ and $\eta \sim 10^{-3}$, this yields density contrasts of order unity. This insures that local emission at $\nu_p(n_g')$ will propagate freely at very low frequencies, $\nu_p(n_g) \gtrsim 100$ Hz until encountering a cloud.

Using (2), a beam can heat a cloud until $n_c = n_g$ if it begins irradiation at a distance L_o and the process ceases at L_f, if

$$L_o = \frac{L* L_f}{L* - L_f}$$

where $L* = \frac{R^2}{ct_h} \left(\frac{R}{R_c}\right)^2$ (kpc)^2 ,

$$t_h = \left(\frac{n_c}{n_g}\right)\left(\frac{T_c}{T}\right)\left(\frac{\phi}{\eta_{-3}}\right)^2 \left(\frac{z}{y}\right) 10^{21} e^{-\mu X} \text{ rec}$$

Here z is the length of the heated zone, with radius of 1 kpc and R_c is the cloud radius. Clearly to achieve heating in the closing time $t_c = 310^{12}(c/v_b) \eta_{-3}^{-1/2}$ sec, we need $\mu X \approx 16$ when $R_c = R$, which is less than the amplification typically invoked for laboratory experiments and Type III bursts (Benford & Smith, 1982).

With this amplification, $L* \sim R^2/10$ kpc and $L_o < R \sim$ kpc. The problem is complicated if $L_o \sim R \sim$ kpc, but the range of $L_o \lesssim 10$ kpc assures that the effect can operate if the beam has $\sim 10^6$ yr to heat the cloud. For these estimates we have taken $(n_c/n_g) \sim 10$, $(T_c/T) \sim 0.1$, $z\phi^2/y \sim 1$. Allowing for lateral inhibition of expansion through the factor $(R/R_c)^2$ lengthens the true heating time, yet there seems ample freedom in the value of μL to accomodate this movement of gas.

In terms of efficiency, we can express the requirement that a jet heat and disperse a cloud of depth d in terms of the ratio of P_r, power radiated at $\sim v_p$, to P_j, the jet power at the working surface. For cloud dispersal,

$$\frac{P_r}{P_j} \geqslant \varepsilon_{-1} \left(\frac{n_c T_c}{n_g T}\right)_1 n_{-3}^{1/2} \left(\frac{d}{L}\right)_{-2} \left(1+\frac{L}{\gamma R}\right)^2 0.03 \ \%$$

The efficiency of jet boring through the ambient gas, n_g, is ε in units of 0.1, and the $1 + L/\gamma R$ arises from the γ-cone geometry. Thus for plausible clouds, $d/L \sim 0.01$, a very mild efficiency $\sim 0.03\%$ is needed. The U.C.I. beam-plasma experiments with $\gamma = 3$ have efficiencies of this order, as do many other experiments in the nonrelativistic regime. Certainly thin clouds, $(d/L)_{-2} \ll 1$, are easy to disperse. To see if the effect is dominant for thick clouds, $d/L \sim 0.1$, requires detailed knowledge of jet conditions beyond our current ability.

This qualitative success implies that electromagnetic tunneling may be an effective agent. The general picture should work for jets of any size. The self-tuning feature of this model arises from the unique nature of plasma waves. Unlike synchrotron radiation, plasma emission is absorbed in very small distances if $v_p(n_c) > v_p(n_g')$. Thus its effects are local and immediate. The emission is of unobservably low frequency, except perhaps in dense, SS433-like environments. Indeed, electromagnetic tunneling may be crucial in the initial setting-up of jets in dense environments, since it could clear a zone in which a nozzle can then be self-consistently made, using beam self-magnetic fields and sidewise external gas pressure.

It seems probable that pre-heating of small clouds can increase the confining pressure around the beam at later times, aiding stability. Magnetic fields in the working surface or in the clouds will have little effect on plasmon emission and subsequent radiation; this appears to be true of laboratory experiments. Thus we expect microscopic collective emission processes can contribute to macroscopic behavior of jets on all astrophysical scales.

*Work supported by AFOSR Grant No. 82-0233

REFERENCES

Benford, G., D.F. Smith, Phys. Fluids 25, 1450 (1982).
Benford, G., D. Tzach, K. Kato, and D. Smith, Phys. Rev. Lett. 45, 1182 (1980).
Cheung, P.Y., A. Y. Wong, C.B. Darrow, and S.J. Qian, Phys. Rev. Lett. 48, 1348 (1982).
Kato, K.G., G. Benford, D. Tzach, Phys. Rev. Lett. 50, 1587 (1983).
Papadopoulos, K., Phys. Fluids 18, 1769 (1975).
Smith, D.F., Adv. Astron. Astrophys. 7, 147 (1970).
Whelan, D.A. and R.L. Stenzel, Phys. Rev. Lett. 47, 95 (1981).

DISCUSSION

Spangler: Given the Lorentz factors of electrons in extended extragalactic sources, and best estimates of the local plasma frequency, your Compton-boosted radiation should be in the observable radio range. Does the absence of spectral "turn-ups" at low frequencies place constraints on your model?

Benford: This process occurs only in a small working surface region, so not much power at very high frequencies comes out. This can easily be swamped by ordinary synchrotron emission from the surrounding hot spot. Also, it is beamed along the jet, not usually toward us. I expect the many more electrons at low γ, giving $\sim$ 100 Hz photons, do most of the cloud heating. A power law electron spectrum folded into eq. (1) yields this conclusion.

Hardee: A very ingenious idea. You solved the problem of heating the cloud, but you created another problem; the acceleration of particles

Benford: I assume reacceleration occurs throughout the jet, as the synchrotron radiation testifies. This occurs independently of the plasma instabilities at the working surface. Indeed, all the particle acceleration and electromagnetic emission comes from jet kinetic energy, through various Alfvén or plasmon waves.

Vasyliunas: Shouldn't the beaming be controlled by the γ of the jet motion and not of the realtivistic electrons (which should be distributed more or less isotropically in the frame of the jet), so that there is no pronounced beaming unless the jet itself moves at relativistic speeds?

Benford: Electrons in the working surface are probably quite anisotropic. Collective emission occurs only where beam-plasma instability drives the plasmon spectrum. By the time electrons turn to the side they may well have left the region of beam-plasma instability; then they join the cocoon. I expect $\gamma \lesssim 10$ electrons do most of the emission anyway, since there are many more of them. Even if you are right, the γ-cone effect is not essential to making the power requirement work.

A UNIFIED THEORY OF CORONAL HEATING

James A. Ionson
Laboratory for Astronomy and Solar Physics
NASA-Goddard Space Flight Center, Greenbelt, MD 20771

This presentation focuses upon the coronal heating problem and reports the results of Ionson's (1984) unified theory of electrodynamic heating. This generalized theory, which is based upon Ionson's (1982) LRC approach, unveils a variety of new heating mechanisms and links together previously proposed processes. Specifically, Ionson (1984) has derived a standing wave equation for the global current, I, driven by emfs that are generated by the $\beta \gtrsim 1$ convection. This global electrodynamics equation has the same form as a driven LRC equation where the equivalent inductance, $L = 4\ell/\pi c^2$, scales with the coronal loop length and where the equivalent capacitance, $C = c^2 \ell / 4\pi v_A^2$, is essentially the product of the free space capacitance, $\ell/4\pi$, and the low frequency dielectric constant, c^2/v_A^2. The driving emf, $\mathcal{E} = vBa/c$, is a formal integration constant associated with the convective stressing of $\beta \gtrsim 1$ magnetic fields. Since the transition from the $\beta \gtrsim 1$ driver to the $\beta < 1$ coronal loop is typically small compared to the "wavelength" of the associated magnetic fluctuation, this integration constant is not sensitive to details of the transition zone. The total resistance, $R_{tot} = L(1/t_{diss} + 1/t_{phase} + 1/t_{leak})$, represents electrodynamic energy "loss" from dissipation, magnetic stress leakage out of the loop and phase-mixing. These three processes have been parameterized by appropriate timescales. Note that $R_{leak} = L/t_{leak}$ and $R_{phase} = L/t_{phase}$ do not result in resistive heating but do participate in limiting the amplitude of the global current, I. This is fairly obvious with regard to magnetic stress leakage but not for phase-mixing. The phase-mixing resistance, R_{phase}, represents coupling between the global current and the local current density. Since the global current is essentially an integration of the local currents, the degree of coherency between the local currents can play an important role in determining the ultimate amplitude of I. The rate at which coherency between the local currents is lost is given by the phase-mixing time, t_{phase}. A loss of coherency implies a corresponding reduction in the amplitude of I. In this sense, R_{phase} measures the phase-mixing contribution to the global current limitation process.

The time averaged coronal heating flux, F_H, is given by

$$F_H = \frac{2\langle I^2 R_{diss}\rangle}{(\pi a^2/4)} \equiv F_{max}\,\mathcal{E} \tag{1}$$

where the factor of two implies two footpoint drivers per loop and where $\pi a^2/4$ represents the cross sectional area of a loop within the corona. The maximum flux, F_{max}, of energy available to the coronal portion of a loop is given by

$$F_{max} = 4\left(\frac{B}{B_{\beta\geq 1}}\right)\left(\frac{4 v_A^{\beta\geq 1}}{v_A}\right) v_A^{\beta\geq 1} \left(\tfrac{1}{2}\rho\, v_{tot}^2\right)_{\beta\geq 1} \tag{2}$$

which can be interpreted by noting that the first factor of four accounts for two footpoints per loop and two excitation polarizations; the $B/B_{\beta\geq 1}$ term results from magnetic expansion into the corona; and the third term represents a transmission coefficient which in this case is given by the Fresnel relation, $4v_A^{\beta\geq 1}/v_A$. The fourth term, $v_A^{\beta\geq 1}(1/2\rho v_{tot}^2)$, represents the maximum available flux of electrodynamic energy within the $\beta\geq 1$ region. The fraction of F_{max} that actually does enter the loop, heating the contained coronal plasma is denoted by an electrodynamic coupling efficiency, $\mathcal{E}$. The electrodynamic coupling efficiency is a convolution of the normalized convection velocity spectrum with a frequency dependent velocity-magnetic field interaction parameter, $e(\nu; t_A, Q_{tot}, Q_{diss})$, i.e.,

$$\mathcal{E} \equiv 2\int_0^\infty \left(\frac{\langle v_\perp^2\rangle_\nu}{v_{tot}^2}\right)_{\beta\gtrsim 1} e(\nu;t_A,Q_{tot},Q_{diss})\,d\nu \tag{3}$$

$$e(\nu; t_A, Q_{tot}, Q_{diss}) = \frac{Q_{tot}^2}{Q_{diss}\left[1 + \left(\nu t_A - \frac{1}{\nu t_A}\right)^2 Q_{tot}^2\right]} \tag{4}$$

with the coronal loop quality factors, $Q_{diss}=2\pi t_{diss}/t_A$ and $Q_{tot}=2\pi/(t_A/t_{diss}+t_A/t_{phase}+t_A/t_{leak})$. The velocity-magnetic field interaction parameter peaks at the resonance frequency, $\nu_{res}=t_A^{-1}$, with a maximum value $e_{max} = Q_{tot}^2/Q_{diss}$ which can be thought of as a measure of the magnitude of the magnetic stressing rate spectrum with respect to the $\beta\geq 1$ mechanical energy spectrum. The characteristic width of $e(\nu; t_A, Q_{tot}, Q_{diss})$ about this maximum can be thought of as an interaction bandwidth, $(\Delta\nu)=\pi/t_A Q_{tot}$.

A general form for the normalized velocity spectrum that parameterizes the details is assumed, viz.,

$$\frac{\langle V_\perp^2 \rangle_\nu}{V_{tot}^2} = \left(\frac{t_c}{\pi}\right) \left[\frac{1}{1 + (\nu t_p - \frac{1}{\nu t_p})^2 (\frac{t_c}{t_p})^2}\right] \quad (5)$$

where t_c represents the correlation time and where t_p^{-1} is the frequency at which the spectrum peaks. Using the above parameterized form for the convection spectrum in equation (3) results in the following general expression for the electrodynamic coupling efficiency, $\mathcal{E}$:

$$\mathcal{E} = \left(\frac{t_c}{t_A}\right)\left(\frac{Q_{tot}}{Q_{diss}}\right) \begin{cases} [1 + (t_p/t_A - t_A/t_p)^2 (\frac{t_c}{t_p})^2]^{-1} & ; \quad \begin{array}{l} Q_{tot} > 1 \\ \text{HIGH-QUALITY COUPLING} \end{array} \\ \\ \frac{Q_{tot}}{(t_c/t_A) + Q_{tot}} & ; \quad \begin{array}{l} Q_{tot} < 1 \\ \text{LOW-QUALITY COUPLING} \end{array} \\ \\ 1 & ; \quad \begin{array}{l} t_c < t_A, t_p, t_A Q_{tot} \\ \text{STOCHASTIC COUPLING} \end{array} \end{cases} \quad (6)$$

Equation (6) for the electrodynamic coupling coefficient highlights three general coupling categories.

If $Q_{tot} > 1$, than a coronal loop responds to external driving as a high quality resonance cavity in the sense that the velocity-magnetic field interaction parameter is sharply peaked bout the resonance frequency. Therefore, a high quality coronal loop with $Q_{tot} > 1$ interacts only with a small portion of the convection spectrum centered about the resonance frequency. This, however, can still lead to significant power absorption because the corresponding magnetic stressing rate is quite large within $(\Delta \nu)$. Specifically, the maximum value of the velocity-magnetic field interaction parameter, $e_{max} > 1$.

If $Q_{tot} < 1$ then there is no preferred excitation frequency because $t_A \Delta \nu > 1$. A large interaction bandwidth does not necessarily imply a large absorption of power because the magnetic stressing rate is small within $\Delta \nu$ since $e_{max} < 1$. It's also important to note that although $\Delta \nu$ is large compared to t_A^{-1} it could be small compared to the correlation width, t_c, of the convection spectrum. In this regard, the maximum value of $\Delta \nu$ is given by t_c^{-1} and the loop interacts with the entire convection spectrum. In fact, when t_c is smaller than all other timescales, both the high and low quality coupling categories degenerate into a third general category describing stochastic coupling.

Each of the above general representations embodies three special cases. These special cases highlight the dominant "loss" process responsible for limiting the magnetic stressing rate with respect to the available convection energy. As discussed earlier, $e_{max}=Q_{tot}^2/Q_{diss}$, represents a measure of the magnetic amplitude limitation process. The three special cases of interest are "dissipation limited" when $Q_{tot} \simeq Q_{diss}$, phase-mixed limited" when $Q_{tot} \simeq Q_{phase}$ and "leakage limited" when $Q_{tot} \simeq Q_{phase}$. This classification scheme consolidates a variety of new heating mechanisms and also illustrates that existing mechanisms are actually special cases of a much more general formalism. Specifically, Alfvenic surface wave heating is a high-quality, phase-mixed limited mechanism, resonant electrodynamic heating is a high-quality, dissipation-limited mechanism, dynamical dissipation is a low-quality dissipation-limited mechanism and stochastic magnetic pumping is a stochastic, dissipation-limited mechanism.

An application of this theory to solar coronal loops indicates that the total quality is essentially equal to the dissipative quality (i.e., dissipation limited); and that the general coupling category changes as loops age and hence increase in length. Specifically, young active region loops and possibly bright points are overdamped systems (i.e., $Q_{diss}<1$), heated at a rate that depends upon the dissipation time t_{diss}, while older active region and large scale loops are underdamped systems (i.e., $Q_{diss}>1$), heated at a rate that is quantitatively independent of the dissipation time. Furthermore, it appears that active region loops are heated by electrodynamically coupling to $\beta \gtrsim 1$ p-mode oscillations, while large scale loops are heated by coupling to the solar granulation.

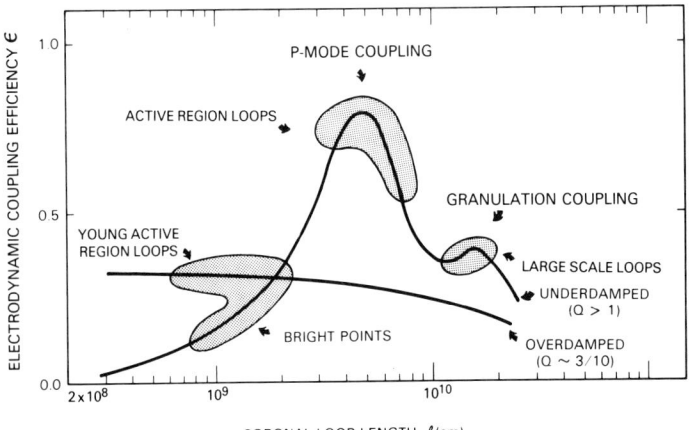

Ionson, J. A., 1982, Ap. J., 254, 318.
Ionson, J. A., 1984, Ap. J., Jan. 1.

DISCUSSION

Rust: I do not think there are definitive studies to indicate that coronal loop lengths are "quantized" as indicated by your theory. Probably the visible loop lengths are determined by the magnetic field distribution, not by the heating mechanism. That is, coronal pictures suggest that any closed loop will be heated.

Ionson: The theory does not predict quantized coronal loop lengths. The observations do, however, as indicated by the blocks of data illustrated in the last figure. In other words, coronal pictures do not suggest that any closed loop will be heated in a thermally stable sense.

Bratenahl: Your driving spectrum did not include the supergranulation time scale $\sim$ 20 hours. It is a strong peak in the overall spectrum.

Ionson: Such long period power only resonates with loops with lengths larger than the solar radius. There are very few loops with such large sizes.

Krishan: How do you make contact with the heating theories invoking flow of current in thin sheaths?

Ionson: The physics of thin sheaths is absorbed within t_{diss} which does not appear in the final result for the heating flux, F_H. Therefore, in many cases one can predict heating rates without delving into the details of current sheaths.

SESSION III

LABORATORY EXPERIMENTS ON RECONNECTION IN CURRENT SHEETS

A. Bratenahl and P. J. Baum
Institute of Geophysics and Planetary Physics
University of California
Riverside, CA 92521

ABSTRACT. Laboratory reconnection experiments dedicated to problems of space and astrophysics are briefly reviewed with the purpose of demonstrating that such experiments can provide important insights of considerable value to the development of reconnection theory. Moreover, many of these insights are of a kind not likely to be perceived either when working directly with space observations or while pursuing a course of pure theoretical reasoning without reference to laboratory results.

1. INTRODUCTION

1.1 History and Motivation

At the Jet Propulsion Laboratory in Pasadena in 1963, one of us (AB) became intensely interested in the pioneering theoretical work on reconnection of Dungey (1953) and the solar flare model of Sweet (1958a,b). A laboratory facility was developed in order to test the validity of some of the basic assumptions and approximations that were used.

Sweet (1958a), for example, assumed that a flat neutral current sheet would form where two bipolar sunspot field systems were pressed together. Note, a neutral current sheet implies, topologically, the complete absence of a weak transverse magnetic field crossing the field reversal plane. Until a possible tearing mode develops (Furth et al., 1963), there are no x-points and therefore no reconnection, only field annihilation via resistive diffusion. Sweet's hypothesis was based on the traditional MHD assumption that solutions at very high conductivity can be approximated by solutions at infinite conductivity to which is simply added the effect of resistive diffusion. However, based on the well known asymptotic paradoxes in ordinary hydrodynamics due to the effect of viscosity (Birkhoff, 1950), AB was suspicious that analogous asymptotic paradoxes might occur in MHD. Thus, just as ordinary viscosity can introduce wakes and boundary layers, so resistivity might introduce neutral points, shock waves, etc. In other words, Sweet's assumption might be invalidated.

1.2 Reconnection Experiments

The laboratory facility, designed to test these and other theoretical ideas related to reconnection became known as the Double Inverse Pinch Device (DIPD) (Fig. 1a). In simplest terms, two insulation-covered conducting rods carry parallel currents of increasing strength. The rod return currents, forced to flow through the preionized argon, develop expanding cylindrical sheets (inverse pinches). Upon colliding at the center of the device, the two cylinders merge into a single expanding oval. The resulting magnetic field line topology (Fig. 1b) has a figure eight-shaped separatrix surface which partitions the field lines into cells according to their linkage with the source current rods, i.e., with either one of the rods alone, the "parent" cells, or with both, the "daughter" cell. On the axis of the device, the separatrix intersects itself along a line called the separator, forming a locus of x-type neutral points in this 2-D magnetic field configuration.

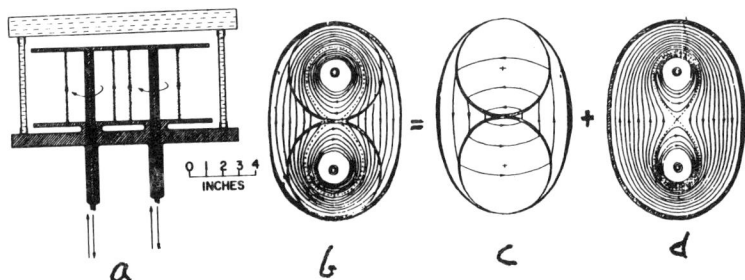

Figure 1. *DIPD with its currents and magnetic fields: (a) the device producing a pair of inverse pinches; (b) the total field and current system; (c) the induced separator current system and its field; (d) the source current and field. (b), (c) and (d) were calculated using the coax approximation.*

In the vicinity of the separator, a secondary current system is induced, which develops into a hyperbolic pinch with extensions trailing off downstream in the daughter cell. These extensions were later identified (Bratenahl and Yeates, 1970) as slow mode shocks in the Petschek (1964) configuration.

The significance of the separator current system is this: The field of this current system (Fig. 1c) when vectorially added to the field of the source current system (Fig. 1d) increases the flux in the parent cells and reduces the flux in the daughter cell. Thus, relative to the minimum energy potential field (Fig. 1d) of the source currents alone, the parent cells build up a store of excess magnetic flux, and the daughter cell is left deficient in flux. The induced separator current circuit is closed by a return current on the outer oval which flows in the opposite direction from that of the source current return. The total current flowing on the outer oval is thus reduced by an amount which is consistent with the reduced field in the daughter cell.

We may note further, that at least in the DIPD, it turned out that Sweet's neutral current sheet does not develop and there may well exist asymptotic paradoxes in MHD, as originally suspected.

We wish to emphasize the fact that in the basic DIPD design concept, the strategy was to set up conditions to investigate the interpenetration of two independent field systems that differed in the topological connectivity of their field lines as in Sweet's interaction between two bipolar sunspot fields. For reasons of experimental simplicity, however, the model actually used was based on the 2-D example of Dungey (1958, Fig. 4.2). Furthermore, in order to avoid unnecessary interference from wall effects, the design strategy mandated the inclusion of the entire field system and plasma dynamics within the test volume, i.e., no field lines should be permitted to intersect the chamber walls.

The first preliminary results of the DIPD program were reported in 1965 (Bratenahl and Hirsch, 1966). In 1969, the project was moved from JPL to UC Riverside and there the work continued until 1980.

In the meanwhile several other laboratory experiments relating to magnetic reconnection joined the effort. The first of these, in about 1967 at the Lebedev Institute, Moscow, is the work of A. G. Frank and co-workers. Their TS device was designed for the specific purpose of investigating the collapse of a current system into a pinch sheet current at an x-type magnetic neutral line and the possibility of its subsequent disruptive breakup through dynamic instability (Frank, 1976). This pair of experimental objectives provide a direct test of a major part of the theoretical effort of Syrovatskii and theorist colleagues at the Lebedev Institute to determine if magnetic energy, gradually stored in the build-up of the pinch sheet, could be converted rapidly and efficiently into particle kinetic energy upon the disruption of the sheet (Syrovatskii, 1976; Somov and Syrovatskii, 1976; Gerlakh and Syrovatskii, 1976; Bulanov and Syrovatskii, 1976; Frank, 1976, and references in this valuable collection of papers published together as Volume 74 of Proc. P. N. Lebedev Physics Inst., N. G. Basov (ed.) entitled <u>Neutral Current Sheets in Plasmas</u> (see also the posthumous review: Syrovatskii, 1980)).

Note that the motion of dynamic collapse of a current system in the vicinity of an x-type neutral line into a pinch sheet "discharge" originated with Dungey (1953, 1958) and represents a local aspect of a problem which, as we shall see, is complementary to, in fact, coupled to its global aspect, the topological conversion of field line connectivity as depicted in Dungey (1958, Fig. 4.2). Since the latter aspect is the formal basis of the DIPD and the former is the basis for TS, one might suppose the two experiments represent investigations of complementary aspects of the same problem. The truth turns out to be otherwise, however, and, in fact, the intercomparison between DIPD and TS provides much more insight into the complexity of reconnection than might be expected. This is a classic case of "the whole can be much greater than the sum of its parts." It is our purpose here today to make just such a comparison.

Following a description of the basic design and operation of TS, we shall return to the history of reconnection experiments, making, however, only brief mention of the various other projects. More details will be found in a recent review (Baum and Bratenahl, 1980).

TS-1 (and its subsequent versions, TS-2 and TS-3) is basically quite simple in design concept (Fig. 2). An ordinary z-pinch system is fitted with two pairs of external conductors carrying D.C. current in opposite directions. The resulting transverse quadrupole field has an x-type neutral line on the axis of the pinch glass-lined tube. Instead of an induced pinch current as in the DIPD, in TS the rapidly increasing current is directly driven by means of an external capacitor bank. The various versions (TS-1, TS-2, and TS-3) represent different methods of producing the plasma, and variations in plasma and field parameters.

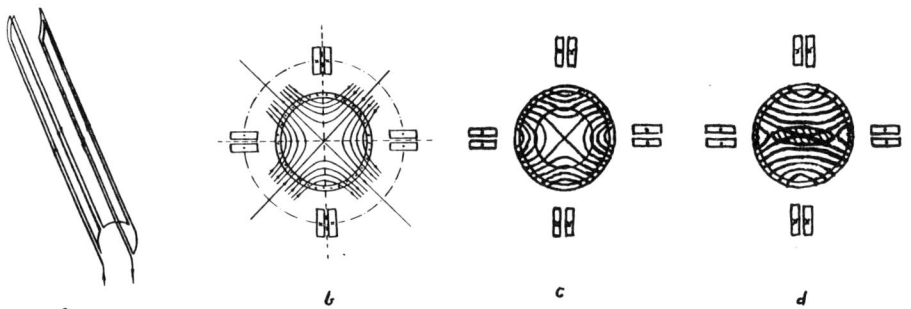

Figure 2. TS: (a) the D.C. quadrupole field conductors; (b) the quadrupole field; (c) converging cylindrical wave and its effect on (b); (d) the developing elliptical pinch current channel. Figs. 2a,b are from Frank (1976, Fig. 7a,b).

The resulting evolution and ultimate form of this directly driven current system is different in nearly all respects from the inductively driven system in the DIPD. Nevertheless, both devices eventually develop an instability leading to the violent tearing or rupturing of their pinch current sheets (Bratenahl and Yeates, 1970; Frank, 1976; Kirii, 1977). We shall return to this intercomparison and its interpretation in Section 3.2.

Continuing now with the historical development, the next to join the family of laboratory reconnection experiments, beginning in about

1969 at the University of Tokyo, was a device somewhat similar in appearance to the DIPD in the sense that the magnetic field is completely contained within the test chamber but operationally the same as the TS concept of a D.C. background field defining a separator line, and a directly driven separator current system (Ohyabu and Kawashima, 1972; Ohyabu et al., 1974). This device also produces the current sheet rupture phenomenon.

Another directly driven reconnection experiment was carried out at MIT (Overskei, 1976; Overskei and Politzer, 1976), while in the meantime, the only other inductively driven device was the "pancake pinch" reported by Dailey (1972).

At the present time, by far the largest and most sophisticated reconnection experiment is that of Stenzel and co-workers at the University of California, Los Angeles, as reported here yesterday (Stenzel, 1985, see also references therein). The UCLA device (we shall call it UCLAD) has an inductively driven separator current system, similar to the DIPD. However, the device is about 10 times larger; it operates in the low density collisionless regime, requiring a hot cathode source to furnish current carriers; the source current rods are replaced by a pair of flat plate conductors; a D.C. axial bias field is needed to confine the plasma more or less to within the space between the plates. As might be expected, UCLAD performs in a way similar to DIPD although there are some noticeable differences, possibly related to the D.C. axial bias field (see Sec. 3 regarding singular lines). In the DIPD, flow is essentially laminar, in UCLAD it is somewhat turbulent. Spontaneous current rupture does not seem to occur (Bratenahl and Baum, 1983; Stenzel, Gekelman and Wild, 1983).

2. THE PROBLEM OF INTEGRATING EXPERIMENT WITH THEORY.

The purpose of the experiments under discussion here is to provide assistance in the development of theory. Unfortunately, there has been widespread skepticism concerning applicability or relevance of laboratory experiments to space and astrophysical problems: the impossibility of accurate scaling; the so-called "wall-effects"; plasma regimes, etc., etc. This basic prejudice persists today, though rapidly diminishing, thanks in large measure to the efforts of Bengt Sonnerup (1979; 1983) who has been actively stirring up interest in a much stronger integrative interaction between theorists and experimentalists. But, the problem of achieving this goal, is also further complicated by some of the concepts and disciplinary approaches that experimentalists must employ and which seem out of place to theorists. The experimentalist is bound to think very often as an electrical engineer: with circuits (albeit plasma conductor circuits which, obviously, behave quite differently from ordinary wires, Sec. 3.1); with inductances and inductive coupling between circuits; with nonlinear resistive impedances; with modes of energy input from outside the global system; with EMF's; with loads; with Faraday's law in integral form, etc., etc.

The experimentalist must also deal with real plasma behavior: for example, plasma reacts to inductive (rotational) electric fields by attempting to cancel them out with their own self-generated space charge (irrotational) fields. This latter process has an extremely important consequence as we shall see in Sec. 3.

It might be helpful here to remind ourselves of the very different relationship, on the one hand, between observations and the development of theories to explain them, and on the other hand, the relationship between the theories and laboratory experiments for the purpose of testing and perhaps even guiding their development. An adequate reconnection theory should be consistent with both space as well as laboratory reconnection phenomena, differences in scaling and wall effects notwithstanding.

2.1 Scaling, Wall Effects, and Global Boundary Conditions.

As to scaling, the best that can be done follows Podgorny's "principle of limited simulation" (Podgorny and Sagdeev, 1970): compared to unity, dimensionless parameter ratios that are large or small in space should be made, correspondingly as large or small as possible in the laboratory, ratios of order one in space should be of order one in the laboratory.

As to wall effects, it turns out that if due consideration is given them, they need not be as detrimental as is generally believed. The study of wall effects, in actuality, is the study of the influence of global boundary conditions (GBC). Contrary to widespread belief, such conditions are always present in the astrophysical context although in theoretical development they are nearly always ignored, inadequately accounted for or for mathematical convenience, replaced by arbitrarily chosen boundaries. This tradition, which even includes much of Syrovatskii's work goes back to Dungey and Sweet not withstanding both of these pioneers in the subject realized that reconnection "is in principle a global process within a topologically complex structure" to quote Vasyliunas (1975) in his thorough going review of this so-called "restricted problem" mode of analysis.

Perhaps the most important of all insights we have been able to glean from our experience with laboratory reconnection experiments is that the restricted problem, despite quite logical reasoning, is incompletely posed (Forbes and Speiser, 1979) and can not furnish unambiguous answers to some of the more important questions raised, for example, by Sonnerup (1979) in his excellent summary of the present status of reconnection theory. The reason behind this, as we shall shortly see, has to do with the strong coupling between, on the one hand, the overall field topology and its time-dependence which introduces an essential nonlocal character to reconnection processes (Sec. 3), and on the other hand, the local conditions prevailing on the separator line which play a role of comparable importance.

A prime example of GBC is that which is determined by the solar photosphere and lower chromosphere: the permanent boundary separating the very high mean beta regime below from the low mean beta regime above (Dubov, 1971). Observational evidence suggests that down below, the high density plasma confines the magnetic field to very high field fibers, leaving most of the space occupied by essentially field-free plasma (Zwaan, 1978). In contrast, up above, strong currents must satisfy the force-free condition $\nabla \times B = \alpha B$ or else be confined to sheet current pinches, leaving most of the space occupied by nearly curl-free (potential) magnetic field. There is some strong observational evidence (Pneuman et al., 1978) for this latter condition.

Crossing the boundary from below, the tiny but intense field fibers fan out (Gabriel, 1976; Kopp and Kuperus, 1968; Dubov, 1971) to produce a continuous field above. The fibers, however, define the footpoints of field lines making up topologically distinct tubular flux cells. As these are shuffled about by the powerful convective motion below, EMF's are generated (Baum, et al., 1978, 1979; Bratenahl andd Baum, 1982) which drive the lively solar activity above, as Ionson (1985) mentioned yesterday. On this basis, we expect reconnection in pinch current sheets to be by far the dominant dynamic dissipation mode in the solar atmosphere. Indeed, we have been lead to propose, following Babcock (1961), a two-level solar dynamo model in which this ubiquitous reconnection dynamics becomes the essential factor in the alpha-process (Bratenahl et al., 1980).

3. THE NONLOCAL ASPECT OF RECONNECTION

If, under any of the following conditions, a nonzero line integral of an electric field exists: (i) around a closed magnetic field line, (ii) along the length a field line, both ends of which pass through a rigid conductor; (iii) along the length of a null line of either x- or o-type; or (iv) along the length of a separator as defined in a reconnection geometry, then such a line possesses singular properties (Syrovatskii, 1977). For instance, in infinite conductivity (ideal) MHD there would be "jump-like singularities in the magnetic field and correspondingly infinite current density" (Syrovatskii, 1978a), and a complete breakdown in the frozen-in flow condition. At such lines, conditions at arbitrarily high conductivity do not connect asymptotically to conditions at infinite conductivity. Here, at last, is theoretical support for one of our suspicions that lead to the DIPD program.

However, the most important aspect of singular separator lines, enters the picture under realistic conditions of finite conductivity. The fact is, the line integral of E in question is the Faraday integral satisfying

$$\int \vec{E} \cdot d\vec{\ell} = - \frac{d\phi}{dt} \tag{1}$$

in which ϕ is the magnetic flux encircled by the closed circuit path of the separator current (for example, the daughter cell in the DIPD). In the Coulomb gauge (Cragin and Heikkila, 1981),

$$\vec{E} = -\frac{\partial \vec{A}}{\partial t} - \vec{\nabla}\phi, \qquad (2)$$

the first term on the R.H.S. is the induction (rotational) electric field, E^i and the second term is the electrostatic (space charge) electric field E^s. Along a singular line, although the condition

$$\vec{E} = \vec{E}^s + \vec{E}^i \simeq 0 \qquad (3)$$

may hold locally due to cancellation of E^i by an equal but opposite E^s, the nonlocal conditions:

$$\int \vec{E} \cdot d\vec{\ell} = \int \vec{E}^i \cdot d\vec{\ell} \neq 0 \qquad (4)$$

$$\int \vec{E}^s \cdot d\vec{\ell} \equiv 0 \qquad (5)$$

must prevail for integrals extended over the whole length of such lines whenever $d\phi/dt \neq 0$ as in eq. (1) (Heikkila et al., 1979). This obviously implies that if eq. (3) holds over most of a singular line, then over the remaining portion, E_i^s must be reversed in direction, i.e., must be in the same direction as E^i and be large enough to satisfy eq. (5) (Baum and Bratenahl, 1980b). If this situation should occur in actuality, then we must be dealing either with a double layer or, equivalently with a large electrode sheath drop. The latter situation was clearly observed in the DIPD by Beeler (1979), in association with a sudden, explosively fast reconnection event we had previously, but quite erroneously, attributed (Bratenahl and Yeates, 1970; Baum et al., 1973) to the sudden development of anomalous (turbulent) resistivity (more about this in Sec. 3.3).

Since the nonlocal aspect of reconnection is related to inductive electric fields E^i, $= -\partial A/\partial t$, its importance arises in time-dependent processes. Unfortunately, time-dependent reconnection has been a very much neglected subject in theoretical work. In truth, however, steady state reconnection may be said to be a special (perhaps degenerate) case of the general problem in which, not only must the process compress, heat and accelerate plasma, but it must also produce a measurable redistribution of magnetic flux between the distinct field line mapping cells defined by the separatrix. The latter process known as flux transfer, does not occur in steady state reconnection (Alfven, 1976). Flux transfer involves Faraday's law in integral form and cannot be properly handled without taking into account the overall field line topology and the form of the current circuits as well.

Flux transfer is caused by changes in the strengths or geometrical arrangement of the various externally driven source currents. These changes on the global boundary produce an EMF along the separator which, in magnitude and direction is, to a good approximation, the line integral of $\vec{E}^1$ that would have been induced there in the total absence of any plasma.

In effect, the system acts like a transformer (Kaburaki, 1975; Bratenahl and Baum, 1982). Consider the DIPD (Fig. 3). The primary consists of the two source currents acting together, and the secondary is the induced pinch sheet current circuit. The EMF driving the secondary is the rate of change of flux coupling primary to secondary, expressed as the product of the primary current and the mutual inductance, between primary and secondary: $I_p M_{ps}$.

The line integral of E on the separator is now the effective "IR" drop or load. It is also, by equation 3, the flux transfer rate. However, the effective "R" need not be a resistance in the usual sense, since, as already noted, it may include a double layer or sheath drop ϕ. This may be expressed as $\int \vec{E} \cdot \vec{d\ell} = I_s R_s + \phi$. Actually, the problem is somewhat more complicated because we are not dealing here with ordinary wire conductors threading a nonconducting medium but current flow in a continuous conducting medium. A special process of development of build-up of the sheet current pinch in this situation will shortly be considered.

Nevertheless, the field of this secondary current circuit as noted in Section 1.2, adds to the primary or source field in the parent cells and subtracts from that field in the daughter cell. In effect, some flux is withheld from the transfer process and stored. The amount of this stored flux is the product of separator current and the self inductance of the current path, $I_s L_s$.

On the basis of the preceding, we can write

$$\dot{I}_s L_s + (I_s R + \phi_s) = \dot{I}_p M_{ps} \quad (6)$$

$$\dot{I}_p L_p + I_p R_p = \Sigma - \dot{I}_s M_{ps}, \quad (7)$$

in which Σ is the externally applied EMF driving the primary circuit. Important facts to note are these: (i) the flux storage rate $\dot{I}_s L_s$ is the difference between the EMF, ($\dot{I}_p M_{ps}$) which drives the secondary (separator circuit) and the flux transfer rate, ($IR + \phi_s$), so that a flux storage reservoir capacity is defined by the asymptotic state, $\dot{I}_p M_{ps} - (I_s R_s + \phi_s) = 0$. Thus the integrity of the reservoir, as it is filled, depends on the ability of the conduction mechanism to support the required current without failure, i.e., either a sudden increase in R_s (anomalous resistivity) or a sudden increase in ϕ_s (double layer or sheath) or both. But if the conduction mechanism can support the required saturation current, the flux transfer rate from a completely filled reservoir is identical to that which would occur in the total absence of a conducting plasma medium. This might appropriately be called the Yeh-Axford theorem (Yeh and Axford, 1970). Note carefully, this concept could hardly have been rigorously derived from the restricted problem approach.

For a given $\dot{I}_p M_{ps}$ (neglecting ϕ_s), the flux storage capacity increases with conductivity and the stored energy capacity $I_s L_s/2$, as the squares of the conductivity. Thus, at high conductivity and significant EMF's we can expect catastrophic flux tranfer events due to one or another form of conduction mode instability.

3.1 Build-up of Pinch Current Sheets.

As electromagnetic energy (Poynting flux) enters a plasma system through some portion of its global boundary, current is initially induced to flow over the path of least inductance, which is the layer of plasma adjacent to that energy-entrance portion of the boundary. Then, as the Lorentz force accelerates the plasma layer to the local drift velocity V_d = E/B and becomes thereby electrostatically polarized, the current path advances into the interior as an elementary hydromagnetic wave, accelerating the successive layers of plasma through which it propagates. Each subsequent increment of current propagates into the plasma in a similar fashion, the ensemble of all such waves constituting a system.

In a laboratory device, the nature of the rigid (non-plasma) boundaries impose conditions on $E_{\parallel}$ which must be considered. Where the boundary is a time-varying current-carrying conductor, $E_{\parallel} = E^i + E^s = 0$. An exposed conductor boundary therefore cannot serve as an entrance window for electromagnetic energy. Only if the conductor is completely covered with a solid dielectric (insulator) can that portion of the boundary permit the entrance of electromagnetic energy. In this case, the surface of the dielectric, easily and quickly acquires charge from the plasma sufficient to cancel the "E^s-effect" of induced surface charge on the conductor, and the necessary condition $E_{\parallel} = E^i \neq 0$ for energy entrance is established.

If the field topology is such as to lead to the development of a (singular) separator line, or if one is already present in the system, the elementary current carrying waves gradually build up, through a nonlinear process, a current channel structure whose cross-sectional form depends significantly on the amplitude of the waves coming in from each direction. It also depends on the somewhat lower amplitude outgoing waves, if present, moving toward portions of the global boundary that act as energy sinks. All this, of course, depends on details of the initial structure of the magnetic field, and the changes in it that are taking place consistent with the changing source currents. In other words, it depends intimately on the initial conditions and global boundary conditions.

3.2 Intercomparison between DIPD and TS.

The development of Petschek shocks evidently must depend in a very sensitive way upon the details of the plasma flow field that is

developed by the passage of the small amplitude wave system (Sec. 3.1).
In the DIPD, this wave system diverges from the two glass-insulated
source current rods and where they run into each other at points other
than precisely at the separator, the Lorentz forces from both wave
systems cooperate to accelerate the plasma into a pair of strong outflow
structures. With no adverse interference of incoming waves in the
outflow sectors, standing waves can develop nonlinearly into slow mode
shocks (much like the bow waves of two speed boats about to collide
head-on). For this to occur, however, the outflow velocity must develop
selfconsistently to a value greater than the slow mode wave speed. The
overall flow, development in the DIPD is laminar.

In contrast to the foregoing, the TS experiments produce radially
inward propagating current-carrying waves (Gerlakh and Syrovatskii,
1976) which are uniformly induced over the inner surface of the
cylindrical glass-insulated outer boundary. As these waves converge
upon the central axis (as in a z-pinch), the interaction term in the
Lorentz force (that which depends on the D.C. quadrupole field)
increases the compressive pinch from the inflow sectors and weakens that
force in the outflow sectors so that the current channel takes on the
forms of a flattening ellipse. In this case the outflows are
essentially blocked by net inward pinch forces. No Petschek shocks are
possible under these circumstances. It should be noted, however, that
this collapse of the current system into a pinch current sheet met
precisely one of the stated objective of the TS experiments. The second
objective, namely the instability of the sheet pinch was met in TS-3
through reduced initial density and increased field. In TS-3 the pinch
sheet ruptures impulsively into two magnetic islands through some form
of tearing mode. The rupture occurring at the separator line (Frank,
1976).

It is an important fact, however, that the qualitatively different
pinch structure that develops in the DIPD also ruptures in an impulsive
event, as we shall now describe in greater detail.

3.3 Impulsive Flux Transfer Events.

One of the earliest observations in the DIPD was the occurrence of
a sudden and major change in the magnetic field and current. On the
basis of detailed investigation, the phenomenon was given the
descriptive identification: impulsive flux transfer event. IFTE is
analogous to the bursting of a dam (Figs. 3b, 4a,b). The induced
separator current system, consisting of the central hyperbolic pinch
together with the attached Petschek shocks is observed to break up and
fly away in all directions with little or no change in the total current
flow. Large amplitude compressive blast waves propagate downstream in
the daughter cell while large amplitude expansive fast mode waves
propagate upstream toward the source current rods in the two parent cells
(Fig. 4a).

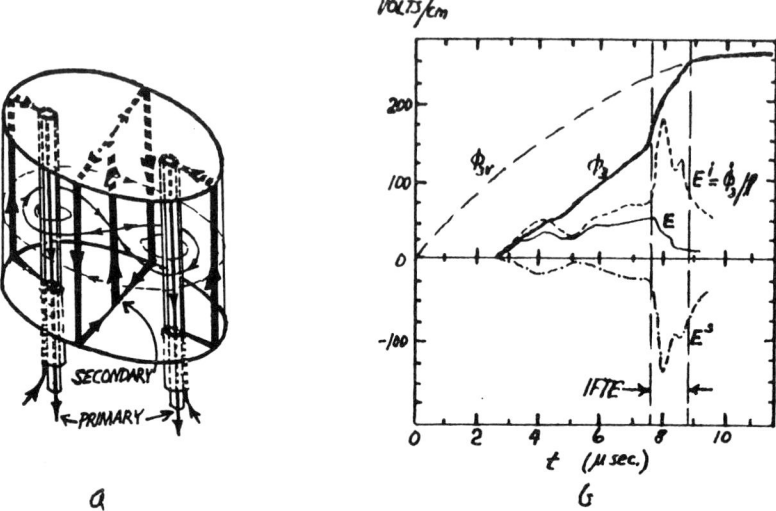

Figure 3. Flux and electric field relationships in the DIPD: (a) schematic view of the transformer basis of Eqs. (6) and (7); (b) long dash; $M_{ps} \int I_s dt$, Eq. (7), heavy solid line, daughter cell flux, $I_s L_s$, short dash $E^i(t)$; dash dot, $E^s(t)$; thin solid, $E(t)$.

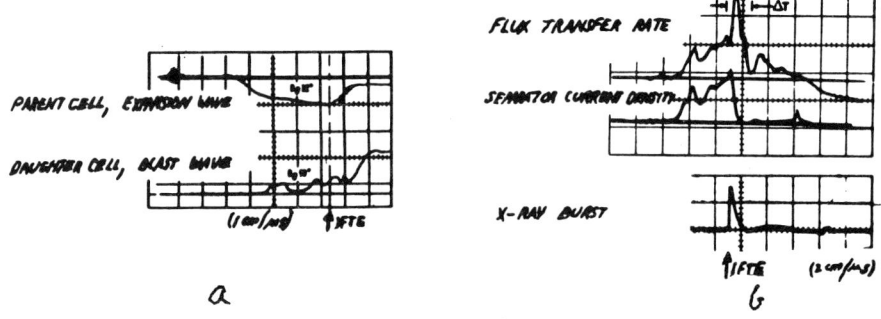

Figure 4. IFTE: (a) magnetic field behavior showing upstream expansion waves and downstream blast waves; (b) $E^i(t)$, $j_s(t)$, and x-ray burst.

The constancy of the total current during IFTE means that the growth of current in the propagating waves is at the expense of the current in the original pinch current system. By Ampere's law, this precisely identifies the rapid intrusion of transferred flux into the daughter cell and the corresponding rapid withdrawal of flux from the parent cells.

IFTE is caused by a conduction mode instability. Until quite recently, all the evidence seemed to indicate a sudden transition to anomalous resistivity: first, the initiation of runaway carriers, evidenced by observation of x-rays produced at the anode end of the separator; and then followed by ion-acoustic noise, accompanying the impulsive event. However, Beeler (1979) discovered that anomalous resistivity was not the principal factor at all. By an ingenious method of separately measuring E^i, E^s, and E at points distributed over 80% of the separator (Fig. 5), he was able to show that: (1) there is no measurable change in resistivity, $\eta = E/j$; (2) that E^s and E^i are large but nearly cancel each other out over the region of measurement. Now recalling eqs. (3), (4) and (5), and the comments following in Sec. 3, it becomes quite clear that the conduction mode instability must result in, or take the form of, the sudden development of a large space charge sheath at one of the electrodes. This is analogous to a double layer in its effects (eq. 6), including the x-ray production through energization in the large potential drop ϕ_s. (Energetic ions have also been observed).

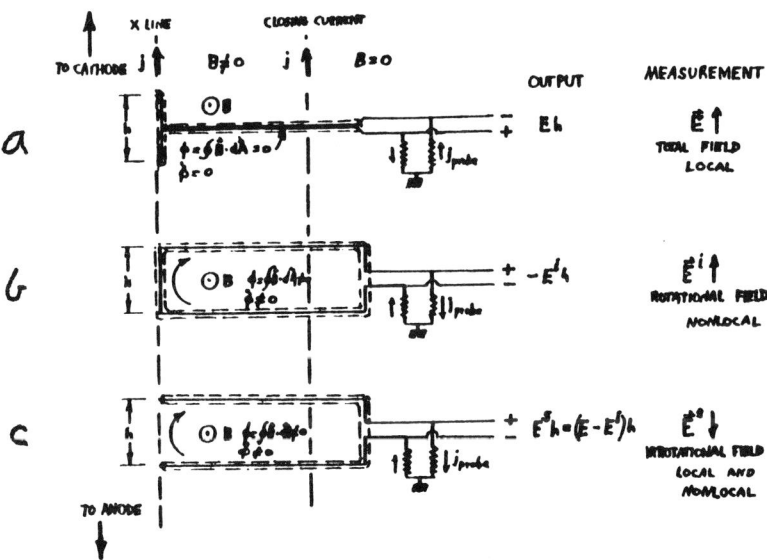

Figure 5. Beeler (1979) method of separately measuring E, E^i and E^s: (a) double probe with close-spaced leads; (b) flux look probe; (c) double probe with separated leads. This system works only for B-fields with 2-D translational symmetry.

Note that an impulsive increase in resistive impedance ($I_s R + \phi_s$) on the separator current channel due to either R or ϕ_s produces an impulsive increase in E^i (Figs. 3b,) and by eq. (4), the flux transfer rate. The resulting large amplitude wave system is the (nonlinear) inverse of the build-up mechanism of Sec. 3.1. Note further, this process provides a very plausible setting for the current interruption solar flare model (Baum et al., 1978; Baum and Bratenahl, 1982) of Alfven and Carlquist (1962) and Carlquist (1982).

Obviously, during the preceding current buildup, rapid Petschek-type reconnection, including flux transfer, was taking place. Evidently also, this proces was not fast enough to compete with the simultaneous flux and energy storage buildup as represented by the ever-increasing pinch current system, I_p. Then, bang! Could this be a more realistic version of Sweet's solar flare model? We certainly can believe it is. It goes "bang" because the demand for current carriers inevitably grows faster than the rate of supply. Considering the inhomogeneous structure of the solar atmosphere: higher density plasma in thin current sheets increasing in strength with time and which are separated by large spaces consisting of lower density plasma in nearly curl-free field, the DIPD process simulation may not be so far off base.

In contrast to the DIPD, it should be noted that in TS, because the quadrupole background field circuit (constant in time) is not inductively coupled to the pinch current circuit, no flux transfer takes place during the process of current build-up. Indeed, parent and daughter cells are not even completely defined within the test volume. However, current rupture into a pair of magnetic islands may certainly be identified as an IFTE, and, as in the case of the DIPD, it too is caused by some form of conduction mode instability. (The precise cause is not known to us at this writing).

This raises an important issue: the resolution of a total B-field into the contributions of the major current circuits, some of which change while others remain constant, may provide a much more advantageous strategy for analysis of reconnection than the not so reliable but traditional strategy of "moving field lines, frozen into the flow." This is especially true because of the presence of singular lines (e.g., separator lines) where the freezing-in concept breaks down anyway.

4. TOWARD A BETTER UNDERSTANDING OF SOLAR ACTIVITY

Why, might you ask, this preoccupation with Sweet's interaction between simple bipolar structure rather than the popular notion among theorists of twisted force-free fields? We would answer simply as follows: There appears to be no reasonable mechanism, either observationally suggested or theoretically proposed within constraints imposed by observations, capable of systematically twisting high field flux fibers or bundles of them to the extent required by the force-free field concept beyond occasional occurrences. On the other hand, based on both topological arguments involving flux conservation as well as overwhelming abundance of direct observational evidence, all new flux erupts upward through the photospheric boundary as bipolar structures (Sheely et al., 1975). The low mean beta condition above does not restrict large currents to flow only in force-free field structures satisfying $\nabla \times B = \alpha B$. The concept that all strong currents in the solar atmosphere must be of this force-free type is merely an assumption which seems to overlook the other possibility. In fact, pinch current sheets, in otherwise very nearly curl free field, are also entirely compatible with the low mean beta regime, and on the basis of evidence are much more likely to be the dominant currrent structure in the solar atmosphere, and yet, we know almost nothing about the structural transformation of a 3-D separator-separatrix system (Sweet, 1978; Baum and Bratenahl, 1980) in the presence of a sheet current pinch. Note, the current on such a separator is strictly field aligned, nevertheless, is not a force-free structure in the traditional sense.

A recent paper in Solar Physics (Fig. 18) (Machado et al., 1983), is quite relevant here. We believe this is the first time the 3-D separator-separatrix structure, originally proposed by Sweet (1958), has been explicitly considered other than in several papers by our colleagues and ourselves (Bratenahl and Baum, 1976b; Baum et al, 1978a, 1979; Baum and Bratenahl, 1980, 1982).

5. SUMMARY

Among the numerous insights derived from laboratory reconnection experiments, we believe the following to be the most significant: (i) "Reconnection," as the term ought to imply, but unfortunately through long usage in a different sense, does not seem to, is fundamentally the time-dependent process of topological conversion of field line connectivity or flux transfer which, however, must draw upon electromagnetic energy to compress, heat, accelerate, and otherwise energize any plasma that gets in its way. Steady state reconnection might more properly be called "merging," as in the "confluence of two streams of plasma:" it is a special limiting case of the more general one, in as much as topological conversion through flux transfer is entirely missing. (ii) Topological conversion necessarily involves a nonzero line integral of E^i along the separator, and this introduces a nonlocal aspect into the problem which tightly couples the global boundary conditions to the local condition in the neighborhood of the separator. Since steady state reconnection must be regarded as an asymptotic state, derivable from the general case, it must surely contain information uniquely determined by GBC and initial conditions. The restricted problem approach is therefore incompletely posed. (iii) The failure of the assumption of a neutral sheet formation by Sweet (1958a) illustrates the dangers of deriving finite conductivity solutions by simply adding resistive diffusion effects to solutions based on ideal MHD. (iv) The hydromagnetic wave process by which pinch current sheets are built up (or destroyed, as in IFTE) depends intimately on the way energy enters the system on its global boundaries, as well as the presence of any preexisting magnetic fields. Therefore, analysis in terms of the superposition of fields belonging to the individual current circuits provides a more powerful approach than the unreliable methods of "moving field lines, frozen into the flow." (v) The unquenchability of E^i along singular field lines and the difference in plasma response to E^i vs E^s, provides an obvious mechanism to produce double layers, and thereby hoists a cautionary flag against the customary and unqualified assumption that anomalous (turbulent) resistivity is the only form or even the most important form of over-current instability.

6. CONCLUSIONS

We have tried to make it clear that laboratory experiments can provide valid testing and guidance in the development of reconnection theory even when that theory is directed toward understanding phenomena in astrophysics. A valuable resource already exists which theorists should now take greater advantage of. We are convinced that important ideas have been presented. For instance, the time has come to seriously challenge the unqualified assumption that the force-free field structure obeying $\nabla \times B = \alpha B$ is mandated for strong currents in the solar atmosphere. An alternative structure, the pinch current sheet is not only permissable, but on the basis of observations seems much more likely.

ACKNOWLEDGMENTS

Without the support at various times of NASA, C.I.T., AFOSR, and NSF, the DIPD program would never come into existence and survived over many years. The production of this paper was supported in part by travel funds made available through the auspices of I.A.U. Symposium 107. It represents a last minute substitution, expressing the authors' views on a subject of mutual interest to Mrs. Ann G. Frank and co-workers at the Lebedev Institute Moscow. Had Frank been able to attend this Symposium as originally planned and expected, her views on this subject might well have been differently expressed.

REFERENCES

Alfven, H.: 1976, J. Geophys. Res. 81, pp. 4019-4021.
Alfven, H. and Carlquist, P.: Solar Phys. 1, pp. 220-228.
Babcock, H.W.: 1961, Astrophys. J. 133, pp. 572-587.
Baum, P.J. and Bratenahl, A.: 1974, Phys. Fluids 17, pp. 1232-1235.
Baum, P.J. and Bratenahl, A.: 1976, Solar Phys. 47, pp. 331-334.
Baum, P.J. and Bratenahl, A.: 1980, Adv. Electron. Electron Phys. 54, (Adademic Press, Inc.), pp. 1-67.
Baum, P.J. and Bratenahl, A.: 1980b, Rept. 80-01, Inst. Geophys. Planet. Phys., Univ. of Calif., Los Alamos, NM.
Baum, P.J. and Bratenahl, A.: 1980a, Solar Phys. 67, pp. 245-258.
Baum, P.J. and Bratenahl, A.: 1982, in R.E. Lingenfelter, H.S. Hudson, D.M. Worrell (eds.), Gamma Ray Transients and Related Astrophysical Phenomena, Amer. Inst. of Physics, NY, pp. 433-442.
Baum, P.J., Bratenahl, A., and White, R.S.: 1973, Phys. Fluids 16, pp. 226-230.
Baum, P.J., Bratenahl, A., Kao, M., and White, R.S.: 1973, Phys. Fluids 16, pp. 1501-1504.
Baum, P.J., Bratenahl, A., and Kamin, G.: 1978, Ap. J. 226, pp. 286-300.

Baum, P.J., Bratenahl, A., Crockett, G., and Kamin, G.: 1979, Solar Phys. 62, pp. 53-67.
Beeler, R.G., Jr.: 1979, Ph.D. Dissertation, Univ. of Calif., Riverside, CA.
Birkhoff, G.: 1950, Hydrodynamics, A Study in Logic, Fact and Similitude, Princeton University Press.
Bratenahl, A. and Baum, P.J.: 1976a, Solar Phys. 47, pp. 345-360.
Bratenahl, A. and Baum, P.J.: 1976b, Geophys. J. R. Astr. Soc. 46, pp. 259-293.
Bratenahl, A. and Baum, P.J.: 1982, EOS Trans., Am. Geophys. U. 63, p. 1063.
Bratenahl, A. and Baum, P.J.: 1983, J. Geophys. Res. 88, pp. 503-505.
Bratenahl, A. and Hirsch, W.: 1966, Bull. Am. Phys. Soc. 11, p. 580.
Bratenahl, A. and Yeates, C.M.: 1970, Phys. Fluids 13, pp. 2646-2709.
Bratenahl, A., Baum, P.J., and Adams, W.M.: 1980, in M. Dryer and E. Tandberg-Hanssen (eds.), Solar and Interplanetary Dynamics, IAU Symp. 91, pp. 29-32.
Bobrova, N.A. and Syrovatskii, S.I.: 1979, Solar Phys. 61, pp. 379-387.
Bulanov, S.V. and Syrovatskii, S.I.: 1976, in N.G. Basov (ed.), Neutral Current Sheets in Plasmas, Proc. (Trudy) P.N. Lebedev Physics Inst., Vol. 72, Consultants Bureau, NY, pp. 87-106.
Carlquist, P.: 1982, Astrophys. and Space Sci. 87, pp. 21-39.
Cragin, B.L. and Heikkila, W.J.: 1981, Rev. Geophys. and Space Phys. 19, pp. 223-229.
Dailey, C.L.: 1972, TRW Interim Sci. Prog. Rept. for AFOSR Contract, TRW Corp., El Segundo, CA.
Dubov, E.E.: 1971, Solar Phys. 18, pp. 43-59.
Dungey, J.W.: 1953, Philos. Mag. 44, pp. 725-738.
Dungey, J.W.: 1958, Cosmic Electrodynamics, Cambridge University Press.
Forbes, T.G. and Speiser, T.W.: 1979, J. Plasma Phys. 21, pp 107-126.
Frank, A.G.: 1976, in N.G. Basov (ed.), Neutral Current Sheets in Plasmas, Proc. (Trudy) P.N. Lebedev Physics Inst., Vol. 74, Consultants Bureau, NY, pp,. 107-163.
Furth, H.P., Killeen, J., and Rosenbluth, M.N.: 1963, Phys. Fluids 6, pp. 459-484.
Gabriel, A.H.: 1976, Philos. Trans. R. Soc. London Ser. A. 281, pp. 339-352.
Gerlakh, N.I. and Syrovatskii, S.I.: 1976, in N.G. Basov (Ed.), Neutral Current Sheets in Plasmas, Proc. (Trudy) P.N. Lebedev Physics Inst., Vol. 74, Consultants Bureau, NY, pp. 73-86.
Heikkila, W.J. and Pellinen, R.J.: 1977, J. Geophys. Res. 82, pp. 1610-1614.
Heikkila, W.J., Pellinen, R.J., Falthammar, C.-G., and Block, L.P.: 1979, Planet Space Sci. 27, pp. 1383-1389.
Ionson, J.A.: 1985, IAU Symp. 107 (this volume).
Kawashima, N., and Ohyabu, N.: 1971, Inst. Space and Aero. Sci., University of Tokyo, Rept. No. 464, Vol. 36, No. 6, pp. 175-185.
Kaburaki, O.: 1975, Tohoku University Science Rept. Ser. 1, Vol. 43, pp. 141-149.
Kirii, N.P., Markov, V.S., Frank, A.G., and Khodzhaev, A.Z.: 1977, Sov. J. Plasma Phys. 3, pp. 303-306.
Konopinski, E.J.: 1981, Electromagnetic Fields and Relativistic Particles, McGraw-Hill Book Co.

Kopp, A.K. and Kuperus, M.: 1968, Solar Phys. 4, pp. 212-223.
Machado, M.E., Somov, B.V., Bobrova, M.G., and De Jager, C.: 1982, Solar Phys. 85, pp. 157-184.
Ohybu, N. and Kawashima, N.: 1972, J. Phys. Soc. Japan 33, pp. 496-501.
Ohyabu, N., Okamuru, S., and Kawashima, N.: 1974, Phys. Fluids 17, pp. 2009-2013.
Overskei, D.: 1976, Ph.D. Dissertation, Mass. Inst. of Tech., Cambridge, MA.
Overskei, D. and Politzer, P.: 1976, Phys. Fluids 19, p. 683.
Petschek, H.E.: 1964, in W.N. Hess (ed.), AAS-NASA Symposium on the Physics of Solar Flares, NASA SP-50, U.S. Govt. Printing Office, Washington, DC, pp. 425-439.
Pneuman, G.W., Hansen, S.F., and Hansen, R.T.: 1978, Solar Phys. 59, pp. 313-330.
Podgorny, I.M., and Sagdeev, R.Z.: 1970, Sov. Phys. Uspekhi 98, pp. 445-462.
Priest, E.R. and Raadu, M.A.: 1975, Solar Phys. 43, pp. 177-188.
Sakurai, T. and Uchida: 1977, Solar Phys. 52, pp. 397-416.
Sheeley, N.R., Bohlin, J.D., Bruekner, G.E., Purcell, J.D., Scherrer, V.E., and Towsey, R.: 1975, Ap. J. (Lett.) 196, pp. 129-131.
Sonnerup, B.U.O.: 1979, in L.T. Lanzerotti, C. F. Kennel, and E.N. Parker (eds.), Solar System Plasma Physics, Chapt. III.1.2, pp. 45-108.
Sonnerup, B.U.O.: 1983, Report of the Working Group on Reconnection. To be published in Proc. NASA Solar Terrestrial Physics Workshop, Berkeley Springs, WV.
Somov, B.V. and Syrovatskii, S.I.: 1976, in N.G. Basov (ed.), Neutral Current Sheets in Plasmas, Proc. (Trudy) P.N. Lebedev Physics Inst., Vol. 74, Consultants Bureau, NY., pp. 13-71.
Stenzel, R.L.: 1985, IAU Symp. 107 (this volume, see also extensive references herein).
Stenzel, R.L., Gekelman, W., and Wild, N.: 1983, J. Geophys. Res. 88, p. 507-508.
Sweet, P.A.: 1958a, Nuovo Cimento Suppl. 8, pp. 188-196.
Sweet, P.A.: 1958b, in B. Lehnert (ed.), Electromagnetic Phenomena in Cosmical Physics, IAU Symp. No. 6, Cambridge Univ. Press, London and New York, pp. 123-134.
Sweet, P.A.: 1969, Ann. Rev. Astron. Ast. 2, pp. 149-176.
Syrovatskii, S.I.: 1976, in N.G. Basov (ed.), Neutral Current Sheets in Plasmas, Proc. (Trudy) P.N. Lebedev Physics Inst., Vol. 74, Consultants Bureau, NY, pp, 1-11.
Syrovatskii, S.I.: 1977, Sov. Phys. Lebedev Inst. Rep. (Engl. Transl.) 5, pp. 9-12.
Syrovatskii, S.I.: 1978a, Solar Phys. 56, pp. 3-12.
Syrovatskii, S.I.: 1978b, Solar Phys. 58, pp. 89-94.
Syrovatskii, S.I.: 1981, Ann. Rev. Astron. Ast. 19, pp. 163-229.
Syrovatskii, S.I., Frank, A.G., and Khodzhaev: 1973, Sov. Phys. Tech. Phys. 18, pp. 580-586.
Van Hoven, G.: 1976, Solar Phys. 49, pp. 95-116.
Vasyliunas, V.M.: 1976, Rev. Geophys. Space Phys. 13, pp. 303-336.
Withbroe, G.L. and Noyes, R.W.: 1977, Ann. Rev. Astron. Astrophys. 15, pp. 363-387.

Yeh, T. and Axford, I.: 1970, J. Plasma Phys. 4, pp. 207-339.
Zwaan, C.: 1978, Solar Phys. 60, pp. 213-240.

DISCUSSION

Priest: What are the values of the magnetic Reynolds number and plasma beta in your experiment and A. Frank's experiment?

Bratenahl: In the DIPD the magnetic Reynolds number is $\sim$16 when based on the distance from the source current rods to the separator line. This being greater than one, the principle of limited simulation is satisfied. In the TS series of A. Frank the Reynolds number is probably larger. The beta in both experiments is much less than unity when averaged over the system.

Kundu: I believe in one of your viewgraphs, you showed an X-ray burst in time coincidence with inductive flux. My question is, what is the energy range of the X-ray burst and what is the time scale?

Bratenahl: The X-ray burst is thick target bremsstrahlung due to the impact on the copper anode of $\sim 2(10)^3$ electrons in the energy range 1 - 2.5 keV. The burst duration $\Delta t \sim 0.25$ μsec is estimated from the observed pulse width of 0.4 μsec and the instrumental width of ~ 0.3 μsec. The burst occurs at the onset of the IFTE which has a duration of ~ 1 μsec (Baum et al., 1973).

Smith: What is the role of ion-acoustic turbulence in impulsive flux transfer events?

Bratenahl: Evidence developed by Beeler (1979) makes it now quite certain that ion-acoustic turbulence is not sufficiently intense to cause a measurable increase in resistivity anywhere over 80% of the length of the separator. On the contrary, the evidence indicates the sudden development of a large electrode sheath drop (similar to a double layer). On the basis of X-ray data (Baum et al., 1973), we can be pretty sure it is an anode sheath. However, the sudden appearance of the turbulence may be related to the sudden development of the sheath.

Mullan: Under what conditions do Petschek shocks form?

Bratenahl: This is a very important question, and at present there is no definitive answer. However, we do know that the presence of shocks of any kind in a quasi-steady flow requires special conditions both upstream and downstream in order to satisfy the jump conditions. The flow must be able, self-consistently, to evolve asymptotically into such a state from some initial state via a transient process. Boundary and initial conditions hold the key. To develop Petschek shocks, a sufficiently large outflow velocity must be able to develop, unimpeded by adverse pressure or Lorentz stresses (to avoid choking as in nozzle flow). In the DIPD, at low enough initial pressure, Petschek shocks are observed (Bratenahl and Yeates, 1970) but not at higher pressures (Beeler, 1979). In the TS experiments of Frank et al., the boundary conditions impose adverse stresses on the outflow and Petschek shocks do not form.

Sonnerup: Is the transition from a compressed flat current sheet to a potential field in an IFTE accomplished by a wave front (a double layer) traveling from the anode along the separator?

Bratenahl: The answer is both yes and no: The sheath remains attached to the anode, but, as I have already explained, IFTE is like the bursting of a dam. Large amplitude blast waves propagate downstream forming the front of transferred flux previously stored upstream, and large amplitude fast mode expansive waves travel upstream as the stored flux is withdrawn in the transfer process. The corresponding changes in field strength must also propagate as wave structures along the separator and away from the sheath region, but this three-dimensional aspect remains to be explored.

Stenzel: Do you have any direct experimental evidence for the existence of a double layer during IFTE's?

Bratenahl: No, on the contrary, the evidence (Beeler, 1979) points to the sudden development of an electrode sheath. But let us not forget that such a sheath produces effects similar to a double layer, especially when in the presence of a strong inductive electric field.

Vasyliunas: The principle of gauge invariance implies that electrostatic and induced electric fields cannot be separately measured as physical quantities; they are mathematical constructs, possibly convenient, but not physically significant.

Bratenahl: I am most grateful that you raise this very fundamental issue. My last figure shows how Beeler (1979), with three different probes did, in fact, separately measure the electrostatic and induced electric fields and their vector sum, the total field. All three, of course, are local quantities, but separate resolution of the electrostatic and induction contributions to the total field requires non-local path integral measurements of appropriate paths that require prior knowledge of the magnetic field line structure together with the assumption of negligible displacement current. However, this fact in no way can be taken to imply that the local quantities, thus obtained, are mere mathematical constructs that cannot be separately measured as valid physical quantities. These quantities, of course, correspond to the Coulomb gauge, but for mathematical convenience, the two contributions to the total field may be transformed, say into the Lorentz gauge (Cragin and Heikkila, 1981).

I would strongly urge those of you here who doubt that the vector potential A has physical significance to consult Konopinski (1981, p. 158).

COMPUTER SIMULATION OF RECONNECTION IN PLANETARY MAGNETOSPHERES

J. Birn
University of California, Los Alamos National Laboratory
Los Alamos, NM 87545

Abstract. The earth's magnetosphere provides an ideal opportunity to model reconnection in well known geometries that are close enough to the idealized analytic models to make a comparison of the computer models with analytic theory meaningful. In addition more detailed, even three-dimensional, models can be used for a comparison with extended data from in situ observations. The computer studies have basically confirmed the reconnection picture that was based on two-dimensional steady state models and linear analytic theory. The three-dimensional models in particular have also added a lot more information on the reconnection process and the structure of flow, magnetic fields, and currents including many features that are consistent with observations and empirical models of geomagnetic substorms.

1. INTRODUCTION

Modeling reconnection in the Earth's magnetosphere has several advantages that cannot be met anywhere else. On the one hand, the underlying equilibrium structures are well known from observation and can be described, in some cases even to a high degree of sophistication, by self-consistent analytic models. They are close enough to simplified structures, such as, for instance, a plane current sheet, that form the base of most analytic work on reconnection, to make a comparison with such analytic work meaningful. The equilibrium or quasi-equilibrium structures, however, also contain small deviations from the simple geometries, which might be important in influencing the reconnection process. An example is the small magnetic field component perpendicular to the current sheet or plasma sheet in the geomagnetic tail. There is no doubt that there can be a stabilizing effect of such a normal component, and as we will see later it also influences the dynamic evolution by the asymmetry that is associated with its presence.

On the other hand, there is the great advantage that the results from reconnection models can be compared with in situ satellite measurements, which are provided even in much more detail than any present theory can cope with.

Although reconnection processes have been suggested to occur also in other than the earth's magnetosphere (e.g., Nishida, 1983), all computer studies of magnetospheric reconnection have been devoted to processes in the earth's magnetosphere, mainly because of the above mentioned detailed knowledge about those processes from observations but also because the earth might serve as a representative example for any magnetospheric reconnection. Also, most of the present reconnection models are highly idealized and therefore general enough to have applications not only in magnetospheres but possibly also in stellar atmospheres and other, even extragalactic, objects.

Reconnection in the earth's magnetosphere is a consequence of the interaction with the solar wind plasma as demonstrated by Fig. 1. At the frontside reconnection leads to a transfer of magnetic flux from the solar wind and the region of closed magnetospheric field lines into the region of open field lines that have only one foot connected to the earth. The occurrence of front side reconnection is strongly favored by a southward component of the interplanetary magnetic field. Without additional reconnection in the tail the magnetic flux on open field lines must increase in time and become stored in the lobes of the geomagnetic tail. On the average (but not necessarily at each instant of time) this flux transfer must be compensated by a transfer of flux from the nightside lobes back into the solar wind and into the closed field line region. The widely accepted view is that this occurs by a nonsteady process in the near tail at about 15 R_E, in so called magnetospheric substorms, but possibly also by a more steady reconnection process in the far tail beyond the moon's distance. Before lobe field lines can reconnect (Fig. 1c) in the near tail, the closed plasma sheet field lines in between them must reconnect first. This leads to somewhat shorter more dipolar-like field lines earthward from the reconnection region and to detached closed loops forming a so-called plasmoid tailward from it (e.g., Hones, 1979). This plasmoid moves tailward and leaves a very narrow plasma sheet behind (Fig. 1b). This process seems to occur quasi-periodically with a reformation of the original plasma sheet in between.

2. THE MODELS

The optimum model would include the entire magnetosphere using the earth's dipole and the ionosphere on one side and the unperturbed solar wind on the other side as boundary conditions. There are indeed such models which will be discussed in Section 3, in two and even three space dimensions. The drawback of most of these models is that resistivity, which is necessary to allow for reconnection, is provided by numerical effects rather than by a physical interaction. Still they have given relevant results on the global shape of the magnetosphere and reconnection sites.

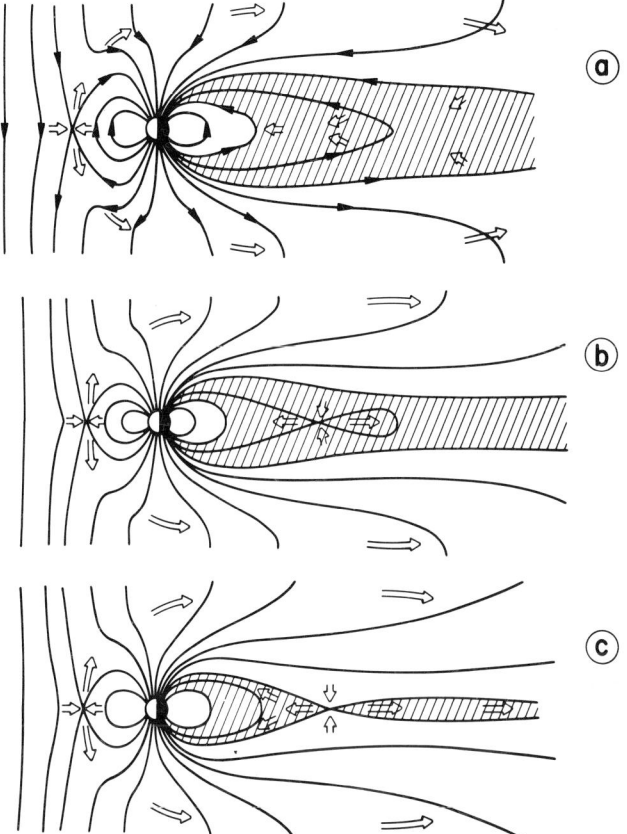

Figure 1. Schematics of reconnection regions of the earth's magnetosphere. The figure shows magnetic field lines in the noon-midnight meridian plane and flow vectors (open arrows). a. Frontside reconnection with transfer of magnetic flux from the solar wind and the front side magnetosphere into the lobes of the night side magnetotail. b. New reconnection in the tail with the formation of a magnetic island ("plasmoid") that subsequently moves tailward. c. Later stage of magnetotail reconnection with transfer of magnetic flux from the lobes into the closed nightside region and the solar wind.

On the other side there are models dealing with more localized regions, such as the magnetotail, or with current sheets in general. These models can give a more detailed knowledge of structural changes, flows, currents, and so on generated by the reconnection process. These models do not include the interaction with the solar wind, or account for it only by some assumed magnetospheric response at the simulation boundary inside the magnetopause. None of the models so far includes the interaction with the ionosphere and a possible feedback from ionospheric currents and electric fields.

In this paper we will concentrate on the results of the most advanced, primarily three dimensional, nonlinear time dependent computer models of reconnection. We will not deal with the models of laboratory experiments with their mostly toroidal or cylindrical geometry which produce some peculiarities.

The mean free path for binary collisions in the magnetosphere is extremely long, much longer than any other length in the magnetosphere. The plasma can therefore be treated as collision free. The best method of simulation, at least in principle, would include the full ion and electron dynamics. Such models (e.g., Katanuma and Kamimura, 1980) are, however, until now restricted to simulation system sizes of several Debye lengths only, i.e., of the order of several hundreds of meters. This is much smaller than the typical equilibrium scales, which are a few R_E for the plasma sheet half-width and roughly 1000 km for the magnetopause thickness. Hybrid codes that include the full ion dynamics but treat electrons as a fluid (e.g., Terasawa, 1981) can at present deal with scale sizes of a few ion Larmor radii (from several hundreds to a few thousands of km) which is comparable to the thickness of the magnetopause but still somewhat smaller than the plasma sheet thickness, at least before thinning associated with the dynamic phase of substorms has occurred. These codes, however, loose the electrostatic effects from individual electron motion.

The very large scale structures and dynamics finally can be dealt with only by fluid or MHD approaches. The microscopic collisonless process that leads to the deviation from ideal MHD with frozen-in fields is usually represented by some more or less ad hoc scalar resistivity. This model can be justified by the following view of the initiation of reconnection:

A driving force leads to a gradual compression of the current sheet until a stability limit for a microscopic instability is exceeded. The microinstability leads to wave turbulence and through wave particle interaction to an anomalous resistivity which initiates the large scale reconnection process. The mostly discussed candidate for this microturbulence is the lower-hybrid drift instability because of its relatively low threshold (Huba et al., 1978).

An alternative is the possibility that the collisionless instability itself is of large spatial scale. Again a driving force is needed to compress the current sheet and in particular reduce a normal magnetic field component until ions become non-adiabatic in the center and are no longer tied to the field lines (Schindler, 1974; Galeev and Zelenyi, 1977; Goldstein and Schindler, 1982).

In either case, deviations from ideal MHD are important only in very localized regions inside the current sheets. It is therefore plausible that the large scale spatial pattern of magnetic field and flow velocity is governed by ideal MHD regardless of what localized process enables the growth of the reconnection mode. This is also a possible justification of discussing reconnection within ideal MHD models, where resistivity is purely numerical. The situation might be comparable to the theory of shocks where useful jump conditions can be derived without knowledge of the details and the dissipation mechanism within the shock.

In this paper we will focus on the large scale structures and dynamics of the reconnection process and therefore discuss mainly the results from the MHD models.

3. GLOBAL MODELS

The first attempt to model a two-dimensional magnetosphere in a global way including a (line) dipole field for the earth and the streaming solar wind was made by Leboeuf et al. (1978). As a consequence of a discontinuity in their solar wind boundary condition representing a southward rotation of the magnetic field they reported enhanced reconnection at the front side and in the tail and the formation and subsequent tailward motion of a region of closed magnetic loops corresponding to the plasmoid in phenomenological substorm models (e.g., Hones, 1979). The resistivity in the so-called MHD particle code by Leboeuf et al., however, was purely numerical corresponding to a very low magnetic Reynolds number $R_m \approx 3 - 10$ and a very short diffusion time $\tau_D = R_m \tau_A$ where τ_A is the typical MHD time scale of e.g. Alfvén waves. As a consequence the resulting magnetosphere, which was of the type suggested by Dungey (1961), had a very short tail. It differed only slightly from the simple superposition of a line dipole and a homogeneous interplanetary field which is expected to result for infinitely fast magnetic diffusion. The model also suffered somewhat

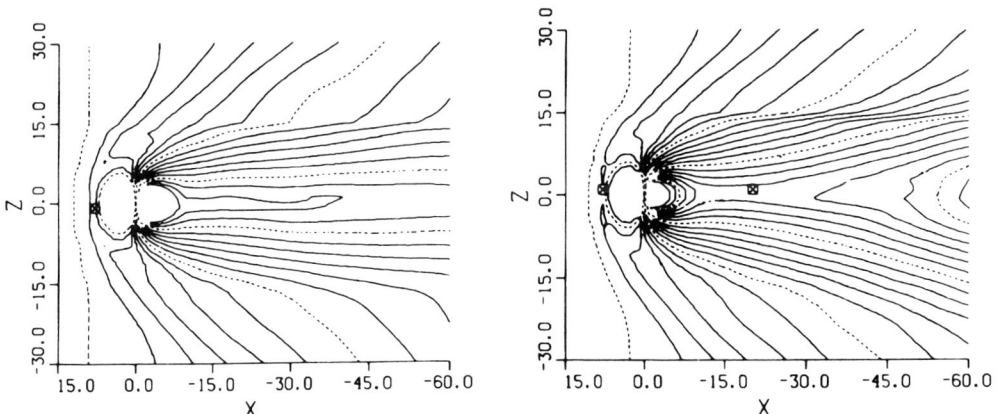

Figure 2. Magnetic field lines in the noon-midnight meridian plane in the global simulation by Lyon et al. (1981). a. After a southward magnetic field has been incident for one hour. b. 20 minutes later at peak of tail reconnection.

from the fact that periodicity in the z direction (north-south) was assumed which meant that instead of the solar wind interaction with a single magnetospheres actually the interaction with a lattice of magnetospheres was treated.

Subsequent 2D global simulations (Lyon et al., 1980, 1981; Brackbill, 1982) tried to reduce the numerical resistivity. Figure 2 shows results from a simulation by Lyon et al. (1981). Following a southward turning of the magnetic field in the incident solar wind, steady reconnection at the frontside was found leading to tailward convection of magnetic flux stretching the tail and increasing its field strength. About an hour later reconnection in the tail started as a spontaneous process. The resistivity in this model allowing for reconnection was still purely numerical.

More recently, Brackbill (1982) presented a 2-D global simulation using the current density dependent resistivity model of Sato and Hayashi (1979). Because of the inherent numerical resistivity the maximum representable Reynolds number R_m was about 50. Figure 3 shows a sequence of evolution in the presence of southward interplanetary magnetic field showing front side reconnection and subsequently tail deformation and tail reconnection.

Leboeuf et al. (1981) were the first to make their global model three-dimensional. With a southward solar wind magnetic field they found again a Dungey type magnetosphere with X-type neutral lines at the front and in the tail. But because their numerical resistivity was still as high as in the 2-D model it is not surprising that their

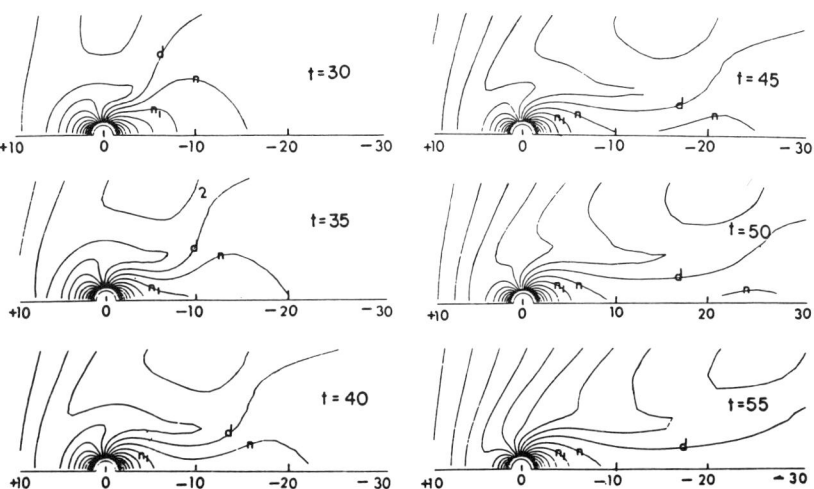

Figure 3. Magnetic field lines in the noon-midnight meridian plane in a global simulation by Brackbill (1982) for time instants indicated in the figures. Time units are R_E/V_s where V_s is the solar wind speed.

resulting configuration was close to the superposition of dipole field and uniform interplanetary field.

Shortly after other global 3-D models followed with reduced numerical diffusion. Brecht et al. (1981) reported results similar to their 2-D simulations with reconnection still based on numerical resistivity. The basic reconnection picture was found again. However, the magnetic field configuration in the tail plasma sheet was found to be more complicated with multiple neutral lines moving tailward while others formed newly closer to the earth. This result is similar to a result by Birn and Hones (1981) using a model of the tail only, which will be discussed in more detail in the following section.

4. MODELS OF LOCAL RECONNECTION

Global models of the magnetosphere including the discontinuities of the bow shock and the magnetopause almost unavoidably contain a large amount of numerical diffusion which makes the study of the dynamic reconnection process and its dependence on physical interaction mechanisms difficult (fortunately, many large scale characteristics seem not to depend much on the details of the interaction mechanism). More local models which deal with only one of the possible reconnection sites usually have less problems with numerical diffusion. They gain more insight into the details of the structures around the reconnection region, however, loose the interaction with the solar wind except for the possibility of prescribing boundary conditions which might reflect this interaction. They usually start from equilibrium configurations which are mostly simple one-dimensional sheets and can be described analytically. The resistivity models, however, vary.

In a 2-D simulation, Ugai and Tsuda (1977) initiated reconnection in a plane current sheet by locally enhancing the resistivity. The maximum resistivity corresponded to $R_m = 10$ with a background value of $R_m = 1000$. They found a pattern of fast flow and a quasi-steady state (Tsuda and Ugai, 1977) similar to analytic steady state models (Petschek, 1964) using open boundary conditions that allowed for free outflow and inflow.

Sato and Hayashi (1979) tried to model the generation of anomalous resistivity by a current dependent resistivity of the form

$$\eta = \begin{cases} \alpha(j-j_c)^2 & \text{for } j \geq j_c \\ 0 & \text{otherwise} \end{cases}$$

They started also from a one-dimensional plane current sheet using a 2-D code. Because the current density had to exceed the threshold j_c before resisitivity was generated and reconnection could be initiated, a

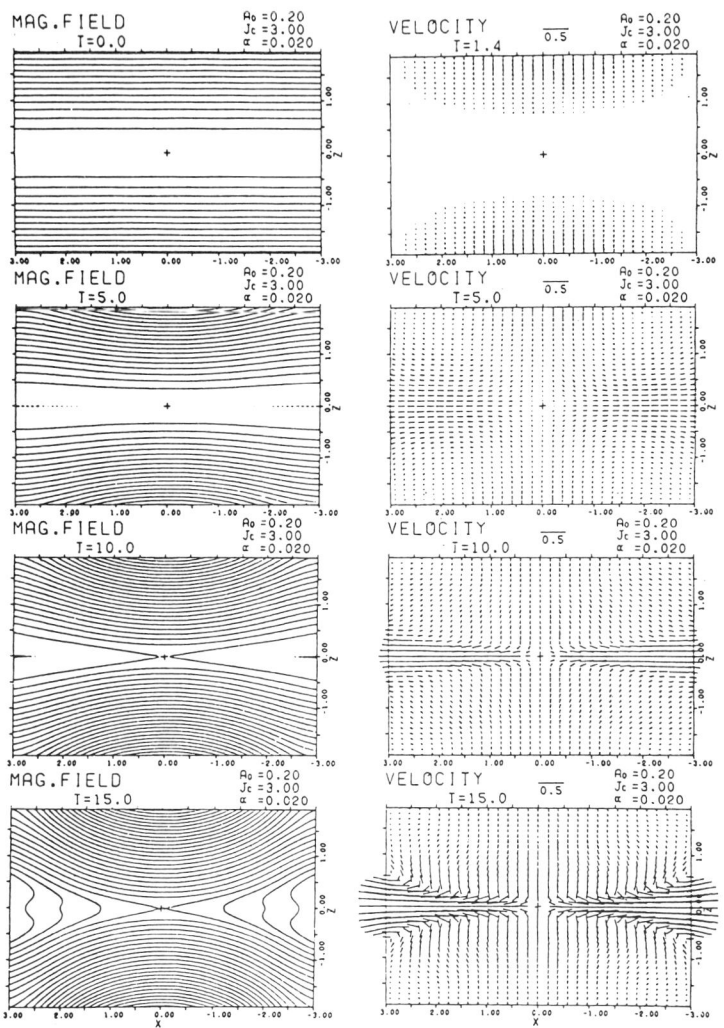

Figure 4. Computer plots of reconnecting magnetic field lines (left) and plasma flow vectors (right) in a two-dimensional model by Sato and Hayashi (1979). All variables are normalized by combinations of the initial maximum magnetic field, density, and plasma sheet half width.

driving force was necessary to gradually compress the current sheet and increase the current density. Sato and Hayashi assumed a nonuniform inflow with a maximum speed of initially typically 20% of the Alfvén speed at the boundaries parallel to the current sheet. The perpendicular boundaries were treated as "free" or "open." Sato and Hayashi found again the typical flow pattern around an X-type neutral point (see Fig. 4) including narrow current layers that were identified as the slow shocks existing in Petschek's (1964) model. They also found that the electric field at the X-point saturated at a level that

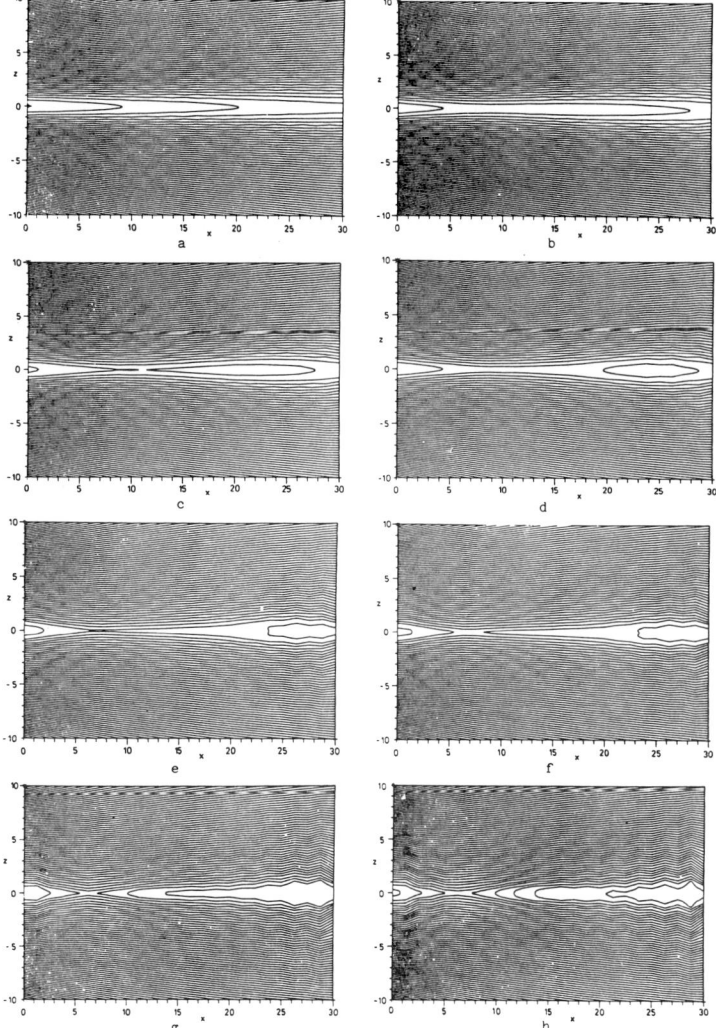

Figure 5. Magnetic field lines in a two-dimensional simulation of tail reconnection by Birn (1980) for selected times: (a) t = 0, (b) t = 110, (c) t = 135, (d) t = 150, (e) t = 165, (f) t = 170, (g) t = 180, and (h) t = 185; all times normalized by L/V_A where L is the characteristic plasma sheet half width and V_A the Alfvén speed.

depended on the inflow velocity, i.e., on the external electric field. This is why they called their reconnection process externally driven.

Different results concerning the role of an external driving mechanism were reported by Birn (1980) and Ugai (1982) using two-dimensional resistive models and by Terasawa (1981) using a two-dimensional hybrid particle simulation code which treated electrons as a fluid. Terasawa started from a Harris plane sheet equilibrum (Harris, 1962) using periodic boundary conditions in the direction along the sheet. He found a phase of nonlinear more rapid growth of the tearing mode before saturation, which was similar to the explosive mode proposed by Galeev et al. (1978). Ugai's model was a continuation of the earlier 2-D calculations by Ugai and Tsuda using a more sophisticated model of localized resistivity which included a threshold for rise and decrease of anomalous resistivity. As the initial drift velocity exceeded the onset threshold slow reconnection started immediately. After some time interval characterized by a decay of the initial resistivity a fast "explosive" increase of reconnection occurred even when the open boundary parallel to the current sheet was replaced by a rigid wall.

Birn (1980) started from a more realistic initial tail configuration including a finite normal magnetic field component and flaring of the lobe field lines (Fig. 5a). After applying constant resistivity with R_m = 500 he found the growth of a tearing instability leading to thinning of the plasma sheet and to the formation of neutral points and a plasmoid moving tailward (Figures 5a-h). A pattern of strong flow was set up around the X-point similar to the Petschek model (Fig. 6). He also found sheets of enhanced current density similar to those found by Sato and Hayashi (1979) related to the slow shocks postulated by Petschek. Similar results were recently obtained by Forbes and Priest (1983) starting from a plane sheet configuration but introducing asymmetry in the x coordinate along the sheet through asymmetric boundary conditions with line-tying at the near earth side and open boundaries elsewhere.

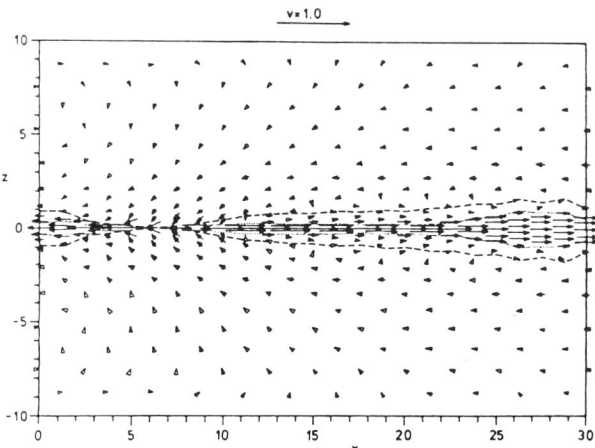

Figure 6. Velocity pattern at t = 180 for the same simulation as in Figure 5. The dashed line represents the magnetic field separatrix through the X-point; the dotted line represents maxima of electric current density for x = const. (from Birn, 1980).

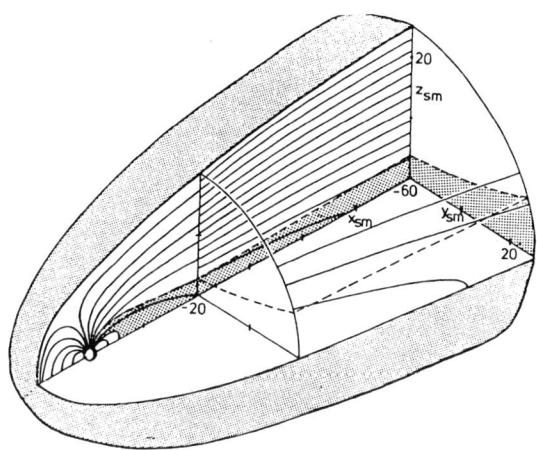

Figure 7. Self-consistent three-dimensional magnetotail equlibrium configuration by Birn (1979) used as initial configuration for the three-dimensional reconnection simulation by Birn and Hones (1981). The figure shows maganetic field lines in the midnight meridian x,z plane and close to the magnetopause boundary. The plasma sheet is indicated by dark shading and the dashed line. The near earth part (not used in the simulation) is not calculated and added only for better illustration (from Birn, 1979).

Subsequently Birn's model was made three-dimensional (Birn and Hones, 1981) starting from a self-consistent three-dimensional tail equilibrium (Birn, 1979; Fig. 7). Again the growth of the tearing mode was initiated by the onset of (constant) resistivity (R_m = 200). Plasma sheet thinning and the formation of a neutral line and the tailward moving plasmoid were observed again with the picture in the midnight meridian plane very similar to the 2-D model (Figs. 5 and 6).

The third dimension, however, enabled new results not present in the 2-D models. This is demonstrated by Figure 8 which shows flow velocity vectors and the neutral lines (dotted lines) in the equatorial x,y plane. Reconnection and strong flows do not occur across the entire tail as they would in a 2-D model but are restricted to a center region where also the thinning takes place. This is a consequence of the 3-D equilibrium with plasma sheet thickening and correspondingly an increase of the normal magnetic field component B_z toward the dawn and dusk flanks of the magnetotail.

The pattern is also more complicated than in 2-D. The region of strong tailward flow does not exactly coincide with the region where B_z has changed sign, which is enclosed by the neutral line. The neutral line itself is more complicated and can become multiple similarly as reported by Brecht et al. (1981).

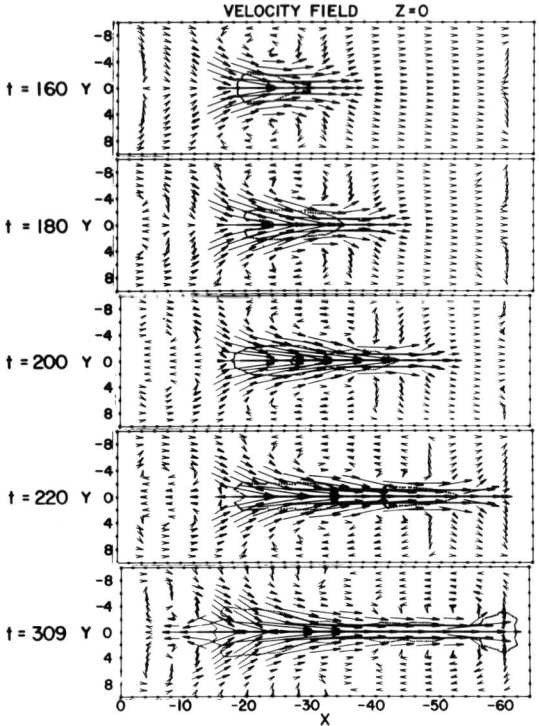

Figure 8. Flow vectors and magnetic neutral lines (dotted lines) in the equatorial x,y plane in a three-dimensional magnetotail reconnection simulation by Birn and Hones (1981). Times as shown on the left are normalized in the same way as for Figure 5. The length of the flow vectors is proportional to the speed with the maximum vectors corresponding to about 0.7 in units of the typical Alfvén velocity.

Of special interest is the current pattern in and around the reconnection region and around the plasmoid. Before we discuss this it is useful to visualize the 3-D structure of field lines in and around the plasmoid as shown in Figure 9. The field lines are represented above the equatorial plane as seen from the tail toward the earth. The centerward draping of the field lines causes a shear of the magnetic field and thereby field aligned currents, in particular close to the separatrix surface, i.e., the field lines originating from the X-line.

The currents are shown in Figure 10 by vectors of current density in the equatorial x,y plane and by projections in a cross-section of the tail. Figure 10a shows a deviation of the current vectors from the original cross-tail direction. The deviation earthward from the main X-line is earthward on the dawnside and tailward on the dusk side. Oppositely directed deviations are found tailward from the X-line. It

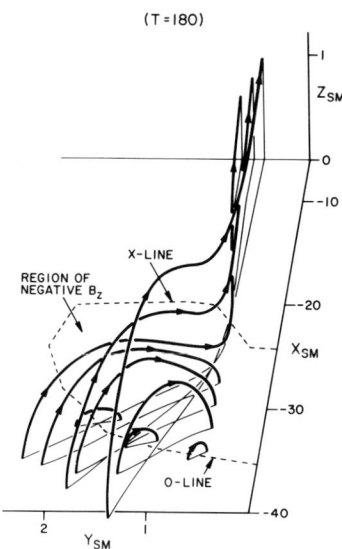

Figure 9. Three-dimensional representation of magnetic field lines computed by the 3-D model of Birn and Hones (1981) for a time about 12 minutes after reconnection begins. Projections of the magnetic field lines into the x,y plane are shown by light lines (from Hones et al., 1982).

looks as if the current tends to flow around the center diffusion region. A similar deviation is found in the cross-sections of constant x at some distance from the X-line. The current tends to flow around the plasmoid. The current density inside the plasmoid is reduced and can even become oppositely directed (Fig. 10b).

Similar results on current deviation are recently reported by Sato et al. (1983) in a model of 3-D driven reconnection. The model was similar to their previous 2-D models already mentioned. The initial configuration was again a plane current sheet. The dependence on the other directions was introduced, and the reconnection was driven by a nonuniform inflow velocity at the boundaries parallel to the sheet with a maximum at some center point. Figure 11 shows the current density vectors in the equatorial x,y plane and some parallel plane above. One can see, even more pronounced than in the model of Birn and Hones, the oppositely directed currents at some distance from the X-line which is in the center of the frame. The currents get partly concentrated right at the X-line and tend to flow around it at higher latitudes.

Although there is some similarity of the two 3-D models discussed above in the changes of current flow, the field aligned currents that are found in both models earthward from the X-line have opposite directions flowing toward the earth on the dawn side and away on the

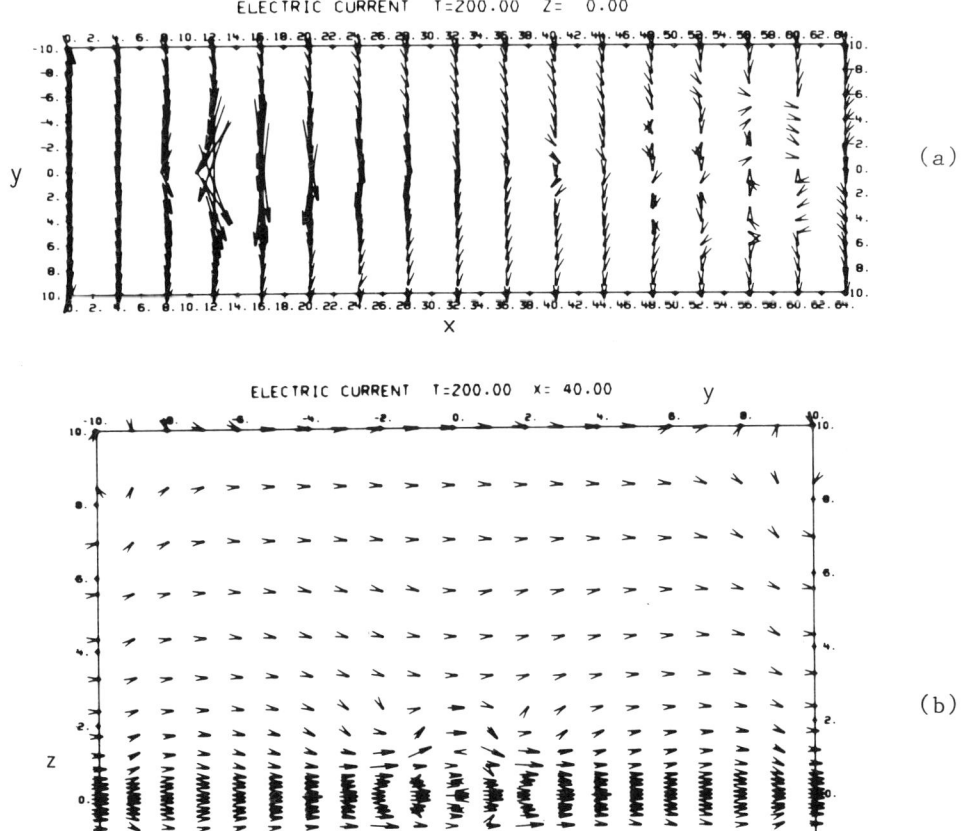

Figure 10. Current density vectors (a) in the equatorial x,y plane and (b) projections in the plane x = 40 for a three-dimensional magnetotail reconnection simulation of Birn and Hones.

dusk side in the model of Sato et al. and the opposite way in the model of Birn and Hones. This need not be contradictory as the two approaches were quite different. Both field aligned current systems are realized in the magnetosphere, and it is indeed conceivable that the outer system which has the signatures of that found by Sato et al. is indeed externally driven while the inner system corresponding to that found by Birn and Hones is caused by the internal dynamics.

5. SUMMARY

We have presented a variety of models of the dynamic evolution of the magnetosphere and sheet like configurations as they are present at the magnetopause and in the magnetotail and probably in many other astronomical objects as well. It seems that the typical reconnection pattern predicted by the 2-D steady state theory (e.g., Petschek, 1964)

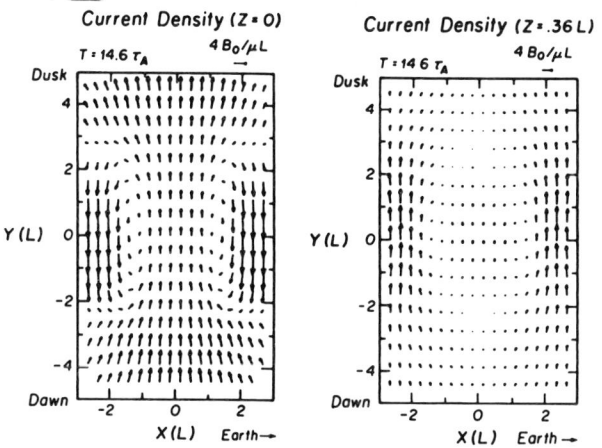

Figure 11. Current density vectors in the equatorial x,y plane (z = 0) and a parallel plane (z = 0.36 L) at t = 14.6 τ_A in a three-dimensional simulation of driven reconnection by Sato et al. (1983).

can be found for a variety of different initial configurations, boundary conditions, and resistivity models, even though most of the models did not reach a steady state. Although the basic view of reconnection formed by Petschek's steady state model and the linear tearing theory (Furth et al., 1963) is thus very well confirmed, the models added also more information about the nonlinear phase of evolution and about the 3-D structure in more realistic geometries especially for the geomagnetic tail. Many of the features that were found are consistent with basic features of phenomenological substorm models, e.g., plasma sheet thinning, neutral line formation, plasma acceleration to speeds of the order of the Alfvén speed, current diversion, and the generation of field-aligned currents. The models have also shown that reconnection may be a localized phenomenon in the sense that it occurs only in a limited region of a current sheet configuration where the normal magnetic field component is weakest. Although this result stems from a magnetotail model, it might also be some clue for the understanding of the patchy type of magnetopause reconnection that is believed to occur in the so-called flux transfer events (Russell and Elphic, 1978).

Improvements are needed to include the particle effects in more realistic geometries including for instance a normal magnetic field component B_z in the magnetotail current sheet. In this context it is useful to point out that it is necessary for such simulations also to include the lobe regions around the plasma sheet where B_z changes sign because a configuration with the same sign of B_z everywhere would be stable (Birn et al., 1975).

Acknowledgments. This work was supported under U. S. Department of Energy contract W-7405-ENG-36.

REFERENCES

Birn, J.: 1979, J. Geophys. Res. 84, pp. 5143-5152.
Birn, J.: 1980, J. Geophys. Res. 85, pp. 1214-1222.
Birn, J. and Hones, E. W., Jr.: 1981, J. Geophys. Res. 86, pp. 6802-6808.
Birn, J., Sommer, R., and Schindler, K.: 1975, Astrophys. Space Sci. 35, pp. 389-402.
Brackbill, J. U.: 1982, Los Alamos Nat. Lab. preprint LA-UR-82-483, submitted to Geophys. Res. Lett.
Brecht, S. H., Lyon, J. G., Fedder, J. A., and Hain, K.: 1982, J. Geophys. Res. 87, pp. 6098-6108.
Dungey, J. W.: 1961, Phys. Rev. Lett. 6, 47-48.
Forbes, T. G. and Priest, E. R.: 1983, J. Geophys. Res. 88, pp. 863-870.
Furth, H. P., Killeen, J., and Rosenbluth, M. N.: 1963, Phys. Fluids 6, pp. 459-484.
Galeev, A. A. and Zelenyi, L. M.: 1977, Sov. Phys. JETP 43, pp. 1113-1123.
Galeev, A. A., Coroniti, F. V., and Ashour-Abdalla, M.: 1978, Geophys. Res. Lett. 5, pp. 707-710.
Goldstein, H. and Schindler, K.: 1982, Phys. Rev. Lett. 48, pp. 1468-1471.
Harris, E. G.: 1962, Nuovo Cimento 23, pp. 115-121.
Hones, E. W., Jr.: 1979, in "Dynamics of the Magnetosphere," ed. by S. -I. Akasofu, D. Reidel, Dordrecht-Holland, pp. 545-562.
Hones, E. W., Jr., Birn. J., Bame, S. J., Paschmann, G., and Russell, C. T.: 1982, Geophys. Res. Lett. 9, pp. 203-206.
Katanuma, I. and Kamimura, T.: 1980, Phys. Fluids 23, pp. 2500-2511.
Leboeuf, J. N., Tajima, T., Kennel, C. F., and Dawson, J. M.: 1978, Geophys. Res. Lett. 5, pp. 609-612.
Leboeuf, J. N., Tajima, T., Kennel, C. F., and Dawson, J. M.: 1981, Geophys. Res. Lett. 8, pp. 257-260.
Lyon, J. G., Brecht, S. H., Fedder, J. D., and Palmadesso, P.: 1980, Geophys. Res. Lett. 7, pp. 721-724.
Lyon, J. G., Brecht, S. H., Huba, J. D., Fedder, J. A., and Palmadesso, P. J.: 1981, Phys. Rev. Lett. 46, pp. 1038-1041.
Nishida, A.: 1983, Geophys. Res. Lett. 10, pp. 451-454.
Petschek, H. E.: 1964, AAS-NASA Symp. on the Phys. of Solar Flares, NASA SP-50, pp. 425-439.
Russell, C. T. and Elphic, R. C.: 1978, Space Sci. Rev. 22, pp. 681-675.
Sato, T. and Hayahi, T.: 1979, Phys. Fluids 22, pp. 1189-1202.
Sato, T., Hayashi, T., Walker, R. J., and Ashour-Abdalla, M.: 1983, Geophys. Res. Lett. 10, pp. 221-224.
Schindler, K.: 1974, J. Geophys. Res. 79, pp. 2803-2810.
Terasawa, T.: 1981, J. Geophys. Res. 86, pp. 9007-9019.
Tsuda, T. and Ugai, M.: 1977, J. Plasma Phys. 17, pp. 451-471.
Ugai, M.: 1982, Phys. Fluids 25, pp. 1027-1036.
Ugai, M. and Tsuda, T.: 1977, J. Plasma Phys. 17, pp. 337-356.

DISCUSSION

Vasyliunas: Your initial state is an equilibrium for zero resistivity. When a non-zero resistivity is introduced, the state is no longer one of equilibrium and will evolve in time; shouldn't that be called simply a time evolution of a non-equilibrium state and not an instability?

Birn: The application of a finite resistivity indeed causes the initial equilibrium configuration to diffuse slowly. This diffusion, however, occurs on a time scale much longer than the rise time of the instability that develops out of the perturbations introduced by the diffusion. The instability therefore is clearly distinguishable from the slow, quasi-steady diffusion.

Steinolfson: The intial state in your 3-D simulation was not in dynamic diffusive equilibrium. As a result, the reconnection was effectively driven (initiated) by the diffusion.

Birn: As I pointed out in my answer to Dr. Vasyliunas, the instability developed from a (slow) diffusion of the initial equilibrium state. However, since this evolution did not require a driving mechanism at the boundary, I would not call this "driven".

Bratenahl: 1) I'm interested in the 3-D field line figure showing the field line shear or rotation as you go up in z. You attributed this shear to field aligned currents. I did not see such currents in your slide showing currents. My comment is that you can get such rotational shear in a potential field, as in my model here. 2) My colleague Peter Baum modified the Schmidt-Harvey field line integrating code to map only through field lines lying on the separator and separatix structures. This can be useful.

Birn: Thank you for your comment.

Mullan: What is the ultimate fate of a plasmoid which forms in the magnetotail following reconnection?

Birn: I don't know. The simulations cannot tell what happens after the plasmoid has reached the tailward boundary of the system. Observationally, the present ISEE-3 mission through the distant tail might give more information about the ultimate fate of the plasmoid.

Wu: I would like to follow Dr. Steinolfson's question a little further, since your initial solution is not an equilibrium solution. I assumed that you also used artificial numerical dissipation in your code. As we all know that all numerical solutions are approximate solutions, then how do you identify whether your reconnection results are due to physical conditions or numerical effects?

Birn: By running the computer code with an without resistivity, I could prove that the reported results are indeed due to the physical resistivity and not to numerical dissipation. The run without resistivity showed no significant changes of the intital configuration.

Wu: What is the resolution of your grid size in the reconnection region?

Birn: I used a variable grid size in the most critical direction perpendicular to the neutral sheet with an innermost grid spacing of about 0.05 in units of the initial plasma sheet half width.

Drake: What boundary conditions do you use on the cross tail current in your 3-D simulation of the tail?

Birn: The velocity was assumed to be zero at that boundary; all quantities which did not vanish, including the cross tail current density, were assumed to have vanishing derivatives perpendicular to the boundary.

Sonnerup: Is the B_y component in the plasmoid predicted by your model consistent with the observed sign and magnitude of B_y in the plasmoid?

Birn: The sign of the B_y component in the plasmoid predicted by my model agrees indeed with observations in the geomagnetic tail; the magnitude of B_y in my model, however, is somewhat smaller than in some of the observations.

Lui: The field-aligned current system in Sato's simulation is consistent with disruption of the cross tail current while in your model it is not. How do you explain this?

Birn: The field aligned current system in my model is realized by the inner ring of "region 2" field-aligned currents found at the earth, which increases with increasing activity. The outer system ("region 1") is probably closer related to a driving mechanism at the boundary and, therefore, to Sato's model. This view is supported by the fact that "region 1" parallel currents are observed to be strong, even during quiet times.

Vasyliunas: Field-aligned (Birkeland) currents are carried by Alfven waves and therefore also involve velocity perturbations; a boundary condition V=0 may thus suppress J in the model, which may account for the discrepancies between various models.

Birn: I agree with your comment on the generation of "region 1" field-aligned currents.

OBSERVATIONAL EVIDENCE FOR MAGNETIC RECONNECTION IN MICROWAVE SOLAR BURSTS

M. R. Kundu
Astronomy Program
University of Maryland
College Park, MD 20742

In this paper, we first discuss a set of 6 cm observations made with the NRAO Very Large Array (VLA) (spatial resolution ~ 2") that pertain to changes in the coronal magentic field configurations that took place before the onset of an impulsive burst observed on 14 May 1980. We also discuss a second set of 6 cm VLA observations (spatial resolution 18" arc) where several interacting loops were involved in triggering the onset of an impulsive burst observed on June 24, 1980, 19:57:00 UT. Both sets of observations are examples of magnetic reconnection process being involved in accelerating microwave emitting electrons.

In the first case, the burst appeared as a gradual component on which was superimposed a strong impulsive phase (duration ~ 2 minutes) in coincidence with a hard X-ray burst. Soft X-ray emission (1.6 - 25 Kev) was associated with the gradual 6 cm burst (before the impulsive burst), as is to be expected. There was a delay of hard X-ray emission (> 28 Kev) relative to 6 cm emission (~ 10 sec delay from 6 cm maximum to hard X-ray start and ~ 20 sec delay from 6 cm maximum to hard X-ray maximum.). The preflare region at 6 cm showed intense emission with peak T_b ~ 1.3×10^7 K and degree of polarization p ~ 65% extended along a neutral line situated approximately in the east-west direction. A gradual burst source of intense emission with T_b ~ 4×10^7 K and p ~ 50 - 80% appeared initially. The most remarkable feature of the 6 cm burst source evolution was that an intense emission (T_b ~ 1.4×10^8 K; p ~ 60%) extending along the north-south neutral line (line of zero polarization at 6 cm), possibly due to reconnections, appeared, just before the impulsive burst occurred. This north-south neutral line must be indicative of the appearance of a new system of loops. In the 20 seconds preceding the impulsive peak (T_b ~ 1.1×10^9 K; p ~ 40%) the arcade of loops (burst source) changed and ultimately developed into two strong bipolar regions or a quadrupole structure whose orientations were such that near the loop tops the field lines were opposed to each other. This quadrupole field configuration is reminiscent of the flare models in which a current sheet develops at the interface between two closed loops. The impulsive energy release must have occurred due to magnetic reconnection of the field lines connecting the two oppositely polarized bipolar regions (Kundu et al 1982; Velusamy and Kundu 1982). After the impulsive phase was over, the gradual burst still continued for another 10 minutes, with the magnetic field configuration being very similar to what existed in the gradual phase prior to the impulsive event. The important changes in the coronal field configuration, and hence magnetic field reconnections, are schematically represented in Figs. 1 and 2. The reconnection process accelerates electrons to energies of the order of 100 keV or higher, which are responsible for the microwave and possibly, the hard X-ray bursts. The delayed hard X-ray emission, assuming it to be nonthermal, must be attributed to the fact that not enough electrons of energy > 28 Kev were able to reach a thick target region to produce observable X-ray emission at the onset of 6 cm impulsive burst. The hard X-ray spectrum may be either thermal (exponential) or nonthermal (power law). In the thermal interpretation, impulsive phase plasma temperatures of 1-2 x 10^8 K and a plasma density of ~ 1 x 10^9 cm^{-3} are implied, if the X-ray emitting volume is comparable to the microwave emitting volume (Holman, this issue). The thermal interpretation implies delayed heating of the flare plasma to ~ 10^8 K.

The observations of the second burst provides a good example of interacting loops (Kundu et al 1984). The 6 cm burst source was complex, consisting initially of two oppositely polarized bipolar sources separated E-W by ~ 1.5 arc. The first brightening occurs in one component at 19:57:10 UT, located at the same position as the burst that occurred at 19:51:05 UT. The western component is much weaker at this time. It then brightens up at 19:58:05 UT, just at the onset of the impulsive rise of the burst and is accompanied by changes in its polarization structure. It then decays and by 19:59:05 UT, it appears to split into two weak sources separated E-W by ~ 12" arc. The eastern component brightens up at 19:58:15 UT and then decays until 20:00:15 UT. This brightening is accompanied by significant polarization changes, including reversal of polarization. A third component appears approximately midway between the eastern and western component at 19:58:45 UT during the peak of the associated hard x-ray burst (Figs. 3 and 4). The appearance of this source is again associated with polarization changes, in particular the clear appearance of several bipolar loops; its location overlaps two opposite polarities implying that it might be situated near the top of a loop. The third source reaches maximum intensity at 19:59:05 UT and by 19:59:15 UT it disappears. At the time of maximum intensity the burst source appears to lie at the interface between two oppositely polarized loops. Clearly, in this set of observations we are dealing with interaction between multiple loop structures and the resultant formation of current sheets between two oppositely polarized loop structures. The magnetic field reconnection process that ensues must be responsible for the acceleration of electrons responsible for impulsive microwave emission.

These two sets of observations provide the first observational evidence for magnetic reconnection in microwave flares.

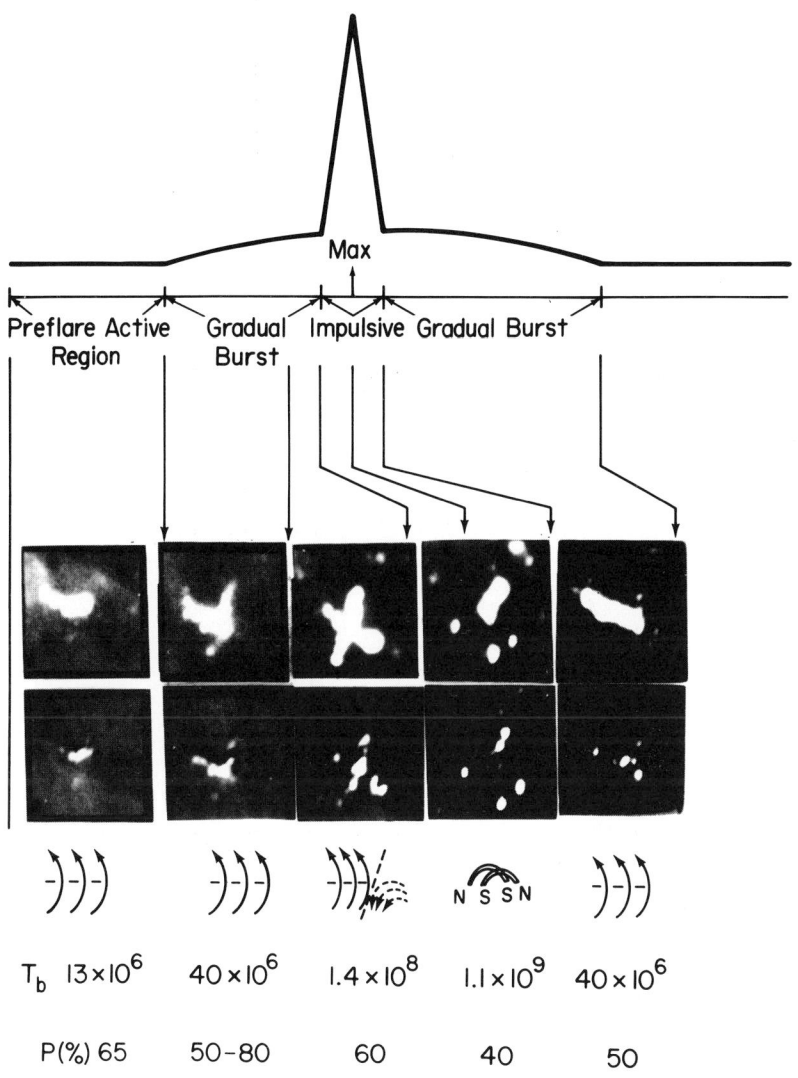

Fig. 1. Preflare active region and burst source maps for the 14 May 1980 burst at 6 cm. Each map was synthesized from data taken during a time interval appropriate to the observed 6 cm flux. Preflare 6 cm map 18:06–18:45 UT, $T_b(max) \sim 13 \times 10^6$ K; gradual phase of burst, 18:59–19:14, $T_b(max) \sim 40 \times 10^6$ K; last 5^m before impulsive phase, 19:14–19:19, $T_b(max) \sim 1.4 \times 10^8$ K; peak of the impulsive phase 19:19:55–19:20:05, $T_b(max) \sim 1100 \times 10^6$ K (Note the remarkable quadrupole structure); gradual phase of burst, 19:21–19:30, $T_b(max) \sim 40 \times 10^6$ K.

References

Kundu, M. R.: 1981, Proc. SMY Workshop, Crimea, p. 24.
Kundu, M. R., Schmahl, E. J., Velusamy, T.and Vlahos, L.: 1982, Astron. Astrophys. 108, 188.
Kundu, M. R., Machado, M., Erskine, F. T., Rovira, M. G., Schmahl, E. J.: 1984, Astron. Astrophys. (in press).
Velusamy, T. and Kundu, M. R.: 1982, Astrophys. J., 258, 388.

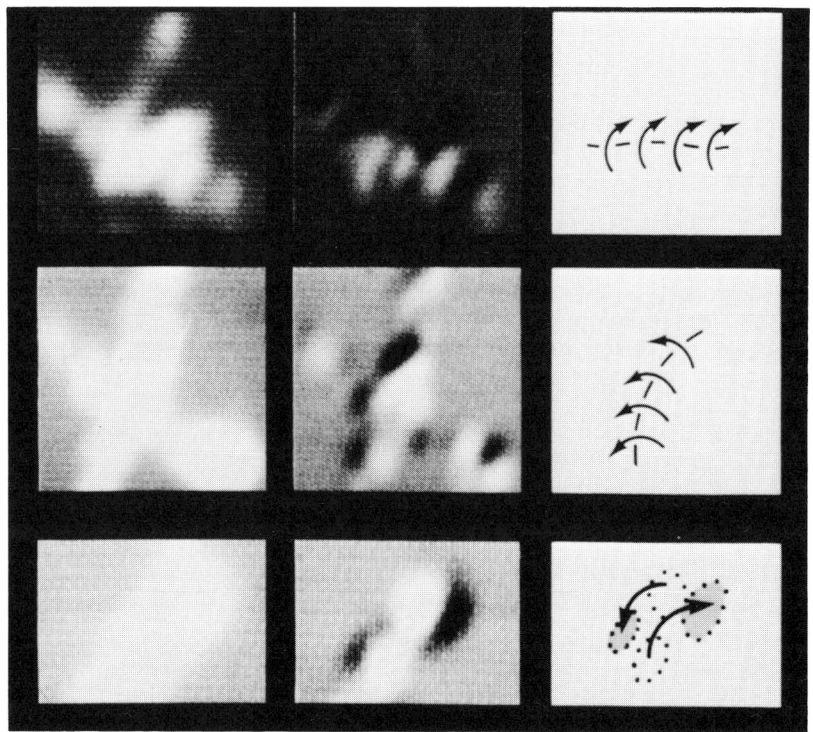

Fig. 2 Total intensity (left) and polarization (middle) maps at three selected periods (from top to bottom) - gradual phase, just before onset and at peak of impulsive phase burst shown in Fig. 1. The inferred magnetic field lines are also shown (right).

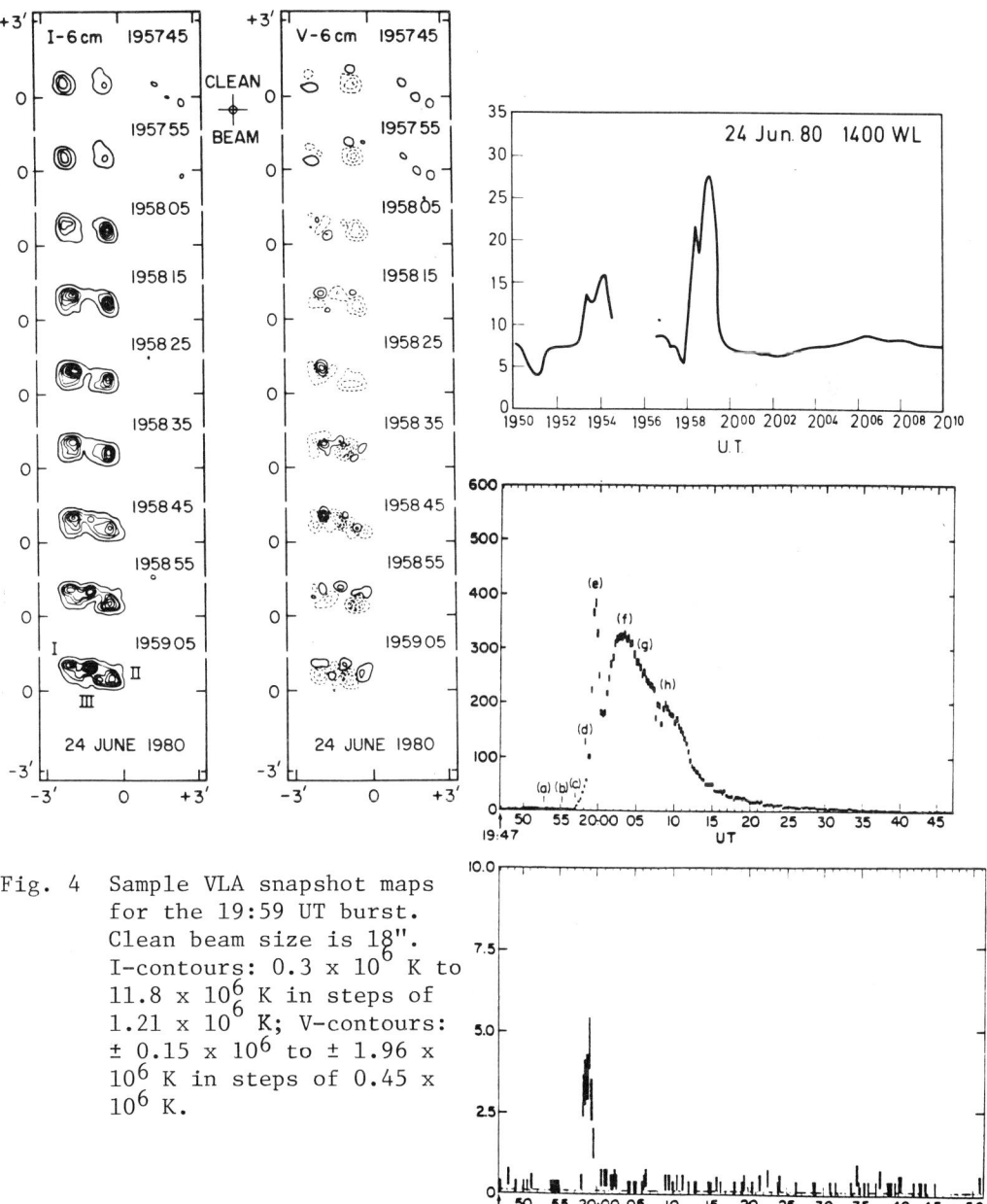

Fig. 4 Sample VLA snapshot maps for the 19:59 UT burst. Clean beam size is 18". I-contours: 0.3×10^6 K to 11.8×10^6 K in steps of 1.21×10^6 K; V-contours: $\pm 0.15 \times 10^6$ to $\pm 1.96 \times 10^6$ K in steps of 0.45×10^6 K.

Fig. 3 Top: Amplitude (in units of sfu, 1 sfu = 10^{-22} w m^{-2} Hz^{-1}) vs. time of 24 June 1980, 19:50 and 19:59 UT bursts at 6 cm; (middle and bottom): total emission light curves (in cts s^{-1}) of the same events in soft (3.5 -8.0. Kev) and hard (22.0 -30.0 Kev) X-rays.

DISCUSSION

Bratenahl: Is there a photospheric magnetogram available? The photospheric magnetic field should not change very much on a time scale of seconds.

Kundu: Yes, we do have photospheric magnetograms for the region in which the flare occurred. They of course do not change on time scales of seconds, but then we do not have photospheric magnetograms during the flare.

Mullan: You have shown examples of reconnection during flares. Do you see examples of reconnection outside flares? In other words, do you see non-thermal emission associated with "normal coronal heating"? Can you distinguish between coronal heating and flaring?

Kundu: I believe we may see reconnections outside flares, but they may not be easy to detect, because they may occur over a long time scale. We have evidence for the existence of non-thermal particles in non-flaring active regions, because we see brightness temperatures as high as 10 million degrees in active regions before flares. However, it is not clear how we can relate such non-thermal particles to coronal heating.

Krishan: Flares can occur in single loops as well as in interacting loops. The spatial position of the flaring region, i.e., the position of magnetic reconnection, will be very different. In fact, one should see at least three flaring regions simultaneously, at the points of interaction as well as at the neutral line of the individual loops.

Kundu: I believe that when flares occur as a result of interacting loops, we shall see mainly one flaring region, namely at the region of interaction. If two loops flare up simultaneously we may observe two regions near their tops, but whether or not we shall see the third region will depend on how strong the interaction is.

ACCELERATION OF RUNAWAY ELECTRONS AND JOULE HEATING IN SOLAR FLARES

Gordon D. Holman
Astronomy Program, University of Maryland
College Park, MD 20742

D.C. electric fields provide the simplest and most direct means of accelerating electrons out of a thermal plasma. Most solar flare models result in the production of D.C. electric fields. On the other hand, microwave and hard X-ray observations of flares provide specific requirements for the number and energy of energetic electrons produced during a flare, and the timescales involved in accelerating them. The microwave emission from flares is understood to be gyrosynchrotron radiation from electrons with energies of 100 keV or greater. The hard X-ray emission ($\gtrsim$ 25 keV) can be interpreted as being either thick-target bremsstrahlung from non-thermal electrons (thin-target radiation may also contribute to the X-ray emission, but the process is less efficient), or thermal bremsstrahlung from hot, impulsively heated plasma. Hence, it is of interest to study the electric field acceleration of "runaway" electrons and the simultaneous Joule heating of the thermal plasma in light of these results from flare observations, without recourse to a specific flare model. Some of the results of such a study are summarized here.

When a thermal plasma is subjected to an electric field, a fraction of the current-carrying ($J = \sigma E$) electrons, those with a velocity greater than a critical velocity, v_c, will be freely accelerated out of the thermal distribution (Dreicer 1959). The velocity v_c is determined by the temperature (T_e), density (n) and resistivity ($\eta = 1/\sigma$) of the plasma, and by the electric field strength (E). As this tail of "runaway" electrons is produced, new particles are supplied to the runaway region ($v \geq v_c$) by collisions. (This assumes that $v_c > v_e$, the thermal electron velocity. Otherwise, the runaway region includes the entire distribution of thermal electrons.) The runaway electrons are further accelerated by the electric field as long as they remain in the current channel.

1. It is of interest to estimate the maximum number of electrons that can be supplied by the original distribution of thermal electrons, N = $n(v > v_c) V_J$, where V_J is the volume of the acceleration region, without considering the additional particles made available by collisions. The

volume of the current channel and, hence, the acceleration region is not arbitrary, since the total allowable current is limited by the induction magnetic field associated with it. From Ampere's law, the magnetic field associated with a current sheet of total current $I = Jw\delta r$ (w is the width of the sheet, δr its thickness) is $B_J = (2\pi/c)(I/w) = (2\pi/c)(J\delta r)$. Requiring $B_J \lesssim B$, the maximum magnetic field strength in the acceleration region, yields $\delta r \lesssim (c/2\pi)(B/J)$. With $V_J = A\delta r$ and $J = nev_D$, where A is the area of the current sheet and v_D is the mean drift velocity of the current-carrying electrons, one finds that $V_J \lesssim (c/2\pi e)(BA/nv_D)$. The __maximum__ number of electrons that can be accelerated is obtained when $v_D = v_c = v_e$, giving

$$N \lesssim \frac{BA}{2\pi e}\left(\frac{c}{v_e}\right) = 4.3 \times 10^{29} \left(\frac{B}{100\ G}\right)\left(\frac{A}{5.3 \times 10^{17}\ cm^2}\right)\left(\frac{T}{10^7\ K}\right)^{-1/2} \text{ electrons.} \quad (1)$$

Note that this result is independent of the density and the resistivity of the plasma. The sheet area chosen here corresponds to a 10" x 10" sheet on the sun, as seen from the earth. It should also be noted that a much smaller limit is obtained if the current is assumed to be confined to a cylindrical volume, rather than to a sheet geometry.

The number of electrons required to produce a typical microwave flare on the sun is $10^{31} - 10^{32}$. Maximum observed magnetic field strengths are on the order of 1000 G. Hence, under extreme conditions it may be possible to generate an observable microwave flare with the original distribution of thermal electrons. In general, however, additional time is required for more electrons to be supplied from the thermal plasma. An important corollary of this result is that *the number of accelerated electrons will, in general, depend upon the thermal collision frequency and, therefore, the resistivity of the plasma.*

2. Interpreting the ≥ 25 keV X-ray emission from a flare to be thick-target, non-thermal bremsstrahlung requires a minimum of 10^{35} electrons/sec to be accelerated. The current associated with an electron flux of $\dot{N}$ electrons/sec, $I = e\dot{N}$, is limited by the induction field it generates, $B_J = (2\pi/c)(I/w)$. Requiring $B_J \leq B$ yields the result

$$\dot{N} \leq \frac{cwB}{2\pi e} = 1.0 \times 10^{30} \left(\frac{w}{10^9\ cm}\right)\left(\frac{B}{100\ G}\right) \text{ electrons/sec.} \quad (2)$$

Hence, since observed magnetic field strengths are limited to $B \lesssim 1000$ G, *the simple acceleration of runaway electrons cannot produce a large enough electron flux to explain the bulk of the observed hard X-ray emission as non-thermal bremsstrahlung.* **This result agrees with a similar result obtained by Spicer (1983).**

Note that the timescale for the generation of N runaway electrons is $t_N \equiv N/\dot{N}$. Hence, the upper limit on $\dot{N}$ yields a minimum timescale for the generation of the number of electrons required for an observable microwave source:

$$t_N \geq 10 \left(\frac{N}{10^{31}}\right)\left(\frac{w}{10^9 \text{ cm}}\right)^{-1} \left(\frac{B}{100 \text{ G}}\right)^{-1} \text{ sec.} \qquad (3)$$

3. The timescale, t_a for the acceleration of an electron in a D.C. electric field from the critical velocity v_c to a velocity v, is limited by the distance over which this acceleration can occur. For the simplest case of a constant electric field, the acceleration distance is $x_a = (1/2)(v_c + v)t_a$. To accelerate an electron to an energy of 100 keV in a distance of 10^9 cm or less requires $t_a \lesssim 0.1$ sec. The electric field strength required to accomplish this is $E = (m/e)(v-v_c)/t_a \geq 3 \times 10^{-7}$ statvolt/cm. If $E < E_D$, the Dreicer field, is also required, so that the entire thermal electron distribution does not run away, the following constraint on the density and temperature in the current channel is obtained:

$$0.18 \left(\frac{n}{10^9 \text{ cm}^{-3}}\right)\left(\frac{T}{10^7 \text{ K}}\right)^{-1} \left(\frac{\ln \Lambda}{23}\right) > 1, \qquad (4)$$

where $\ln \Lambda$ is the Coulomb logarithm and the collision frequency is assumed to be classical. Hence, an acceptably high density or a low temperature in the current channel will maintain $E < E_D$. Alternatively, a high (anomalous) resistivity will maintain $E < E_D$.

4. A flare that is well suited for testing the applicability of a simple D.C. electric field acceleration, Joule heating model is the May 14, 1980 (1920 UT) flare, discussed in the preceding paper by Kundu. The 6 cm VLA maps show that interaction among several magnetic loops is involved, with the possible formation of a current sheet between two of the loops. The impulsive 6 cm microwave emission peaks ~ 10 sec before the onset of the hard X-ray flare. The microwave and hard X-ray data can be interpreted as follows:

The microwave emission increases from one-tenth its peak value to its peak value in ~ 30 sec. This timescale can be related to either t_a or to t_N. The results of sections 2 and 3 indicate that t_N is more likely to be relevant, with $t_a \ll t_N$. Since $t_N(N = 0.1 N_{peak}) = 0.1 t_N(N = N_{peak})$ and $t_N(N = N_{peak}) - t_N(N = 0.1 N_{peak}) \simeq 30$ sec, $t_N(N = N_{peak}) \simeq 33$ sec.

The hard X-ray emission, in light of the result of Section 2, is taken to be thermal. The electron temperature deduced from the X-ray spectrum is greatest at the onset of the (X-ray) flare. Since the onset of the hard X-ray flare follows the microwave peak by ~ 10 sec, the Joule heating time is $t_J \simeq t_N + 10$ sec $\simeq 43$ sec.

The timescale for the production of runaways is $t_N \equiv N/\dot{N} \simeq N/\gamma n v_J$, where γ is the runaway production rate (Kruskal and Bernstein 1964). This can be written as (a more detailed presentation will be published elsewhere)

$$t_N \simeq 160 \left(\frac{N}{10^{32}}\right)\left(\frac{A}{5.3 \times 10^{17} \text{cm}^2}\right)^{-1}\left(\frac{B}{100 \text{ G}}\right)^{-1}\left(\frac{T}{10^7 \text{K}}\right)^{1/2}\left(\frac{\nu}{20 \text{sec}^{-1}}\right)^{-1}/f\left(\frac{v_c}{v_e}\right) \text{ sec,} \quad (5)$$

where ν is the thermal collision frequency in the current channel and $f(v_c/v_e) \leq 1$ is an exponential function of v_c/v_e:

$$f(v_c/v_e) = 4.66 \, (v_c/v_e)^{11/4} \exp[-2^{1/2}(v_c/v_e) - (1/4)(v_c/v_e)^2]. \quad (6)$$

This timescale is consistent with the inferred timescale of ~ 33 sec.

The fundamental timescale for Joule heating is $\tau_J \equiv nkT/(J \cdot E)$. The actual timescale ($t_J$) will be longer than this, however, since the volume of gas to be heated will generally be larger than the volume of the current channel. The actual heating time can be estimated to be $t_J \gtrsim \tau_J/\varepsilon$, where $\varepsilon \equiv V_J/V_X$, the volume of the current channel divided by the volume of the X-ray emitting gas. A thermal fit to the X-ray spectrum at the time of the temperature maximum gives an emission measure of $EM \equiv n_X^2 V_X \simeq 3 \times 10^{44}$ cm^{-3}. The Joule heating time can be shown to be

$$t_J \gtrsim 3.5 \times 10^4 \left(\frac{A}{5.3 \times 10^{17} \text{cm}^2}\right)^{-1}\left(\frac{B}{100 \text{ G}}\right)^{-1}\left(\frac{EM}{3 \times 10^{44} \text{cm}^{-3}}\right)\left(\frac{n_X}{1 \times 10^9 \text{cm}^{-3}}\right)^{-1}$$

$$\cdot \left(\frac{T}{10^7 \text{K}}\right)^{1/2}\left(\frac{\nu}{20 \text{sec}^{-1}}\right)^{-1}\left(\frac{v_c}{v_e}\right)^2 \text{ sec} \quad (7)$$

(n_X is the average electron density in the X-ray emitting volume, n is the density in the current channel).

In order to achieve the inferred heating timescale of ~ 43 sec through Joule heating, *either the resistivity of the plasma in the current channel must be anomalous, or the density in the X-ray emitting region (and, therefore in the current channel as well) must be much greater than 10^9 cm^{-3}.* With the other parameters remaining unchanged from those used in equations (5) and (7), a collision frequency that is 2×10^4 times greater than the classical thermal (Coulomb) collision frequency will give the required timescales. This also requires $v_c/v_e \sim 5$ and $v_D \simeq 1.5 \, c_s$, were c_s is the ion sound speed. Hence, the ion acoustic or the ion cyclotron instability is a likely source of the anomalous resistivity. If the resistivity remains classical, a plasma density in the X-ray emitting region and current channel of $\sim 10^{11}$ cm^{-3} will give the required timescales. This also requires $v_c/v_e \sim 4$ and $v_D \simeq 3c_s$, however. Since both the ion acoustic and the electrostatic ion cyclotron instabilities set in near or below this drift velocity, *anomalous resistivity is likely to play a significant role in the development of the flare*.

This work was supported in part by NASA Grant NGR 21-002-199, NASA Contract NSG 5320, NSF Grant ATM 81-03809, and NGL 21-002-033.

REFERENCES

Dreicer, H.: 1959, Phys. Rev. 115, pp. 238-249.
Kruskal, M., and Berstein, I.B.: 1964, Phys. Fluids 7, p. 407.
Spicer, D.S.: 1983, Adv. Space Res. 2, No. 11, pp. 135-137.

DISCUSSION

Vlahos: The restriction of the maximum number of electrons accelerated by an electric field E to 10^{31} is an artifact of the oversimplified model. For the model you used, for example, what if the region is split into many small volumes with electric fields in opposite directions, or if the field is localized and the return current flows from the boundaries to produce a total current that is small, but copious non-thermal particles?
Or if $E(t) \sim \cos\omega t$, it accelerates electrons in both tails that carry no current!

Holman: Splitting the acceleration region into many small, oppositely directed current channels is indeed a possible way of getting around the induction field limit on the electron flux. This requires at least 10^4 individual current channels to produce a ($\geq$ 25 keV) hard X-ray burst, however. The plausability of producing this situation, and its stability, requires further study.
I do not see how "return current flow from the boundaries" can result in a small total current, but "copious non-thermal particles". If the current is reduced, so is N, and the nonthermal electron flux is not large enough.
A time oscillating electric field does not necessarily avoid the induction field limitation since, although electrons are accelerated in both directions, they are not cospatial, there is a current, and an induction B field is still generated. Cospatial, oppositely directed beams would avoid the inducation field limitation, but this arrangement will not be produced by simple electric field acceleration.
The model used here is intentionally simple, since its purpose is to test the applicability of the D.C. electric field acceleration of electrons to solar flares. The limitations that I have discussed are within this context. On the other hand, this simple model does appear

to adequately explain at least one class of solar flares, such as the May 14 flare. In this sense the model is not at all over simplified.

Vasyliunas: In computing the magnetic field associated with the runaway electron current, you have assumed in effect that the return current is outside the sheet. Could some of the return current (which includes any displacement current effects present) be within the sheet itself, thus reducing the calculated B?

Holman: Since the current is driven by an electric field (which is necessary to accelerate the runaway electrons), the return current cannot be within the sheet itself. Hence, the return current cannot reduce the calculated induction magnetic field.

Bratenahl: There may be another process (other than this Joule theory), namely, acceleration in double layers - does this help the problem? Have you looked into it?

Holman: The formation of double layers is expected to be driven by the interruption of a macroscopic current, with the electrons accelerated by the enhanced electric field within the double layer. This process is limited by the induction B field associated with the current and, therefore, is also subject to this limitation.

Spicer: It makes no difference what flare mechanism one uses in an inductive circuit to produce a flare, because an inductive circuit responds to an enhanced impedance in the form of double layers, anomalous resistivity, or reconnection by producing a backward emf so as to attempt to keep the net magnetic flux constant (it, of course, does not succeed). However, by doing so $\frac{dN}{dt} = I/|e|$ will not increase, but decrease. In addition, the particles accelerated by the emf may be relativistic, but the number of relativistic electrons will not exceed $\frac{dN}{dt} = I/|e|$ (Spicer 1983).

Degaonkar: We find the difference in time of peak emission of hard X-rays and microwaves varies from a few millisecond to a few seconds. On what mechanism or process does this depend? Does it have bearing on the acceleration of particles or the distance between the sources?

Holman: These are certainly important questions, for which I do not claim to have a general answer. For the May 14 flare, the time difference between the hard X-ray and microwave peaks is taken to be the difference between the Joule heating time, t_J, and the runaway production time, t_N (as is required by the acceleration mechanism). The microwave and hard X-ray sources are essentially cospatial. Shorter time delays in other flares may be related to an entirely different process. Fully non-thermal models for the hard X-ray and microwave emission are constrained by the limit on the accelerated electron flux (Equation 2, and Spicer 1983), however.

D. Smith: Have you accounted for the change in the number of runaways which occurs in the presence of ion-acoustic turbulence?

Holman: Yes. When the plasma resistivity is higher, a higher electric field strength is required to accelerate a given number of runaways. This electric field strength is consistent with the other parameters in the acceleration region.

THE COALESCENCE INSTABILITY IN SOLAR FLARES

T. Tajima and F. Brunel, University of Texas;
J.-I. Sakai, Toyama University and University of Maryland
L. Vlahos and M. R. Kundu, University of Maryland

ABSTRACT

The non-linear coalescence instability of current carrying solar loops can explain many of the characteristics of the solar flares such as their impulsive nature, heating and high energy particle acceleration, amplitude oscillations of electromagnetic and emission as well as the characteristics of 2-D microwave images obtained during a flare. The plasma compressibility leads to the explosive phase of loop coalescence and its overshoot results in amplitude oscillations in temperatures by adiabatic compression and decompression. We note that the presence of strong electric fields and super-Alfvenic flows during the course of the instability play an important role in the production of non-thermal particles. A qualitative explanation on the physical processes taking place during the non-linear stages of the instability is given.

1. Introduction

Direct observations in soft x-rays (Howard and Svestka, 1977) of interconnecting coronal loops suggest that loop coalescence may be a very important process for energy release in the solar corona. It was suggested (Tajima, Brunel and Sakai, 1982) that the most likely instability for impulsive energy release in solar flares is the coalescence instability (Tajima 1982; Brunel, Tajima and Dawson, 1982). In the present article we examine the existing observational and theoretical results together with a global energy transfer model and conclude that the merging of two current carrying solar loops can explain many of the known characteristics of solar flares.

2. Observations

In this section we briefly outline several recent observational results, which agree well with the model of two interacting loops which will be presented in the following section.

The hard x-ray emission from solar flares results from the interaction of energetic electrons with protons and ions, with electron energies ranging from 10 to hundred keV's in the upper part of the chromosphere (Brown, 1971). The microwave emission, on the other hand, is interpreted as gyro-synchrotron emission resulting from the gyration of the energetic electrons around the magnetic field. The microwave emission is closely correlated with hard x-ray emission (Kundu, 1961). With the availability of SMM and Hinotori experiments and ground based observations using the Very Large Array (VLA), it has been possible to obtain spatially resolved two-dimensional images of the microwave burst sources and of hard x-ray sources with energies less than 30 keV. The main conclusions that emerge from the analysis of the existing observations are that the microwave emission is often confined on the upper part of a closed magnetic loop (Marsh and Hurford, 1980; Kundu et al., 1982) and hard x-ray emission is mainly emitted from the foot points of the loop (Duijveman et al., 1982; Hoyng et al., 1983). The time structure of the impulsive emission is usually "spiky" and is characterized by pulses of short duration (Kiplinger et al., 1983; Kaufmann et al., 1984).

The two-dimensional maps which were obtained with the VLA during an impulsive flare observed on May 14, 1980 probably provide the first direct evidence of a coalescence of coronal loops (Kundu et al., 1982; Kundu, this issue). The 6 cm burst appeared as a gradual component on which was superimposed a strong impulsive phase (duration ~ 2 minutes) in coincidence with a hard x-ray burst. Soft x-ray emission (1.6 ~ 25 keV) was associated with the gradual burst (before the impulsive burst, as is to be expected. There is a delay of hard x-ray emission (> 28 keV) relative to 6 cm emission (~ 10 sec delay from 6 cm max to x-ray start and ~ 20 sec delay from 6 cm max to x-ray max). The preflare region showed intense emission with peak T_b ~ 10^7K extended along a neutral line situated approximately in the east-west direction. A burst source of intense emission with T_b ~ 4×10^7K, appeared initially. The most remarkable feature of the burst source evolution was that an intense emission extending along the north-south neutral line (line of zero polarization at 6 cm), possibly due to reconnection, appeared just before the impulsive burst occurred. This north-south neutral line must be indicative of the appearance of a new system of loops. In the 20 seconds preceding the impulsive peak (T_b ~ 1.1×10^9K) the arcade of loops (burst source) changed and ultimately developed into two strong bipolar regions or a quadrupole structure whose orientations were such that near the loop tops the field lines were opposed to each other. The impulsive energy release must have occurred due to magnetic reconnection of the field lines connecting the two oppositely polarized bipolar regions.

Observations of gamma-rays and high-energy neutrons are a relatively new and useful diagnostic of relativistic particle acceleration in solar flares. Gamma-ray lines are the products of nuclear reaction between flare accelerated protons and nuclei with the ambient solar atmosphere. Narrow line emission results from the de-excitation of nuclear levels in solar atmospheric nuclei, such as C, O, Ne, Mg, Si and Fe, and from neutron capture and positron annihilation. Broad-band nuclear emission results from the superposition of such lines and de-excitation radiation from excited heavy nuclei in the energetic particle population. Of particular significance is the 4-7 MeV band in which the nuclear lines C, N and O produce the bulk of the observed emission. The strongest narrow line from disk flares in the 2.223 MeV line results from neutron capture on H in the photosphere. The underlying gamma-ray continuum in solar flares is produced by bremsstrahlung from relativistic electrons. Gamma-ray lines and continuum are formed from the interaction of MeV electrons and GeV ions in the low chromosphere or upper photosphere (average plasma density $\simeq 10^{14}$ cm^{-3}). An interesting result that can be explained with the present flare model is the appearance of a "double peak" in the amplitude profile of the June 7 and 21, 1980 observation of gamma-ray emission (Forrest et al., 1981; Nakajima et al., 1982). These oscillations are present in the electron and ion temperature profiles on the numerical simulation presented below and can be explained from the dynamics of the coalescence instability.

3. The coalescence instability

It is well known that the annihilation of magnetic energy and its conversion into kinetic energy by the tearing instability (Furth et al., 1963) are too slow to account for the impulsive energy release in solar flares. Many authors (Sweet, 1958; Parker, 1963; Petschek, 1964) have proposed fast magnetic reconnection mechanisms. Recently Tajima (1982) found that the reconnection rate for a compressible plasma with weak toroidal magnetic field B_t is much larger (by a factor of $10^2 \sim 10^3$) than that for a nearly incompressible plasma with large B_t and that the sharp transition in reconnection behavior takes place when the poloidal field B_p (created by the field aligned current J_t) exceeds approximately B_t. Brunel et al. (1982) found further that when the plasma is compressible a faster second phase of reconnection sets in after one Alfven time of the Sweet-Parker first phase with reconnected flux

$$\psi = \psi_{SP}(t_A) \cdot (t/t_A)^{\rho_i/\rho_e}, \qquad (1)$$

where ψ_{SP} is the Sweet-Parker flux

$$\psi_{SP}(t) = \cdot^{1/2} B_p(y=a)(\rho_i/\rho_e)^{1/2}(v_{Al}L)^{1/2}, \qquad (2)$$

ρ_i and ρ_e are densities inside and outside of the current channel ($\rho_i \geq \rho_e$), t_A the Alfven time ($t_A = a/v_{Al} = a\sqrt{4\pi\rho/B_p}$), η the resistivity, a the current channel width, and L the length of reconnecting region. According to the above theory, the annihilation of the magnetic flux proceeds much faster ($\psi \propto t^{\rho_i/\rho_e}$) than the

Sweet-Parker rate ($\psi \propto t$), where for compressible plasmas $\rho_i/\rho_e >$ 1. When the plasma has a strong toroidal field and the plasma is incompressible, the reconnection rate reduces to the Sweet-Parker rate even for $t > t_A$. The theory is in good agreement with the computer simulation results of Brunel et al., 1982 and Tajima, 1982. Nevertheless the reconnection process (before the coalescence instability starts) in itself, however fast it is, is not responsible for the large magnetic energy conversion into particle energy, but rather the change is in the magnetic geometry before and after the reconnection process. Indeed only a small fraction of the total poloidal magnetic energy is released through the reconnection process which necessarily takes place at the x-point, i.e. the field null point, where not much magnetic energy is available in the first place.

It is the non-linear development of the <u>coalescence instability</u> of the current filaments (loops) that can release a large amount of magnetic energy (Wu et al., 1981; Leboeuf et al., 1982). Although the coalescence instability is of ideal magnetohydrodynamic (MHD) nature in the linear stage and the growth rate for compressible plasmas is somewhat smaller than that for incompressible cases (Pritchett et al., 1979), the non-linear development of this instability involves field line reconnection and therefore is of nonideal MHD nature. Since the reconnection rate drastically differs by two or three orders of magnitude (Tajima, 1982), the non-linear coalescence time differs by two or three orders of magnitude for case $B_p \gtrsim B_t$ and case $B_p < B_t$. In fact the recent study (Bhattacharjee et al., 1983) confirmed that the rapid coalescence occurs in a compressible plasma as characterized by Eq. (1). We studied the coalescence of two current filaments in detail using analytical and computer simulation techniques.

a. The simulation model
A plasma configuration which is unstable against the tearing and subsequent coalescence instabilities has been studied by a fully self-consistent electromagnetic relativistic particle simulation code (Langdon et al., 1976; Lin et al., 1974).

b. Short description of time sequence
When the linear stage where the energy release is small is past and two magnetic islands (i.e. current filaments) approach, the islands are squashed. The plasma near the contact area of islands is squeezed and has a high density, which leads to fast reconnection according to Eq. (1). Because of this, the total flux reconnection of two islands into a coalesced island takes place only within 1 ~ 2 Alfven times according to the simulation of Tajima et al., 1982. The magnetic energy contained in the island fields is explosively released into ion and electron kinetic energy as seen in Fig. 1. The ion temperature shown in Fig. 1 sharply increases over the non-linear coalescence stage in 1 ~ 2 Alfven times. Significantly, there appear amplitude oscillations in the electron and ion temperature. This temperature oscillation behavior can be attributed to the overshooting of coalescing and colliding of two current blobs.

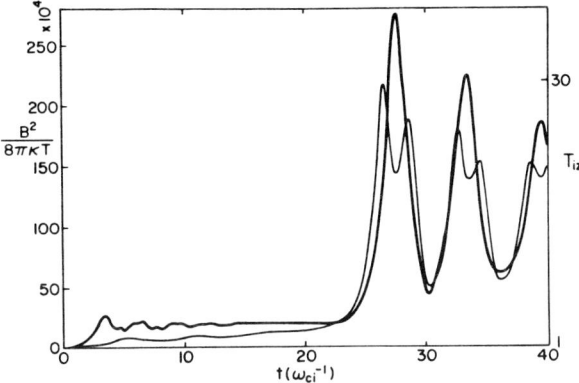

Figure 1. Magnetic field energy and ion temprature oscillations before and after the coalescences of the current loops.

c. Velocity distributions - heating

Once two current blobs coalesce, they are bound by the common magnetic flux and the coalesced larger island which also vibrates, with a smaller amplitude. Within the coalesced island the colliding of two plasma blobs causes turbulent flows which dissipate their energy quickly into heat, thereby reducing the amplitude of the temperature oscillations. As a result, the momentum distributions of plasma electrons and ions [shown in Fig. 2(b) in Tajima et al., 1982] exhibit the intense bulk heating and acceleration of the tail. The heating in the poloidal direction is due to adiabatic compression and decompression of the coalesced loops. The eventual bulk heating is a result of turbulent dissipation of counter streaming instabilities (either the Buneman instability or the modified two stream instability). The temperature in the poloidal direction was increased in our simulation by a factor of 60. The heating in the toroidal direction is due to heating/acceleration by the inductive toroidal electric field which is several times the classical Dreicer field. The momentum distribution of ions in the toroidal direction shows three regimes, the first being the bulk, the second an exponential section, and the third a flat distribution up to the relativistic regime with a relativistic factor $\gamma \sim 2$, where $P_o^2/2M \simeq 10 \times$ (bulk temperature). The momentum distribution of electrons in the toroidal direction shows two regimes, the first being the bulk, the second a flat distribution up to the relativistic region.

d. Temperature pulsations - double peaks

The double peaks in the time development of the temperature (Fig. 1) occur just before ($t = t_1 \sim 27 \omega_{ci}^{-1}$) and after ($t = t_2 \sim 29 \omega_{ci}^{-1}$) the maxima of magnetic fields ($t = t_2$). In Fig. 2 schematic sequential pictures of plasma dynamical behaviour during coalescence are shown.

At $t = t_1$, the magnetic (jxB) acceleration of ions becomes maximum so that the magnetic flux behind the colliding plasma blobs as well as plasma blobs is strongly compressed. This plasma compression causes the first temperature peak at $t = t_1$. After this maximum acceleration phase ions acquire super-Alfvenic velocities along the direction in which the loops collide so that they detach from the magnetic flux against which ions have been compressed. This results in an expansion phase ($t = t_2$) of ions and hence in an adiabatic cooling of the plasma as the magnetic fields obtain maximum values. The process reverses after the maxima of the magnetic fields at $t = t_3 \sim 29\ \omega_{ci}^{-1}$, which gives rise to the second peak of the temperature.

e. High energy tail acceleration − electrostatic fields

The high energy tail particle acceleration of ions and electrons is probably due to a combination of localized electrostatic field acceleration across the poloidal magnetic field (Sagdeev and Shapiro, 1973) and magnetic acceleration of the poloidal to toroidal directions. As shown in Fig. 2(a), electrons are magnetized and are carried away with the magnetic flux, while ordinary ions are accelerated by the JxB force. On the other hand the high energy ions are dragged by the charge separation created near the piled flux.

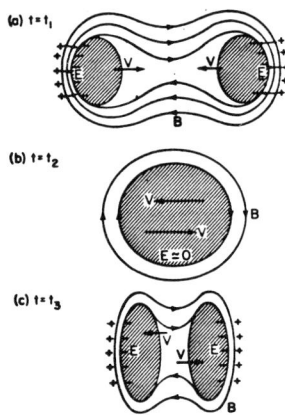

Figure 2. Schematic sequential of pictures of the plasma dynamics during the coalescence.

Ions (as well as electrons) acquire super-Alfvenic velocities upon coalescence. The difference of motions between ions and electrons around $t = t_1$ causes a strong localized shock-like electrostatic field, E, which propagates with a velocity V_p. In Fig. 3, the density distribution and shock-like electrostatic fields just before the coalescence ($t = t_1$) are shown. This $V_p \times B$ acceleration causes the formation of high energy particles in the toroidal direction. By this acceleration process, ions and electrons are accelerated to

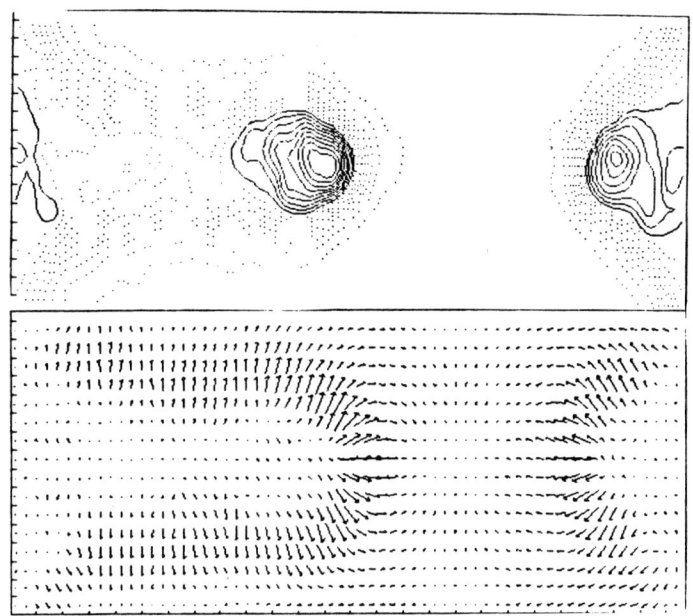

Figure 3. The density and electrostatic field distribution just before the coalescence.

relativistic energies in opposite directions along the toroidal magnetic field.

The temperature anisotropy between the poloidal and toroidal component of the velocity distribution that was observed in our simulation may give rise to the onset of the Alfven-ion cyclotron instability (Tajima et al., 1980). This instability may create large amplitude Alfven waves that are traveling away from the coalescence region as well as ion cyclotron resonance heating by Alfven waves with frequencies near the cyclotron frequency.

f. <u>The role of B_p/B_t on the growth</u>

Our simulations show that the non-linear coalescence time for case $B_p > B_t$ is more rapid by two or three orders of magnitude than the case of $B_p \lesssim B_t$. When the toroidal magnetic field is stronger than the poloidal field, the super-Alfvenic plasma flows cause a plasma rotational motion around the toroidal magnetic field, rather than the counter streaming flow. The result of plasma incompressibility leads to weaken the double peak structure in the temperature oscillations.

g. <u>Flare energetics</u>

The flare loop slowly expands after it emerges from the photosphere as the toroidal field curvature of the loop makes the

centrifugal motion. However, its relative stable configuration
suggests that the quiet loop is in some kind of equilibrium such as
the Walter-Taylor equilibrium which is force-free stable. In time the
toroidal current J_t builds up increasing the poloidal magnetic field
B_p. As the poloidal field B_p reaches a critical value which is of the
order of B_t in magnitude, the adjacent flare current loop can coalesce
rapidly, being facilitated by the fast reconnection process governed
by Eq. (1) (the faster second phase). A process similar to this that
can happen upon increased toroidal current J_t (or more twisted
magnetic tube) is the (global) kink instability. This instability
creates a local section of parallel currents, which may also coalesce
rapidly. Such a fast coalescence of flare current loops proceeds
explosively once in its non-linear regime in a matter of 1-2 Alfven
times, releasing more than 1/10 of the poloidal magnetic energy into
(ion) kinetic energy. Since the flare loop magnetic field (100G) with
current rod size (a = 10^8 cm), $W_c \sim 0.5 \times 10^{20} \ln(L/a) \sim 1.5 \times 10^{20}$
erg/cm and the energy available in length d $\sim$ L($\sim 10^9$ cm) is E = 1.5 $\times$
10^{29} erg for a = 10^8 cm, d = L = 10^9 cm and E = 1.5 $\times$ 10^{31} erg for a =
10^9 cm, d = L = 10^{10} cm. Release of ion energy, therefore, is
$E_{ion} \sim \frac{1}{6} E$ is in between 2 $\times$ 10^{28} erg and 2 $\times$ 10^{30} erg due to the
coalescence. This amount of energy is in the neighborhood of the
energy released in a solar flare (Sturrock, 1980). For magnetic field
$\simeq$ 100G the Alfven time is of the order of 1 sec, which is
approximately the time scale for the explosive coalescence. The time
scale of the impulsive phase is observed to be several seconds, and is
in good agreement with the above theoretical estimate. The sudden
nature of the impulsive flare phase (Sturrock, 1980) is thus explained
by increasing field aligned current and the faster second phase
reconnection in the course of coalescence.

4. Radiation signatures from the coalescence of two solar loops

In order to relate the observations with the physical picture of
the coalescence instability reported above, one must analyze the
energy transfer in a 3-D (global) magnetic topology that connects the
interacting loops and follow the evolution of the energetic particles
and the hot plasma away from the energy release volume. In Figure 4
we present the magnetic topology of the interacting loops during the
coalescence. As was pointed out in the previous section the by-
products of the interaction are: (a) the generation of hot electron
and ion distributions: (b) the existence of run-away tails on the ion
and electron distributions, and (c) the presence of MHD and
electrostatic waves emitted away from the energy release volume. In
what follows we will summarize how each of these components will
evolve in time and locate the most likely place for the emission of
the microwave, x-ray, γ-ray and meter wave-length bursts associated
with the coalescence instability.

a. Energetic electrons and ions stream away from the energy
release region following the magnetic field lines. The majority of
these electrons and ions will reach the chromosphere in a fraction of

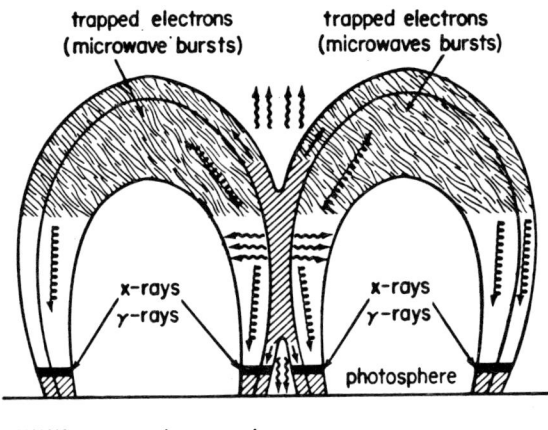

Figure 4. A schematic model for the energy transfer and location of different radiation signatures that results during the coalescence of two solar loops.

a second. The interaction of these particles with the chromosphere will result in bremsstrahlung hard x-ray, gamma-ray continuum and gamma-ray lines. A fraction of the high energy electron population will be trapped in the upper part of the interacting loops and emit gyro-synchrotron radiation. We expect that if the VLA resolves these structures we observe two sources with two different neutral lines. Each radio source will be split into two oppositely polarized sources. This picture is indeed in good agreement with observations, in particular the May 14 event (Kundu et al., 1982). Energetic electrons and ions can also be accelerated by the electrostatic waves excited from the unstable currents inside the energy release volume. Electrostatic waves propagating away from their excitation region will be absorbed in different places inside the merged loops. For example, if the drift velocity of the currents during coalescence exceeds the ion speed, the energy release region becomes a source of lower hybrid waves which propagate across the magnetic field with phase velocity along the field lines $\gtrsim 3v_e$ and which can easily accelerate electrons up to 100 keV energies (Tanaka and Papadopoulos, 1983). These waves will progressively propagate further towards the center of the merged loops since the accelerated electrons escape from the interaction region and locally reduce the damping. The net result is that a relatively large volume around the energy release region can be affected by the electrostatic waves. Excitation of electrostatic ion-cyclotron waves will have the same effect on the ions. We thus suggest that the non-thermal tails co-existing with the hot plasma inside the energy release region and the waves driven by the unstable currents inside it will be responsible for the prompt acceleration of the trapped and precipitated population of energetic electrons and ions that will radiate the microwave, hard x-ray and gamma-ray emission.

b. The hot (several keV) electron and ion components will expand along the field lines forming heat waves propagating along the field lines. It is possible that under certain conditions (see Brown, Melrose and Spicer, 1979 or Vlahos and Papadopoulos, 1979) the heat conduction will be anomalously inhibited due to the formation of ion acoustic fronts in the interface of the hot plasma with the surrounding material. At the same time the chromosphere will also be heated by the precipitated electrons and ions. The result is another hydromagnetic heat wave expanding in the opposite direction. This expansion phase lasts only a few$_0$ seconds and unless one of the interacting loops is larger $L \gtrsim 10^{10}$ cm, it will not be observed with the presently available instruments. The mixture of the hot coronal plasma with the dense and hot chromospheric plasma will be responsible for the appearance of two post-flare loops that will slowly cool off during the decay phase of the flare.

c. MHD waves: Finally a significant amount of energy is emitted in the form of hydromagnetic waves which may or may not steepen to form shocks. These waves are travelling away from the interaction region with velocity (v_s) larger than the local Alfven speed (v_A). They can fill a volume V ($V \simeq (V_s \tau)^3$) which can be as large as 10^{27} cm^3 in a few seconds. the interaction of these waves with the ambient plasma and the pre-accelerated electrons will increase the number and average energy of the energetic particle population. Most of these electrons and ions will end up in the chromosphere but a fraction of them will escape in the interplanetary space.

Although in the discussion above we qualitatively analyze the coalescence of two isolated loops, one must keep in mind that in most cases many such collisions take place simultaneously or nearly simultaneously, lasting sometimes several minutes. In these cases the MHD waves excited from each individual collision join together to form a large volume with turbulent wave activity that accelerates electrons and ions to very high energies with good efficiency. The hydromagnetic waves excited from the coalescence instability can destabilize another solar loop which, as it expands, forms a coronal transient (Sakai 1982) and a shock that drives type II and/or type IV bursts.

5. Summary

The hot plasma, non-thermal particles and waves (electrostatic and MHD) which are present during the coalescence instability can explain several of the observed characteristics of solar flares. The interactions of solar loops in the manner described here are quite common in the sun, but due to the fact that only if B_p/B_t is small the coalescence instability is rapid, not all of them produce impulsive, energetic flares.

Acknowledgements

This work was supported at the University of Texas by NSF grant ATM82-14730 and DOE grant DE-FG05-80ET53088, and at the University of Maryland by NASA grant NAG W-81 and NSF grant 81-03089.

References

Bhattacharjee, A., Brunel, F. and Tajima, T.: 1983, IFSR #93, The University of Texas, Austin.
Brown, J.C., Melrose, D.B. and Spicer, D.S.: 1979, Ap. J., 228, 592.
Brunel, F., Leboeuf, J.N., Tajima, T., Dawson, J.M., Makino, M. and Kamimura, T.: 1981, J. Comput. Phys. 43, 268.
Brunel, F., Tajima, T. and Dawson, J.M.: 1982, Phys. Rev. Lett., 49, 323.
Chupp, E.L., Forrest, D.J. and Suri, A.N.: 1974, in Solar Gamma-, X-, and EUV Radiation (edited by S. Kane) (Reidel, Holland) p. 341.
Duijveman, A.P., Hoyng, P. and Machado, M.E.: 1982, Solar Physics, 81, 137.
Forrest, D.J., Chupp, E.L., Rayan, J.M., Reppin, C., Rieger, E., Kanbach, G., Pinkau, K., Share, G., and Kinzer, G.: 1981, to be published in the Late Volumes of the 17th International Cosmic Ray Conference; Paris, France.
Furth, H.P., Killeen, J. and Rosenbluth, M.N.: 1963, Phys. Fluids, 6, 459.
Gold, T. and Holye, F.: 1960, Monthly Notices Roy. Astron. Soc., 120, 8.
Howard, R., Svestka, Z.: 1977, Solar Phys., 54, 65.
Hoyng, P., Marsh, K.A., Zirin, H. and Dennis, R.B.: 1983, Ap. J., 285, 865.
Kaufman et al.: 1984, Solar Phys., 91, 354.
Kiplinger, A.L., Dennis, B.R., Emslie, A.G., Frost, K.J. and Orwig, L.E.: 1983, Solar Phys,
Kundu, M.R.: 1961, J. Geophys. Res., 66, 4308.
Kundu, M.R., Schmahl, E.J., Velusamy, T. and Vlahos, L.: 1982, Astron. Astrophys., 108, 188.
Kundu, M.R.: this issue.
Langdon, A.B. and Lasinski, B.F.: 1976, Methods in Computational Physics, Academic Press, New York, Vol. 16, p. 327.
Leboeuf, J.N., Tajima, T. and Dawson, J.M.: 1982, Phys. Fluids, 25, 784.
Lin, A.T., Dawson, J.M. and Okuda, H.: 1974, Phys. Fluids, 17, 1995.
Marsh, K.A. and Hurford, G.T.: 1980, Ap.J. (Lett.), 240, 611l.
Nakajima, H., Kosugi, T., Kai, K., and Enome, S.: 1983, Nature, (in press).
Parker, E.N.: 1963, Ap. J. Suppl. Ser., 77, 177.
Petschek, H.E.: 1964, in Symposium on the Physics of Solar Flares, ed. by W.N. Hess, NASA, Washington, p. 425.

Pritchett, P.L. and Wu, C.C.: 1979, Phys. Fluids, 22, 440.
Sagdeev, R.Z. and Shapiro, V.D.: 1973, JETP Lett., 17, 279.
Sakai, J.: 1982, Ap. J., 263, 970.
Sturrock, P.A. ed.: 1980, Solar Flares: Monograph from Skylab Solar Workshop II, Colorado Associated University Press, Boulder.
Svestka, Z.: 1976, Solar Flares, D. Reidel Publ. Co., Holland.
Svestka, Z. and Howard, R.: 1981, Solar Phys., 71, 349.
Sweet, P.A.: 1958, Nuovo Cimento Suppl., 8, (Ser. 10), 188.
Tajima, T.: 1982, in Fusion Energy - 1981, (International Atomic Energy Agency, ICTP, Trieste, 1982) p. 403. Also IFSR#4...
Tajima, T. and Dawson, J.M.: 1980, Nuclear Fusion, 20, 1129.
Tanaka, M. and Papadopoulos, K.: 1983, Phys. Fluids, 26, 1697.
Vlahos, L. and Papadopoulos, K.: 1979, Ap. J., 233, 717.
Wu, C.C., Leboeuf, J.N., Tajima, T. and Dawson, J.M.: 1983, Phys. Fluids, submitted for publication.

DISCUSSION

Sato: I think the initial condition is essential to the result. I wonder if your initial configuration is force-free equilibrium, or non-force-free equilibrium, or non-equilibrium?

Tajima: Although the initial start up of the run is not an exact equilibrium, the stage of the island formation is. These two islands are unstable to coalesce. Remember it takes a long time for two islands to coalesce. But once they approaches a little bit, they do so very rapidly. We also tried runs starting from the Faddeev equilibrium (MHD particle model), in which a rapid coalescence was observed.

van Hoven: This model does not have access to more magnetic energy than a tearing model, it only claims to have faster access to this store.

Tajima: The point is that a global magnetic reconfiguration in a short time scale is needed. Tearing models may be said to be studying some important aspects of this, but it concentrates only on the minor part of the magnetic energy or a very early episode of magnetic energy conversion. What I want to emphasize is that we should more boldly go into the crux of the matter.

van Hoven: How hard is this system driven (how far away is it from equilibrium)? What is v_D/v_{Te}?

Tajima: The original magnetic island positions were essentially equilibrium positions. At that point $v_{D_z} = 0$. Of course as a result of coalescence v_{D_z} becomes non-zero. I do not recall at this moment if $v_{D_z} > v_{Te}$ or not.

van Hoven: What is the spatial resolution in the direction perpendicular to the line of centers?

Tajima: I have 32 grid points in that direction with the subtracted dipole interpolation.

Migluolo: Is the total magnetic energy decreasing?

Tajima: I have tried two sets of investigations. First, the kinetic model which I primarily discussed today has a coupling to the external circuit, i.e. the total effective plasma current $J_z + \frac{1}{4\pi}\dot{E}_z$ for the uniform component kco is returned via the external circuit. Thus the total energy in the system is not constant. The magnetic field increases. Second, the MHD particle model has no coupling to the external circuit and thus the total energy conserves and the magnetic energy decreases.

Migliuolo: Is the work being done to maintain current systems at the boundaries?

Tajima: As I mentioned in the above, in the kinetic model it is an open system.

Sturrock: As I understand it, your model involves two current filaments which merge, keeping the same total current. The magnetic energy will <u>increase</u> in this process. Hence your model cannot represent a solar flare which is generally agreed to involve the <u>release</u> of magnetic energy. By contrast, if two twisted flux tubes merge by reconnection, the total current is found to decrease.

Tajima: Magnetic energy increases. When the inductance of the loop is large, the current (effective total) has a tendency to be kept constant. If disrupted, an inductive electric field will drive currents, which may give rise to double layer type structure. We will be discussing this process in the future (Wagner, Tajima, Akasofu, and Alfvén). So, I do not believe that the only model of solar flares is that of decreasing current.

Vlahos: Although in this model the current is kept constant, one can repeat the experiment by fixing the total initial current. In this case the coalescence will quickly be damped out after the first few encounters.

Tajima: Results from our other starts (i.e. the current is allowed to decrease) using MHD particle code also show relatively prolonged oscillations, although they are much less pronounced.

Spicer: Would you clarify the time scales you are using, because an ion gyro-period is quite short for solar conditions which, according to your view graphs, is the time scale during which these processes occur? Also, would you comment on what you mean when you say that the ions are unmagnetized? This either means you are treating a process which occurs in a time shorter than an ion gyro-period, or turbulance exists that unmagnetized the ions. Finally, the Modified Two Stream Instability is very hard to excite, while the Lower Hybrid Drift instability is much easier, probably the most likely instability to be excited.

Tajima: The time scale of the amplitude oscillations is the compressional Alfvén time. The Alfvén time scale is well separated from the ion cyclotron time scale albeit that the scale separation is much smaller than nature's. Only <u>hot</u> ions are unmagnetized over the distance of the accelerating structure. In fact this hot ion component is the very agency which set up charge separation in the magnetized shock that is responsible to accelerate high energy tails. The turbulent heating time scale is several Alfvén times and I clearly see counter streaming ions which become turbulent as they stream through each other.

Vasyliunas: The energy release time should be bounded by the travel time τ_A of an Alfvén wave across the system. In your simulations, the time scale is of the order of $10 \, \omega_{ci}^{-1}$: (ω_{ci} = ion gyrofrequency). If $\tau_A \sim \omega_{ci}$, this implies that the system is of relatively small size, but for small systems the energy release is relatively easier -- the entire difficulty in astrophysical applications comes from the enormous scale sizes.

Tajima: The energy release time is 1 ∿ 2 Alfvén times. Because of the numerical simulation nature, we used much smaller mass ratio than 1800. The relevant dynamical time scale of energy release, which is clearly tied to the compressional Alfven time, has to be recognized. We have to use our brains to properly interpret the raw data. Also, remember we did carry out MHD particle simulations in which fast coalescense occurred. For more detail, see Bhattacharjee, Brunel and Tajima.

3D SIMULATION OF EXTERNALLY DRIVEN RECONNECTION

T. Sato
Institute for Fusion Theory, Hiroshima University, Hiroshima
730, Japan, and Institute of Geophysics and Planetary Physics,
University of California, Los Angeles, CA 90024, U.S.A.

A 3D magnetohydrodynamic simulation is presented. The essential conclusions obtained by the previous 2D model, such as the slow shock formation and the strong plasma jet generation, are reconfirmed. In addition, several new findings pertinent to three dimensionality are obtained. Among them, particularly interesting and important is the generation of field aligned currents at the slow shock associated with a local interruption of the neutral sheet current. It is also interesting to observe the generation of super-magnetosonic flows with the Mach number of 2.

1. INTRODUCTION

The solar flare and the magnetospheric substorm are the most familiar, naturally occuring energy release phenomena. Among others, magnetic reconnection is the most efficient and explosive energy conversion mechanism in plasmas.

In the earth's magnetosphere, there are two places where reconnection is likely to occur; they are the dayside magnetopause and the magnetotail neutral sheet. On the dayside no such a violent energy release as we observe on the nightside is observed. Provided reconnection actually takes place on both sides, therefore, the energy conversion rate of reconnection must be largely dependent upon circumstantial conditions.

In the past we studied extensively the reconnection process for several different conditions and compared their results in detail by using 2D simulation codes (1-5). The conclusion obtained by these studies is that the energy conversion rate is certainly largely dependent on the external conditions. More specifically, energy conversion occurs most efficiently and rapidly in such a situation that an open-ended Harris type high β neutral sheet is locally compressed by an external force. The conversion rate is rather independent of resistivity but largely dependent on the strength of the driving force (1).

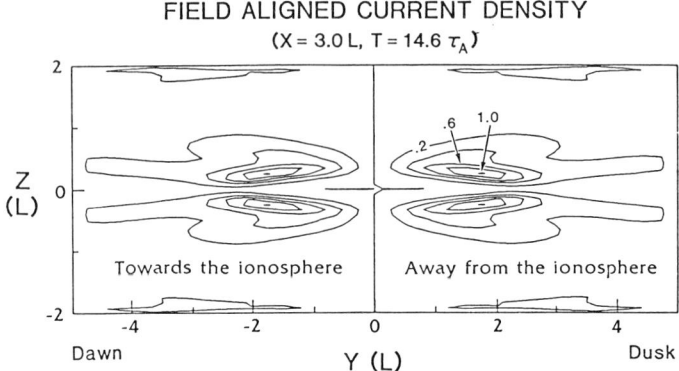

Fig. 1 Contour map of the field aligned current generated by 3D driven reconnection.

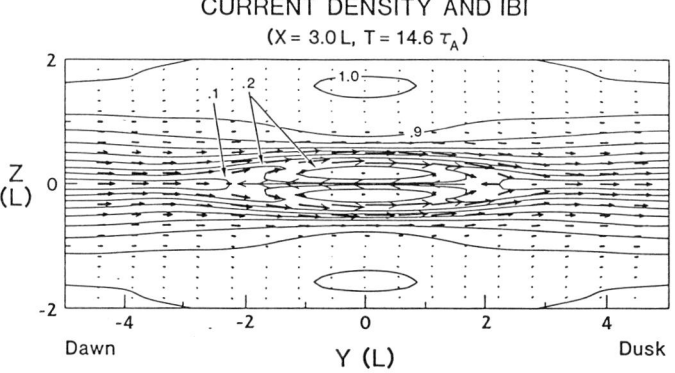

Fig. 2 Neutral sheet current vectors on the same cross section as that of Fig. 1. The solid lines represent the contours of the magnetic field strength.

2D models were sufficient to reveal the slow shock structure and the plasma acceleration. Actual magnetospheric substorms, however, are associated with other complicated features that can never be explained by a 2D model. As such, we know the existence of localized field aligned currents which are presumably connected to strong westward ionospheric currents called "westward auroral electrojets" (6), and the three-dimensional deformation of the magnetic field.

These facts provide us with a sufficient motive force to study the externally driven reconnection in 3D dimensions (7).

2. SIMULATION MODEL AND RESULTS

A rectangular box nested by equally spaced grid points in three directions (x,y,z) is provided wherein a Harris type equilibrium is set up; the neutral sheet is placed at z=0, and the anti-parallel magnetic field is parallel to the x axis. A plasma with a certain amount of magnetic energy density is continuously injected through the two boundary planes parallel to the neutral sheet ("input" boundaries, $z = \pm L_z$). On the two output boundaries ($x = \pm L_x$) and on the side boundaries ($y = \pm L_y$), we impose a free boundary condition that the normal derivative of any variable vanishes. The plasma injection pattern on the input boundaries is assigned in such a way that the neutral sheet is most strongly compressed at the origin (x=y=z=0) of the simulation box.

The most striking feature observed is the local interruption of the neutral sheet current and associated generation of field aligned currents. Fig. 1 shows a contour map of the field aligned current density on a cross section close to the output boundary. It is clearly seen that two pairs of field aligned currents are generated in a narrow band at the slow shock front, namely, at the sharp edge of the plasma sheet: The upper pair closes in the northern hemisphere and the lower one in the southern hemisphere in the earth's case. The numbers attached to the contour lines represent the current density normalized by the initial maximum neutral sheet current. It turns out that the maximum field aligned current density becomes comparable to the neutral sheet current density, this indicating a complete interruption of the neutral sheet current at the edge of the plasma sheet. In fact, the neutral sheet current is locally interrupted. The neutral sheet current density vectors are shown in Fig. 2 for the same plane as that in Fig. 1. Along the second row from the midplane (the neutral sheet plane) the vectors are seen to disappear in the range $-2 \leq Y \leq 2$, this being coincident with the location of the field aligned current divergence and convergence. Incidentally, the solid lines represent the equi-contours of the magnetic field strength. In this respect, it is interesting to remark that the field aligned current generation is associated with $\underline{j}_{NS} \cdot \nabla B$ where $\underline{j}_{NS}$ is the neutral sheet current. This is in good agreement with the previous prediction (8) based on the theory of field aligned current generation (9).

An important implication of such a strong field aligned current generation by 3D externally driven reconnection is that electrons could be accelerated along the field lines as a result of some current driven instability, possibly by a double layer.

Another interesting feature observed is the generation of a super-magnetosonic flow. The distributions of the outflow speed and the magnetosonic speed along the x axis are shown in Fig. 3. Beyond a certain distance from the x point the outflow becomes super-magnetosonic and the Mach number reaches roughly 2. We also wish to remark that the outflow has a strong parallel component near the edge of the plasma sheet. As predicted in the earlier literature (8), a super-magnetosonic flow along the mirror field

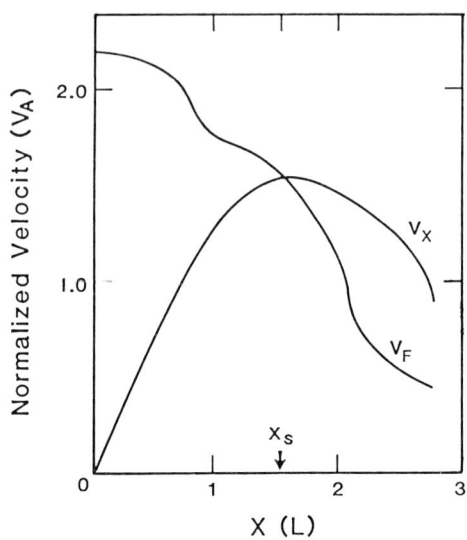

Fig. 3 The distributions of the generated outflow speed (v_x) and the fast magnetosonic speed (v_F) along the line passing the x point on the neutral sheet.

would also be a potential energy source of field aligned acceleration of electrons. We have recently developed a theory that can adequately accelerate electrons up to the directed energy of proton flows.

In conclusion, externally driven reconnection can generate localized strong field aligned currents as well as the super-magnetosonic plasma flows in a reasonably fast time scale, say, 10 τ_A (Alfvén transit time). These macroscopic energies would be potential energy sources of parallel acceleration of electrons.

REFERENCES

1) Sato, T. and Hayashi, T., Phys. Fluids, 22, 1189, 1979
2) Sato, T., J. Geophys. Res., 84, 7177, 1979.
3) Sato T. and Walker, R.J., J. Geophys. Res., 87, 7453, 1982.
4) Sato, T. and Hasegawa, A., Geophys. Res. Let., 9, 52, 1982.
5) Sato, T., Oda, Y., Otsuka, S., Katayama, K. and Katsurai, M., to be published in Phys. Fluids.
6) Kamide, Y., Space Sci Rev., 31, 127, 1982.
7) Sato, T., Hayashi, T., Walker, R.J. and Ashour-Abdalla, M., Geophys. Res. Lett., 10, 221, 1983.
8) Sato, T., in Magnetospheric Plasma Physics, edited by A. Nishida, Center for Academic Publ., Japan and D. Reidel Publ. Co., Tokyo, 1982.
9) Hasegawa, A. and Sato, T., in Dynamics of the Magnetosphere, edited by S.-I. Akasofu, D. Reidel Publ. Co., Boston, 529, 1979.

DISCUSSION

Vasyliunas: What boundary conditions were imposed at the x = const. boundaries, especially the earthward ones?

Sato: Free boundary conitions are imposed. So field-aligned currents can flow out. I suppose that driven reconnection can be a current-generator, in other words, the impedance of the reconnection region would be much larger than the ionospheric resistance. Therefore, I don't think that the inclusion of the ionosphere would alter the conclusion so much. However, certainly it is an interesting thing to develop a model that can include the finiteness of the field line and the terminating ionosphere.

Krishan: While calculating the jets, have you included gravity? This must be done before you can apply your work to double ribbons on the Sun.

Sato: No. Explicitly the earth's magnetosphere is considered. Inclusion of gravity may somewhat destroy the symmetry. But I do not think that gravity would make any essential difference.

Birn: What were the boundary conditions of your 3-D model in the y direction?

Sato: A free boundary condition is imposed; that is, the normal derivative vanishes.

Steinolfson: Is there an asymmetry along the tail in your initial state?

Sato: No. A symmetric configuration is taken, since I am interested in a rather general feature.

Bratenahl: Your last two figures refer to structures as "fast shocks", but actually these are fast mode expansion fronts because they go from sub-Alfvenic to super-Alfvenic, is it not so?

Sato: I have not checked quantitatively the Rankine-Hugoniot shock relation, but the observed feature that the outflow decreases after it passes the super-magnetosonic point seems to indicate it to be a fast shock. The fact that the tail current direction reversed in the super-magnetosonic region indicates that the outflow is self-blocked, thus leading to shocks.

OPENING OF THE MAGNETIC FIELD LINES IN A FAST ROTATING MAGNETOSPHERE, WITH AN APPLICATION TO JUPITER

J. J. ALY
Service d'Astrophysique - CEN Saclay
91191 Gif-sur-Yvette Cedex, FRANCE

In this Communication, we consider a simple model of magnetosphere around a fast rotating Jupiter-like object possessing a spin-aligned dipolar moment μ. In this model, low-energy plasma released by inner sources located beyond the corotation radius diffuses outward through closed lines, forming a thin equatorial disk in which there is a quasi-static balance between the centrifugal force and the magnetic tension. At some critical radius r_0, however, the magnetic field is no longer strong enough to hold the matter, and blows open, the matter escaping freely (Fig. 1) (this model has been introduced by Hill and Carbary, 1978).

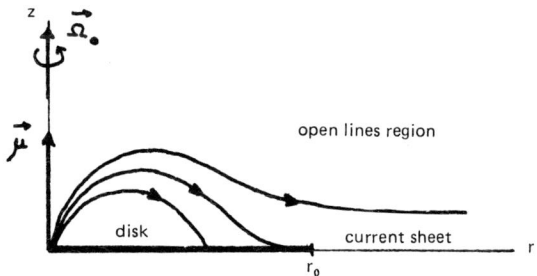

Figure 1: Geometry of the model

The complete solution of the whole set of equations of this model is well beyond the scope of this Paper, in which we just would like to present the solution of the following restricted sub-problem, which can be solved analytically: for given functional forms for the quantities characterizing the plasma in the closed region (angular frequency $\Omega(r)$ and flux-tube mass-contents $n(r)$ for $r < r_0$), what is the "self-consistent" value of r_0, and what is the structure of the magnetic field (B_r, B_ϕ, B_z) in the region outside the disk and the central object (we will assume this region to be force-free, and the currents flowing through it to have a negligible effect on the poloidal component $\vec{B}_p$ of $\vec{B}$).

We first consider the following problem for the (quasi-vacuum) potential $A = \int_0^\infty c(k) e^{-k|z|} J_1(kr) dk$ of that part of $\vec{B}_p$ created by the disk currents: to find the function $c(k)$ such that: i) for $z = 0^+$, $r < r_o$, the B_r component of the field is given by $B_r = + 2\pi r \Omega^2(r) \eta(r) = \chi(r)$ (from the radial component of the momentum balance equation) and then $\int_0^\infty k c(k) J_1(kr) dk = \chi(r)$; ii) for $z=0$, $r>r_o$, the z-component of the field is zero (open field) and then $\int_0^\infty c(k) J_1(kr) dk = -\mu/r^2 + \phi/2\pi r$, where the constant of integration ϕ is equal to the flux on the open lines; iii) at $(z=0, r=r_o)$, the field is continuous (this condition will determine the value of ϕ).

Solving this problem by standard techniques (Sneddon, 1966), one gets a formal solution for the poloidal structure of the field. For this solution to be physical, it is necessary that the components B_r and B_z of the field keep the same sign on $\{z=0^+, r<r_o\}$ ($B_r > 0$, $B_z < 0$ for $\mu > 0$) and that the surface mass density $\sigma = -\eta B_z > 0$ (this is obvious from the momentum balance equation above, and the fact that the plasma diffuses). Then we choose $\chi(r) > 0$ and $\eta(r) > 0$, and look for a condition on these functions for $B_z(r) < 0$. A necessary condition writes

$$\int_0^1 \frac{x}{(1-x^2)^{1/2}} \frac{d}{dx} [x^2 \chi(r_o x)] \leq \frac{2\mu}{r_o^3} \tag{1}$$

(1) is also sufficient if we restrict our attention to the large class of function χ for which $\frac{d}{dr}[r\frac{d}{dr}(r^2\chi)] \geq 0$. The maximum value of r_o for which (1) holds has clearly to be interpreted as the opening radius. If we take for instance Hill and Carbary's assumptions ($\Omega=\Omega_o=$ angular frequency of the central object, $\eta=\eta_o$), one gets $r_o = (\mu/2\pi\eta_o\Omega_o^2)^{1/4}$.

The toroidal component of the field is easily computed with the assumptions above. At the surface of the disk, one has, because of angular momentum conservation,

$$B_\phi = \frac{2\pi \dot{M}}{B_z} \frac{d}{dr}(r^2 \Omega) \tag{2}$$

($\dot{M}$ is the rate at which plasma is released by the inner sources), and this value can then be transported along the lines of the zero-order poloidal structure by using the well-known force-free relation rB_ϕ = constant along a line.

To apply the results presented in this Paper to Jupiter's magnetosphere, it is necessary to do some specific hypotheses on Ω and η, as do Hill and Carbary. Using their scaling and their value for $\hat{M}$, one gets $r_o \simeq 35\ R_J$. Choosing, instead of $\Omega = \Omega_o$, $\Omega = \Omega_o a^2/(r^2+a^2)$ to take into account the observed lack of strict corotation, one gets a larger value for r_o (one can take e.g. a equal to the "corotation breaking" radius introduced by Hill (1979)).

The research summarized in this Paper has been completed during a stay of the author at the University of Illinois at Urbana-Champaign, supported in part by a CNRS-NSF exchange grant, and by NASA grant NSG 7653 and NSF grant PHY 80-25605.

REFERENCES

Hill, T. W., 1979, J. Geophys. Res., **84**, 6554.
Hill, T. W. and Carbary, J. F., 1978, J. Geophys. Res., **83**, 5745.
Sneddon, I. N., 1966, Mixed Boundary Value Problems in Potential Theory, Wiley, New York.

DISCUSSION

Sturrock: Your assumption that the system has linear translational symmetry has the drawback that an open field system would have infinite energy per unit length. Similarly, the assumption that $j = \mu B$, where μ is a constant, involves an infinite total current in the system. Hence it would be better to relax these two assumptions although this would, of course, make the problem much more difficult.

Aly: That's right, considering a translational invariant configuration makes any energetic considerations impossible to apply to actual situations, as the energy per unit of length of the open field is indeed infinite (the energy of the sequence of force-free field $\{A_\mu\}$ increases as Log μ when $\mu \to \infty$). However, it is possible to prove an asymptotic result, similar to the one I have presented here, for other 2-D configurations for which the energy of the open field, corresponding to the given fixed value of the normal component B_n of the field on the boundary, is finite. In that case, you find that the energy of the force-free field increases monotonically with the parameter μ which measures the degree of shearing of the structure, and approaches asymptotically the energy of the open field when $\mu \to \infty$. This proves indirectly that the energy of the open field is the least upper bound for the energy of the force-free configurations having the same given B_n on the boundary and satisfying the conditions I have quoted in my paper.

Birn: In your theorem, did you have to assume that a solution A_μ exists for all values of μ?

Aly: Yes, the asymptotic result I have presented rests on the assumption that the sequence $\{A_\mu\}$ does exist for all values of μ (actually, it can be also applied if $\{A_\mu\}$ exists only for large values of μ, $\mu > \mu_1$ say). At the present time, the existence of $\{A_\mu\}$ is still a conjective, there is not yet a complete proof of this point. However, it is worth noticing that we are sure that $\{A_\mu\}$ exists in some cases, as we know explicitely some particular sequences, as the one you have computed yourself with Goldstein and Schindler.

QUASI-STATIC EVOLUTION OF FORCE-FREE MAGNETIC FIELDS AND A MODEL FOR TWO-RIBBON SOLAR FLARES

J. J. Aly
Service d'Astrophysique – CEN Saclay
91191 Gif-sur-Yvette Cedex, FRANCE

Abstract. We show that a sheared 2-D force-free field can evolve in a quasi-static way towards an open configuration, and apply this result to a qualitative theory of two-ribbon solar flares.

I. INTRODUCTION

A popular model for the main phase of a two-ribbon solar flare involves the reconnection of a magnetic field which has been previously blown open by some unspecified mechanism (Kopp and Pneuman, 1976). In this communication, I would like to show that such an open field can be produced (or at least closely approached) by slowly shearing, during a long enough period of time, the feet of the coronal field lines on the photosphere.

II. QUASI-STATIC EVOLUTION OF 2-D FORCE-FREE FIELDS

To show this result, let us consider a quite simple model in which the corona is taken to be the half-space $\{z > 0\}$ and the coronal field $\vec{B}$, which is force-free, is assumed to be invariant by the translations parallel to the x-axis. Then we can write (Birn and Schindler, 1981) $\vec{B} = \vec{B}_p + B_x \hat{x} = \nabla A \wedge \hat{x} + B_x(A)\hat{x}$, where the potential $A(y,z)$, which can be used to label the field lines $\mathcal{C}$, as well as their projection $\mathcal{C}_p$ onto the plane $\{x=0\}$, obeys the equation

$$-\Delta A = \frac{d}{dA} \frac{B_x^2(A)}{2}. \qquad (1)$$

On the other hand, we have the following relation between $\vec{B}$ and the difference $\xi(A)$ between the x-positions on $\{z=0\}$ of the left and right feet of $\mathcal{C}(A)$ (Fig. 1)

$$\xi(A) = B_x(A) \int_{\mathcal{C}_p(A)} \frac{ds_p}{B_p} = -B_x \frac{d\Sigma}{dA} \qquad (2)$$

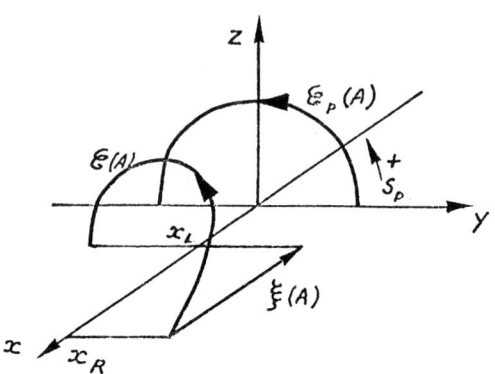

Fig. 1 Geometry of the model

where $d\Sigma$ is the area between $\mathcal{C}_p(A)$ and $\mathcal{C}_p(A+dA)$.

Let us now start from a potential configuration $A_o(y,z)$ without critical points ($\nabla A_o \neq 0$), and apply to the feet of its lines on $\{z=0\}$ a given velocity field, parallel to the $\hat{x}$-axis. Then the coronal field is brought by electromagnetic coupling into an evolution that we suppose to be sufficiently slow for a quasi-static approximation to apply. If the embedding plasma is perfectly conducting, $\xi(A,t)$ can be assumed to be known and we will take it hereafter to be of the form $\xi(A) = \mu\zeta(A)$, where ζ is a given function and μ ($0 \leq \mu < \infty$) will be considered as a mere parameter.

If we assume that there is a continuous sequence of solutions $A_\mu(y,z)$ of (1)-(2) corresponding to these conditions, with the topology of the lines of A_μ being independent of μ, then, when $\mu \to \infty$, A_μ approaches asymptotically an open field, in which all the current is concentrated in an infinitesimally thin sheet.

The proof of this statement rests on the following fundamental inequality

$$\int_{\{z>0\}} B_x^2 \, dy\,dz = -\mu \int \zeta(A) B_x(A) dA \leq \left[\int_{-\infty}^{+\infty} B_z^2(y,0) dy \int_{-\infty}^{+\infty} y^2 B_z^2(y,0) dy\right]^{1/2} \quad (3)$$

from which it can be deduced, with the help of (2), that $\lim_{\mu\to\infty} B_x(A) = 0$ for almost all values of A, and $\lim_{\mu\to\infty} |d\Sigma/dA| = \infty$ for almost all values of A for which $\zeta(A) \neq 0$. Then the field expands indefinitely and it approaches a configuration with open lines in which the current must necessarily be concentrated in singularities. Those can be shown to form

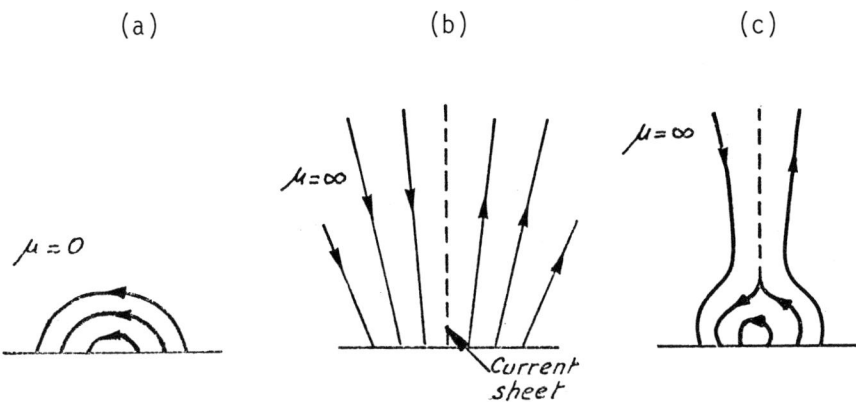

Fig. 2 Totally (b) and partially (c) open configurations which can be approached by large shearing of a potential field (a). See text for details.

a current sheet going to infinity and starting either from (0,0) at which $B_z(0,0) = 0$ (if $\exists a > 0$, $\zeta[A(y,0)] \neq 0$ for y ϵ] 0, a[(Fig. 2b), or (if $\exists$ b > 0, $\zeta[A(y,0)] = 0$ for yϵ[0,b]), from the line separating the region where $\zeta(A) \neq 0$ from the region around (0,0) where $\zeta = 0$ (Fig. 2c).

It is worth to note, to support our assumption on the existence of the sequence A_μ, that an explicit example having $\xi[A(y,0)] = \mu|y|$, has already been computed (Birn and Schindler, 1981).

III. APPLICATION TO THE THEORY OF TWO-RIBBON SOLAR FLARES

Let us now return to the problem of the origin of the two-ribbon flares, for which we suggest the following simplified scenario, in which the previous asymptotic result is included along with ingredients already proposed by other authors. At first, there is a preflare phase, which can be divided into two stages: in the first one, the magnetic structure, which is supposed to be sheared by photospheric motions parallel to the inversion line, does not change very much, but a large amount of energy is stored in it; in the second one, the field expands and is brought into a quasi-open configuration. This configuration however is resistively unstable and a tearing mode develops. That the sheared magnetic structure is not destroyed by such an instability before reaching the quasi-open configuration may be due either to the fact that the shearing process is fast enough for the tearing mode to not fully develop in the mildly sheared field, or to the existence of a stabilizing effect (line tying?...) whose effectiveness decreases when the field is stretched out. The tearing mode initiates a large scale reconnection process, which proceeds at a rising neutral point and releases the free magnetic energy stored in the configuration. This process is responsible for the appearance of the hot coronal flare loops and of the two receding bright H_α ribbons on the photosphere, as in Kopp and Pneuman's model. However, as our field is not really open, reconnection ceases after some period of time, and one is left with a configuration with almost zero shear but with a suspended weakly twisted flux tube (magnetic island). In accord with Svestka's interpretation (1983), the gigantic stationary arche, recently observed above two active regions where two ribbon flares occurred, may be identified with such a tube, assumed to be in equilibrium.

The results reported in this paper have been obtained during a stay of the author at the University of Illinois at Urbana, and the University of California at Berkeley, supported in part by a CNRS-NSF exchange grant, and by NASA grant NSG 7653 and NSF Grant PHY80-25605 at Illinois.

REFERENCES

Birn, J. and Schindler, K., 1981, Chap. 6 in "Solar Flare Magnetohydrodynamics," E. R. Priest, Ed., Gordon and Breach, New York.
Kopp, R. A. and Pneuman, G. W., 1976, Solar Phys., 50, 85.
Svestka, Z., 1983, Space Science Rev., 35, 258.

SOLAR MICROWAVE EMISSION IN ACTIVE REGIONS

B. Lokanadham, P. K. Subramanian and M. Sateesh Reddy
Department of Astronomy, Osmania University, Hyderabad

and

B. M. Reddy and D. R. Lakshmi
National Physical Research Laboratory, New Delhi, India

Abstract: Multi-frequency Observations of Solar Microwave bursts recorded during solar maximum period 1980-81 are analysed and compared with x-ray data for studying the nature of microwave emissions from active regions. Most of the microwave burst spectra showed that the spectral index below the peak frequency is always less than 2

The magneto-ionic conditions of the burst sources and the electron energies as obtained from these multi-frequency observations of the bursts showed that the centimetric and x-ray observations are satisfactorily explained, if the emitting regions are dense, hot and compact associated with strong magnetic fields of a few hundred gauss, suggesting that the thermal gyroresonance process is the most likely emission mechanism involved in the emission of microwave and x-ray radiations from the active regions of sun.

1 INTRODUCTION

The solar microwave bursts have been extensively studied by several investigators and established their characteristics (Kundu 1965). With the advent of the recent developments of high resolution radio interferometric techniques and space instruments, much work has been done on the structure of solar active regions (Kundu etal 1977; Pye etal 1978; Feldman and Doschek 1978; Lang and Willson 1980). The recent skylab x-ray and EUV data have provided an additional wealth of information on solar flare phenomenon. (Reeves etal 1976; Vorphal etal 1977). It has been realized that a coordinated study of centimetric and x-ray observations of solar flares are of particular important for understanding the physical mechanisms involved in the active regions of microwave emission (Schmahl 1980). The aim of this paper is to make such a comparative study of centimetric observations of active regions and the simultaneous x-ray observations carried out during solar maximum period 1980-81 in order to understand the emission mechanisms.

2 MICROWAVE OBSERVATIONS AND SOURCE SPECTRA

The nature of these active regions can be investigated either by means of intensity and polarization measurements at several frequencies or by comparing radio measurements at a single frequency with simultaneous observations at other wavelengths (Kakinuma and Swarup 1962).

A first step in this direction is to construct a spectrum of a microwave burst from its multi-frequency radio observations and to make a statistical analysis of such spectra obtained from the study of a number of events recorded during solar maximum period. A statistical analysis of the low frequency part of microwave spectra makes it possible to estimate the importance of different obsorbing mechanisms and a better understanding of the physical conditions of the microwave source (Schoechlin and Magun 1979).

Microwave solar burst observations have been carried out simultaneously at frequencies of 2.8, 10 and 19 GHz. The 10 GHz microwave Dicke-type radiometer was operated at Japal-Rangapur Observatory of Osmania University and the other two at Physical Research Laboratory, Ahmedabad. These observations form a part of the international solar maximum year program for solar flare studies. Fig. 1 shows simultaneious recordings at these frequencies of a typical solar burst observed on 4 JUNE 1980. These are compared with the x-ray data obtained from SMS mission and H_α photographs. It is seen from the figure that the time of occurrance of x-ray burst is coinciding well with that of microwave bursts suggesting, that both the radiations are originating from the same solar active region.

The multifrequency observations of the microwave solar bursts carried out in India during the solar maximum period and the international Solar Geophysical data are used to construct radio spectra of the sources as shown in Fig. 2. A typical microwave spectrum is characterized by the parameters:
Spectral index $\alpha = d(\log s)/d(\log f)$ where s is flux in s.f.u. and f is frequency in MHz, maximum flux density s_{max} and the peak frequency f_{max} (Schoechlin and Magun 1979).

The magnetic fields associated with burst source and the electron energies involved in the production of microwave bursts are estimated from the source spectrum, since a radio specturm has a maximum at a frequency $f_{max} \simeq 4 f_H$, where f_H is the gyrofrequency (Takakura 1967). For the event of Fig. 1 the magnetic field strength of the source has been estimated as ~ 500 gauss. The energy of electrons producing the microwave burst is ≈ 1.5 MeV, which is mildly relativistic. The angular size of the burst source has been determined using the relation given by Kellerman (1966) and modified to solar conditions as

$$f_{max} = 24 \, H^{1/5} \, (s_{max}/\varphi^2)^{2/5}$$

In the present case the source size $\varphi \approx 4$ arcsec.

3 STATISTICAL ANALYSIS OF MICROWAVE SPECTRA

From the observations recorded by the above radiometers combined with that of solar geophysical data of 1980-81, about 60 active events have been analysed. The radio spectra of these events have been plotted similar to that in Fig. 1 and the corresponding statistical parameters of the Low frequency part of the spectra are determined, The distribution of spectral indicies as shown in Fig. 3 indicates the peak occurrence of α is in the range of 1.0 to 1.5. The most probable value of α has been estimated as 1.3.

Following the method as discribed in section 2, the magetic fields of the energetic sources have been determined. Most of the sources are associated with magnetic fields in the range ~ 400 to 1000 gauss. The energy range of electrons is of the order of 1.5 to 2 MeV. The source sizes emitting the x-ray and microwave emissions are estimated as 4 - 8 arcsec with brightness temperatures in the range of 1 to 2 x 10^6 K. The values of all these parameters as determined from radio spectra are in good agreement with high resolution measurements of the active regions of sun (Allissandrakis and Kundu 1975, Pick 1980, Velusamy and Kundu 1980).

4 INTERPRETATION

The above estimations of the source parameters - the magnetic field strengths, electron energies, source sizes and brightness temperatures indicate that both the x-ray and microwave emissions can be satisfactorily explained, if the emitting regions are dense, hot and compact associated with magnetic fields of a few hundred gauss.

The estimated radio brightness temperature, 1 to 2 x 10^6 K is of the same order of the maximum temperature of EUV and x-ray loops. The region with this temperature is optically thick and it cannot be explained due to thermal bremsstrahlung absorption. Hence, the source of absorption is mostly due to gyroresonance absorption (Zheleznyakov 1962, Kundu 1965) which satisfies the present estimation of magnetic fields from the condition $f_{max} = 4f_H$. This is in good agreement with the results obtained by Schmahl (1980) from simultaneous x-ray and centimetric observations of active regions, where he found that the 3rd and 4th harmonics are highly optically thick layers. The present low value of spectral index 1.3 may then be explained mainly due to this gyroresonance absorption.

Thus, it is concluded from the present estimated values of the magnetic fields, electron energies, spectral indicies, source sizes and brightness temperatures as obtained from the radio spectra of microwave bursts, points out to the thermal gyroresonance process as the most

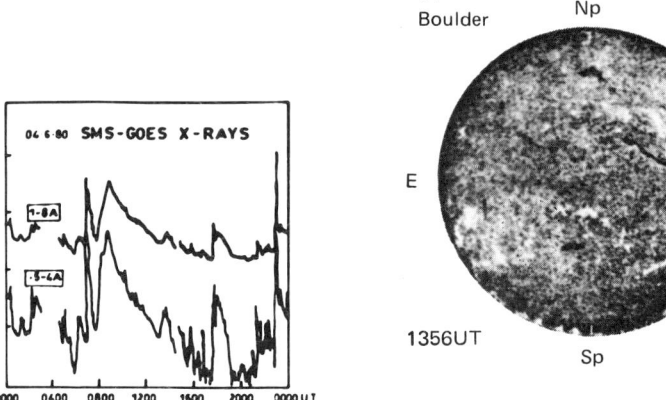

Hα Photograph on 04.6.80

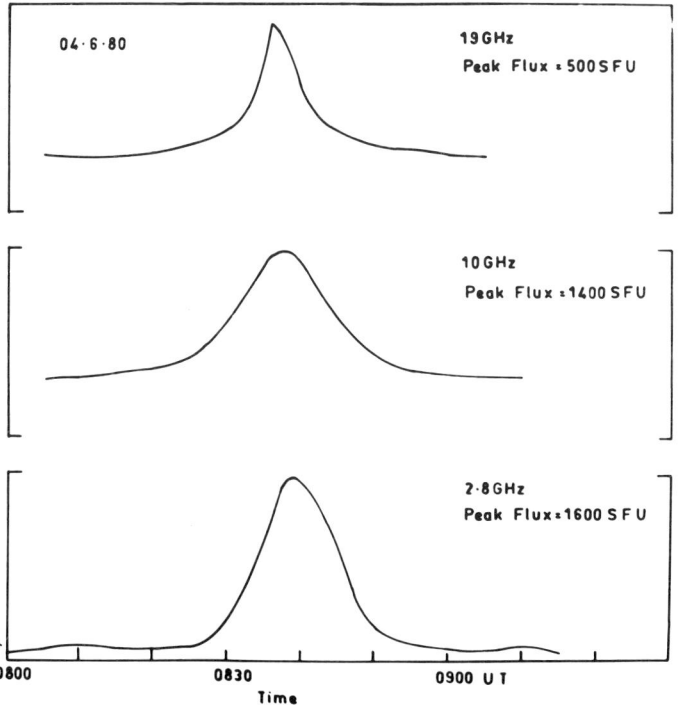

Fig.1 Intense solar microwave burst recorded at different frequencies and simultaneous X-ray observations

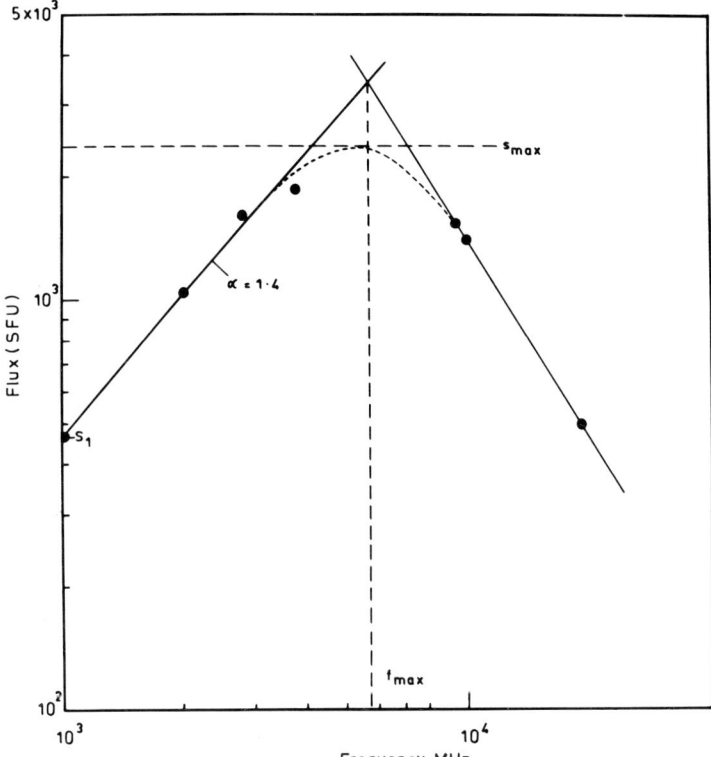

Fig.2 A Typical microwave spectrum of the burst observed on 4 June 1980 0840 UT

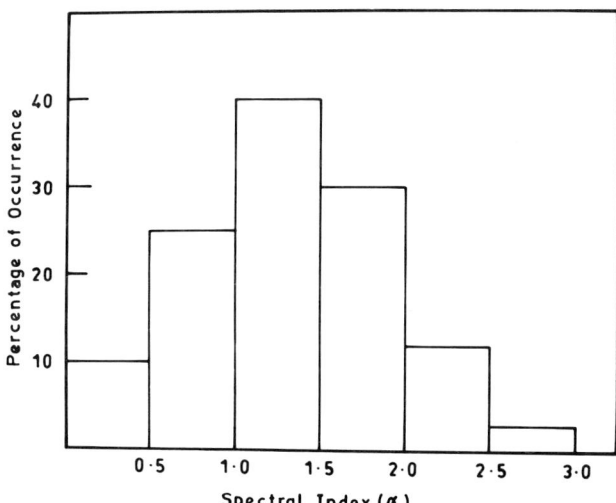

Fig.3 Distribution of spectral index

likely emission mechanism for the bright radio components associated with active regions.

5 ACKNOWLEDGEMENTS

Dr. B. Lokanadham thanks the Vice-Chancellor, Osmania University for providing a travel grant and Prof. M.R. Kundu for arranging living expenses for the stay at University of Maryland during the IAU Symposium 107. The authors would like to thank Prof. R.V. Bhonsle, Physical Research Laboratory, Ahmedabad for his keen interest in the progress of this work.
This work has been supported by University Grants Commission, New Delhi through the Research Project No. F-23-1189/79(SR II/III).

REFERENCES

Allissandrakis, C.E. and Kundu, M.R.: 1975, Solar Phys. 41, 119
Feldman,U. and DOSCHEK, G.A.: 1978, Astr. Ap. 65, 215
Kakinuma, T. and Swarup, G.: 1962, Ap.J 136, 975
Kellermann, K.I.: 1966, Ap.J 146, 621
Kundu, M.R.: 1965, Solar Radio Astronomy (Inter Science, New York)
Kundu, M.R., Alissandrakis, C.E., Bregman, JD., and Hin, A.C.:1977
    Ap.J 213, 278
Lang, K.R., and Willson, R.F.: 1980, Radio Physics of Sun
    (ed. M.R. Kundu and T.E. Gergerly), 109
Pick, M.: 1980, Phil. Trans. R.Soc.Lond., A 297, 587
Pye, J.P., Evans, K.D., Hutcheon, R.J., Grassimenko, M., Davis, J.M.,
    Krieyer, A.S., and Vesecky, J.F.: 1978, Astr. Ap.65, 123
Reeves, E., Timothy, J., Foukal, P., Huber, M., Noyes, R., Schmal, E.,
    Vernazza,J. and Withbroe,G.: 1976, Progress in Astronautics
    Aeronautics, V.48(ed. M.Kent, E.Stuhlinger, and S.T.Wu)
    New York, AIAA
Schoechlin, W and Magum,A.: 1979, Solar Phys. 64, 349
Schmahl,E.: 1980, Radio Physics of Sun (ed. M.R.Kundu and T.E.Gergely),
    71
Takakura, T.: 1967, Solar Phys. 1, 304
Velusamy, T., and Kundu, M.R.: 1980, Radio Physics of the Sun
    (ed. M.R. Kundu and T.E. Gergely), 105
Vorphal, J., Tandberg - Hanssen,E. and Smith, J., Jr.: 1977, Ap.J.212,
    550
Zheleznyakov, V.V.: 1962, Astron. Zh., 39, P5 (Soviet Astron., 6 P.3)

SESSION IV

CURRENT SHEETS IN SOLAR FLARES

E. R. Priest
Applied Maths Dept.
St. Andrews University
Scotland

ABSTRACT   Until recently magnetic reconnection in solar flares was discussed simplistically in terms of either a spontaneous tearing mode instability or a driven Petschek mode. Now the subtle relationship between these two extremes is much better understood. Current sheets may form and reconnection may be initiated in many different ways. There are also a variety of nonlinear pathways from a reconnection instability and several types of driven reconnection.

In solar flares current sheets may be important as new flux emerges from below the photosphere and also as a magnetic arcade closes down after being blown open by an eruptive instability. Numerical simulations of these sheets will be described, including new features such as the presence of a fast shock in Petschek's mechanism and impulsive bursty reconnection due to secondary tearing and coalescence.

1. INTRODUCTION

The collapse of the magnetic field near an X-point to form a sheet of intense current is a universal phenomenon in a cosmic plasma. Such current sheets are important as sites where the magnetic energy may be converted to kinetic energy, heat and fast particle energy. Also, they allow the global topology of the magnetic field to be changed as magnetic field lines break and reconnect.

A few years ago the emphasis in laboratory plasmas was on the tearing mode instability, whereby time-dependent linear reconnection grows spontaneously. The emphasis in cosmic plasma research was instead on the Petschek regime, in which steady nonlinear reconnection is driven at a fast rate. Now, however, we have entered an age of sophisticated numerical experimentation on time-dependent nonlinear reconnection. This has linked the two previous strands of thought and has produced many unexpected results.

In the next two sections my aim is to summarise some of the lessons

learnt while preparing a major review of the MHD of current sheets
(Priest, 1984) and also to mention some of the distinctive properties
of current sheets in solar flares.

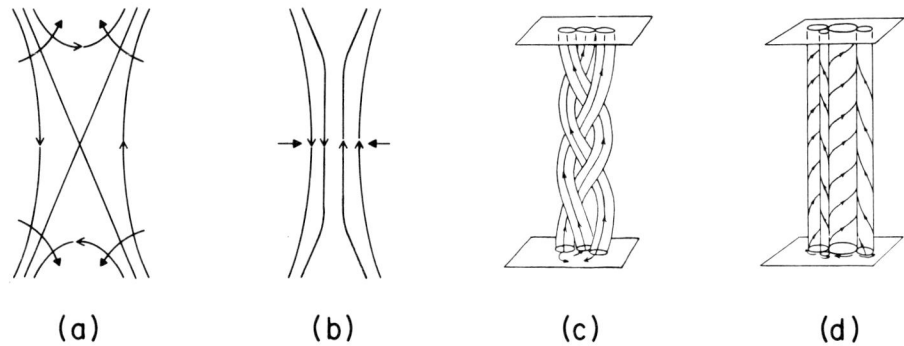

Figure 1. Current sheet formation.

2. GENERAL PROPERTIES OF MAGNETIC RECONNECTION

2.1. Current sheets may form in different ways.

Because the magnetic Reynolds number (vL/η) is so large throughout most
of a cosmic plasma, where η is the magnetic diffusivity, the magnetic
field is frozen very effectively to the plasma except in intense con-
centrations of current such as sheets. These may form if the region
near an X-point (or separator) collapses (Dungey, 1953; Syrovatsky,
1966), which occurs when, for example, topologically separate flux
systems are pushed towards one another (Figure 1a, b). Also, Parker
(1972, 1983) has pointed out that a closely packed set of flux tubes
will tend to create sheets. When the tubes are braided (Figure 1c) he
conjectures that no equilibrium is possible so that magnetic nonequilibrium ensues (Syrovatsky, 1978; Tsinganos, 1982; Rosner and Knobloch,
1982). When the tubes are twisted (Figure 1d) with neighbouring tubes
possessing the same sense of twist an equilibrium does exist but there
is a current sheet at the interface between them.

Current sheets may also be created by ideal magnetic instabilities
such as the classical kink instability (Figure 2a) studied in the solar
context by Raadu (1972), Hood and Priest (1979), Van Hoven et al. (1981)
and, in its nonlinear regime, by Sakurai (1976). Other ideal modes are
the eruptive instability (Figure 2b, §3.3) and the coalescence instability (Finn and Kaw, 1977; Pritchett and Wu, 1979), which drives neighbouring magnetic islands together in a tearing layer (§3.3).

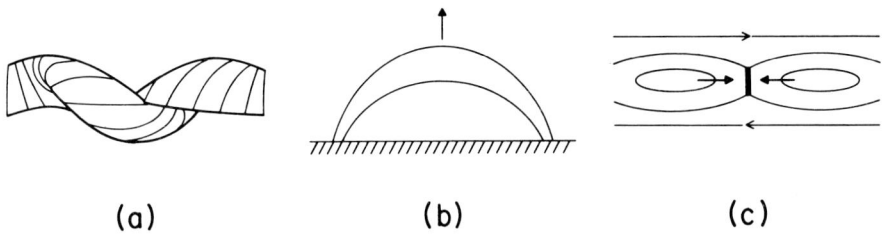

Figure 2. Ideal magnetic instabilities.

2.2. Reconnection may be initiated in different ways

Magnetic reconnection in a current sheet may occur spontaneously by a resistive instability such as a tearing mode (Furth et al. 1963). Alternatively, it may be driven locally by a sudden enhancement of the magnetic diffusivity (η) due to, for example, the onset of microturbulence (Ugai and Tsuda, 1977). Furthermore, reconnection may be driven from outside when topologically separate flux systems are pushed together (Sato and Hayashi, 1979).

2.3. Regimes of driven reconnection

When separate flux systems are forced together at a steady speed v (at large distances from the neutral point), the resulting type of reconnection depends on the value of v. If the reconnection rate is smaller than the overall diffusion speed, i.e.

$$v < \frac{\eta}{L},$$

where L is the overall length-scale of the system, we have a state of <u>very slow (vacuum) reconnection</u> (Figure 3(I)), in which the field diffuses through a series of potential fields. If instead

$$\frac{\eta}{L} < v < \frac{v_A}{R_m^{1/2}},$$

where $v_A = B/(\mu\rho)^{1/2}$ is the Alfvén speed at large distances and $R_m = L v_A/\eta$ is the large-scale Lundquist number, the result is a <u>slow (Sweet-Parker) mode</u> (Figure 3(II)) with a narrow diffusion region (shaded) of length L where the magnetic field slips through the plasma.

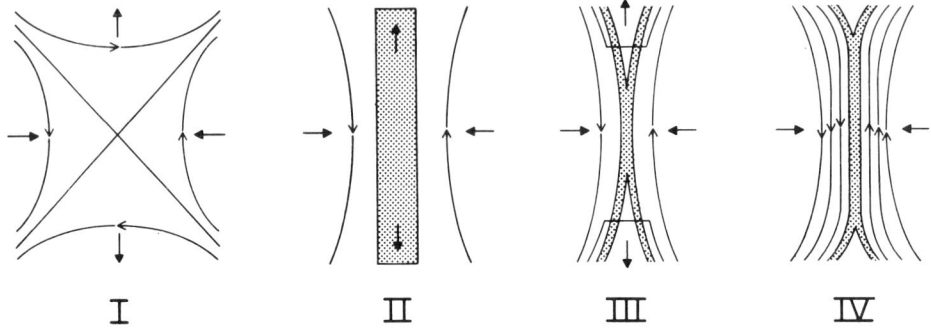

Figure 3. Reconnection regimes: (I) very slow, (II) slow, (III) fast, (IV) supercritical.

When the distant inflow speed (v) lies in the range

$$\frac{v_A}{R_m^{1/2}} < v < v_{max},$$

where $v_{max} = \pi v_A / (8 \log R_m)$ lies typically between 0.01 v_A and 0.1 v_A, one finds <u>fast (Petschek) reconnection</u>, in which the current sheet bifurcates into a pair of slow magnetoacoustic shock waves (Figure 3 (III)). The central diffusion region is just a Sweet-Parker current sheet, which is insignificant energetically but acts as a source for the slow shocks. The shocks create hot fast jets of plasma with typically $\frac{3}{5}$ of the inflowing magnetic energy being converted into kinetic energy and $\frac{2}{5}$ into heat. They form simply because the inflow is supersonic with respect to the slow-mode speed.

When the inflow speed is so large that

$$v_{max} < v$$

a fast steady state is impossible and we have <u>supercritical reconnection</u> with <u>flux pile-up</u> (Figure 3 (IV)). The magnetic flux is being brought in faster than it can reconnect and so it piles up outside the diffusion region, gradually increasing its length until there is no room for the slow shocks. This is a highly time-dependent regime with fast-mode waves propagating back towards the source and must not be confused with the Sweet-Parker mode.

In practice the nonlinear development of a tearing mode in a cosmical plasma is likely to lead to regime III (§3.1), unless reconnecttion is inhibited by severe constraints. For example, in a tokamak there is a strong magnetic field component along the X-line as well as conducting walls and restrictions on the wavelengths that can fit in the torus, with the result that the tearing soon slows down to a simple diffusion (the Rutherford regime). Ideal magnetic instabilities, on the other hand, (Figure 2) occur on Alfvénic times and are likely to drive very fast reconnection in regime IV, which needs to be explored in more detail.

3. SOLAR FLARES

3.1. Important extra effects in the solar atmosphere

Steinolfson and Van Hoven (1983, 1984) have stressed that the Lundquist number ($S = a\,v_A/\eta$) and dimensionless wavelength (λ/a), where a is the current sheet width, are both much larger in the solar atmosphere than is normally considered in tearing mode calculations. For $S \geq 10^6$ they find that at $\lambda/a=2$ the tearing has the usual linear growth rate ($\sim S^{2/5}$) and the nonlinear growth is very slow (at "constant-ψ"). However, at larger wavelengths ($\lambda/a=20$) the linear tearing is much faster ($\sim S^{2/3}$) and the perturbation extends far from the sheet with a much faster non-linear growth than before (at non-constant-ψ).

Another feature of modelling the solar atmosphere is that a realistic <u>energy balance</u> should be employed including optically thin radiation. For the Sweet-Parker or Petschek mechanism this leads to a beta-limitation, since a steady-state diffusion region becomes impossible when the plasma beta is too low (Milne and Priest, 1981). Furthermore, the tearing mode becomes coupled to the faster radiative mode, which exhibits a surprisingly high level of reconnection and magnetic energy conversion (Van Hoven <u>et al</u>., 1983).

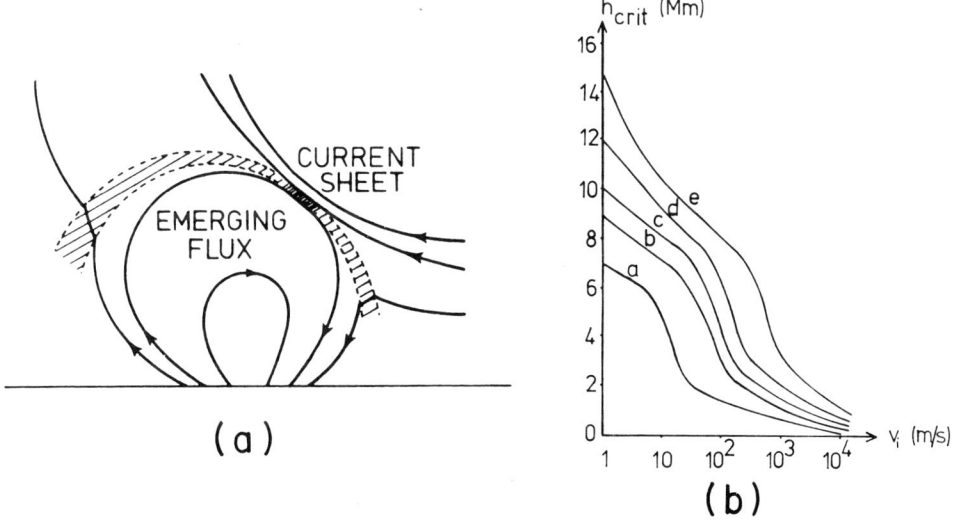

Figure 4. (a) Flux emergence, (b) Critical sheet height.

Finally, it is important to incorporate the stabilising effect of photospheric line tying, since the footpoints of coronal magnetic field lines are usually anchored in the dense photosphere. It has been suggested that this may even make the resistive modes completely stable in a loop (Mok and Van Hoven, 1982) or an arcade (Hood, 1983) unless there is a reversal in the axial field component.

3.2 Emerging flux

According to the emerging flux model (Heyvaerts et al., 1977), as new magnetic flux emerges from below the photosphere it creates a current sheet at the interface with the overlying field at a height h (Figure

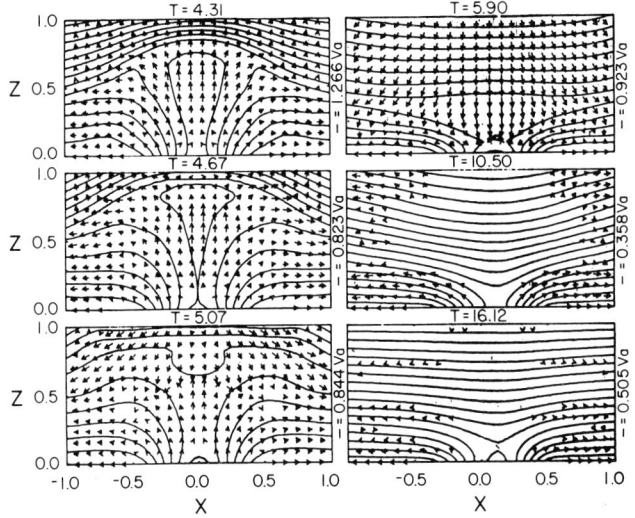

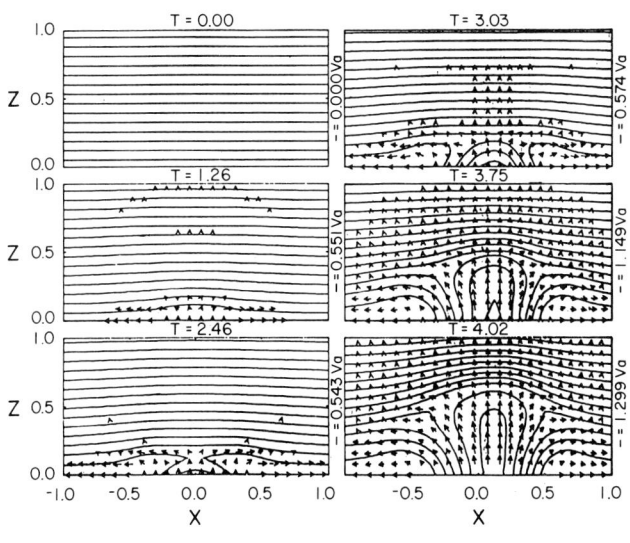

Figure 5. Numerical simulation of emerging flux (Forbes and Priest).

4a). The type of flare depends on the magnetic environment into which the new flux emerges. Usually, an X-ray bright point is produced or a small flare if a lot of flux emerges. If the emergence is near the sheared magnetic field of an active-region (or plage) filament, a large two-ribbon flare may result, with the emerging flux triggering the release of energy stored in the overlying sheared field. The model also applies to horizontal (rather than vertical) motion of the new flux (e.g. satellite sunspots), because all that is needed to produce current sheets is a relative motion of the two flux systems. It is suggested that the onset of the flare occurs when the sheet reaches a critical height h_{crit} such that the current density becomes large enough to trigger microturbulence. Its variation with emergence speed (v) and field strength (B) is calculated by solving the energy balance in the sheet (Heyvaerts and Priest, 1976; Tur and Priest, 1978; Milne and Priest, 1981) as shown in Figure 4b, where a,b,c,d,e refer to field strengths of 10^3, $10^{2.5}$, 10^2, $10^{1.5}$ and 10 Gauss, respectively.

Some preliminary results of a numerical experiment on emerging flux by Forbes and Priest are shown in Figure 5, with time measured in units of the Alfvén travel time over unit distance and the initial magnetic field being horizontal. The boundary conditions are free-floating on the sides and top, whereas on the base an emerging flux region of oppositely directed field is modelled by imposing the normal magnetic field component and the mass flux normal to the field lines. For this case the magnetic Reynolds number of 400 and the emergence speed is rather large (one eighth of the Alfvén speed). In the first six frames it can been seen how the flux emerges and reconnects. Even though no more new flux is forced through the base after t = 4, the flux continues to rise

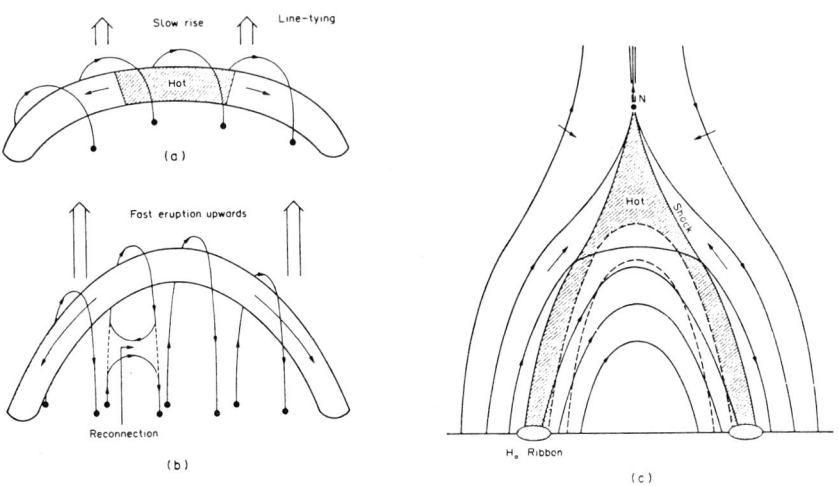

Figure 6. Overall behaviour of large flare.

through its own inertia. In the second six frames the flux pinches off near the base and tears to form a plasmoid which is ejected upwards out of the top of the box. By the last frame the configuration has quietened down to an almost potential state. As well as being important in solar flares, the formation and acceleration of plasmoids may be occuring in X-ray bright points and EUV Brueckner jets, and it has even been speculated that they may represent a large part of the solar wind (Pneuman, 1983).

3.3 Large (two-ribbon) flares

For most major flares one observes in the preflare phase a large flux tube (active-region filament) starting to rise slowly for typically half an hour (Figure 6a). This slow rise may be caused by an ideal eruptive instability (Hood and Priest, 1980; Migliulo and Cargill, 1983) or it may be the result of magnetic nonequilibrium when the flux tube can no longer remain in equilibrium under the combined action of magnetic tension and magnetic buoyancy (Parker, 1979; Browning and Priest, 1983).

As the flux tube rises it stretches out the overlying arcade of magnetic field lines until they start to reconnect by the tearing mode, thereby no longer holding down the tube. This represents the flare onset and the beginning of a much more rapid eruption (Figure 6b). During the main phase the filament has disappeared from view and the reconnection continues, as shown in a section across the arcade in Figure 6c. As the open field closes back down the neutral point and the Petschek shocks rise, creating hot rising loops of plasma ($\approx 10^7$K) with two separating ribbons of emission at their footpoints (Kopp and Pneuman, 1976; Cargill and Priest, 1982).

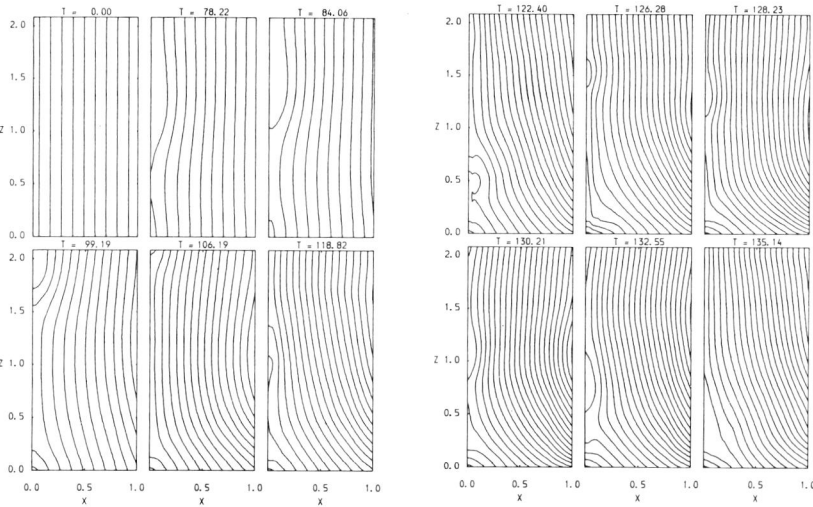

Figure 7. Line-tied reconnection (Forbes and Priest, 1983).

A numerical simulation of this reconnection process in stretched-out field lines has included the line tying of the footpoints at the base of the numerical box (Forbes and Priest, 1982a, 1983). Boundary conditions along the upper and right-hand edges are free-floating and along the left-hand edge they are symmetry conditions, since only the right-hand part of the configuration is shown in each frame of Figure 7. The initial state just consists of vertical field lines and the magnetic Reynolds number ($R_m = v_A a/\eta$) is 300, while the ambient plasma beta is 0.1 and the width (a) of the initial vertical current sheet is 0.1 of the box width. A flux-corrected algorithm is used with a variable grid and time is measured in units of the Alfvén travel time (a/v_A).

For the first three frames of Figure 7 the behaviour was as expected. The sheet tears near the base and in the nonlinear development the magnetic field continues to close down while the X-type neutral point rises and a plasmoid is ejected. From a single frame during this development (Figure 8a, for $R_m = 10^3$) the presence of two slow Petschek shocks can be seen extending up and down from the neutral point. They show up much more clearly in the current density or plasma velocity plots than in the magnetic field line sketches. A new feature is the presence of a fast magnetoacoustic shock wave, as predicted by Yang and Sonnerup (1976). It rapidly slows down the jet of plasma that is being squirted down towards the obstacle of closed magnetic field lines near the base and therefore degrades some of the kinetic energy released in the Petschek mode further into heat.

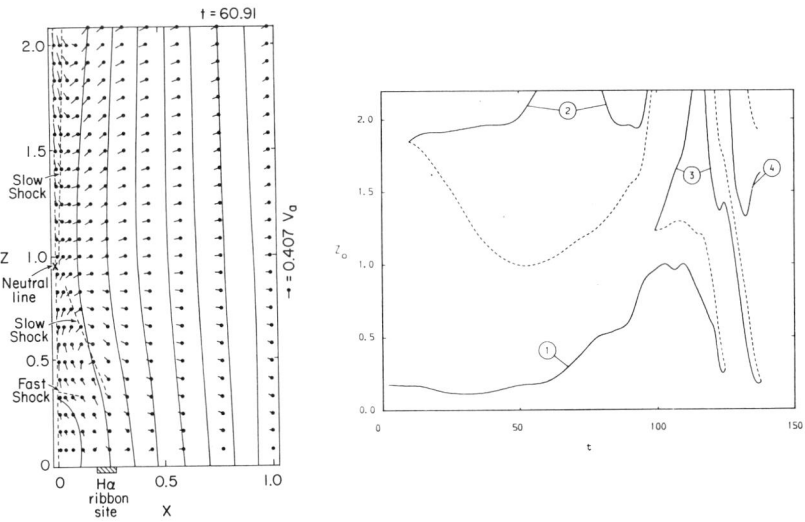

Figure 8(a). Magnetic field lines and flow velocity vectors.
(b). Creation and annihilation of neutral point pairs. (Forbes and Priest, 1983.)

The subsequent development (t>90) was a surprise. The current sheet thins and tears again, creating a pair of X- and 0-points. Reconnection at the upper X dominates and the 0 is shot down and coalesces with the lower X extremely rapidly by the <u>coalescence instability</u>. Meanwhile a new pair of neutral points is created and the process repeats. This relatively slow secondary tearing and much faster coalescence produces an <u>impulsive bursty mode</u> of reconnection and is found to occur more easily as R_m, β^{-1} and the box width are increased.

4. CONCLUSION

It is important to increase greatly the contact (both theoretical and observational) between those working on reconnection in the laboratory, the magnetosphere and the solar atmosphere, so that we can understand better the fundamental behaviour of current sheets. In so doing, we should seek to recognise the distinctive properties of reconnection in these diverse plasmas in order to discover the real similarities and differences rather than being over-prejudiced by our own speciality (see e.g. Forbes and Priest (1982b) for a comparison of geomagnetic substorms and solar flares). It will be especially interesting to continue the present crop of numerical experiments on reconnection and to compare with analytical models so as to appreciate more of the basic physical mechanisms at work in this beautiful process. Clearly, current sheets will continue to intrigue and surprise us in years to come.

ACKNOWLEDGEMENTS

I am most grateful to those at Durham, New Hampshire, who made my summer so stimulating and enjoyable. This work was supported by the NASA Solar Terrestrial Theory Program under grant number NAGW-76.

REFERENCES

Browning, P. and Priest, E.R.: 1983, submitted.
Cargill, P.J. and Priest, E.R.: 1982, Solar Phys. 76, 357.
Dungey, J.W.: 1953, Phil. Mag. 44, 725.
Finn, J.M. and Kaw, P.K.: 1977, Phys. Fluids 20, 72.
Forbes, T.G. and Priest, E.R.: 1982a, Solar Phys. 81, 303.
Forbes, T.G. and Priest, E.R.: 1982b, Planet. Space Sci. 30, 1983.
Forbes, T.G. and Priest, E.R.: 1983, Solar Phys. 84, 169.
Furth, H.P., Killeen, J. and Rosenbluth, M.N.: 1963, Phys. Fl. 6, 459.
Heyvaerts, J., Priest, E.R. and Rust, D.M.: 1977, Astrophys. J. 216, 123.
Hood, A.W. and Priest, E.R.: 1979, Solar Phys. 64, 303.
Hood, A.W. and Priest, E.R.: 1980, Solar Phys. 66, 113.
Hood, A.W.: 1983, submitted.
Kopp, R. and Pneuman, G.W.: 1976, Solar Phys. 50, 85.
Migliulo, S. and Cargill, P.: 1983, submitted.
Milne, A.M. and Priest, E.R.: 1981, Solar Phys. 73, 157.
Mok, Y. and Van Hoven, G.: 1982, Phys. Fluids 25, 636.
Parker, E.N.: 1972, Astrophys. J. 174, 499.
Parker, E.N.: 1983, Astrophys. J. 264, 635.
Pneuman, G.W.: 1983, Astrophys. J. 265, 468.
Priest, E.R.: 1984, Rep. Prog. Phys.

Pritchett, P.L. and Wu, C.C.: 1979, Phys. Fluids 22, 2140
Raadu, M.A.: 1972, Solar Phys. 22, 425.
Rosner, R. and Knobloch, E.: 1982, Astrophys. J. 262, 349.
Sakurai, T.: 1976, Pub. Astron. Soc. Japan 28, 177.
Sato, T. and Hayashi, T.: 1979, Phys. Fluids 22, 1189.
Steinolfson, R.S. and Van Hoven, G.: 1983, Phys. Fluids 26, 117.
Steinolfson, R.S. and Van Hoven, G.: 1984, Astrophys. J.
Syrovatsky, S.I.: 1966, Soviet Astron. 10, 270.
Syrovatsky, S.I.: 1978, Solar Phys. 58, 89.
Tsinganos, K.C.: 1982, Astrophys. J. 259, 832.
Tur, T.J. and Priest, E.R.: 1978, Solar Phys. 58, 181.
Ugai, M. and Tsuda, T.: 1977, J. Plasma Phys. 17, 337.
Van Hoven, G., Ma, S.S. and Einaudi, G.: 1981, Astron. Astrophys. 99, 232.
Van Hoven, G., Steinolfson, R.S. and Tachi, T.: 1983, Ap. J. Letts.
Yang, C.K. and Sonnerup, B.U.O.: 1976, Astrophys. J. 206, 570.

DISCUSSION

Vasyliunas: In your Figure 3, there should be an additional regime IIIa, between III and IV, which might be called the Sonnerup regime, where plasma is being pushed together faster than v_{max} but is also being sucked out the sides, creating a slow-mode expansion ahead of the reconnection region and allowing steady reconnection of essentially arbitrarily high rates. This could occur, for example, at the dayside of the earth where the flow goes both toward and around the magnetosphere.

Priest: I agree that one can have a slow mode expansion Petschek (or Sonnerup) mode or a fast mode expansion Petschek mode. Indeed, it is possible to have a modified Petschek mode with both slow and fast mode expansion effects in the inflow region and with a quite different scaling from the normally quoted value.

It is the inflow boundary conditions that determine which of these modes occurs. However, if reconnection is driven locally by a resistivity enhancement or by a tearing mode, I would expect the fast mode expansion Petschek regime to be the relevant one. There are great difficulties comparing numerical experiments with the Petschek or Sweet-Parker modes. For example, in numerical simulations the magnetic field may not be steady or uniform at large distances; there may be external sources of waves and currents; numerical diffusion may be so high that it smooths out the shocks over large regions. Just comparing with the η scaling is too simplistic, and it is preferable to use the presence of slow mode shocks as the criterion for the Petschek mechanism - such shocks are best detected in maps of the current density contours rather than magnetic field line plots.

Migliuolo: What is the width of the current sheet in the simulation by Forbes and Priest?

Priest: The width (a) of the current sheet (within regions of uniform field) has been varied from 0.5 to less than .03 of the width (w) of the numerical box. The interesting feature is that the new regime of impulsive bursty reconnection is absent at low values of magnetic Reynolds number (R_m) or w/a or high values of external plasma beta (β). But, as R_m, w/a or β^{-1} are increased the impulsive bursty regime becomes more evident. In this regime the reconnection rate (as measured by the electric field at the dominant X-point) is significantly higher. Also, this new regime could develop in theory from either regime II, III or IV when the sheet length becomes large enough for tearing (Bulanov et al., 1978).

Migliuolo: Why is diffusion in neutral sheets limited by the value of β?

Priest: When the plasma beta is too low the optically thin radiation can no longer be balanced by the other terms in the energy balance. It arises because of the maximum in the radiative loss curve.

Van Hoven: What is the vertical scale in the line-tied reconnection plots (Fig. 7)?

Priest: The numerical box is twice as high as it is wide.

Van Hoven: Would you clarify the effects of the distance to the free boundaries on the "bursts" of X- and O-point movement?

Priest: The "free" boundary conditions (setting e.g. $\partial \rho/\partial n = 0$, $\partial v/\partial n = 0$) are freer than fixed conditions (ρ = const, v = const), but they are still not completely free since they do inhibit reconnection to a certain extent. Therefore, as the boundaries are taken further away from the sheet the impulsive bursty reconnection becomes more possible.

Drake: In the case in which the inflow velocity initially exceeds the Alfvén velocity, it seems to me that the magnetic flux will simply build up (increasing local B) until the Alfvén velocity exceeds the inflow velocity so that the Petschek model can be applied to the whole process.

Priest: When $v > v_{max}$ (not v_A) I agree that the propagation of fast mode waves in the inflow region will tend to increase B, but the result is a subtle interaction between these waves and the sources of magnetic field. This highly time-dependent process has just not been looked at in detail yet.

Drake: In all simulations of resistive instabilities with high magnetic Reynolds numbers, the magnetic reconnection rate is never as large as the Petschek rate. The maximum reconnection rates are seen in simulations of the coalescence instability and kink mode (ideal modes) in which the Sweet-Parker model seems to describe the scaling (Biskamp and Park et al.). Petschek reconnection rates have only been seen when a constant velocity external driver has been imposed from the outside. In my opinion the real question is whether the plasma itself can ever internally generate reconnection rates as fast as predicted by Petschek.

Priest: The important feature is the assumed boundary conditions. If these are free-floating, as in our simulations or those of Tsuda, for instance, then the tearing mode does develop nonlinearly into the Petschek mode. If instead you impose fixed or periodic conditions, then it evolves much more slowly into the Rutherford regime. In my view the solar flare is likely to be caused by an ideal instability (such as the kink or eruptive mode) which then drives fast reconnection.

ENERGY DISSIPATION MECHANISMS IN FLARE STARS

D.J. Mullan
Bartol Research Foundation of The Franklin Institute
University of Delaware
Newark, Delaware 19711 U.S.A.

Flare stars derive their name from intermittent increases in luminosity which have certain characteristics reminiscent of solar flares (e.g. enhanced strengths of emission lines in the stellar spectrum during the outbursts). When a flare star is observed in a filter which transmits, say, the violet part of the visible spectrum, the increase in luminosity during a flare may range from noise level up to perhaps 100 times the quiescent brightness. During a flare, certain spectral features of the quiescent star (e.g. molecular bands) remain visible, indicating that the flare occupies only a fraction of the visible disk. Thus, analagous to a solar flare, a stellar flare is confined to a single active region. However the total power is large enough to affect the integrated light from the stellar disk. In contrast, the largest solar flare ($E_{tot} \approx 10^{32}$ ergs) has a rate of energy release ($L \approx 10^{29}$ erg/sec) which is so small that a distant observer would record such a flare as a luminosity increase of less than $10^{-4} L_{sun}$. However, even apart from the flares themselves, it has become apparent in recent years that flare stars in their "quiescent state" provide some extreme contrasts with the sun.

As a result, models which have been proposed for solar flares may be testable in stellar flares over a broader range of physical conditions than is possible in the sun. In order to quantify this statement, we will first summarize the relevant physical conditions in the atmosphere of a flare star. (See Part I below.) We will find that even outside flares, the dissipation of mechanical energy in the atmospheres of flare stars is in a certain sense qualitatively different from what is observed in the sun, and is therefore of interest in its own right. In Part II, we will discuss energy dissipation in flares. However, we shall find that, at least in the very coolest flare stars, the distinction between the flaring and non-flaring states of the star becomes difficult to define.

PART I. CHARACTERISTICS OF FLARE STARS

Physical parameters

Flare stars are cool dwarfs, with spectral class mainly of type M. In young clusters (e.g. Orion, Pleiades), flare stars of spectral type K are also observed, but flares on stars of spectral class G or earlier have very rarely been detected (even if one believes the scattered and unconfirmed reports; cf. Kunkel, 1975). The exceptional case of the sun has already been noted: if this G2 dwarf were at a distance of even the nearest star, its classification as a flare star would be highly unlikely.

Effective temperatures, T_{eff}, of flare stars range from ~4000 K to 2200 K (Pettersen, 1980). The prototype flare star, UV Ceti, has T_{eff} = 2950 K. For purposes of scaling various physical quantities from the sun (T_{eff} = 5780 K), the flare stars can be considered as having $T_{eff} \approx 0.5\, T_{eff}$(sun).

The absolute visual luminosities M_V of many flare stars are known with some precision, because the stars are close enough to be susceptible to astrometry. Thus, UV Ceti has M_V = 15.85. In order to determine total luminosities of these stars, photometry must extend to wavelenths of several microns. For the coolest flare stars, the bolometric corrections exceed 5^m (Pettersen, 1982). For UV Ceti, the bolometric luminosity is some 800 times less than the solar luminosity.

Thus, the radius of UV Ceti is smaller than the solar value by a factor of 7 ± 1. Other flare stars have radii which are somewhat larger than this: the mean value of log R/R(sun) in a sample of 35 flare stars (Pettersen, 1980) is -0.49, i.e. the mean radius is 2.2 times larger than that of UV Ceti.

Reliable masses of some flare stars are now available (cf. Mullan, 1976). In the case of UV Ceti, in particular, the mass is accurately known, 0.108 M(sun). Along the lower main sequence, the mass-radius relation can be described fairly well by $M \sim R^\alpha$, where α is very close to unity (Hoxie, 1973). Hence, the gravitational acceleration g in the atmospheres of the flare stars is essentially proportional to R^{-1}: g exceeds the solar value by factors of 3-6 among the flare stars.

Because of the reduced effective temperature, molecules (especially H_2) form in the upper photospheres of flare stars. Dissociation of these molecules gives rise to a new convection zone high in the photosphere of these stars (in addition to the hydrogen ionization zone, which lies beneath the photosphere). As a result, convection sets in at a much higher level in the atmosphere than in the sun. This new convection zone creates a remarkable signature in the broad-band colors of cool dwarfs: this signature has been detected by Mould and Hyland (1976). Hence, the convection zone which is nearest to the chromosphere and corona of a flare star is composed predominantly of molecular hydrogen, with a mean molecular weight μ which is therefore larger than in the solar convection zone by a factor of 2.

The reason for discussing the above quantities in detail is that we wish to estimate the scale-height of the convection zone which is nearest to the corona. The pressure scale height H_p is proportional to $T_{eff}/\mu g$, and so, combining the above estimates, we find that in the flare stars, H_p is smaller than H_p(sun) by factors of order 12-24.

Convective time scales

For purposes of coronal heating, it is important to estimate the period at which the mechanical power generated within the convection zone is a maximum. To first order, this maximum can be characterized by the time-scale of convective overturning. With typical cell dimensions of order L, the convective overturning time-scale is $t_c \approx L/v_c$, where v_c is the convective flow velocity. In the sun, the granules have typical sizes of $L \approx 10^3$ km and convective flows are typically in the range of 1-3 km/sec. (Theoretical models of solar convection also predict maximum convective velocities in the solar convection zone of about 3 km/sec; cf. Mullan, 1971). Hence, the overturning time-scale is about 500 seconds in the sun. Actually, the mean lifetimes of the granules is also about 400-500 seconds (cf. Leighton, 1963), which indicates that convection cells in the turbulent solar convection zone persist for about one overturning time and then lose their identity. (The Reynolds numbers of the solar convection zone are too high to allow the formation of steady hexagonal cells of Benard type.)

In red dwarfs, spectral line broadening shows no evidence for flows in the atmosphere which are faster than 1-2 km/sec. Model convection zones for such stars also suggest (Mullan, 1971) that the maximum convective velocities are not much less than in the sun, $v_c \approx 2$ km/sec. Hence, in red dwarfs, the convective velocities may be less than solar values by a modest amount, say 1.5.

We can now estimate the period at which the convection zone power is maximum in red dwarfs:

$$\frac{t_c \text{ (RD)}}{t_c \text{ (sun)}} = \frac{L \text{ (RD)}}{L \text{ (sun)}} \frac{v_c \text{ (sun)}}{v_c \text{ (RD)}}$$

It is usually assumed that the dimensions of convection cells scale with the pressure scale height. Adopting this assumption, we find t_c (RD) $\approx (0.06-0.13)\ t_c$(sun). This leads to

$t_c(RD) \approx 30\text{--}70$ seconds

This result will be of significance when we discuss the heating of loops in the coronas of flare stars.

Starspots

A prominent characteristic of most flare stars is that their surfaces are not uniformly bright. Dark spots are inferred to exist on the surface, at times covering up to 10% or more of the visible disk. More typically, the spots may occupy a few percent of the surface. This is to be compared with a maximum reduction of order 0.1% in the case of large sunspots. The origin of large dark spots on flare stars is not definitively known. However, by analogy with the solar case, the spots are believed to be caused by localized clumpings of vertical magnetic field lines. An indirect argument in favor of this interpretation can be made on the basis of the rotation periods of spotted stars: in a sample of 16 stars which are known to be chromospherically active and for which rotation periods are known, the six stars which are known to be definitely spotted are the six stars with the shortest rotational period (Vogt, 1983). Short rotation periods ($P \leq 7$ days), or equivalently, fast rotation velocity ($V \gtrsim 5$ km/sec) appear to be the single best criterion for defining the group of spotted stars, and such stars are also expected to have the most efficient dynamos for creating magnetic fields. Thus, the possibility that starspots on flare stars are associated with magnetic fields (as in the sun) appears plausible. Magnetic field strengths in spots in red dwarfs have not yet been measured. However, the upper limits which are available on the measured field strengths are such that fields of order 10 kilogauss could exist in the spots without contradicting any of the current data on flare stars (Mullan, 1979). The best evidence at present for the presence of fields in flare stars is the occurrence of high degrees of circular polarization in the radio emission from flares (Spangler et al., 1975; Gibson, 1983), but these data are not suitable for estimating the field strength. An indirect method of estimating the magnetic field strength on the surface of BY Draconis (based on broad-band optical linear polarization) suggested that indeed fields of order 10 kilogauss might have been present at the time the polarization was observed (Mullan and Bell, 1976).

The question arises: what is the ultimate disposition of the missing flux from the dark spots? In the presence of a vertical magnetic field, the normal convective motions cannot occur freely. The convective efficiency is therefore reduced, although it cannot be suppressed entirely, since the electrical conductivity is not infinite. Residual convection interacts with the field and causes emission of MHD waves of various modes. The joint effect of reduction in convective efficiency and emission of MHD waves cool the spot. Some years ago, a self-consistent model of a spot, in which the convective efficiency is reduced by an amount which depends on the electrical conductivity, and this reduced efficiency is compensated by emission of Alfven waves (Mullan, 1974a, 1974b), was calculated for both sunspots and starspots. In these models, the missing flux was carried to a large extent by the Alfven waves. These waves are strongly reflected off the steep density gradient near the surface of the spot, and so, in a sunspot, for example, the missing flux is effectively trapped inside the sun, under normal conditions. If, however, the Alfven waves could leak into the corona, their mechanical energy would be so large compared with the normal mechanical energy flux which heats the corona that the corona over a sunspot would appear to be in a continual state of "flaring". During certain solar flares, the reflection conditions over the umbra of a spot are altered, and at such times, the leakage of umbral Alfven waves into the corona may contribute significantly to the energy budget of the flare (Mullan, 1981). For the most part, however, in the case of a sunspot, the missing flux remains trapped inside the sun (Willson et al., 1981).

A small fraction of the missing flux may appear in the form of a faint bright ring around a sunspot (Fowler et al., 1983). The emission in the ring amounts to several percent of the missing flux. The maximum intensity in the ring seems to be too bright to be due to reduction in convective efficiency alone (Fowler et al., 1983). It is possible that the ring is a manifestation of Alfven waves leaking out of the umbral walls. This would be in accord with the proposal that a sunspot is the site of a large flux of wave energy, capable of contributing significantly to coronal heating and/or flaring, if conditions are favorable.

In the case of starspots, it has been pointed out by Gershberg (1983) that the missing flux is close to the X-ray flux emitted by the stellar corona. He suggests, therefore, that the MHD waves in a starspot may have somehow found a way to leak much more efficiently into the stellar corona than in the case of a sunspot. Once the waves reach the corona, they dissipate rapidly and heat the corona. In fact, the quiescent coronae of red dwarfs are observed to be hotter than the mean solar corona (Golub, 1983). In this view, therefore, the corona of a flare star is essentially in a state of continual "flaring" due to the efficient leakage of Alfven waves from the starspot into the corona. In this regard, it is worth noting that some of the statistical properties of stellar flares can be understood if stellar flares also are powered by the missing flux from starspots (Mullan, 1975b). This raises the important question as to the distinction (if any) between coronal heating and flaring. We will see below that in the case of the coolest flare stars, this distinction may become difficult to make.

There is no quantitative estimate at present of the change in bolometric luminosity of a flare star when a starspot is present on the surface. Thus it is not known to what extent the missing flux of a starspot is stored beneath the surface.

Chromospheres

The most prominent characteristic in the quiescent optical spectra of flare stars is strong emission in the Balmer lines of hydrogen. In the case of a star as hot as the sun (T_{eff} = 5780 K), the radiation from the photosphere is so strong that the formation of, say, the H_α line has a large (if not dominant) contribution from this radiation. The remainder of the solar H_α line is formed by collisional processes in the chromosphere. However, in the case of red dwarfs, the effective temperatures are so low that the radiation field from the photosphere makes a negligible contribution to the formation of the Balmer lines. (In the case, e.g. that the stellar energy distributions can be represented as black bodies, the intensity at the Balmer limit in a flare star is weaker than in the sun by a factor of order 600.) As a result, the Balmer lines in flare stars are formed essentially entirely by collisions in the chromosphere, and therefore the Balmer lines in flare stars become good diagnostics of the physical conditions in the stellar chromospheres.

The modelling of Balmer lines in red dwarfs by means of a full non-LTE treatment has been reported by Kelch et al. (1979) and Cram and Mullan (1979). In these papers, the chromosphere is parametrized by imposing a temperature rise from the top of the photosphere (where the temperature is T_{min}) up to a point (the "top" of the chromosphere) where the hydrogen is beginning to ionize rapidly, at temperatures of 8000-9000 K. The mass loading at the latter point, m_0, is the main parameter which characterizes these model chromospheres. In order to drive the first three Balmer lines into prominent emission features (as they are observed to be in flare stars), it turns out that $\log m_0$ must exceed -4.5 to -4.0 (Cram and Mullan, 1979). Converting to pressures, $p_0 = m_0 g$ (where $\log g \approx 5$ in red dwarfs), we find p_0 (i.e. the pressure at the "top of the chromosphere") to be greater than 3 dyn/cm. By comparison, the quiet sun has $p_0 \approx 0.2$ dyn/cm^2. Hence, even if the Balmer emission in flare stars is coming from the entire surface of the star, the chromospheric densities are at least 10 times higher than in the sun. However, in practice, the Balmer emission is probably concentrated in a few active regions across the stellar disk. Therefore, the true pressures in the active regions are even higher than those estimated above. In fact, certain features of stellar flare spectra (Cram and Woods, 1983) suggest that flares occupy only a small fraction (order 10%) of the visible disk. Active regions may therefore occupy only 10% of the visible surface also. We conclude that the enhancement by a factor of $\Delta N \sim 10$ in densities in the upper chromosphere mentioned above are lower limits to the true enhancements: the latter, ΔN, are likely to be in the range from 10-100. If this scaling holds throughout the chromosphere, then flare stars are characterized by the chromospheres which are 10-100 times denser than the solar chromospheres.

These dense chromospheres require enhanced mechanical energy to heat them. Linsky et al (1982) have estimated the required energy fluxes by summing over the radiative losses in as many spectral lines as were accessible to them, including emission lines of MgII, CaII, FeII, and Balmer lines. If we express by F_{ch} the fraction of the photospheric flux which is required to balance the radiative losses in the various chromospheric lines, then for red dwarfs with active chromospheres F_{ch} turns out to be 3-10 times larger than the solar value. Thus, flare stars are more efficient than the sun at converting thermal into mechanical energy for the heating of their chromospheres, by factors of 3-10. This estimate of course says nothing about the source of the mechanical energy: it does not even specify the direction from which the energy comes. We shall see below that the energy which heats the chromosphere in an M dwarf may be provided from above.

Coronae

The fraction of the photospheric flux which is required to account for the radiation from the quiescent corona of a flare star, F_{cor}, has been found to be unexpectedly high: values as high as 10^{-3}-10^{-2} have been reported. The value of F_{cor} seems to reach a peak of $\sim 10^{-2.2}$ among stars with B-V $\approx$ 1.5 (i.e. spectral type M0 - M2), falls off at later spectral types, to less than $\sim 10^{-3.1}$ at B-V $\approx$ 2.0 (at spectral type ~M6; for conversion of B-V colors to spectral type, cf. Mould and Hyland, 1976). (These values of F_{cor} have been derived from X-ray fluxes given by Golub (1983) and bolometric luminosities provided by Pettersen (1983).) By comparison, the solar value is $F_{cor}(sun) \approx 5 \times 10^{-6}$ (Withbroe and Noyes, 1977) (quiet sun). It appears therefore that flare stars are at least 100 times more efficient at heating their coronae than the sun is. This is a striking characteristic of flare stars, and it has the effect that M dwarfs contribute appreciably to the X-ray background from the sky (Rosner et al., 1981).

Spectral lines which are formed in the transition region between chromosphere and corona have also been observed by Linsky et al. (1982) in the spectra of flare stars. These lines also show that the transition regions in flare stars are heated with an efficiency which is some 100 times greater than in the sun. This has led Linsky et al to conclude that the process which heats transition regions and coronae in red dwarfs is different from that which heats chromospheres.

In the sun, the mechanical energy which is deposited in the chromosphere is typically 10 times greater than the mechanical energy which is deposited in the corona (Withbroe and Noyes, 1977). In flare stars, the situation is qualitatively the opposite: the mechanical energy deposited in the corona appears to be almost an order of magnitude greater than that which is deposited in the chromosphere. In fact, it has been suggested (Cram, 1982) that the chromospheres of M dwarfs may be heated almost entirely by X-rays from the overlying coronae. Apparently, in the flare stars, the corona is more effectively coupled to the source of mechanical energy than the chromosphere is. This is an important distinction which may be important in understanding the dissipation of energy in the atmospheres of flare stars.

PART II. FLARES

In principle, a flare represents a short-lived localized enhancement in the rate of dissipation of mechanical energy, relative to the rate of dissipation in the "quiescent" atmosphere. The result is an increase in radiation from all levels of the atmosphere, including X-rays from the corona, emission lines from the transition region and chromosphere, and most significantly, "white-light" optical continuum. The latter feature is much more common in stellar flares than in solar flares.

Time-scales

When a stellar flare is observed through a broad-band optical filter, the flare appears typically as a rapid rise in brightness followed by a slower decline to normal light. All of the parameters of a flare light curve are quite variable from one flare to the next, such as the rise time, the decay time, the maximum amplitude, the total energy, and the time interval which has elapsed since the preceding flare. Despite these irregularities, there are some statistical trends which have been established. For example, the rate of occurrence, R, of a flare whose maximum brightness in a particular filter (say the U filter) is equivalent to a star of magnitude m_U behaves as $R = \exp a(m_U - m_0)$. Here, m_0 is a parameter which increases in the fainter flare stars, which a is almost constant from one star to the next. According to this formula, faint flares may be occurring very frequently in a star, but still remain undetectable against the statistical fluctuations which are intrinsic to our measurements of the "quiescent" photon flux from the star. This is an important consideration, and we shall return to it subsequently.

If we consider flares regardless of amplitude, then it appears that, in general, the time intervals Δt between successive flares can be fairly well described by a Poisson distribution (Oskanyan and Terebizh, 1970), i.e. and flares can be considered to be in general random events, with no connection between any one flare and either the preceding or following flare. An important exception to this general behavior, however, appears at the shortest time intervals, $\Delta t = \lesssim 10$ minutes. At such intervals, there is observed to be (in some of the flare stars at least) an excess (by a factor of ~ 2) over the numbers of flares which would be expected if the Poisson distribution were extrapolated to these intervals. The cause of this excess is not known with certainty, but sympathetic flaring is one possible explanation for it (Mullan, 1975a): if this is correct, then the upper limit of about 10 minutes on Δt would correspond to the time required for a disturbance (presumably an MHD wave) to propagate from the site of the first flare to a flare site at the maximum possible distance, i.e. at the antipodal point. With stellar radii of $(1-2) \times 10^{10}$ cm, this interpretation would suggest MHD propagation speeds of 500-1000 km/sec in the flare stars. Now flare disturbance waves sweeping out through the solar corona after certain flares ("Moreton waves") also travel at speeds of 500-1000 km/sec (Uchida, 1968): such speeds are probably characteristic of the Alfven speed in the corona. This is the first indication that despite the higher densities which characterize the upper atmospheres of the flare stars relative to the sun (ΔN = 10-100), the Alfven wave speeds might be rather similar. This would require that the coronal magnetic fields be stronger than in the sun by factors of $\Delta N^{1/2} \sim 3-10$. Such enhancements are consistent with the discussion of surface fields given above, where the strengths in starspots were estimated to be of order 10 kilogauss or more, i.e. at least 3 times as strong as in sunspots.

As regards the decay times in stellar flares, Kunkel (1975) has found that the time scale $t_{0.5}$ for the flare light to decay by 0.5 magnitudes is related to the absolute magnitude of the star:

$$\log_{10} t_{0.5} = 1.084 - 0.119 M_V$$

Where $t_{0.5}$ is in minutes. Hence, in the faintest flare stars (M = 16-19), $t_{0.5} \approx 4-10$ seconds. The rise times are even shorter. These are average time-scales: in individual flares, the light curve may evolve so rapidly that even time constants of one second are not sufficient to resolve the peak of the curve (Moffett, 1974). Optical flares on the sun evolve on considerably longer time-scales, of order one minute or so. Very rapid variations in optical emission are a striking characteristic of stellar flares.

Energies

As regards total energies of flares, these can be derived for specific band-passes, e.g. the U filter, E_U. The frequency of occurrence ν of flares of energy E_U has been derived by Lacy et al. (1976): $\log \nu \propto -\beta \log E_U$. The numerical values of β lie in an interesting range: they are close to unity, especially among the fainter stars. The significance of this is that the total power radiated by a flare star in the form of flares is dominated by large flares if β is less than unity, and by small flares if β is greater than unity. In the former case, one can distinguish, meaningfully, between flaring and non-flaring behavior. However, in the case of the faintest flare stars, the star's output of energy in "flares" is dominated by the smallest events, which can hardly be distinguished from the noise. Such stars are in a sense in a state of continual flaring, and therefore it becomes increasingly difficult among the faintest stars to draw a line of demarcation between what we would call <u>bona-fide</u> "flaring" and what we would call more or less continuous heating of the upper atmosphere by mechanical energy release.

The maximum energy recorded in an optical flare is 6×10^{34} ergs in YZ CMi (Kunkel, 1969). Flares with total energies greater than 10^{34} ergs in the U-filter alone have also been reported for YY Gem (Lacy et al., 1976). Major uncertainty surrounds the correction of these estimates for energy radiated by the flare outside the band-pass of observation: the uncertainties are at least as large as an order of magnitude (Mullan, 1976). The reason for the uncertainties is that the true nature of flare radiation has not yet been identified with certainty. Different investigators have proposed to fit various spectral features with radiation from a black body, from recombining hydrogen plasma, from deep in the photosphere where negative hydrogen ions can form, and radiation from optically thin coronal plasma. Unfortunately, predictions of (e.g.) flare colors by all of these models agree fairly well with observations, and so no discrimination on the basis of flare colors is reliable. Moreover, there is no estimate whatever of the mechanical or magnetic energy carried away from stellar flares by blast waves. [In solar flares, these components often dominate the energy budget of the flare.] Thus, the above estimates are certainly lower limits on the total energy released in a large stellar flare.

The source of this energy has not yet been definitely determined. One suggestion which has recently been advocated as a source of energy in certain stellar flares is the release of energy when matter falls on to the surface of the star. The arguments in favor of this model are based on a short-lived increase in absorption which is observed just prior to certain flares. The increase in absorption has been reported at optical wavelengths (Giampapa et al., 1982) and also at X-ray wavelengths (Haisch et al., 1983). The interpretation is based on an analogy with a class of solar flares called "disparitions brusques" where material which was originally suspended above the surface in a prominence is disturbed (for some reason), and falls to the surface, releasing its original potential energy in the process. The enhanced absorption observed just prior to the stellar flare is attributed to the prominence material before it releases its energy.

Now, the suspension of prominence material above a stellar surface requires a magnetic field. Hence, in a sense, the infall model of stellar flares is a magnetic model. However, other stellar flares (which show no evidence for pre-flare absorption) in all likelihood derive their energy from magnetic fields, but more directly, probably by converting magnetic free energy into fast particles or hot gas. As was mentioned above, the existence of magnetic fields in stellar flare plasma is most obviously suggested by intense (almost 100%) circular polarization of flare radio emission.

Magnetic fields

A convenient classification of magnetic energy release in solar flares has been proposed by Spicer and Brown (1981), involving electric currents which flow either parallel to the field ($j_\parallel$) or perpendicular to it ($j_\perp$). In both cases, flare onset occurs when something happens locally to enhance the Joule dissipation rate, j^2/σ, by a large factor. For example, if turbulence sets in, the conductivity σ will drop rapidly from its classical (Spitzer) value to a much lower value σ_t. Whatever the details are, the resulting effect is the conversion of magnetic energy into fast electrons and hot gas. It is useful to keep these ideas in mind also in attempting to interpret stellar flares.

For example, conduction of heat away from the flare site towards the surface of the star forces the "top of the chromosphere" downwards into deeper layers. This effect can explain rather well the spectroscopic characteristics (line strengths, line widths and continua) of certain optical flares (Cram and Woods, 1982). On the other hand, if the fast electrons are the dominant agent for transporting flare energy away from the site of primary energy release, they may create shock waves which propagate upwards (into free space) and downwards (into the dense photosphere). Because of the denser gas in flare star atmospheres, the downward propagating shock wave can create an appreciable optical thickness in the visible part of the spectrum ($\tau_{4500A} \approx 1$) before dissipating. As a result, this shock-heated and shock-compressed material can act as an efficient source of "white-light" continuum in certain stellar flares (Livshits et al., 1981). The faint photospheric background against which a stellar flare is observed (T_{eff} = 3000 K, rather than 5780 K as in the sun) also helps to make it easier to detect the continuum in stellar flares than in solar flares.

Accepting the magnetic source of stellar flare energy, the properties of stellar flare plasma allow one to place lower limits on the field strengths in the flare region. Stern et al. (1983) and Haisch (1983) have done this for a large X-ray flare in a Hyades star: the lower limits on the field are 350-700 gauss. If we apply analogous arguments to a large solar flare ($E = 10^{32}$ ergs, volume = 10^{29} cm^3), the lower limit on the flare field (assuming complete annihilation of the field) is 160 gauss. However, it is known that the fields existing near flare sites in the sun (in spots) are larger than this by factors of 10-20. If similar factors apply to stellar flare case, the flare fields of 350-700 gauss may be consistent with surface fields of 5-14 kilogauss. Thus, these flare fields may not seriously constrain the values of field strength on the surfaces of red dwarfs.

X-ray data on stellar flares have provided important information on the temperatures of flare plasma (Haisch, 1983); it is remarkable that temperatures of a few times 10^7 K are recorded, close to the values observed in solar flares, despite the fact that the total energies in stellar flares may exceed those in solar flares by more than an order of magnitude. If flares are powered by magnetic energy, the thermal energy density of the flare plasma, $2NkT$, should be comparable to the pre-flare magnetic energy density, $B/8\pi$ (Moore and Datlowe, 1975). Then the similarity of temperatures in solar and stellar flares suggests that B^2/N values are similar, i.e. the Alfven speeds at the sites of stellar flares are apparently not very different from those at the sites of solar flares. This confirms the conclusion derived above on the basis of sympathetic flares, i.e. the Alfven speed in flare star coronae may be similar to the solar value.

Given the central role of magnetic fields in stellar flares, the higher densities which are characteristic of flare star atmospheres will affect the rate of energy release. Consider, for example, the general class of flare models in which the flare energy comes from dissipation of the current in a loop (cf. Spicer and Brown, 1981). In these models, the flare begins when the parallel current exceeds a critical value, $j_c = Nev_c$ where v_c is a critical velocity which depends on the temperature. When $j > j_c$, the conductivity becomes turbulent, σ_t, and the Joule dissipation rate, j_c^2/σ_t, becomes locally very large. This is thought to be the cause of primary flare energy release. Now, the temperatures in solar and stellar coronal loops are not very different (about 2×10^6 K): hence, v_c is about the same in both solar and stellar flares. Thus, $j \propto N$, and therefore the rate of flare energy release, j_c^2/σ_t is proportional to N^2/σ_t. Since $\sigma_t \sim N^\gamma$, [where $\gamma \approx +0.5$ (cf. Rosner et al. 1978)] the rate of Joule dissipation at the site of primary flare energy release in flare stars is larger than in the sun by factors of order $\Delta N^{1.5} \approx 10^{1.5} - 10^3$. These large increases relative to the sun are likely to be an important factor in understanding why stellar flares evolve much more rapidly than solar flares.

Another advantage of the detection of X-rays from stellar flares is that it has allowed the possibility of estimating certain properties of the sources of the X-rays. With certain assumptions about flare cooling, Haisch (1983) has estimated that the X-rays are emitted by loops with lengths which range from $(2-6) \times 10^9$ cm up to $>6 \times 10^{10}$ cm. Using different reasoning, Kodaira (1983) has estimated loop lengths of order 10^{10} cm in a particular flare on YZ CMi. Loops of length 10^{10} cm are comparable in extent to the radius of flare stars. Hence, the footprints of such loops on the surface of the star must also be comparable in extent to the stellar radius: such footprints would be suitable candidates for some starspots. Flaring loops in the sun are typically an order of magnitude shorter than the loop lengths mentioned above (typically 10^9 cm). In view of the similarity in Alfven speeds in the coronas of the sun and flare stars (see above), this leads us to expect that the resonance time-scales of coronal loops in flare stars might be an order of magnitude longer than in the sun.

Periodicity or quasi-periodicity

Some indication in support of this comes from recent radio data. Linsky et al. (1983) have detected a radio flare in the star L726-8A (dM5.5e) with the usual characteristics of large flux (10 mJy) and a high degree of circular polarization. However, the interesting feature was the presence of quasi-periodic structure in the flare emission, with periods of 50-60 seconds, extending over about 6 cycles. The authors suggest that this might be caused by Alfven waves oscillating along a large loop (of length $(1-2) \times 10^{10}$ cm). An independent example of periodic behavior following a flare in the optical region was reported by Rodono (1974): there, the light level fluctuated with a very regular peiord (≈ 13 seconds) for several dozen cycles. It is not clear that this periodicity is attributable to the same source as in the radio flare of Linsky et al. (1983), but the strict periodicity is striking.

It has been known for many years that flare radio emission from the sun also shows periodic or quasi-periodic behavior after certain flares (Abrami, 1970; de Groot, 1970; Rosenberg, 1970; McLean et al. 1971). The periodic behavior is most clearly observed at low frequencies (100-300 MHz), corresponding to source heights of 0.2-0.3 R(sun) above the photosphere. At such frequencies, the observed periods are well defined, and from event to event span the range 1.7-3.1 seconds. At higher frequencies of observation, the variations are not as strictly periodic: moreover, the quasi-periodicities span a range which is wider than the above range (Maxwell and Fitzwilliam, 1969). The physical interpretation of the quasi-periodic behavior at higher frequencies may be different from that of the strictly periodic behavior at lower frequencies (Maxwell and Fitzwilliam, 1969). For example, the lack of strict periodicity at higher frequencies may be due to confusion of many loops of varying sizes which lie within the beam width when the latter is observing deep levels of the atmosphere, whereas at lower frequencies, which are sensitive to higher lying loops, individual loops may be detectable, standing in isolation above the background. However, even at microwave frequencies, strict periodicity is occasionally observable: a well-documented case is that reported by Gaizauskas and Tapping (1980) who observed a complex active region (non-flaring) in which a source persisted at a period of 2.5 seconds for many hours.

Coupling with convection zone

Comparing the 50-60 second quasi-periodicity in the stellar radio flare reported by Linsky et al. (1983) and the 13-second periodicity reported in the optical (Rodono, 1974) with the typical solar periods reported above, it appears that stellar flares favor periods which are longer than in the sun. The increase in the stellar periods relative to the sun covers a wide range ~ (4-40), with a mean which may be about an order of magnitude longer than in the sun. This is consistent with loops on flare stars being about an order of magnitude longer than in the sun, provided the Alfven wave speeds are comparable.

The significance of this result can be seen by noting that the preferred periods of mechanical energy generation in the convection zones of flare stars are $t_c \approx$ 30-70 seconds (see above), and this range overlaps with the estimates of 13-60 seconds for the preferred periods of oscillation of coronal loops, t_A. Hence, coronal loops in these stars may be almost ideally matched to the convection zone as regards period. Ionson (1983) has given the following formula for the efficiency of coupling, ϵ, between convection zone and coronal loops:

$$\epsilon = \frac{t_c}{t_A} \frac{1}{\left[1+(\frac{t_c}{t_A} - \frac{t_A}{t_c})^2\right]}$$

In certain flare stars, ϵ must be close to unity. In contrast, in the sun, where $t_c \approx$ 500 seconds, while $t_A \approx$ 2.5 seconds, Ionson's formula leads to $\epsilon \approx 0.005$. Hence, the efficiency of coupling to the corona is larger by a factor of order 10 in certain flare stars then in the sun. As we noted above, the efficiency of coronal heating in red dwarfs (as evidenced by the coronal X-rays) is indeed larger than the solar value by factors of order 10^2.

This leads us to conjecture that as one examines stars which lie further and further down along the main sequence, one is observing a better and better matching of the source of mechanical energy (in the convection zone) with the load where that energy is eventually deposited (coronal loops). To the extent that this conjecture is true, the corona of the certain red dwarfs are fed with such high levels of mechanical energy that they may be considered to be in a state of quasi-continual flaring. In such stars, the boundary between coronal heating and flaring may be quite indistinct. The statistical data of Lacy et al. (1976) on optical flares provides support for this idea: among the cooler red dwarfs, the stars emit most of their "flare" power in small events.

Ultimately, as one examines stars farther along the main sequence, t_c may continue to even smaller values, while t_A may continue to increase. In that case, ϵ would have a maximum value when $t_c = t_A$, but would than decrease rapidly ($\sim (t_c/t_A)^{-3}$) at later spectral types. This may help to explain why the coronal heating efficiency, F_{cor}, has a rapid decline at the latest spectral types (Golub, 1983), after passing through a large maximum at B-V = 1.5. The chromospheres in the coolest M dwarfs would therefore not have access to the X-ray heating which seems to heat the active chromosphere stars (Cram, 1982). Such chromospheres would therefore be quite weak: there is observational support for this prediction (cf. Giampapa, 1983).

REFERENCES

Abrami, A.: 1970, Solar Phys. 11, p. 104.
Cram, L.E.: 1982, Ap. J. 253, p. 768.
Cram, L.E. and Mullan D.J.: 1979, Ap. J. 234, p. 579.
Cram, L.E. and Woods, D.T.: 1982, Ap. J. 257, p. 269.
De Groot, T.: 1970, Solar Phys. 14, p. 176.
Fowler, L. et al.: 1983, Solar Phys. 84, p. 33.
Gaizauskas, V. and Tapping, K.F.: 1980, Ap. J. 241, p. 804.
Gershberg, R.E.: 1983, in Rodono and Byrne (1983).
Giampapa, M.: 1983, in Proc. IAU Symp. No. 102, p. 187.
Giampapa, et al.: 1982, Ap. J. (Letters) 252, p. L39.
Gibson, D.M.: 1983, in Rodono and Byrne (1983).
Haisch, B.M.: 1983, in Rodono and P. Byrne (1983).
Haisch, B.M. et al: 1983, Ap. J. 267, p. 280.
Hoxie, D.: 1973, Astron. Astrophys. 26, p. 437.
Ionson, J.A.: 1983, Ap. J. (In press).
Kelch, W. et al.: 1979, Ap. J. 229, p. 700.
Kodaira, K.: 1983, in Rodono and Byrne (1983).
Kunkel, W.E.: 1975, in "Variable Stars and Stellar Evolution:, ed.
   V. Sherwood and L. Plaut (Dordrecht:Reidel), p. 47.
Lacy, C.H.: 1976, Ap. J. Suppl. 30, p. 85.
Leighton, R.B.: 1963, Ann. Rev. Astron. Ap. 1, p. 19.
Linsky, J.L. et al.: 1982, Ap. J. 260, p. 670.
Linsky, J.L. et al.: 1982, Ap. J. (Letters) 263, p. L79.
Livshits, M.A. et al.: 1981, Solar Phys. 73, p. 269.
Maxwell, A. and Fitzwilliam, J.: 1973, Astrophys. Letters 13, p. 237.
McLean, D.J. et al.: 1971, Nature 234, p. 140.
Moffett, T.J.: 1974, Ap. J. Suppl. 29, p. 1.
Moore, R.L. and Datlowe, D.: 1975, Solar Phys. 43, p. 189.
Mould, J. and Hyland, A.: 1976, Ap. J. 208, p. 399.
Mullan, D.J.: 1971, Monthly Not. Roy. Astron. Soc. 154, p. 467.
Mullan, D.J.: 1974a, Ap. J. 187, p. 612.
Mullan, D.J.: 1974b, Ap. J. 192, p. 149.
Mullan, D.J.: 1975a, Astron. Astrophys. 40, p. 41.
Mullan, D.J.: 1975b, Ap. J. 200, p. 641.
Mullan, D.J.: 1976, Irish Astron. J. 12, p. 161; ibid. 12, p. 277.
Mullan, D.J.: 1979, Ap. J. 231, p. 152.
Mullan, D.J.: 1981, Solar Phys. 70, p. 381.
Mullan, D. J. and Bell, R. A.: 1976, Ap. J. 204, p. 818.
Oskanian, V.S. and Terebizh, V.J.: 1971, Astrofizika 7, p. 83.
Pettersen, B.: 1980, Astron. Astrophys. 82, p. 53.
Pettersen, B.: 1983, in Rodono and Byrne (1983).
Rodono, M.: 1974, Astron. Astrophys. 32, p. 337.

Rodono, M. and Byrne, P.B. (editors): 1983, "Activity in Red Dwarf
   Stars", Proc. IAU Colloq. No. 71. (Dordrecht:Reidel).
Rosenberg, P.: 1970, Astron. Astrophys. 9, p. 159.
Rosner, R. et al.: 1978, Ap. J. 222, p. 317.
Rosner, R. et al.: 1981,, Ap. J. (Letters) 249, p. L5.
Spangler, S. et al.: 1974, Ap. J. (Letters) 194, p. L43.
Spicer, D.S. and Brown, J.C.: 1981, in The Sun as a Star, ed. S.
   Jordan (Washington: NASA SP-450), p. 413.
Stern, R.A. et al.: 1983, Ap. J. (Letters) 264, p. L55.
Uchida, Y.: 1968, Solar Phys. 4, p. 30.
Vogt, S.S.: 1983, in Rodono and Byrne (1983).
Willson, R.W.: 1980, Science 211, p. 700.
Withbroe, G. and Noyes, R.W.: 1977, Ann. Rev. Astron. Ap. 15, p. 363.

DISCUSSION

Sturrock: Since the Alfvén speed is about 500 km/s and the length of a flare loop is about $10^{10} - 10^{11}$ cm, I would expect the time scale of energy release to be $10^2 - 10^3$ s. However, the time curves show a rise time of about 1 sec. How can one uderstand this discrepancy?

Mullan: The rising part of a flare light curve represents the rapid release of energy at the "spark point" where the flare was initiated. This flare emission is mainly optical continuum (in the so-called "spike flares" where the rise time is very short), and is distinguished from the flare emission later on, which is mainly emission lines. The continuum can be interpreted consistently as thermal bremsstrahlung from the coronal flare plasma, while the emission lines emerge from denser gas (cf. Mullan, Ap.J. $\underline{210}$, 1976). Thus, the later emission in the flare represents the after effects of the "spark", after the "conflagration" has spread out over the entire volume of the available flux tube. The conflagration time scale is indeed 100-1000 seconds, as you point out. But the rapid rise time scale represents only emission from a localized region at the "spark point" itself, and is therefore characterized by a much more rapid time evolution. However, the energy emitted in the rise phase is only a small fraction of the total flare energy: the main energy release comes about when the entire available volume is engulfed by the spreading conflagration.

Kundu: I have two comments: 1) When you drew the analogy between solar radio oscillations with a periodicity of approximately 2.5 sec and red dwarf flare star oscillations with a periodicity of a few tens of seconds, you should remember that in the example from Culgoora, there was a shock wave (type II burst) which was responsible for compressing the plasmoid (type IV) and producing oscillations. In the optical stellar flare that you showed from Rodono, was there any evidence of a shock in the flare? How are the oscillations produced in the stellar flare? 2) The other radio oscillations that you referred to (paper by Gary et al. 1982) - I believe it is the paper in which they invoked masering to explain the stellar variability. One must not forget that in the case of solar masers, we have millisecond time structure, whereas with the VLA we have only a time constant of 10 sec. So, in spite of a high brightness temperature, they conceivably could not have been observing "stellar masers"!

Mullan: 1) In Rodono's optical flare data (on an early type flare star, of spectral type K), there is no evidence for or against the existence of a shock wave. In stellar flares, observers in radio bands have occasionally discussed the presence of type II bursts on the basis of time lags between features on the "light curves" of the flare at different frequencies. But no such data were available in conjunction with Rodono's oscillations. In the optical stellar flare, there is no clear answer to the mechanism which creates the oscillations. However, their highly regular behavior suggests a rather strict time-keeping mechanism, and the only one which comes to mind is oscillations in a flux tube which preserves its identity and gross structure intact for the duration of the flare oscillations. 2) You are correct in saying that the oscillations in the radio flare from L726-8A were reported in a paper where Gary et al. favor a maser interpretation. However, they did discuss the possibility that the oscillations might involve bouncing MHD waves on a closed flux tube, and, although Gary et al. finally viewed the latter interpretation with less enthusiasm than they viewed the maser interpretation, my own prejudice is that their data are quite consistent with the MHD bouncing wave interpretation. Note also that the star observed by Gary et al. was of considerably later spectral type (M6) than Rodono's star (K), and this would be consistent with our claim that the bounce time, t_A, increases as the spectral type becomes later.

Rust: In trying to use the rise times for solar and stellar flares to deduce properties of the chromospheres, you make two errors:
1. Flares, as you admit, happen in the corona. The chromospheric flare happens only because heat or particles or photons from the corona excite the chromosphere. We learn nothing about the density of the chromosphere from the rise time of the flare. 2. Observations of white-light flares show rise times of a few seconds, not one minute as you claim (see papers by Zirin and by Rust and Hegwer). Thus solar and stellar optical flare rise times are about the same.

Also, pay attention to the fact that there is no single density relationship between the corona and the chromosphere. Coronal holes have a density perhaps as low as 10^{-7} cm^{-3}, but active region loops may have $10^{11} cm^{-3}$. Yet, under each, the chromosphere is essentially the same. You cannot scale density from the chromosphere into the corona.

Mullan: 1) The first "error" is not an error: you have misinterpreted as coronal bremsstrahlung (Mullan, Ap.J. 210, 1976), and is therefore a signature of the stellar corona. Admittedly, the chromosphere of the star contributes to the flare later, but at maximum continuum emission, we are seeing the corona. The rise time therefore is a measure of a coronal process, and is therefore sensitive to the density in the corona. The latter is proportional to the density in the chromosphere (see below). 2) Your admission that solar white light flares have rise times of a few seconds, taken in conjunction with the observation that rapid stellar flares can evolve on times of less than one second, still allows us to conclude that the evolution of stellar flares is faster than solar flares. Thus, qualitatively, my conclusion is unchanged. Quantititatively, the ratio may be less extreme than I quoted. However, the origin of white light emission in solar flares is not yet known with certainty. Since stellar optical continuum is coronal in origin, we should strictly compare the stellar rise times with the rise time for the thermal plasma in the solar corona, i.e. the time-scale on which the soft X-rays (not the impulsive X-rays which are chromospheric in origin due to precipitating electrons) rise. The soft X-rays generally show a rather gradual behavior, with time-scales which are longer than a few seconds. Thus, even in a quantitative sense, my estimates of the ratio of relevant time-scales may not be far off the true mean value.

I strongly disagree with your final statement. The chromosphere in an active region is certainly different from the chromosphere in a coronal hole. The densities are higher, the macro-turbulent broadening is larger (especially if the active region is young), and the energy requirements differ by more than an order of magnitude. The work of Linsky, Shine, Withbroe, and Noyes, and many others has shown this to be the case. In an active region, the calcium emission cores are much stronger and wider than they are in a quiet region (such as a coronal hole), and Linsky and co-workers have systematically examined how the pressure at the "top" of the chromosphere must be varied in order to fit the observed calcium emission in various features. The results is that the pressure is lowest in quiet regions, increases with increasing activity, and is highest in flares. Exactly the same sequence of pressures can be derived completely independently for the coronal gas from X-ray pictures (such as the AS&E photos during Skylab): the coronal gas pressure is lowest in coronal holes and quiet regions, and increases in active regions (especially young ones), reaching maximum values in flares. Thus there is a one-to-one correspondence between chromospheric pressures (densities) and coronal pressures (density).

Bell: Have you tried using J-K or some similar IR color in your plot of X-rays versus B-V, as B-V is so much affected by TiO in M dwarfs.

Mullan: No. The X-ray data were plotted by Golub (1983), and he used the most common colorimetric data available in the reference literature, i.e. B-V. As you say, it would be interesting to re-plot the figure with a temperature index which was less susceptible to distortions by (e.g.) molecular bands. However, for many of the stars in Golub's sampel, J-K colors may not be readily available.

A UNIFIED TREATMENT OF THE FILAMENT AND FLARE INSTABILITIES*

Gerard Van Hoven
Department of Physics
University of California
Irvine, California 92717 (U.S.A.)

ABSTRACT

Filaments and flares occur in sheared magnetic structures as a result of radiative cooling and resistive reconnection, respectively. A new integrated theory of these two unstable processes is described, which includes the relevant effects of magnetohydrodynamics and energy transport. The normally dissociated thermal and tearing phenomena are coupled together by a temperature-dependent Coulomb resistivity. As a result, the filamentation and flaring instabilities of a sheared field may coexist, as is familiar from the solar example.

The growth rates and spatial structures of these two modes are detailed here. The much faster radiative instability is shown to provide significant magnetic reconnection, particularly at shorter wavelengths. The long-wavelength reconnection mode is found to be abetted by the resistivity increase caused by the dominance of cooling at the X point, in contrast to its nonradiative behavior. Implications of these results for the development of coronal activity are described.

INTRODUCTION

It is well known, as exemplified by the development of solar activity, that increasing magnetic-field shear or stress gives rise to the formation of filaments (Chiuderi and Van Hoven, 1979) and to flares (Van Hoven, 1979). Theories of the former mechanism (Field, 1965), which is driven by a radiation ouput that decreases with temperature, have usually ignored the resistive magneto- hydrodynamic effects of the resulting, very collisional, relatively low-temperature plasma. Treatments of the latter reconnection instability (Furth, Killeen, and Rosenbluth, 1963), which is catalyzed by finite conductivity, have usually ignored the energy-flux consequences of the resulting magnetic energy release. Since, in astrophysical situations, radiation is a strong effect and the relevant resistivity is the temperature-dependent Coulomb value, these two dynamic processes should be treated in a unified way (Van Hoven, Steinolfson, and Tachi, 1983). This paper describes the outcome of such a coupled-instability study, in which the relevant linearized equations are solved numerically over a range of parameters, and interprets the most important results.

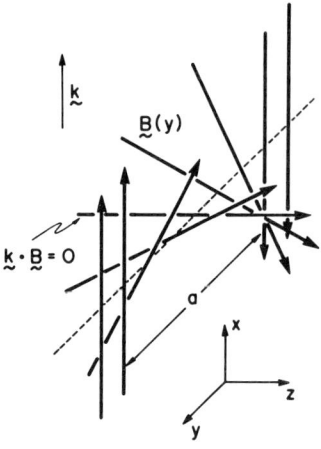

Figure 1. Sheared force-free field.

FORMULATION

We model a current-carrying volume of plasma by the sheared-field form

$$\underline{B}_0 = B_0[\underline{e}_x \tanh(y/a) + \underline{e}_z \mathrm{sech}(y/a)] \quad , \quad (1)$$

shown in Fig. 1, which is force-free and thus consistent with an initially uniform temperature and density. We describe the temporal development of this system by using the resistive MHD equations, relating flow velocity to magnetic induction,

$$\rho \, d\underline{v}/dt = \underline{J} \times \underline{B} \quad (\mu_0 \underline{J} = \underline{\nabla} \times \underline{B}) \quad (2)$$

$$d\underline{B}/dt = (\underline{B}\cdot\underline{\nabla})\underline{v} + (\eta(T)/\mu_0)\nabla^2 \underline{B} \quad , \quad (3)$$

along with the incompressible energy-transport equation

$$K'\rho dT/dt = \underline{\nabla}\cdot[\underline{B}\kappa_\| B^{-2}(\underline{B}\cdot\underline{\nabla})T] - R\rho^2 T^r + \eta(T)J^2 \quad . \quad (4)$$

The notation here is the same as that in Van Hoven, Steinolfson and Tachi (1983), except that K' is Boltzmann's constant per unit mass, divided by $\gamma-1$. These equations are solved by assuming small-amplitude perturbations of the form (dictated by symmetry) $T_1(\underline{r},t) = T_1(y,t)\exp(ikx)$, for example, where $T_1 \ll T_o$, the equilibrium value.

If we temporarily ignore the underlined resistive terms in (2) - (4), we obtain two uncoupled dynamic systems. The first, from (2) and (3), describes the frozen-in-field condition, and variations on the hydromagnetic time scale $\tau_{hm} = a(\mu_o \rho_o)^{1/2}/B_o$. The second, from (4), provides unstable, incompressible, radiative cooling at

$$\tau_{ra} \tilde{=} \Omega_\rho^{-1} = [-rR\rho_o T_o^{r-1}/K']^{-1}, \text{ when } r < 0 \text{ so that the}$$

radiation falls with temperature (Hildner 1974). This unstable cooling only occurs, however, in the center of the magnetic-shear layer $y = 0$, where $\underline{B} \cdot \underline{\nabla} = ikB_x = ikB_o\tanh(y/a) = 0$, so that the much stronger parallel (to $\underline{B}$) thermal conduction is ineffective (Chiuderi and Van Hoven, 1979).

Let us now restore the underlined resistive terms to (2) - (4), using the Coulomb value $\eta(T) = \bar{\eta}T^{-3/2}$ (Spitzer, 1962) as the relevant form for astrophysical applications. This added level of fidelity has two important coupling effects, beyond the superposition of (slow) resistive diffusion on the time scale $\tau_{re} \sim \mu_o a^2/\eta$. First, as is well known (Furth, Killeen, and Rosenbluth, 1963; Van Hoven, 1979), Eqs. (2) and (3) now exhibit a new reconnection, or magnetic-tearing, excitation which grows on the hybrid time scale $\tau_{te} \sim \tau_{hm}^\nu \tau_{re}^{1-\nu}$, where $2/5 \leqslant \nu \leqslant 2/3$ (Steinolfson and Van Hoven, 1983). The tearing mode, as with the radiative-cooling perturbation of (4), exists primarily in the layer $y \approx 0$, where the usually dominant first term on the right of (3) is ineffective. This local frustration of the normal frozen-in-field constraint allows the magnetic field to reconnect into a topology from which it can release its magnetic energy (Van Hoven, 1979). [For completeness, the resistive term has no significant influence (for astrophysical conditions) on the radiative phenomena described by (4) alone.]

The second primary effect of a temperature dependent resistivity is the coupling together of the magnetodynamic system (2,3) and the energy transport of (4) through the underlined Ohmic diffusion/heating terms. The consequences of this new interaction form the principal topic of this paper.

RESULTS

The coupled dynamic system (2) - (4) exhibits two, growing, small-amplitude excitations, a (mainly) radiative mode and a tearing-like mode. We solve the linearized equations by using a finite-difference scheme (Killeen, 1970), with the addition of variable grid spacing (Steinolfson and Van Hoven, 1983) to resolve the sharp central gradients which arise in this problem. The growth rate ω is measured from $\partial T_1/T_1 \partial t$, when this quantity becomes uniform throughout the x-y grid, at which time various eigenfunction profiles and topographic plots are produced.

The input parameters and output quantities are expressed in normalized terms. These include the wavenumber $\alpha = ka$, the magnetic Reynolds number τ_{re}/τ_{hm}, and the ratio of the equilibrium radiation to Ohmic heating

$$\varepsilon = R\rho_o^2 T_o^r / \eta_o J_o^2 = -\beta \Omega_\rho \tau_{re}/3r \text{ (with } \gamma = 5/3 \text{ and } \beta \text{ the ratio}$$

of plasma to magnetic pressures).

Growth rates ω are given either in terms of the hydromagnetic time, $\bar{p} = \omega \tau_{hm}$, or of the resistive time, $p = \omega \tau_{re}$, convertible by a factor of S. The magnetic Reynolds number S is artificially scaled (Van Hoven, Steinolfson, and Tachi, 1983) by the coefficient $\bar{\eta}$ of the Coulomb resistivity, with the equilibrium temperature T_o held constant. Thus, $S = S_c \bar{\eta}/\eta$, where the subscript c denotes the classical (or correct) value. Since the parameter $\varepsilon \propto \eta^{-1}$, it also varies as $\varepsilon_c S/S_c$.

An illustrative run of growth rates $\bar{p}$ vs S, for $\alpha = 10^{-1}$ is displayed in Fig. 2. Under conditions of large S ($S_s = S_c$ is the classical solar coronal value), the radiative filamentation) mode grows ~100x more quickly than the tearing (flare) mode, and the character of the eigenfunctions ($\underline{B}, T, \underline{v}$) is quite different, as will be described in what follows. As the value of S drops, the ratio of radiation loss to Ohmic heating falls to unity with ε, and the mode profiles become quite similar. Below $S \approx 10^{7.3}$, where the curves meet, the growth rates are complex.

A more general plot of the growth rate, normalized to the resistive-diffusion time, is shown in Fig. 3, where the α (wavenumber) dependence is emphasized. [The physical conditions here, which can affect the normalized magnitudes, are $T = 10^6 K$, $n = 10^{10}$, $B = 83G$ and $a = 10^7$ cm.] The solid curves specify the growth rates of the tearing mode, and the dashed line segments those of the radiative

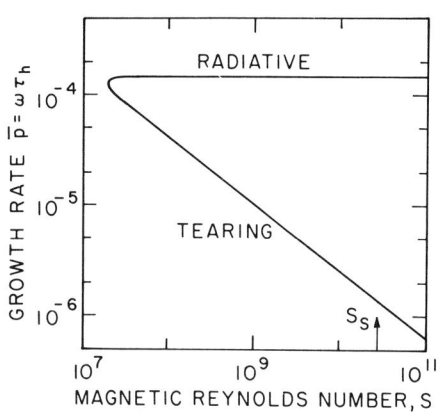
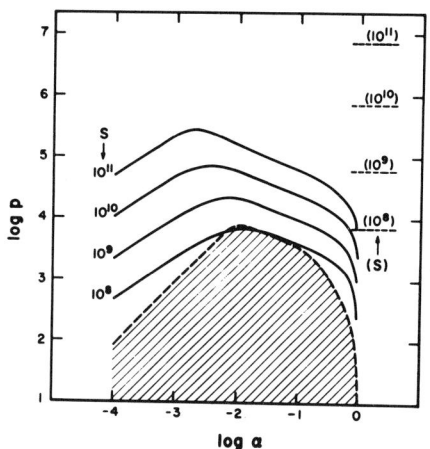

Figure 2. Growth rates at $\alpha = 0.1$.

Figure 3. Growth rates vs wavenumber.

mode (Van Hoven, Tachi and Steinolfson, 1983). The growth of the latter mode is dispersionless over this range of (relatively long) wave-lengths and so is not shown in detail. [For solar coronal parameters, the growth rate is independent of parallel (to $\underline{B}$) heat flow and does not drop until $\alpha \sim 10^2$ where perpendicular thermal conduction becomes important.] The cross-hatched area is the regime of strong radiative-to-tearing mode coupling (as in the $S < 10^{7.3}$ range of Fig. 2), in which the growth rate becomes complex.

For S values above 10^9, the tearing mode growth is unchanged from the normal (non-radiative) case. That is, $p \sim S^{-\nu}$, where $\nu = 2/3$ for (fixed) α to the left of the peak, $\nu = 1/2$ at the peak values of α, and $\nu = 2/5$ to the right (Steinolfson and Van Hoven, 1983). The growth rate of the radiative mode can be expressed as $p \approx -2(\gamma-1) r \epsilon/\beta$, which is equivalent to the well-known result $\omega \approx \Omega_\rho$ (Chiuderi and Van Hoven, 1979).

Some of the more interesting aspects of this coupled reconnection/radiation system arise when one looks at the structure of the eigenfunctions (Steinolfson and Van Hoven, 1984). Figure 4 shows a situation in which each of the two modes (at S_s) has individually (orthogonally) grown to a level in which nonlinearity would just start to become important. [The sum of the nonlinear terms, computed but not used in the dynamic equations, is a fixed fraction of the largest linear term; this measure of equivalency is approximately the same as that of equal linear energy content.] The plots of Fig. 4 refer to the coordinate system of Fig. 1, and show the x-y halfplane (on its side) as seen from the z direction. The horizontal (x) scale is linear and shows one wavelength, and the vertical scale ($y/a = 10^{-2}$ at the top) is expanded near the origin to better show the details in this narrow layer.

The top plot (a) of Fig. 4 shows the magnetic flux function, or B_x-B_y field lines, artificially amplified by a factor of 10^2 so that the fine structure can be appreciated. On the left is the well-known X-point field pattern of the tearing mode (Furth, Killeen, and Rosenbluth, 1963). A new result is shown at the right. At an equivalent energy level, the fast

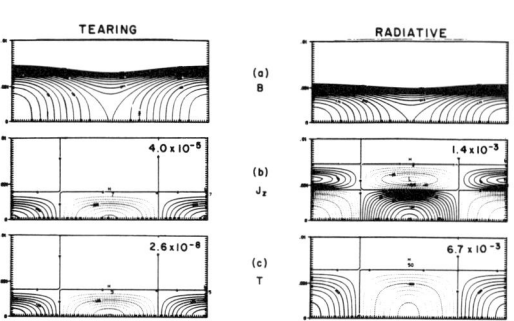

Figure 4. Mode structures at equivalent levels.

radiative mode provides 30% as much magnetic reconnection as the tearing mode, with a similar potential for magnetic energy release. Part (b) displays the z-directed current density which, when multiplied by η_o, provides the parallel (to B) electric field intrinsic to the reconnection process. The peak levels shown (with respect to J_o) are small, at this amplitude, but the radiative-mode $E_\parallel$ field is relatively 35x larger.

Part (c) of Fig. 4 shows the isotherms of the perturbations, with the dotted curves indicating negative values. The strong temperature reduction in the center of the radiative mode was expected, but the moderate cooling of the X point of the tearing mode is significant. This behavior is opposite to that in the absence of radiation (Tachi, Steinolfson, and Van Hoven, 1983), when the locally concentrated Ohmic dissipation heats the plasma and can reduce the level of Coulomb resistivity which catalyzes the reconnection.

DISCUSSION

This paper has described the first unified treatment of the resistive filamentation/radiation and flare/reconnection instabilities. The growth rates and excitation structures of the coupled linear modes in a sheared magnetic field (Van Hoven, Steinolfson, and Tachi, 1983), as determined from numerical computations, have been detailed. [An alternative, approximate, analytic calculation of the growth rates has been given by Steinolfson (1983)].

The growth-rate results show a strong coupling of the modes below a minimum value of the magnetic Reynolds number. However, for solar coronal $\underline{S}$ conditions, the growth rates are similar to their uncoupled values, which are separated by a factor of order 10^2, with the radiative mode being the faster. For $T = 10^6$ K, $B = 10^2$ G, $a = 10^7$ cm and $n = 10^9 - 10^{10}$, the growth time of the radiative mode is 3 - 15 minutes (Van Hoven, Tachi, and Steinolfson, 1983).

Although the growth behavior is not much modified under high-$\underline{S}$, sheared-field, astrophysical conditions, the individual mode structures are significantly changed. The tearing-like mode, as shown in Fig. 4, exhibits a negative temperature perturbation at the X point. This is the result of the strong radiation dominance at high $\underline{S}$, and indicates that the ηJ^2 heating accompanying tearing cannot lower the local Coulomb resistivity. Thus, the magnetic reconnection can proceed to higher levels, and further energy release, without the self-quenching or slowing-down that a thermally reduced resistivity would provide.

What is more important is the fact that the radiative mode has a previously unknown magnetic-reconnection component (Steinolfson and Van Hoven, 1984). This cooling mode is 30x times faster than, and provides 30% of the flux reconnection of, the normal tearing mode. This radiative excitation, therefore, has the potential of providing magnetic energy release on a much shorter time scale than previously believed possible.

The results of this paper show that there is a more fundamental connection between filaments and flares than the fact that they occur at the same magnetic-field site in the central ($\underline{k} \cdot \underline{B} = 0$) layer of Fig. 1. Radiative filament formation can lead directly to magnetic energy release and thus to flares. Normal resistive flare reconnection is also aided by radiative cooling, so that it may go beyond the limitations due to self heating. The final characterization of the connection between these two instabilities must come from nonlinear computations, which are under development.

*This research was performed in collaboration with R. S. Steinolfson, whose contributions are gratefully acknowledged, and was supported in part by the Atmospheric Sciences Section of NSF and the the Solar-Terrestrial Theory Program of NASA.

REFERENCES

Chiuderi, C. and Van Hoven, G.: 1979, Astrophys. J. (Letters) 232, L69.
Field, G.B.: 1965, Astrophys. J. 142, 531.
Furth, H.P., Killeen, J., and Rosenbluth, M.N.: 1963, Phys. Fluids 6, 459.
Hildner, E.: 1974, Solar Phys. 35, 123.
Killeen, J.: 1970, in Physics of Hot Plasmas, eds. B.J. Rye and J.C. Taylor (New York: Plenum), pp. 212-219.
Spitzer, L.: 1962, in Physics of Fully Ionized Gases (New York: Wiley Interscience), pp. 136-145.
Steinolfson, R.S.: 1983, Phys. Fluids 26, .
Steinolfson, R.S. and Van Hoven, G.: 1983, Phys. Fluids 26, 117.
Steinolfson, R.S. and Van Hoven, G.: 1984, Astrophys. J., to appear.
Tachi, T., Steinolfson, R.S., and Van Hoven, G.: 1983, Phys. Fluids 26, .
Van Hoven, G.: 1979, Astrophys. J. 232, 572.
Van Hoven, G., Steinolfson, R.S., and Tachi, T.: 1983, Astrophys. J. 268, 860.
Van Hoven, G., Tachi, T., and Steinolfson, R.S.: 1983, UCI Phys. Rept. No. 83-25, submitted to Astrophys. J.

DISCUSSION

Mahajan: In the presence of tearing modes, sharp gradients can build up. In that case can one justify using the simple Ohm's law $E = \eta J$? Anomalous viscosity could be a very important term, and is shown to be so in Tokamaks. In your range of parameters, what are the effects of viscosity?

Van Hoven: We have included viscosity in other calculations. The effects on the growth of the modes are small (factors of 3) and only occur at the shorter wavelengths.

Sturrock: What solar phenomena do you attribute to this radiatively enhanced reconnection process?

Van Hoven: The answer to this question will come from the nonlinear behavior. Radiative reconnection may initially favor short wavelengths (compressibility effects indicate this) and thereby merely provide the smaller-scale, transverse, magnetic fields observed in prominences. Such excitations would saturate early, as is known from the tearing mode, because of the free energy invested in bending the nearby frozen fields. Then, after a time of continuing shear increase, a long-wavelength radiative reconnection mode would take off, which would have much better access to the stored magnetic energy (again, by analogy with normal tearing) and thus provide the energy release of a flare. These possibilities require, I must emphasize, a nonlinear leap of the imagination.

Priest: If the electric field is 3 times bigger in the radiative mode than in the tearing mode, how do the magnetic energy releases compare?

Van Hoven: The parallel electric field in the radiative mode is 250 times larger (the value given below Fig. 4 did not include the resistivity perturbation). The reconnected fluxes (or island widths) differ by a factor of 3 which, in this linear calculation, is the ratio of the magnetic-energy perturbations. A proper evaluation of the energy release requires a nonlinear computation.

Priest: You are modelling an arcade or loop magnetic configuration by a field that varies with one variable alone and are taking boundary conditions along a second direction. What are the boundary conditions in the third direction? Would not thermal conduction along the field in that direction be important?

Van Hoven: For the formation of a filament, conduction must be suppressed so that one requires the distance (ℓ) to the boundary in the z direction to satisfy $\ell^2 \gg R\ell^2 T^{r-1}/\kappa_{\parallel}$.

Migliuolo: What is the role played by compressibility in radiative and tearing modes?

Van Hoven: The compressibility influence on tearing modes is negligible; the radiative growth is increased by a factor of 2 (for coronal parameters) for ka > 1, before it falls at ka > 10^2 because of perpendicular thermal conduction.

NONLINEAR EVOLUTION OF THE RESISTIVE TEARING MODE

R. S. Steinolfson and G. Van Hoven
Department of Physics
University of California
Irvine, California 92717

ABSTRACT

Numerical solutions of the MHD equations are used to investigate the nonlinear behavior of the tearing instability. The mode evolves from a linearly growing excitation, followed by a period of greatly reduced nonlinear growth. Constant-Ψ solutions evolve much more slowly than comparable nonconstant-Ψ modes with orders of magnitude less conversion of the stored magnetic energy. The nonconstant-Ψ computations indicate a reduction by approximately 20% of the energy in the initial shear layer. For long-wavelength solutions, secondary-flow vortices, opposite in direction to the linear vortices, generate a new magnetic island centered at the initial x-point.

COMPUTATIONAL PROCEDURE

The nonlinear phase of the tearing mode (Furth et al., 1963) is studied in slab geometry using incompressible, constant-resistivity, MHD theory. The initially stationary plasma, with uniform thermodynamic properties, is embedded in a force-free, nondissipating magnetic field. A linear mode, at its maximum linear growth, provides the initial state for the nonlinear computation. We present results for a magnetic Reynolds number S (ratio of the resistive time to the hydromagnetic time) of 10^4 and values of the wavelength parameter $\alpha(2\pi a/\lambda$, where a is the shear scale and λ the disturbance wavelength) of 0.05, 0.13, and 0.50. [A larger parameter range and additional computational results are considered by Steinolfson and Van Hoven (1983b).] The mode with $\alpha = 0.5$ is a constant-Ψ solution in the linear regime, while the other two are nonconstant-Ψ, and the $\alpha = 0.13$ mode corresponds to maximum linear growth (Steinolfson and Van Hoven, 1983a).

The linear theory predicts a chain of x-points and islands in the magnetic field lying along the tearing surface (x-axis in our geometry) at y = 0. We isolate one wavelength of this initial disturbance and do not allow it to interact with adjacent wavelengths. Because of the symmetries involved, our computation only extends from the center of one island (x = 0) to the adjacent x-point (x_{max}) and from the tearing surface (y = 0) to a relatively large distance y_{max}. The distance y_{max} is large enough that the perturbation is essentially negligible and decaying exponentially with y. Symmetry boundary conditions are applied at the remaining three boundaries. An expanding grid is used in the y-direction, with minimum spacing near y = 0, in order to resolve the tearing layer. The nonlinear equations are solved numerically using a fully-implicit, alternating-direction procedure.

NUMERICAL RESULTS

The evolution of two of the modes, as measured by reconnected flux and nonlinear growth, is shown in Fig. 1. The dashed curves represent continued linear growth. The long wavelength mode (α = 0.05) evolves almost identically to the α = 0.13 solution, in terms of these quantities, with somewhat (a few percent) more flux reconnection at the final time. Although the two nonconstant-Ψ modes display comparable nonlinear behavior, they are in sharp contrast to the considerably smaller reconnected flux for the constant-Ψ mode.

The total magnetic energy, per unit distance perpendicular to the tearing plane, removed from the magnetic fields is tabulated in the first row of part A of Table I. By contrast, the energy initially in the shear layer for the α = 0.05 solution is 3.6 x 10^{17} ergs/cm_z (scales inversely with α), which, for this case, means that the shear-layer energy has been reduced by 20%. The second and third rows in part A show the percent of the total energy that was removed from the x- and z-components, respectively. Longer wavelength modes remove more energy from the z-component, while none is removed from the z-component for the constant-Ψ solution (energy actually transfers into this component). The available energy in part A is distributed among the various components as shown (on a percentage basis) in part B. Note that more energy goes into heating as the disturbance wavelength increases, with less into the y-component of the magnetic field.

For all three of the modes, a secondary flow vortex, oriented oppositely in direction to the initial, linear vortex, forms near the x-point, as illustrated in Figure 2(a), which has a nonlinear y-scale. The velocity is parallel to the flux function contours on the left while some of the magnetic field lines near the tearing surface are shown on the right. The dashed flux-function contours indicate a clockwise vortex (the distorted linear vortex), and solid curves represent a counter-clockwise vortex. The two modes not shown in this figure continue evolving for the duration of the calculation with the qualitative spatial behavior in Fig. 2(a). However, for the long wavelength mode, the secondary flows become strong enough to alter the basic magnetic topology and cause the formation of an additional magnetic island centered at the linear x-point [Fig. 2(b)].

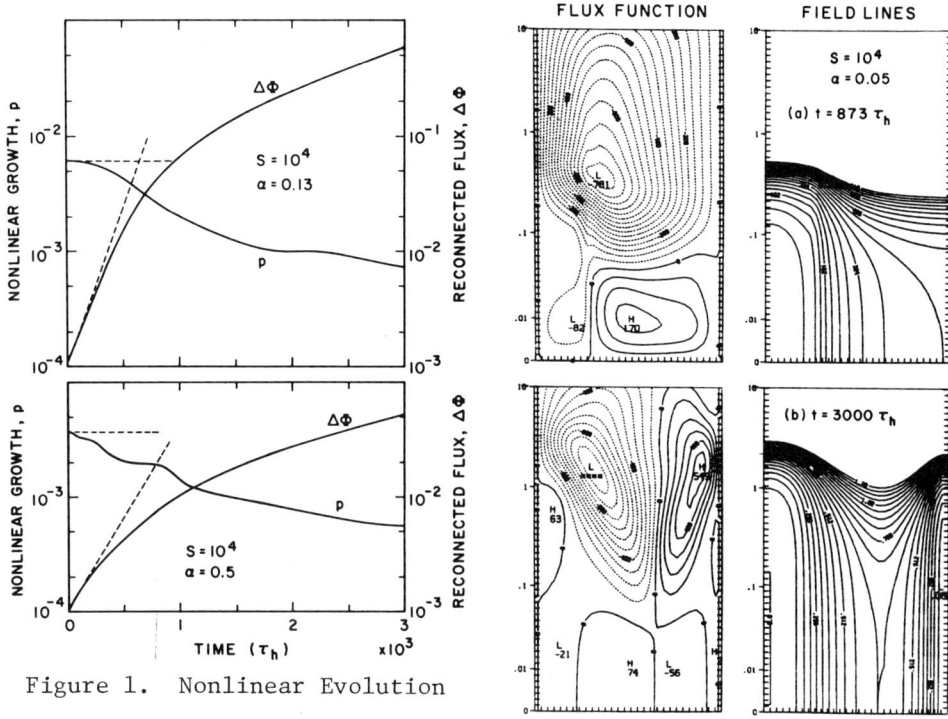

Figure 1. Nonlinear Evolution

Figure 2. Formation of a new magnetic island

α	0.05	0.13	0.5
A. energy source:			
total	7.6×10^{16}	1.7×10^{16}	2.4×10^{13}
magnetic (x)	89.7	97.0	100.0
magnetic (z)	10.3	3.0	0.0
B. energy budget:			
magnetic (y)	8.0	16.2	49.2
magnetic (z)	0.0	0.0	2.6
kinetic (x)	4×10^{-2}	3×10^{-2}	8×10^{-2}
kinetic (y)	6×10^{-4}	5×10^{-4}	3×10^{-4}
kinetic (z)	3×10^{-3}	3×10^{-4}	5×10^{-5}
thermal	92.0	83.8	48.2

Table I. Energy balance. Total magnetic energy (ergs/cm_z) computed for reference B = 37.3 G, a = 10^7 cm; remaining energies given in percent.

CONCLUSION

A primary result in these computations is that the nonlinear evolution generally differs from one region of parameter space to another, and hence, a typical characterization of the evolution in the nonlinear regime is not possible. Some general statements that do apply to all solutions are: (1) The nonlinear spatial distributions of the physical variables differ substantially from the linear behavior; (2) Once nonlinear effects become important, the growth slows considerably from the linear rate; (3) Neither the linear growth rate nor the nonlinear growth is a good predictor of the nonlinear performance of a particular mode in terms of magnetic energy conversion; and (4) More of the stored magnetic energy is converted to thermal energy as disturbance wavelength increases ($>$ 90% at $\alpha = 0.05$).

ACKNOWLEDGEMENTS

This work was supported by the Solar Terrestrial Theory Program of NASA under Grant NAGW-93. Acknowledgement is also made to the Institute of Geophysics and Planetary Physics at the Los Alamos National Laboratory, which provided funding under their grant program, and to the National Center for Atmospheric Research, which is sponsored by NSF, for the use of their computer facilities.

REFERENCES

Furth, H. P., Killeen, J., and Rosenbluth, M. N.: 1963, Phys. Fluids 6, pp. 459-484.
Steinolfson, R. S., and Van Hoven, G.: 1983a, Phys. Fluids 26, pp. 117-123.
Steinolfson, R. S., and Van Hoven, G.: 1983b, submitted to Phys. Fluids.

RESISTIVE INSTABILITIES IN ASTROPHYSICAL CONDITIONS: A CRITICAL DISCUSSION.

P. Batistoni, G. Einaudi, F. Rubini
Scuola Normale Superiore, Pisa, Italy
and
C. Chiuderi, G. Torricelli
Osservatorio Astrofisico di Arcetri, Firenze, Italy.

Abstract. Resistive instabilities have often been indicated as the possible cause of rapid release of energy in astrophysical situations. A correct assessment of the validity of this idea requires a detailed analysis of the theory of resistive instabilities in the regimes of astrophysical interest. In particular, effects as the presence of asymmetries due to current gradients, the influence of geometry and of shear flows must be explicitly evaluated. We have started a program of investigation on this subject, whose preliminary results are reported here.

Magnetic fields are one of the commonest components of the astronomical universe. Their presence constitutes, among other things, a potential source of energy that could be utilized in a number of situations. In many instances they are the only conceivable energy source. However, the exploitation of this reserve is generally made difficult by the fact that astrophysical plasmas are, as a rule, almost ideal electrical conductors. The Alfvén theorem then prevents topological changes in the field structure and makes inaccessible to the system many states with low magnetic energy. The presence of a small, but finite, electrical resistivity relaxes this constraint and allows the transformation of magnetic energy into different forms, like heat or bulk particle motion.

The above considerations explain the interest in resistive instabilities, namely those that either disappear in the infinite conductivity limit, or strongly modify instabilities that may exist also in ideal conditions. Most of our knowledge of this class of instabilities derives from studies of laboratory plasmas and often results valid for fusion devices have been directly applied to astrophysical systems in spite of the vast difference in regimes, boundary conditions and geometrical constraints encountered in the two cases. Another problem that has often prevented a successful application of the theory of resistive instabilities in the astrophysical context is that of the timescale of magnetic energy release. In the case of the solar flares, for instances, the resistive timescales are substantially larger than those on which the phenomenon is observed to develop.

Considering the interest of the problem, the number of still unanswered questions and the fact that from our experience it is hard to predict the behaviour of a given system by simply transferring the results of a completely different one, we have felt the necessity of a detailed analysis of resistive instabilities in regimes of direct astrophysical interest. The first results of such a study are briefly reported here and will form the subject of subsequent papers. All these results refer to the linear stage of the instability.

The linear development of resistive instabilities in plane geometry has been extensively studied since the pioneering FKR paper (Furth, Killeen and Rosenbluth, 1963) both analytically and numerically. Analytic investigations generally rely on boundary layer techniques and often require further approximations, as for instance the so-called "constant-ψ". Numerical investigations have until recently been limited to relatively low S values, where S is the magnetic Reynolds number. An extensive numerical study has been published recently (Steinolfson and Van Hoven, 1983) where the range of S has been increased up to astrophysically relevant values. We have included in our numerical code the effects of gradients of the current density at the singular points that produce asymmetries in the F-profile (F = $\underline{k} \cdot \underline{B}$). The reason for this inclusion derives from the fact that asymmetries are always present in geometries more realistic than the planar one, so that the entity of these effects can be only assessed in the plane case. The field profile was the familiar $\underline{B} = B_0$ (tgh x $\underline{e}_y$ + sech x $\underline{e}_z$), with $\underline{k} = k_y \underline{e}_y + k_z \underline{e}_z$. The equations were Fourier time-analyzed, unlike in Steinolfson and Van Hoven, 1983. Our analysis generally confirms all classical results on symmetric configurations along with the only analytical results for the asymmetric case (Bertin, 1982) valid in the framework of a generalized constant-ψ approach. Our growth rates are lower than those of Steinolfson and Van Hoven, 1983 at small and large values of the normalized wavenumber, α, for large S ($\gtrsim 10^{10}$). The presence of asymmetries modifies the dispersion curve (20-50%) especially at low α's but has little influence on the form of the eigenfunctions.

We have also examined the resistive behaviour of a system with cylindrical symmetry under modes possessing an m = 1 azimuthal symmetry. The equilibrium field represents a current channel surrounded by a potential field. This type of field had already been studied by Chiuderi and Einaudi, 1981 (hereafter referred to as CE) that showed that configurations exist that are absolutely (i.e. for all k's) stable to ideal perturbations. It is, however, also possible to find ideally unstable configurations for low k-values. A first study of the linear tearing-mode in these systems had also been given in CE, by using the same technique employed by Furth et al. (1973). The use of the constant-ψ approximation, implicit in CE, must be considered with caution in these configurations even at relatively large α's (unlike in the planar case) because of the possible existence of an ideal unstable region for $0 \leq \alpha \leq \alpha_c$. In fact, when this happens we may have sizable α's corresponding to a marginal behaviour. For $\alpha \gtrsim \alpha_c$ the resistive modes appear to be of the reconnecting (tearing) type, whereas for $\alpha \lesssim \alpha_c$ a different mode develops, quite similar to the $m = 1$, $n = 1$ internal resistive kink found in tokamaks and studied by Coppi et al. (1976).

The computational method used is essentially a numerical boundary layer with the internal solution computed exactly. A shooting technique is used to find the eigenvalues: this method appears to be fast and accurate enough up to $S = 10^9$, but is not reliable at higher S-values due to the extreme thinness of the resistive layer. There are several distinct aspects of the problem that must be considered. One is the spatial structure of the mode that is strongly influenced by the value of α, particularly for α close to α_c. Crossing α_c from the ideally stable side, we find that the perturbation corresponds to an essentially rigid radial displacement of the internal part of the plasma inside the current channel towards the singular surface. This type of perturbation is possible also in an ideal plasma (internal kink) but saturates at low amplitude due to the build-up of a magnetic counter-pressure. The presence of a finite conductivity allows the continuation of the process. In this situation the perturbed magnetic field (ψ) changes sign, but the displacement does not, unlike the usual reconnecting mode. This different behaviour may prove to be important in the subsequent non-linear stage.

A second aspect refers to the maximum growth rate attainable by these processes. Since the constant-ψ reconnecting and internal resistive kink modes scale differently with S, the concept of fast and slow reconnecting modes has been sometimes introduced in the literature. We would like to point out that, on the basis of different scaling properties only, it is not possible to assess the relative speed of the two processes. For a given S an unstable configuration attains a maximum growth rate for a well-defined value of α: this may correspond either to an internal resistive structure or to a $\psi \neq$ const tearing mode. To decide which is the configuration that evolves faster we must simply compare the maximum growth rates. From our preliminary results, that incidentally confirm the known S-scalings of the tearing and internal resistive modes, we find that at a given S different configurations corresponding to different spatial structures of the mode evolve on essentially the same timescale. This is due to the fact that the only meaningful comparison must involve the maximum growth rates that do not occur at the same value of α for different configurations and S. From our computations it turns out that at $S = 10^9$ the shortest timescale is $\simeq 10^5$ s, regardless of the topology of the operating mode. Since all timescales increase with S this result implies that resistive instabilities based on classical resistivity in a static configuration are not a viable mechanism for solar flares that occur in plasmas with $S \simeq 10^{12}$.

The presence of shear flows in the unperturbed state can modify considerably the growth rates of the resistive modes (Hoffmann, 1975, Pollard and Taylor, 1979, Dobrowolny et al. 1983). The presence of a fluid velocity considerably complicates the numerical problem since all the growing modes are generally overstable. We have produced a numerical code, based on finite different methods, that computes the (complex) eigenvalues by finding the zeros of the determinant of the coefficients of the homogeneous linear system to which the original differential system has been reduced. The eigenvalues, besides of k, depend on β and on the ratio of the spatial scales for the velocity field and the magnetic field. Initial tests of the code have reproduced known results in selected limiting cases.

REFERENCES.

Bertin, G.: 1982, Phys. Rev. A, 25, 1786.
Chiuderi, C. and Einaudi, G.: 1981, Solar Phys., 73, 89.
Dobrowolny, M., Veltri, P. and Mangeney, A.: 1983, Journ. Plasma Phys. (in press).
Coppi, B., Galvao, R., Pellat, R., Rosenbluth, M.N. and Rutherford, P.H.: 1976, Soviet J. Plasma Phys. 2, 533.
Furth, H.P., Killeen, J., Rosenbluth, M.N.: 1963, Phys. Fluids, 6, 459.
Furth, H.P., Rutherford, P.H. and Selberg, H.,: 1973, Phys. Fluids, 16, 1054.
Hoffmann, I.: 1975, Plasma Phys., 17, 143.
Pollard, R.K. and Taylor, J.B.: 1979, Phys. Fluids, 22, 126.
Steinolfson, R.S. and Van Hoven, G.: 1983, Phys. Fluids, 26, 117.

Magnetodynamical Processes in Interacting Magnetospheres
of RS CVn Binaries

Yutaka UCHIDA and Takashi SAKURAI*
Tokyo Astronomical Observatory, University of Tokyo
*Department of Astronomy, University of Tokyo

Abstract : Magnetodynamical processes in RS CVn binaries are discussed in the scheme of Active-Longitude-Belt picture (Uchida and Sakurai, 1983) in which the "photometric wave" is due to a number of spot pairs which emerge, drift across, and submerge in the "active longitude belt" on the K-star. The formation of the corona and the origin of flares in these close binary systems having starspots are interpreted in terms of the reconnections of the magnetic flux tubes of the companion star with the emerging and submerging pairs of spots on the K star. The injection of the hot plasma into the large scale pole-to-spot connections is required to explain the extended corona with large emission measure, and we attribute this to the "sweeping-pinch" mechanism (Uchida and Shibata, 1984) associated with the relaxation of the toroidal component in the twisted magnetic flux tubes which emerge and reconnect with the flux tubes connecting pole and spots.

1. Introduction

The super starspot interpretation of the "photometric wave" (= PW) of RS CVn binaries seems to lead to some paradox. If the drift of PW along the light curve is attributed to the migration of starspots toward the equator on a differentially rotating star, the differential rotation derived from the drift rate of PW is rather small, of the order of $\Delta\Omega/\Omega$ $\sim 10^{-3}$. On the contrary, the differential rotation law scaled from the sun (Durney and Robinson, 1982) suggests a much larger differential rotation in such rapid rotators like RS CVn's, even the tidal effect may tend to bring the stellar atmospheres into the synchronized rotation (Zahn, 1977). The drift curve of PW for SS Boo and V711 Tau, showing reversals of the direction of the migration during a cycle, may also be difficult to interpret in the framework of solar analogy. The very small differential rotation deduced from the almost static super starspot makes the picture of the system rather static because there is no very drastic occurrence expected if relative orientations of everything do not charge. Although it is not impossible that stars in such close binary systems may show rather peculiar behavior, the discrepancies seem to be too large between this and the G5 star, the sun.

Uchida and Sakurai (1983) proposed a different view that the PW is the stellar version of the active-longitude-belt (=ALB) in which individual spot-pairs appear, drift across, and disappear. This model, at a glance, seems to be similar to the multiple spot model of Eaton and Hall (1979) but the latter did not include the notion of such lively activities in the zone. The introduction of ALB picture relaxes the otherwise inevitable restrictions in the somewhat queer (though fascinating) super starspot model, and allows an activated picture in which many pairs of smaller spots appear and disappear after being carried across ALB by appreciable differential rotation.

2. Magnetic Field in ALB Model

It is known that RS CVn systems have coronae and produce flares in X-rays (Walter et al.,1981) and radio (Gibson et al.,1978, Feldman et al.,1978). The coronae have two temperature structure (Swank et al.,1981) and the hotter component might be extending to the size of the binary system. In AR Lac's case, the extended corona is found nonuniform and localized around the leading side (facing the direction of revolution) of the K-star (Walter et al., 1983).

In the solar case, not only flares but also the corona turned out to be magnetic in origin (e.g. Sheeley et al., 1974; Sakurai and Uchida,1977). The presence of radio and X-ray emissions in RS CVn system nicely fits in the situation of stars having starspots, and we discuss the origin of the corona and flares in terms of a magnetic field model.

In order to see what occurs in such a system, we try to visualize the magnetic fields in the binary by taking RS CVn itself as a typical case. In calculating the field configurations, we used Sakurai-Uchida method (1977). An aggregate of starspots whose magnetic field intensity is of the order of 10^3 G (Vogt,1981; Giampapa et al.,1983) are distributed on the K-star. The dipole-like general field is also assumed on the K-star (~ 10 G at the pole) as well as on the F-star ($\sim 10^2$ G at the pole). The computed magnetic field (Figure 1) shows usual active region loop systems as seen on the sun, and also *a loop system connecting the two stars*. These two classes of loop systems could be the reason for the two temperature structure of the corona found by Swank et al. (1981). As the spots are carried by the differential rotation, the geometry of the loop system changes drastically through magnetic field reconnection. For example in Figure 2, the spots moving toward the leading edge of ALB lose their connection with the companion star. The surplus flux of these spots must then be absorbed by the adjacent spots, and as a final consequence large scale pole-to-pole connections appear. This is an example of the "sequential reconnection" mediated by the magnetic field of the companion star. As the new spot pair emerges near the trailing side of ALB while the old pair submerges near the leading

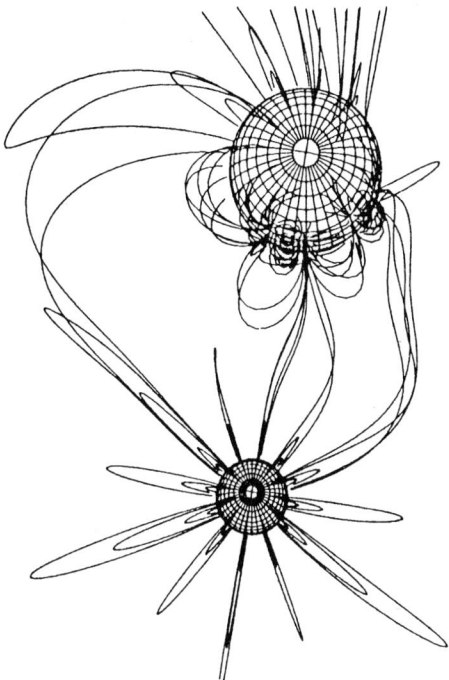

Fig. 1. Magnetic field structure calculated for RS CVn (Uchida and Sakurai, 1983, Fig. 1).

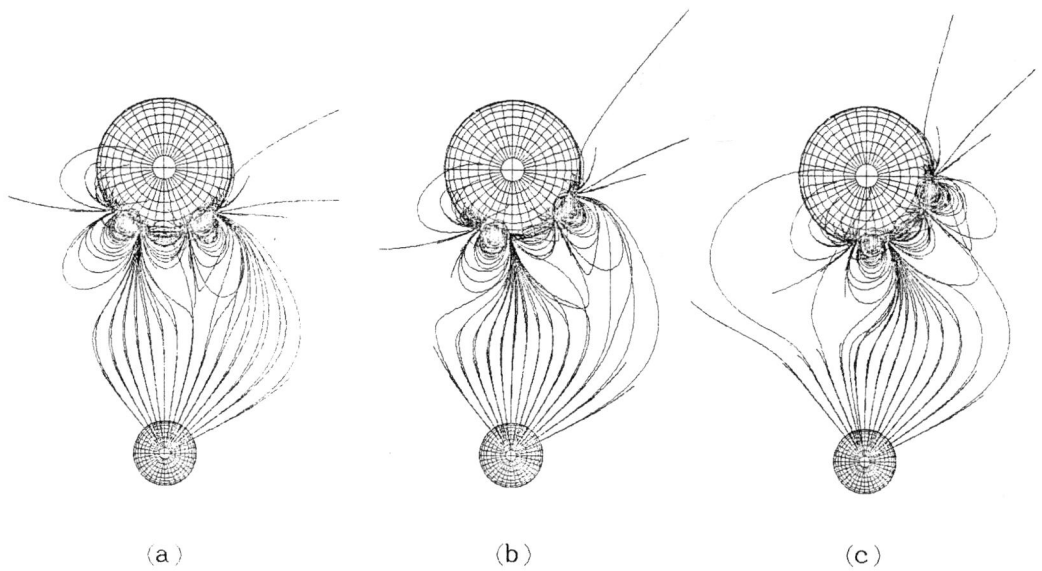

Fig. 2. Changes in the magnetic field structure as the active regions are carried by the differential rotation. The rotation and the revolution of the stars are supposed to be counterclockwise.

side of ALB, the reverse process from Figure 2c→2a can also take place, creating a new pole-to-spot connection.

3. Coronal Heating and Flares in RS CVn Binaries

Keeping in mind the behavior of the magnetic field discussed above, we may consider possible mechanisms of coronal heating in these systems. The two temperature structure of the coronae could result from the existence of two classes of loop systems. The cooler component might be explained by simply scaling up the solar corona. The hotter (and presumably very extended) component, as we believe, may be trapped in the large scale loop system connecting the two stars. Since it has a high temperature ($\sim 10^8$ K) and a large emission measure ($\sim 10^{54}$ cm^{-3}), not only the heating but also a very powerful mechanism of mass supply would be needed. It is possible, for example, that the emergence of a twisted flux tube at the footpoint of the pole-to-spot connection may feed its twist into the large scale connection through reconnection process. The propagation of the twist, which pinches the flux tube and heat the plasma, sweeps up the hot mass along the flux tube and results in the injection of hot and dense plasma into the large scale connections. This process, the sweeping-pinch mechanism, will be discussed elsewhere (Uchida and Shibata, 1984). It is interesting in this context that the extended corona in AR Lac is found above the trailing side of ALB (Walter et al., 1983) where the flux emergence is expected to be most frequent.

The amount of energy and mass provided in this mechanism may be roughly estimated as

$$\dot{W} = 2 \times 10^{33} \times (\frac{B}{1000G})^2 (\frac{a}{10^5 km}) \frac{\Delta\Omega}{\Omega} f \quad \text{ergs/sec}$$

and

$$\dot{M} = 1 \times 10^{18} \times (\frac{n}{10^{13} cm^{-3}}) (\frac{a}{10^5 km}) \frac{\Delta\Omega}{\Omega} f \quad \text{g/sec}$$

where a is the radius, B is the field strength, n is the density in the emerging flux tube, respectively, and f is the filling factor of the spots in ALB. It is also assumed that the drift of the spots by the differential rotation in ALB makes a steady flow with the emergence (and submergence). $\dot{W}$ is sufficient to explain the observed X-ray luminosity, and $\dot{M}$ accumulated for a day or so would provide enough mass to account for the observed X-ray emission measure.

The emergence of new magnetic flux may or may not lead to the catastrophic change in the magnetic configuration, depending on the environment of the emerging flux region. Flares can be identified with the drastic liberation of magnetic energy associated with the global reconfiguration of the magnetic field, probably triggered by the flux emergence. Therefore flares are expected to occur above the leading side of ALB where the magnetic field may be most stressed due to the differential rotation. A flare eventually observed on AR Lac (Walter et al.,1983) seems to lie in this region. Mass flow from the K-star to its companion observed in a flare on UX Ari (Simon et al.,1980) also support this interpretation.

In conclusion, our model predicts that the location of the hot corona and the site of flares are closely related to the relative position of ALB in the binary system, which can be checked by observations.

References

Durney, B.R. and Robinson, R.D.:1982, *Astrophys. J.* **253**, 290.
Eaton, J.A. and Hall, D.S.:1979, *Astrophys. J.* **227**, 907.
Feldman, P.A., Taylor, A.R., Gregory, P.C., Seaquist, E.R., Balonek, T.J., and Cohen, N.L.:1978, *Astron. J.* **83**, 1471.
Giampapa, M.S., Golub, L., and Worden, S.P.:1983, submitted to *Astrophys. J.*
Gibson, D.M., Hicks, P.D., and Owen, F.N.:1978, *Astron. J.* **83**,1495.
Sakurai, T. and Uchida, Y.:1977, *Solar Phys.* **52**, 397.
Sheeley, N.R., Jr., Bohlin, J.D., Brueckner, G.E., Purcell, J.D., Scherrer, V., and Tousey, R.:1974, *Solar Phys.* **40**, 103.
Simon, T., Linsky, J.L., and Schiffer, F.H.,III.:1980, *Astrophys. J.* **239**, 911.
Swank, J.H., White, N.E., Holt, S.S., and Becker, R.H.:1981, *Astrophys. J.* **246**, 208.
Uchida, Y. and Sakurai, T.:1983, in M.Rodono and P.B.Byrne (eds.), 'Activity in Red Dwarf Stars', *IAU Colloq.* **71**.
Uchida,Y. and Shibata,K.: 1984, this volume.
Vogt, S.S.:1981, *Astrophys. J.* **247**, 975.
Walter, F.M., Charles, P.A., and Bowyer, S.:1981, *Astrophys. J.* **244**, 671.
Walter, F.M., Gibson, D.M., and Basri, G.S.:1983, *Astrophys. J.*, in press.
Zahn, J.P.:1977, *Astron. Astrophys.* **57**, 383.

DISCUSSION

Henriksen: [After the computer-animated movie of mass injection into the coronal loop by Uchida and Shibata was shown]. What was the geometry of the initial configuration? What jet velocity was attained? How long does the field continue to drive the jet?

Uchida: As shown in the paper by Uchida and Shibata, the field at t=0 is a potential field with sources at large distances, namely, diverges moderately and converges at large distance again (see Figure 1 in Uchida and Shibata in these Proceedings). The jet velocity attained is roughly the Alfvén velocity. The length of time in which the drive is effective depends on the assumption of $B_\phi(r,z,0)$, and in the calculated case, the drive gradually loses strength with the expansion.

Bratenahl: I am very much interested in your mass ejection mechanism. Have you thought of applying this to spicules on the sun? Don't forget, the usual statement that the base of spicules is the middle chomosphere simply means that observational evidence for them ends there. The base of spicules could well be below the photosphere.

Uchida: Yes. We are thinking of applying this also to the jets of radio galaxies, etc.

"Sweeping Pinch" Mechanism and the Acceleration of Jets in Astrophysics

Yutaka UCHIDA and Kazunari SHIBATA*
Tokyo Astronomical Observatory, University of Tokyo
*Dept. Earth Science, Aichi University of Education

Abstract : A magnetodynamic mechanism of jet formation, in which a packet of the toroidal component of the magnetic field B_φ plays a role, is proposed. Such a packet of toroidal field, produced by the rotational motion in the $\beta = p_g/p_m \gg 1$ region, relaxes itself in the $\beta \ll 1$ region when brought up into such a region, for example, by the process of flux emergence due to magnetic buoyancy. In the $\beta \ll 1$ region, a progressive pinch is caused by this relaxation and the mass is swept out by the pinch near the axis and also by the $j \times B$ force in the twisted field region surrounding the axis.

1. Introduction

Acceleration of jets is pretty common in cosmical objects, ranging from solar flare-surges to the bipolar jets in radio galaxies and quasars. It is getting clearer that the origin of these highly directional flows may be related to the magnetic field combined with the rotational motion in the objects. For example, jets in the solar atmosphere (surges, Brueckner's jets (1980), and so on) are known to originate from magnetic patches and it is suggested that the emergence of the twist of the magnetic field produced in the $\beta \gg 1$ region down in the photosphere by convective motions may play an important role in the formation of these jets along the external field lines (Uchida and Shibata 1983). Also it is likely that the part of the large scale intergalactic magnetic field lines, entangled in the accretion disk of the galaxy or quasar and twisted up by the rotation of the disk, may play a similar role in causing jets along the large scale fields when the magnetic twist is released from the accretion disk.

2. Numerical Simulations

In the following we demonstrate that the jet can actually be accelerated when toroidal component of the magnetic field emerges from the $\beta \gg 1$ region to the $\beta \ll 1$ region. We solve the standard equations of conservations of mass, momentum, and energy together with the induction equation in the MHD approximation in a cyclindrical coordinate (r, φ, z) where z is taken to be antiparallel to the gravity under which the parts of the atmosphere of widely different β-values stratify. In solving the equations, we adopt modified Lax-Wendroff scheme (e.g. Rubin, and Burstein 1967, Shibata 1983) in pseudo three dimension (axisymmetry but allowing v_φ and B_φ) with artificial viscosity. As an example, the unperturbed condition was taken to be that of the solar case, in which the corona ($\beta \ll 1$) and the chromosphere ($\beta < 1$) with temperatures 10^6 K and 10^4 K, respectively, stratify under gravity in hydrostatic equilibrium with the transition layer at around $z_t = 2000$ km above the photosphere. Magnetic field is assumed to be an axisymmetric potential field which diverges moderately with height (these are seen in Figure 1 at $t=0$). We assume the initial distribution of $B_\varphi(r,z)$ in the chromosphere with the radius of the peak, $r_0 = 300$ km as seen in Figure 2 at $t=0$. This corresponds either to the twisted part of the flux tube floated up as the emerging flux appearing in active regions, or the twist of compact loop transferred to a large scale open flux tube through magnetic reconnection (Uchida and Shibata 1983).

The boundary conditions are those of free boundary on the surface of the cylinder (right-hand-side of the rectangular region in the Figures) and the top and bottom surfaces. On the axis (left-hand side of Figures) the symmetry condition with vanishing v_r, v_φ, B_r and B_φ are imposed.

3. Results and Physical Interpretation

Figures 1 and 2 show the behavior of $\mathbf{B}_{//}$ and B_φ in an example in which $\alpha = (B_\varphi/B_z)_{z=0} = 1$ and $\beta = (p_g/p_m)_{z=0} = 1$, respectively, and Figures 3 and 4 show the corresponding velocity components v_φ and $\mathbf{v}_{//}$ induced in this process respectively. Figures 5 and 6 are corresponding changes in log ρ and log T. It is clearly shown that a directed flow or a jet is accelerated as B_φ relaxes itself into the transition zone and into the corona. The magnetic field pattern in Figures 1 and 2 shows that the bunch of B_φ, which was passively produced in the $\beta \gg 1$ region by the rotational motion of the convection and could not relax by itself, now relaxes itself to a new equilibrium with $\mathbf{B}_{//}$ as it is brought up by the flux emergence process into the $\beta \ll 1$ region where the gas pressure can no longer confine the magnetic field. A pinching toward the axis at around the height of the transition layer occurs when B_φ starts relaxing in the upper chromosphere to the transition region (whereas B_φ at larger r expands), and as the front of this twist proceeds into the corona, the pinching sweeps up the material along $\mathbf{B}_{//}$. In the region surrounding the axis, the front drives plasma also through $\mathbf{j} \times \mathbf{B}$ force (cf. Hollweg et al. 1982) as the originally untwisted field lines are pulled into the helical pattern, or the twist proceeds into the region. The pinching can be understood as the process of relaxation of the field to a new equilibrium in which force-free field is to be established after the twist wave front comes into the region and bounces on the borders if the region concerned is a closed finite region. In our present situation of open flux tube, the twist wave escapes out of the region.

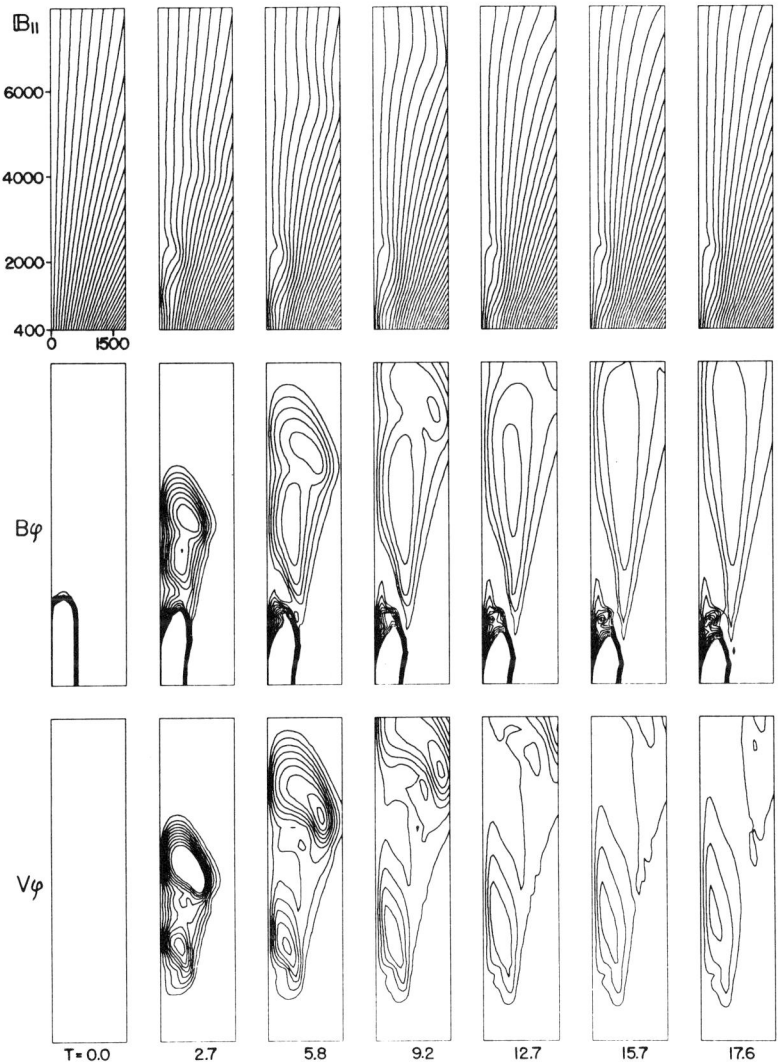

Figure 1, 2 and 3 (from the top to the bottom) : Time variation of the poloidal field $B_{||}$ (Fig.1) in response to the relaxation of the toroidal component B_φ into the $\beta \ll 1$ region (Fig.2 in B_φ/B_0). Figure 3 shows the induced v_φ (in v_φ/v_0). Numbers on the abscissa and ordinate are r and z in km.

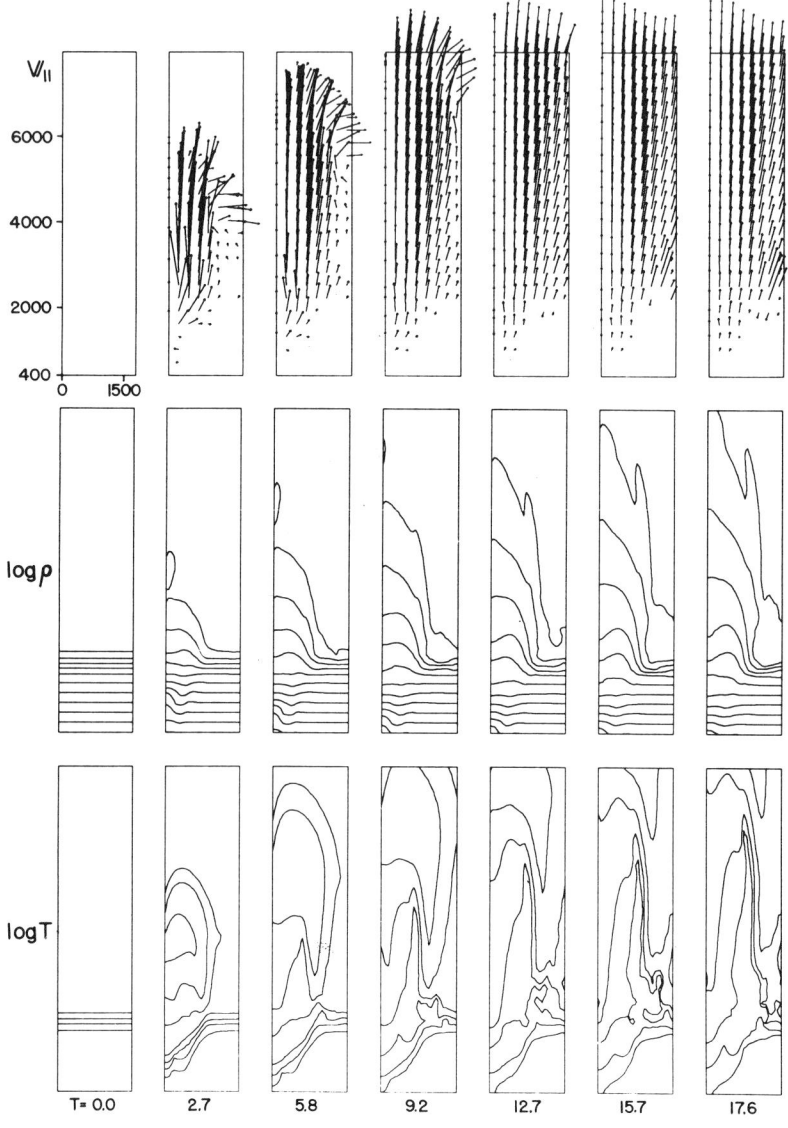

Figure 4, 5, and 6 (from the top to the bottom) : Corresponding evolution of $\mathbf{v}_{\parallel}$ (Fig. 4), $\log \rho$ (Fig. 5), and $\log T$ (Fig. 6). These show the formation of a jet with a high temperature ($>10^7$ K) part at the head progresses roughly with the Alfven velocity, and a sheath of the spinning material with the transition zone temperature follows it.

It is to be noticed that the temperature rises markedly due to strong pinching at the incidence of the twist wave front into the transition region, and this high temperature blob progresses along the field lines as seen in Figure 6. After some time the high temperature part of the jet caused by the sweeping pinch is followed by a spinning sheath with the transition zone temperature. The velocity of the front of the sweeping pinch is of the order of the Alfven velocity, and that of the material of the spinning sheath in this case reaches a good fraction of it, or about 50 % of the Alfven velocity. The velocity of the latter may be brought up to the Alfven velocity if multiple twist wave fronts repeatedly outrun it.

4. Discussion and Prospect for Applications

There seem to be a number of applications of this mechanism in astrophysics. It is readily seen from the above example that the observed properties of jetting phenomena in the sun seem to favor the present mechanism. Often they clearly show helical motions, which is a clear characteristic of our mechanism. Another application on the sun is the dynamical theory of formation of simple-loop flares (Uchida and Shibata 1983). As shown in that paper, the appearances of the superhot region as well as the acceleration of non-thermal particles (Tanaka et al. 1983, Tsuneta et al. 1983) in the solar simple-loop flares may be explained in a natural way in our mechanism.

Other cosmic jets, like quasar's jets with radio lobes, may also be explained by the present mechanism. For example, the rotation of the accretion disk may twist up the part of the large scale intergalactic magnetic field lines, and when the mass is freed from the field line by magnetic reconnection and falls into the central blackhole, the liberated twist wave front sweeps out along the large scale field and accelerates the plasma and particles. In such a picture, the energy comes from the gravitational energy, and the magnetic field plays the role of a converter of this gravitational energy with high efficiency.

The authors acknowledge the assistance of Mr.N.Shibuya, Mrs.H.Suzuki, and Mr.Y.Shiomi in the preparation of this work.

References

Brueckner, G., 1980, Highlights in Astronomy, Vol. 5, ed. A. Wayman (D. Reidel), p 557.
Hollweg, J.V., Jackson, S., and Galloway, D., 1982, Solar Phys., 75, 35.
Rubin, E.L., and Burstein, S.Z., 1967, J. Comp. Phys., 2, 178.
Shibata, K., 1983, to appear in Publ. Astron. Soc. Japan, 35.
Tanaka, K., Nitta, N., Akita, K., Watanabe, T., 1983, in "Recent Advences in the Understanding of Solar Flares", ed. Y. Uchida et al. (Tokyo Astron. Obs.) p 95, also in Solar Phys. 86, special issue, p. 91.
Tsuneta,S.,Takakura,T.,Nitta,N.,Ohki,K.,Makishima,k.,Murakami,T., Oda,M.,and Ogawara, Y., 1983, in "Recent Advances in the Understanding of Solar Flares", ed. Y. Uchida et al. (Tokyo Astron. Obs.) p 333, also in Solar Phys. 86, special issue, p. 313.
Jchida, Y., and Shibata, K., 1983, to be submitted to Solar Phys.

ENERGY RELEASE IN SOLAR FLARES

P. A. Sturrock[1], P. Kaufmann,[2] and D. F. Smith[3]
[1]Stanford University, Stanford, California
[2]Instituto de Pesquisas Espaciais, São Jose dos Campos, SP, Brazil
[3]Berkeley Research Associates, Berkeley, California

In early research on solar flares, attention was focused on the impulsive or flash phase, and it was assumed implicitly that virtually all the energy of a flare is released during that short phase. In recent years, however, it has been realized that the long-lived soft X-ray emission which follows the impulsive phase may require a separate energy-release process, which has been termed the "gradual phase" (Kane 1974). The fact that the impulsive phase is often preceded by soft X-ray emission has also led to the suggestion that there may be a third phase of energy release, which might be termed the "onset phase" (Sturrock 1980). It has long been realized that filament eruptions frequently precede flares, and it has been suggested (Kiepenheuer 1964) that the two should be regarded as parts of the same process. For these and other reasons, it is appropriate to question how many phases of energy release are involved in flares and what are their characteristics.

There is no doubt that some of the soft X-ray emission is due to bremsstrahlung from hot plasma, trapped in coronal loops, which has been evaporated from the chromosphere during the impulsive phase. This causal relationship has the consequence that the time curve of soft X-ray emission is approximately a time-integral of the time curve of the hard X-ray emission (Hudson 1983). On the other hand, as indicated above, detailed study of some flares indicates that the soft X-ray emission is fed by a continued energy release after the impulsive phase has ended. In their analysis of the 1979 March 31 flare, Feldman et al. (1982) adduce evidence indicating that the energy release responsible for hard X-ray emission and the energy release responsible for most of the soft X-ray emission occurred in different magnetic-field systems. For instance, the soft X-ray emission varied in a smooth manner from the onset to the decay of the flare; there was no abrupt change in either the emission measure or the temperature characterizing the soft X-ray emission during the time when the impulsive hard X-ray burst was in progress. This analysis adds further support to the proposition that the impulsive phase and the gradual phase involve distinct and independent processes. Indeed, it is known from H-α and X-ray studies that some flares do not possess an impulsive phase at all.

Although there appears to be a clear case in favor of the existence of two distinct energy-release processes responsible for the impulsive phase and the gradual phase of a flare, it is not so clear that a third process is required to explain the "onset phase". It is possible that, in some flares, such as the one analyzed by Feldman et al. (1982), the "onset phase" is simply the beginning of the gradual phase. It is also possible that, in other flares, the soft X-ray emission which begins early in the flare, before the recognized impulsive phase, is in fact due to early impulsive energy release at a sufficiently low level that it is not detected by its hard X-ray emission, yet is detected through the soft X-ray emission resulting from chromospheric evaporation.

The question now arises as to the distinction between the impulsive and gradual phases of energy release. Adopting the conventional view that energy released during a flare is the free energy of a current-carrying magnetic field which is released by reconnection, there are two classes of explanation of the difference between the impulsive and gradual phases. These two processes may represent either (a) two different modes of reconnection in otherwise similar magnetic-field configurations; or (b) two different current-carrying magnetic-field systems.

Recent research on the tearing-mode instability shows that there are indeed different possible stages in the time evolution of this instability. The work of Carreras et al. (1980) and Carreras et al. (1981) shows that the tearing process may involve the following three stages: (I) the initial phase described by FKR (Furth, Killeen and Rosenbluth 1963) linear theory in which there is a slow but exponential growth; (II) a nonlinear development in which the amplitude grows linearly with time; and, if two or more modes are unstable and if the spatial configuration is such that these modes can interact, a third stage (III) characterized by a very rapid stochastic variation of one or more modes involving growth rates considerably larger than the growth rate of the initial linearized regime. For a variety of reasons, it does not seem possible that one can understand the gradual phase as the manifestation of processes I and II and the impulsive phase as a manifestation of process III. On the other hand, it is possible that some of the rapid variation characteristic of the impulsive phase may be due to the stochastic behavior of the strongly nonlinear regime (III) of the tearing mode.

The analysis of Feldman et al. indicates that the distinction between the impulsive and gradual phases is to be found in the characteristics of different magnetic-field configurations. Of the possible distinctions, it seems most likely that the difference is due to the characteristic length scale of variation of the magnetic field. There are reasons for expecting at least three different length scales to arise in the magnetic-field configurations of active regions.

Observations by Tarbell and Title (1977) and others indicate that, at the photospheric level, magnetic field lines tend to be pulled together into small flux regions with dimensions of order 500 km or less, in which the magnetic field strength is of order 1,000 to 1,500 gauss. In consequence, the magnetic field of the photosphere tends to be aggregated into "knots" in which the flux has values of order $10^{18.4}$ Mx. In consequence, the magnetic field in the corona will comprise an assembly of flux tubes, which could be termed "elementary flux tubes", each originating in a strong-field knot in the photosphere and terminating in another knot. This leads us to the following three characteristic dimensions: (a) the thickness of the current layer separating adjacent elementary flux tubes (perhaps $10^4 - 10^6$ cm); (b) the minor radius of an elementary flux tube (10^8 cm); and (c) the characteristic dimensions of an active region ($10^9 - 10^{10}$ cm). It seems likely that currents of types (a) and (b) are both involved in the impulsive phase of a flare, and that the current of type (c) is responsible for the gradual phase of a flare. Indeed, the evolution of two-ribbon flares indicates that the impulsive phase is associated with activity very close to the neutral line, whereas the gradual phase is associated with magnetic-field lines which meet the photosphere far from the neutral line.

Recent detailed studies of the X-ray and microwave emission from flares indicate that much of the time-structure can be interpreted as due to aggregates of "elementary bursts". The time scale of an elementary X-ray burst is of order 1 s and involves the release of about 10^{28} erg. The time scale of an elementary microwave burst is appreciably less (tens to one hundreds of milliseconds). The energy actually radiated in microwaves is very small, but it represents only a small fraction of the primary energy release. There is sufficient uncertainty in the processes involved that the magnitude of energy involved in each elementary microwave burst may be in the range $10^{24} - 10^{27}$ erg.

The above characteristics are compatible with the proposed scales of fine-structure of the magnetic field of an active region. An elementary X-ray burst may be attributed to the release of the free energy of a single twisted elementary flux tube. An elementary microwave burst may be attributed to energy release of the current sheet separating an adjacent pair of elementary flux tubes or to energy release in one of the "magnetic islands" that one expects to develop in an elementary flux tube during reconnection. The time scale and energy characteristic of the gradual phase are consistent with the attribution of that phase to energy release of the distributed current system of an active region.

In the case that a flare follows the eruption of a filament, it is possible (a) that the eruption is a secondary process following early magnetic-field reconnection; or (b) that the flare is "triggered" by the filament eruption; or (c) that the flare involves the release of energy associated with a current system produced by the filament eruption; or (d) that magnetic-field reconnection and filament eruption form a pair of symbiotic processes. The detailed study of the early stages of some flares favors the last possibility (Moore 1983). A filament involves a magnetic-field configuration which typically has many "legs" tying the filament to the photosophere near the neutral line. Reconnection at the junctions of different legs can simultaneously lead to flare-producing energy release and to eruption of the filament. The continued eruption of the filament can then trigger the release of energy associated with both the elementary-tube structure and the gross structure, producing the impulsive and gradual phases of the flare.

This work was supported in part by NASA Grants NGL 05-020-272 and NAGW-92 and the Office of Naval Research contract N00014-75-C-0673. P.A.S. acknowledges the generous hospitality of the Instituto de Pesquisas Espaciais of Brazil during a recent visit to São Paulo.

REFERENCES

Carreras, B. A., Hicks, H. R., Holmes, J. A., and Waddell, B. V. 1980, Phys. Fluids, $\underline{23}$, 1811.
Carreras, B. A., Rosenbluth, M. N., and Hicks, H. R. 1981, Phys. Rev. Letters, $\underline{46}$, 1131.
Feldman, U., Cheng, C.-C., and Doschek, G. A. 1982, Ap. J., $\underline{255}$, 320.
Furth, H. P., Killeen, J., and Rosenbluth, M. N. 1963, Phys. Fluids, $\underline{6}$, 459.
Hudson, H. S. 1983, private communication.
Kane, S. R. 1974, Coronal Disturbances (ed. G. Newkirk: Dordrecht-Holland: Reidel), p. 105.
Kiepenheuer, K. O. 1964, Proc. AAS-NASA Symposium on the Physics of Solar Flares (ed. W. N. Hess; NASA SP-50; Washington, DC: NASA), p. 323.
Moore, R. 1983, private communication.
Sturrock, P. A. 1980, Solar Flares (ed. P. A. Sturrock; Boulder: Colorado University Press), p. 411.
Tarbell, T. D., and Title, A. M. 1977, Solar Phys., $\underline{52}$, 13.

DISCUSSION

Vlahos: Why do we need large scale magnetic fields in cases that we have a gradual phase? Can this be one phase of the reconnection instability with elementary flux tubes? R. Lin has recently reported an interesting phenomenon in X-ray observation. A "gradual" burst was composed of small "spikes" that only low sensitivity instruments can detect. How does this fit to your suggestion?

Sturrock: I did not propose that the magnetic field structure responsible for the gradual phase necessarily has a larger size than th structure responsible for the impulsive phase. What I did propose is that the characteristic length scales of the current systems are differ ent. It seems unlikely that the impulsive and gradual phases represent two consecutive phases in the development of reconnection in the same magnetic-field and current system. Reconnection even in a large currer sheet will involve the development of small-scale structure - such as "magnetic islands" - which might be related to the small X-ray spikes which occur during the gradual phase.

Mullan: (i) Field "knots" are always present in the photosphere (even apart from flares). Therefore, one expects that interwoven flux tubes (such as you suggest) would always create current sheets in the corona. Can such current sheets heat the "quiet" corona in the way envisioned by R. H. Levine (Ap.J. 192, 1974)? In the latter, the time scale for field collapse in the corona (T = 10^6 K, n = 10^9 cm^{-3}) is ~ 1 second (as you require for "elementary X-ray bursts"). Maybe X-ray bursts are elementary flux reconnection events. (ii) Do you believe that the gradual phase involves essentially the same process as "normal" coronal heating?

Sturrock: I agree that interwoven flux tubes, separated by current sheets, must always be present in the corona. It is quite possible that slow, continued reconnection may play a role in coronal heating. However, the release rate involved in the gradual phase must be considerably more rapid, although it is not as rapid as the impulsive phase.

Tsinganos: What is the vertical extent of the loops? Do you take into account two facts: (i) that the tube expands at larger heights (ii) that the twist is concentrated in these expanded parts of the tube?

Sturrock: The vertical extent may range from a few thousand km up to many tens of thousands of km. We do take account of the variation of cross section and of twist along the length of the flux tube.

Lokanadham: What are the time scales of the microwave spikes observed? Are there observational evidences for any spikes observed on a time scale of seconds and, if so, what is the mechanism involved?

Sturrock: Structures have been observed with a wide range of time-scales, from seconds down to below 100 msec. On times scales of seconds, the time structures are poorly correlated at different frequencies and between microwaves and hard X-rays, so it is unlikely that the structures are to be attributed to fluctuations of the primary energy-release process.

Kundu: There are many examples of the impulsive phase evolving into large loop prominence systems (VLA 6 cm and Nebyama 17 GHz data), which I believe you consider as "gradual phase". So the question is, do you really need two different mechanisms of energy release, (1) for the impulsive phase and (2) for the gradual phase?

Sturrock: If there is no continued energy release after the impulsive phase, there is no need for a "gradual phase" of energy release. However, if there is continued energy release, there clearly is need for another phase.

Rust: It may not be so easy to separate gradual and impulsive phases. Recent Hα and hard X-ray flare observations indicate that filaments (large scale features) rise gradually during the gradual phase onset - consistent with your picture - but also that they move suddenly during the impulsive phase. How can you fit this large-scale, impulsive phase phenomenon into your scheme?

Sturrock: In our picture, the initial rapid reconnection occurring below a filament during the impulsive phase gives rise to a magnetic disconnection of the filament from the photosphere which allows it to erupt. The eruption in turn produces a large-scale current sheet which reconnects during the gradual phase.

AN INTERPRETATION OF THE MILLISECOND TIME VARIATION IN HARD X-RAY SOLAR FLARES

V. Krishan
Indian Institute of Astrophysics, Bangalore 560034, India
and
M. R. Kundu
Astronomy Program, University of Maryland
College Park, Maryland, USA 20742

Abstract

Recent observations of the fast time variability in the hard X-ray emission from solar flares are interpreted. The fast spikes are assumed to be superimposed on the thermal X-ray emission. The rise and fall of a spike are caused by disruptions in the plasma. The rise time represents the impulsive heating time and the decay or fall time represents a quick cooling of the plasma due to the accelerating growth rate of the m=1 tearing mode. The estimated characteristic time durations of the spike are found to be in good agreement with the observed ones.

Introduction

Fast spikes in the hard X-ray solar bursts have been observed with the hard X-ray burst spectrometer (HXRBS) of the solar maximum mission (SMM) by Kiplinger et. al (1983). Spikes with varying temporal structures have been identified. X-ray features of total duration of about 0.25 seconds with approximately equal rise and fall times of 120 milliseconds are seen, although features with faster rise and decay times of 30-50 milliseconds are also present. One finds examples of fast rise and slow decay and vice versa in the complex temporal behavior of hard X-rays. Thus, a spike with a fast 30 millisecond rise time and a relatively smooth 1 second decay time is observed immediately before another spike with a 180 millisecond rise time and a 20 millisecond decay time. One concludes from these observations that a variety of time profiles of spikes can be present in hard X-ray bursts. A study of spike amplitudes shows that intense spikes occur in large events and are not observed in the absence of a gradual hard X-ray burst. The absolute intensity of a typical spike above the associated gradual component is less than the intensity of the gradual burst.

A Theoretical Model of the X-ray Spikes

In this paper we assume a thermal X-ray model and explain the temporal behavior of the hard X-ray emission in terms of fluctuations in the plasma temperature. We present a model of the spikes which is based on disruptions in the plasma (Spicer, 1982) within the framework of the thermal model. We begin with a volume of plasma at temperature T_{eo} which is responsible for the slowly varying thermally generated X-ray emission. It is the time profile of the spikes superimposed on the thermal emission that we are attempting to explain. In the absence of any disruptive phenomenon such as the excitation of instabilities, the plasma temperature would be a smoothly decreasing function of the plasma radius. Such a temperature profile would result from the interplay of ohmic heating and conductive and convective cooling. But disruption causes a significant deviation from this equilibrium. The region near the center of the plasma is at a lower temperature because of the disruption in a state of local nonthermodynamic equilibrium. This region is impulsively heated through anomalous ohmic heating. This represents the rising phase of the spike. An increase in the temperature reduces the anomalous resistivity which then leads to an increase in the toroidal current density and the shear at the singular surface. This enhances the growth rate of the tearing mode until the magnetic island produced by the tearing mode again flattens the temperature and current density profile. This represents the decay phase of the X-ray spike. The details of the calculations of the rise and decay time of a spike will be given elsewhere. Here we give the results. The duration of the spike t_o, the rise time t_r and the decay time t_d are given by (Jahns et. al 1978):

$$t_o = 4 \cdot 3 \times 10^{-23} [2\ell n \frac{\omega_f}{\omega_i}]^{1/3} n^{5/6} T_{eo}^{2/3} (\frac{B_z}{\delta \ell})^{-4/3} \frac{n}{n_b} (\frac{V_{Te}}{V_b})^2 (\frac{V_b}{\Delta V_b}),$$

$$t_r = \frac{\tilde{T}_e^o}{T_{eo}} \times 1.9 \times 10^{-12} n^{3/2} T_{eo} (\frac{n}{n_b}) (\frac{V_{Te}}{V_b})^2 \frac{V_b}{\Delta V_b}$$

and

$$t_d = 0.47 [2\ell n \frac{\omega_f}{\omega_i}]^{-1} t_o$$

where $\tilde{T}_e^o$ is the amplitude of the spike, ω_f and ω_i are the final and the initial widths of the magnetic island respectively, n, T_{eo} and V_{Te} are the density, temperature and thermal velocity of the electrons of the ambient plasma, n_b, ΔV_b and V_b are the density, thermal spread and velocity of the electron beam which provides anomalous resistivity, B_z is the toroidal magnetic field and $\delta \ell$ is the length scale of the magnetic field variation perpendicular to the magnetic field. The numerical estimates of these time scales for several values of the ambient electron density for typical solar flare conditions and the X-ray emitting regions are given below: We use $T_{eo} \sim 2.4 \times 10^8 K$, $\frac{n_b}{n} \sim 10^{-4}$, $\frac{V_{Te}}{V_b} \sim 10^{-1}$, $\frac{\Delta V_b}{V_b} \sim \frac{1}{3}$, $B_z \sim 500$ Gauss, $\delta \ell \sim 5 \times 10^5$ cm,

$J_z \sim 1.8 \times 10^8$ statamp/cm^2 and $\frac{\omega_f}{\omega_i} \sim 5$, $\frac{\tilde{T}_e^o}{T_{eo}} \sim 0.1$. We find for

$n = 4 \times 10^{11}$/cm^3, $t_o = 340$ ms, $t_r = 100$ ms and $t_d = 53$ ms.

$n = 2 \times 10^{11}$/cm^3, $t_o = 191$ ms, $t_r = 35$ ms and $t_d = 29$ ms.

and

$n = 1 \times 10^{11}$/cm^3, $t_o = 107$ ms, $t_r = 13$ ms and $t_d = 16$ ms.

In conclusion, the estimated time durations agree quite well with the observed times.

REFERENCES

Jahns, G. L., Soler, M. and Waddell, B.V.: 1978, Nuclear Fusion 18, 605.
Kiplinger, A. L., Dennis, B. R., Emslie, A.G., Frost, M. J. and Orwig, L.E.: 1983, Ap.J. 265, L99.
Spicer, D.S.: 1982, Space Science Rev. 31, 351.

DISCUSSION

Vlahos: The number of non-thermal electrons that you introduce to increase the resistivity are enough to create the spikes by themselves, and their acceleration controls the characteristics of the spike.

Krishan: One could propose other mechanisms for increasing the anomalous resistivity, like the current going unstable locally and exciting the ion-acoustic instability. Therefore it is not absolutely essential to invoke the presence of a nonthermal population of electrons. Even if the number of nonthermal electrons are sufficient to account for the intensity of the spike, it still does not explain the temporal characteristics of the spike. Instead of pushing the temporal variations under the rug of the unknown acceleration processes, the mechanism I have presented actually accounts for the temporal characteristics of the X-ray spike in a very quantitative way.

OBSERVATIONS OF THE EARTH'S CROSS-TAIL CURRENT SHEET AND THEIR IMPLICATIONS

A. T. Y. Lui
The Applied Physics Laboratory
Johns Hopkins University
Laurel, Maryland 20707

Abstract

Observations of the neutral sheet in the Earth's magnetotail are presented to show different magnetic signatures of the neutral sheet which have been used to infer (1) wave profiles on the neutral sheet surface, (2) magnetic islands embedded in the neutral sheet, and (3) localized turbulent magnetic field regions. The occurrence of these features even at magnetospheric quiet conditions suggests that the above features are intrinsic to the current sheet and may possibly play a role in its stability. There are indications that these features are common to other current sheets in space.

Introduction

One of the most extensively surveyed current sheets in space is in the Earth's magnetotail where the cross-tail current flows in the plasma sheet between the northern and southern tail lobes. The cross-tail current is believed to be driven by the solar wind dynamo action at the magnetopause and a portion of the current is transmitted to the ionosphere via magnetic-field-aligned currents. The cross-tail current extends from a downstream distance of about 10 R_E (R_E = earth radius) to possibly more than 1000 R_E. The typical energies of thermal ions and electrons in this current (plasma) sheet are 5 keV and 1 keV, respectively. The number densities range from <0.1 to ~ 1 cm^{-3}, with plasma beta (ratio of particle pressure over magnetic pressure) varying from ~ 0.1 to $\geqslant 1$. The current densities are about 3×10^5 A/R_E near the earth and decrease exponentially with an e-folding distance of about 60 R_E. A significant portion of the current is conducted in the neutral sheet region at the center of the current sheet.

Observed Neutral Sheet Signatures

The neutral sheet is recognized as the region in which the magnetic field orientation reverses from pointing sunward to tailward or vice versa. The field reversal is often accompanied by a decrease in field magnitude and an increase in the field elevation angle (northward field). There are various departures from this type of classical signature of the neutral sheet. Observed deviations are the following: (1) the field elevation angle may decrease in a rather step-wise fashion at the neutral sheet crossing; (2) the dawn-dusk field component at successive neutral sheet crossings may be alternately duskward and dawnward; (3) the field component normal to the neutral sheet may become southward occasionally at multiple neutral sheet crossings; (4) the field magnitude may be minimum at the start of the neutral sheet transition rather than in the middle; and (5) all field components may show large, rapid fluctuations and the dawn-dusk component may become the largest field component occasionally.

Inferred Features of the Current Sheet

The observed magnetic field signatures of the neutral sheet have been used to infer the structure of the current sheet. For example, the observed feature (1) in the previous section has been used to infer a wave profile of the neutral sheet surface mainly along the tail axis (Speiser, 1973). Indeed, such observed field variations are reproduced very well by the one-dimension equilibrium current sheet solution (Harris, 1962) modified to allow for such a wave profile. In the solar magnetospheric coordinate system, the magnetic vector potential appropriate for this case may be written as (Lui, 1983)

$$A_y = -B_o L \ln [\cosh(\frac{Z - A \sin kX}{L})] + B_n X \qquad (1)$$

where B_o is the field magnitude outside the current sheet, L is the current sheet thickness, B_n is the field component normal to the sheet, A is the amplitude of the wave and k is the wave-number. The waves have periods of a few to about 10 minutes (Pc5 micropulsations). Similar wave modulations are detected at the plasma sheet boundary (Spjeldvik and Fritz, 1981) and may be related to hydromagnetic oscillations of the magnetotail (McKenzie, 1970). The signature in feature (2) is indicative also of a wave surface of the neutral sheet in the dawn-dusk direction (Lui et al., 1978). Feature (3) suggests the presence of magnetic islands embedded in the neutral sheet (Schindler and Ness, 1972). The magnetic vector potential which can be used to describe this field geometry is (Kan, 1979)

$$A_y = -B_0 L \ln \left[\frac{\alpha^2 + 1}{2\alpha^2} \cosh\left(\alpha \frac{Z}{L}\right) + \frac{\alpha^2 - 1}{2\alpha^2} \cos\left(\alpha \frac{X}{L}\right)\right] \quad (2)$$

where the parameter α is related to the locations along the X-axis of the X-type neutral lines (at $X = 2n\pi L/\alpha$, $n =$ integer) and O-type neutral lines (at $X = 2(n+1)\pi L/\alpha$) in the island structure. Features (4) and (5) are suggestive of turbulent field structures (Hruska and Hruskova, 1970; Lui, 1983). These features, which are summarized in Figure 1, are apparently not unique to the earth's cross-tail current since similar features are also found in the current sheets in the solar wind (Smith et al., 1977) and in the Jovian magnetotail (Behannon, 1983).

Dependence on Substorm Activity

Although the different types of neutral sheet signatures which deviate from the classical neutral sheet signature tend to occur more frequently with increasing substorm activity, it is important to note that all these features have been observed even during magnetospheric quiet conditions when no substorm activity is indicated by the AE

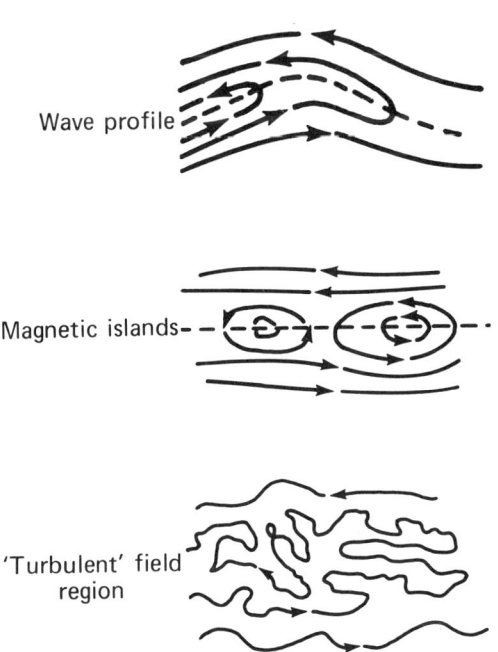

FIGURE 1. A sketch of three basic features in the earth's cross-tail current sheet.

index and global auroral images (see e.g. Lui and Meng, 1979). This result suggests that the structures sketched in Figure 1 are intrinsic features of the current sheet rather than uniquely tied with substorm occurrence, which is consistent with recent theoretical work (Kan, 1979; Zwingmann and Schindler, 1980) that equilibrium tail configuration can be obtained with magnetic island structures embedded within.

Conclusion

Observations indicate that the cross-tail current sheet contains several intrinsic features such as wave profiles, magnetic islands and localized turbulent regions. The fact that similar features are found in the current sheets in the solar wind and in the Jovian magnetotail suggests these features to be common for current sheets in space. The role these features play in the stability of a current sheet is yet to be determined.

References

Behannon, K. W., 1983, "J. Geophys. Res". (submitted).
Harris, E. G., 1962, "Niovo Cimento", 23, 116.
Hruska, A., and J. Hruskova, 1970, "J. Geophys. Res". 75, 2449.
Kan, J. K., 1979, "Planet. Space Science"., 27, 351.
Lui, A. T. Y., C.-I. Meng, S.-I. Akasofu, 1978, "Geophys. Res. Lett"., 5, 279.
Lui, A. T. Y., C.-I. Meng, 1979, "J. Geophys. Res". 84, 5817.
Lui, A. T. Y., 1983, in Magnetospheric Currents, ed. T. A. Potemra.
McKenzie, J. F., 1970, "J. Geophys. Res". 75, 5331.
Schindler, K., N. F. Ness, 1972, "J. Geophys. Res". 77, 91.
Smith, E. J., B. T. Tsurutani, and R. L. Rosenberg, 1977, "Trans. AGU, EOS", 58, 484.
Speiser, T. W., 1973, "Radio Sci"., 8, 973.
Spjeldvik, W. N., T. A. Fritz, 1981, "J. Geophys. Res". 86, 2480.
Zwingmann, W., K. Schindler, 1980, "Geophys. Res. Lett"., 7, 909.

DISCUSSION

Birn: I have two comments: a. The smooth current sheet equilibrium solutions are just a subset of possible solutions. Most of the solutions indeed show wavyness and even island structure. b. Most of the current sheet crossing are due to the motion of the plasma sheet across the satellite and are therefore connected with some kind of dynamic state of the plasma sheet which would explain, at least in part, why the current sheet is seldom found to be "quiescent".

Lui: a) I do not disagree with your statement. One of the points in my presentation is that the initial state of the current sheet is far from smooth and investigation of the current sheet stability for initial states with these structures seems appropriate. b) Plasma sheet motion does not imply the current sheet to be in some kind of dynamic state, e.g. motion induced by the diurnal wobbling of the magnetic dipole axis. Furthermore, the word "quiescent" is in reference to substorm activity and is used to describe the initial state of the current sheet before the dynamic evolution later as a substorm develops.

Vasyliunas: You cannot really distinguish between waves in the current sheet surface and tilts of the whole surface (which are in effect waves of infinite wavelength).

Lui: The interpretation of tilt instead of wave is a possibility, but is a very unlikely one for the following reason. The tilt required to account for some of the observations is more than 30°. If the tail axis is tilted by such an extent, it would mean that the tail deviates substantially from the solar wind flow direction. Since the tail extends at least 100 R_e and possibly > 1000 R_e, this tilt may make the tail present as much an obstacle to the solar wind as the dayside magnetosphere. Furthermore, plasma sheet boundary waves of similar wavelength have been observed by the ISEE satellites.

PLASMA INSTABILITIES GENERATED BY STREAMING PARTICLES

Mats André
Kiruna Geophysical Institute, University of Umeå
S-901 87 Umeå, Sweden

It is well known that particles streaming along the ambient magnetic field in space plasmas may generate waves with frequencies of the order of the local ion gyrofrequency (ion waves), (Kindel and Kennel 1971). In this study we analyze the dispersion relation of these waves numerically and discuss mechanisms for damping and instability. All numerical results in this report are obtained with the computer code WHAMP (Rönnmark 1982a,b), which solves the dispersion relation of linear waves in a homogeneous plasma for a complex frequency as a function of a real wavevector. As a specific example we consider the S3-3 satellite observations of banded electrostatic ion waves, associated with streaming particles (Kintner et. al. 1979, Cattell 1981).

1. DISPERSION SURFACES

In order to compare wave observations with theory, it is important to have an adequate display of all the wave modes which may be excited. Hence, we present dispersion surfaces, i.e. plots of the frequency f versus wavevector components k_z parallel and $k_\perp$ perpendicular to the ambient magnetic field (fig:s 1 and 2). The magnetic field strength in the model is 0.07G, and the plasma consists of 1.5 protons/cm^{-3} with temperature T_p=2eV and an equal number of electrons with temperature T_e=1keV. The proton gyrofrequency is denoted by f_{cp} and the Larmor radius of 2eV protons by ρ_p. Waves that are heavily damped, e.g. because they are close to a harmonic of f_{cp}, are excluded from the figures.

When the electron temperature is decreased, the electromagnetic modes shown in the figures are essentially unaffected. This is also true for the electrostatic waves with small k_z (C and D in fig 1). However, the banded electrostatic modes at large k_z (B) become heavily Landau damped when $T_e \sim T_p$. The ion-acoustic wave (A) also becomes damped when T_e is lowered, and the phase velocity v_{ia} decreases in agreement with the well known relation $v_{ia}^2 = T_e/m_p$, where m_p is the proton mass.

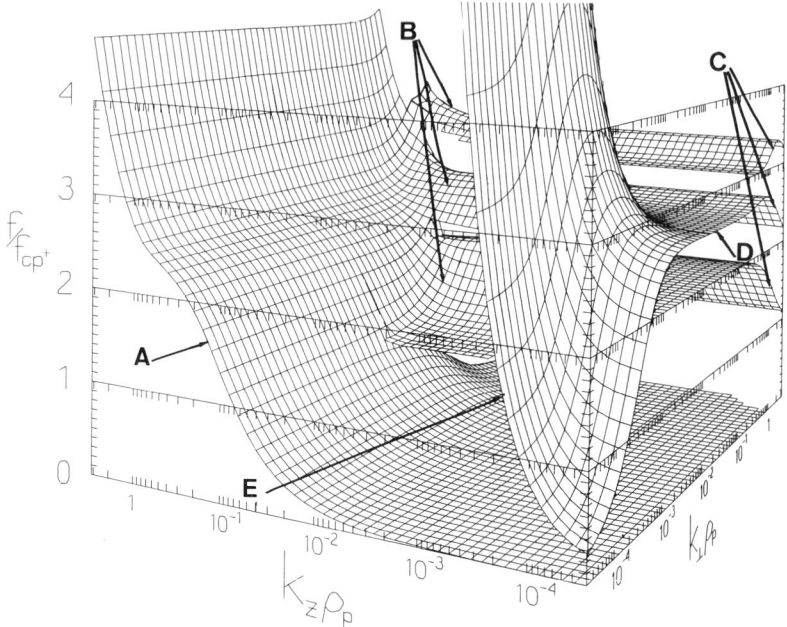

Fig.1 Dispersion surfaces for the first harmonic bands. The plasma model is given in the text. A Ion-acoustic mode (electrostatic). B Banded electrostatic waves, sometimes called Bernstein-like modes or electrostatic ion cyclotron (EIC) waves or, specifically for this plasma composition, electrostatic hydrogen cyclotron (EHC) waves. C Banded electrostatic waves, also called ion cyclotron harmonic (ICH) modes. D Lower hybrid "plateau" (electrostatic for large $k_\perp$, electromagnetic for small $k_\perp$). E Right circularly polarized mode, sometimes called the compressional Alfvén wave. The frequency of this mode approaches zero in the limit of small wavevector.

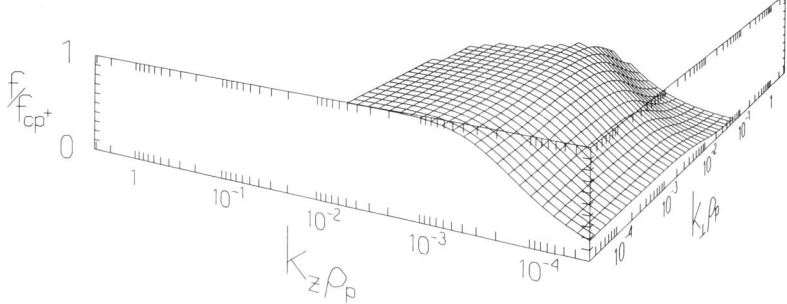

Fig.2 This dispersion surface exist simultaneously with the modes in fig.1, but is shown separately for the sake of clearness. For propagation parallel to the ambient magnetic field this surface represents the left circularly polarized mode, also called the shear Alfvén wave, which approaches zero in the limit of small k_z. This mode gradually becomes electrostatic when $k_\perp$ increases.

2. PARTICLE BEAM INSTABILITIES

Kaufmann and Kintner (1982) study an S3-3 event in which emissions between the first few harmonics of f_{cp} are observed. In their fig. 7 they consider a plasma model including an observed ion beam, and find an electrostatic instability near $1.5 f_{cp}$ ($k_{\perp}\rho_p \sim 1.5$, $k_z\rho_p \sim 0.15$). However, our calculations show that this model gives an additional, strong, broadband instability of ion-acoustic waves, which are not observed. This is not surprising, since the temperature of the only electron component is taken to be rather high (1keV), (Kindel and Kennel 1971, Bergmann 1982). By lowering this temperature to the measured value (250 eV) and including some cool electrons in the model, the ion-acoustic waves can be surpressed. These changes in the model damp also the wave at $f \sim 1.5 f_{cp}$ unless the total density is simultaneously decreased by an order of magnitude. Model 1 presented in Table 1 containes the components observed by Kaufmann and Kintner, a low density 2 eV proton distribution and a mixture of warm and cool electrons. An instability at $f \sim 1.5 f_{cp}$ is obtained in this model, but the ion-acoustic waves are stable.

The model including a measured ion beam does not predict the observed instabilities between higher harmonics of f_{cp}. Currents are often detected during electrostatic wave events on S3-3 (Cattell 1981) and drifting low energy electrons, carrying field aligned currents, are a plausible energy source for ion waves (Kindel and Kennel 1971, Ashour-Abdalla and Thorne 1978, Bergmann 1982). Using Model 2 of Table 1 we find unstable electrostatic waves in two bands, around $1.2 f_{cp}$ and $2.2 f_{cp}$, and no ion-acoustic instability. The observed ion beam is not essential for this current driven instability but rather the fact that not too many cool, damping electrons are present. When the temperature of 40 % of the non-drifting electrons in Model 2 (0.8 cm^{-3}) is lowered to 10 eV, the growthrate in the first harmonic band is not significantly decreased, but the wave at $f \sim 2.2 f_{cp}$ is damped.

In the examples given here we assume that the cool particles have a temperature of a few eV. We present models where banded electrostatic ion waves are generated by proton or electron beams with a drift velocity less than the thermal speed of the cool electrons and also less than the Alfvén speed. When constructing such models some basic facts have to be considered e.g. 1) Too many non-drifting cool electrons may damp all ion waves, 2) The ion-acoustic mode, rather than banded ion waves, may be destabilized when only a few cool electrons (drifting or not) are present, and 3) The growth rate of the ion-acoustic wave is decreased when some of the ions have a temperature of the same order as the warm electrons. Both models in Table 1 include warm protons and cool electrons and hence no ion-acoustic waves are destabilized. In model 1. the ion beam density is a large fraction of the total density and we thus obtain an instability, although cool damping electrons are present. In model 2 the drifting cool electrons generate the instability. Using this kind of models we can get some information about the plasma composition from wave observations.

Model 1

	p	p	p	e	e
T(eV)	2	50	1069	250	10
n(cm^{-3})	1	0.55	0.34	1.59	0.3
V_D	0	3.3	0	0	0

$f/f_{cp} = 1.5$ $k_\perp \rho_p = 0.6$ $k_z \rho_p = 0.2$ $\tau = 0.1$s $D_\perp = 1$ km $D_z = 3$ km

Model 2

	p	p	e	e
T(eV)	10	250	10	250
n(cm^{-3})	2	2	2	2
V_D	0	0	0.7	0

$f/f_{cp} = 1.2$ $k_\perp \xi_p = 1.2$ $k_z \xi_p = 0.1$ $\tau = 0.02$s $D_\perp = 0.1$ km $D_z = 0.3$ km

$f/f_{cp} = 2.2$ $k_\perp \xi_p = 2$ $k_z \xi_p = 0.12$ $\tau = 0.1$s $D_\perp = 0.2$ km $D_z = 2$ km

Table 1. The temperature (T), density (n) and drift velocity along the ambient magnetic field, normalized to the thermal speed of the drifting component (v_D), for protons (p) and electrons (e). The frequencies and wavevectors given correspond to maximum growth in each band. The magnetic field strength is 0.07G and the Larmor radius of 2eV and 10eV protons are denoted ρ_p and ξ_p respectively. The time for one e-folding of wave amplitude is given by τ. Since the instabilities are convective we introduce the distances $D_\perp$ and D_z which show how far the group velocity takes the wave perpendicular and parallel to the ambient magnetic field during one e-folding.

REFERENCES

Ashour-Abdalla M. and Thorne R.M.:1978,J.Geophys.Res. 83,pp 4755-4766
Bergmann R.:1982,UCB Space Sciences Laboratory Preprint No.SSL-82-111-MKJ
Cattell C.: 1981, J. Geophys. Res. 86, pp 3641-3645
Kaufmann R.L. and Kintner P.M.: 1982, J.Geophys.Res. 87,pp 10487-10502
Kindel J.M. and Kennel C.F.: 1971, J. Geophys. Res. 76, pp 3055-3077
Kintner P.M.,Kelley M.C.,Sharp R.D.,Ghielmetti A.G.,Temerin M.,Cattell C
  Mizera P.F. and Fennel J.F.: 1979, J. Geophys. Res. 84, pp 7201-7212
Rönnmark K.: 1982a, Plasma Phys. 25, pp 699-701
Rönnmark K.: 1982b, WHAMP-Waves in Homogeneous Anisotropic Plasmas,
  Kiruna Geophysical Institut Report no. 179.

SESSION V

ANOMALOUS TRANSPORT IN CURRENT SHEETS

J.D. Huba
Geophysical and Plasma Dynamics Branch
Plasma Physics Division
Naval Research Laboratory
Washington, D.C. 20375

ABSTRACT
   A review of several microinstabilities that have been suggested as possible anomalous transport mechanisms in current sheets is presented. The specific application is to a `field reversed plasma´ which is relevant to the so-called `diffusion region´ of a reconnection process. The linear and nonlinear properties of the modes are discussed, and each mode is assessed as to its importance in reconnection processes based upon these properties. It is concluded that the two most relevant instabilities are the ion acoustic instability and the lower-hybrid-drift instability. However, each instability has limitations as far as reconnection is concerned, and more research is needed in this area.

I.   INTRODUCTION

   The subject of anomalous transport in current sheets is of great interest to space plasma physicists, especially as it can impact collisionless reconnection processes. A simple concept of a reconnection process is illustrated in Fig. 1, which depicts a field-reversed plasma. The magnetic field $\tilde{B}$ is in opposite directions on the two sides of the neutral line, and a uniform electric field $\tilde{E}$ is directed into the page. The plasma motion in this configuration is roughly described by Ohm's law, which for the present situation may be written $\tilde{E} + \tilde{U} \times \tilde{B} = \eta \tilde{J}$ where $\tilde{U}$ is the plasma velocity, η the resistivity, and $\tilde{J}$ the current density). Away from the neutral line, the resistivity term is usually small, and the plasma obeys $\tilde{E} + \tilde{U} \times \tilde{B} = 0$, which is sometimes called the frozen-in-field condition. Loosely speaking, this means that particles are tied to a particular magnetic-field line. In this region, far from the neutral line, the plasma and the magnetic field are carried towards the neutral line with a velocity $U_{in} \simeq E/B$. However, the frozen-in-field condition breaks down in the diffusion region, where the magnetic field becomes very weak. The governing equation is $\tilde{E} = \eta \tilde{J}$, and the plasma and magnetic field are decoupled, i.e., no longer tied together. When this occurs the magnetic field can slip through the plasma and reconnect. The plasma and magnetic field then leave the diffusion region with a velocity U_{out} as shown in Fig. 1. In this process $U_{out} > U_{in}$, so that the plasma energy has been increased at the expense of magnetic-field energy.

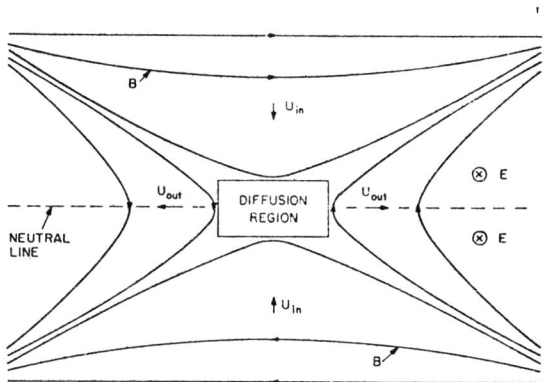

Figure 1: Schematic of a forced reconnection process.

One of the problems in applying this model to collisionless space plasmas (such as the earth's magnetotail) is properly describing the diffusion region. The resistivity η associated with coulomb collisions between particles is very small in space; the plasma is essentially collisionless. What then can balance the electric field in the diffusion region? There are other terms in Ohm's law, such as electron inertia and pressure anisotropy, but these also appear to be quite small (Vasiliunas, 1975). Another explanation is the occurrence of anomalous resistivity in the diffusion region. In this situation, particles scatter off collective electric fields generated by a plasma microinstability, and this scattering process decouples the plasma from the magnetic field. Recent laboratory experiments (Gekelman et al., 1982; Stenzel et al., 1983) in fact report observations of anomalous scattering.

Incorporating microturbulence effects in a reconnection process is a formidable task. Several issues need to be addressed. First, the linear theory of a microinstability needs to be developed appropriate for the plasma and magnetic field configuration of the diffusion region. In this study it is important to determine the relevant plasma conditions needed to excite the instability (e.g., width of the current sheet, electron/ion temperature ratio, etc.). Second, a nonlinear theory of the microinstability in question needs to be developed. Here, it is crucial to determine the level of microturbulence produced by the instability (i.e., saturation energy), and whether or not the turbulence is steady state. Finally, given the linear and nonlinear properties of the unstable waves, this information needs to be <u>self-consistently</u> incorporated into the hydrodynamic flows associated with a reconnection process. Development of such a self-consistent theory of collisionless reconnection is indeed difficult.

In general, plasma theorists have focussed on the first two issues: the linear and nonlinear theories of a microinstability as it applies to the diffusion region. The final issue, incorporating turbulence into reconnection flows, has been ignored. [A notable exception to this is the work of Coroniti and Eviatar (1977).] Although this may be considered, perhaps, a "cop-out" on the part of plasma theorists, the information regarding plasma microturbulence in the diffusion region is still crucial to understanding the overall process. Moreover, the anomalous transport properties of instabilities in the diffusion region can be <u>modelled</u> and incorporated into 2D and 3D MHD simulations of <u>reconnection</u> (Sato and Hayashi, 1980; Ugai, 1983; Sato, these proceedings). Although this is <u>not self-consistent</u>, it does provide insight into the collisionless reconnection process.

In this spirit, the purpose of this paper is to review the various microinstabilities that have been suggested to play a role in reconnection phenomena. Hence, only the linear and nonlinear properties of the instabilities will be discussed. Based upon these properties one can then assess whether or not a particular instability is a viable source of anomalous resistivity for a reconnection process. Finally, it should be noted that a review article of this nature has been published (Papadopoulos, 1979). The present work, in fact, draws heavily from Papadopoulos (1979); however, we attempt to elucidate certain aspects of the problem not emphasized in Papadopoulos (1979), and to present new results that have been obtained in the past four years.

The organization of the paper is as follows. In the next section, we describe the basic plasma and magnetic field configuration under consideration. In Section III, a description of the linear and nonlinear properties of several macroinstabilities is given. In Section IV, a discussion of the relevance of each of these instabilities to a reconnection process is presented. Finally, the last section contains a summary of the important results obtained to date.

II. PLASMA AND FIELD CONFIGURATION

The basic plasma and magnetic field configuration to be considered in this review is shown in Fig. 2. We take a simple, 1D reversed field geometry shown in Fig. 2a. The magnetic field $\underline{B}$ reverses direction at $x = 0$ and is supported by a plasma current $\underline{J}$ which is peaked at $x = 0$. The width of the reversal layer (or current sheet) is given by λ. An important parameter shown in Fig. 2a is x_e and x_i where $x_\alpha = (2\rho_\alpha \lambda)^{1/2}$, where ρ_α is the mean Larmor radius of the α species, and typically $x_e \ll \lambda \leqslant x_i$. This parameter will be discussed shortly. In Fig. 2b, the slab geometry used in the stability analysis is shown. The magnetic field $\underline{B}$ is in the + or − z direction, the density gradient ∇n is directed towards $x = 0$, the magnetic field gradient ∇B is directed away from $x = 0$, and the current $\underline{J}$ is in the y direction. For the purposes of this review, the microinstabilities discussed are driven <u>solely by the cross-field current J</u>. Thus, the wave vector $\underline{k}$ for the instabilities discussed is in the same direction as $\underline{J}$, i.e., $\underline{k} = k\,\underline{e}_y$. Instabilities driven by particle distribution functions which include beams, tails, or temperature anisotropies are not considered.

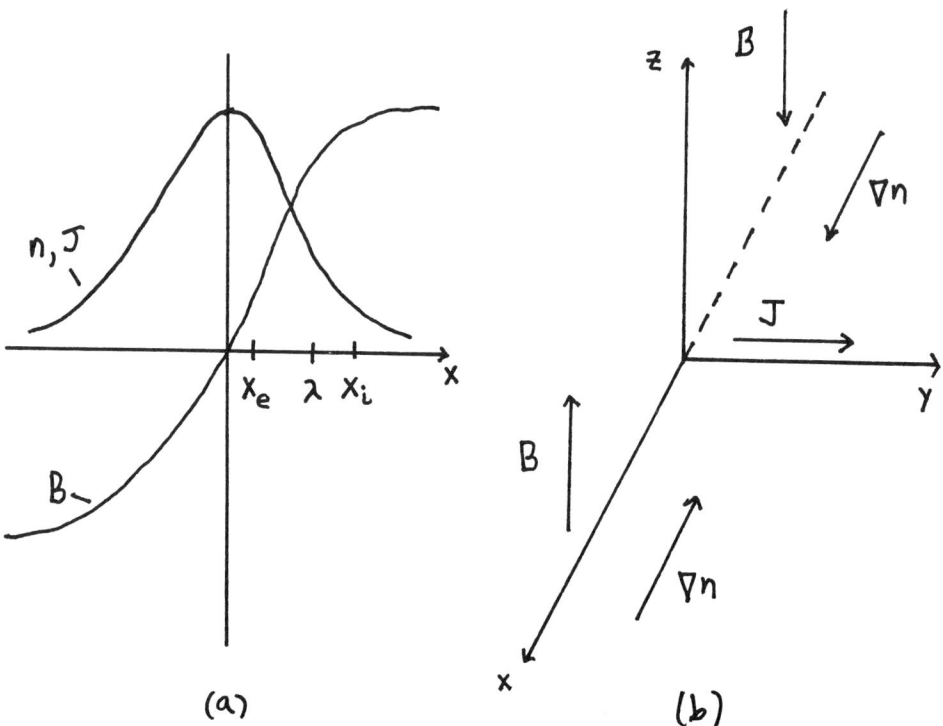

Figure 2: Plasma configuration and geometry.

Finally, we make one further simplifying assumption in the analysis which concerns the parameter $x_\alpha = (2\rho_\alpha \lambda)^{1/2}$ (Hoh, 1966). This quantity indicates the boundary of crossing and non-crossing thermal particles. Thermal particles in the region $|x| < x_\alpha$ cross the "neutral line", i.e., pass through the magnetic null region $B = 0$ at $x = 0$. These particles execute rather complicated orbits not amenable to analysis. On the other hand, thermal particles in the region $|x| > x_a$ do not cross the "neutral line". These particles are magnetized and execute gyro-orbits about $\boldsymbol{B}$. Thus, we consider two regimes: unmagnetized electrons ($|x| < x_e$) and magnetized electrons ($|x| > x_e$). The ions are taken to be unmagnetized which is valid for $|x| < x_i$ or $\gamma > \Omega_i$ where γ is the growth rate of an instability and $\Omega_i = eB/m_i c$ is the ion gyrofrequency. The assumption of unmagnetized electrons (i.e., "straight line orbits") is an over-simplification but is reasonably valid for $|x| \ll x_e$.

III. REVIEW OF MICROINSTABILITIES

A. Unmagnetized Regime ($|x| < x_e$)

1. Buneman instability

The Buneman instability is the classic electron-ion two-stream instability (Buneman, 1959). It is a fluid-like (or hydrodynamic) instability in that it does not involve wave-particle resonances (i.e., $\omega/k \gg v_\alpha$ where $v_\alpha = (T_\alpha/m_\alpha)^{1/2}$ is the thermal velocity of species α). The turn-on condition for instability is roughly $V_d \gtrsim 2v_e$ where V_d is the relative electron-ion drift. In the linear regime at maximum growth one finds that $\omega_r \simeq \omega_{pe}$, $\gamma \simeq (m_e/m_i)^{1/3} \omega_{pe}$ and $k \simeq V_d/\omega_{pe} \lesssim \lambda_{de}^{-1}$ where $\omega = \omega_r + i\gamma$, $\omega_{pe} = (4\pi n_e^2/m_e)^{1/2}$ is the electron plasma frequency, and $\lambda_{de} = v_e/\omega_{pe}$ is the electron Debye length (Krall and Trivelpiece, 1973). Thus, the instability is considered to be high frequency and short wavelength. In the nonlinear regime the instability is saturated by electron trapping which leads to strong electron heating (i.e., $v_e \gtrsim V_d$) (Davidson et al., 1970; Biskamp and Chodura, 1973). In the presence of a steady state electric field, the anomalous resistivity η_{an} is not steady state (i.e., $\eta_{an} \sim$ constant) but is spiky (Papadopoulos, 1977).

2. Ion acoustic instability

The ion-acoustic instability, like the Buneman instability, is driven by the relative electron-ion drift V_d. However, the ion acoustic instability is a resonant (or kinetic) instability and is driven via an electron-wave resonance. The turn-on condition for this instability is somewhat less stringent than that of the Buneman instability when $T_e \gg T_i$. The condition is approximately $V_d \gtrsim (T_i/$

$m_e)^{1/2}$ for $0.2 < T_e/T_i < 5.0$ (Coroniti and Eviatar, 1977). However, when $T_e \lesssim T_i$ the turn-on is comparable to that of the Buneman instability. Linear theory predicts (at maximum growth) that $\omega_r \simeq kc_s \simeq \omega_{pi}$, $\gamma \simeq (m_e/m_i)^{1/2}(V_d/c_s)\omega_{pe}$ and $k \sim \lambda_d^{-1}$ where $c_s = (T_e/m_i)^{1/2}$ is the ion sound speed and $\omega_{pi} = (4\pi ne^2/m_i)^{1/2}$ is the ion plasma frequency (Papadopoulos, 1979).

There have been many nonlinear theories of the ion-acoustic instability proposed (e.g., quasilinear, resonance broadening, nonlinear Landau damping). Rather than discuss any of these theories in detail it will simply be noted that (1) a steady state anomalous resistivity can be achieved (Coroniti and Eviatar, 1977), and (2) near marginal stability, the anomalous collision frequency is roughly $\nu_{an} \simeq 10^{-2}\omega_{pe}$ (Papadopoulos, 1979) so that the anomalous resistivity is $\eta_{an} = 4\pi\nu_{an}/\omega_{pe}^2 \simeq 10^{-1}\omega_{pe}^{-1}$.

B. Magnetized instabilities ($|x| > x_e$)

1. Beam cyclotron instability

The beam cyclotron instability (also known as the electron cyclotron drift instability) (Wong, 1970; Lampe et al., 1972) is a fluid-like (or hydrodynamic) instability that is excited via the coupling of an electron Bernstein wave to an ion mode. The relative electron-ion drift allows the ion mode to be Doppler-shifted so that its frequency matches an electron cyclotron harmonic. The turn-on condition for this instablity is $V_d > \text{Max}\,[c_s, (\Omega_e/\omega_{pe})v_e]$ where $c_s = (T_e/m_i)^{1/2}$ (Papadopoulos, 1979). For the case $T_e \ll T_i$, maximum growth is characterized by $\omega_r \simeq k(c_s + V_d)$, $\gamma \simeq (m_e/m_i)^{1/4}\Omega_e$ and $k \simeq \lambda_{de}^{-1}$ (Lampe et al., 1972). The mode saturates because of turbulent scattering of the electrons which effectively "demagnetize" them and they are unable to maintain coherent gyro-orbits (Lampe et al., 1971). The saturation energy of the instability is relatively small so that a small anomalous collision frequency results: $\nu_{an} \simeq (V_d/v_e)^3\Omega_e$ (Papadopoulos, 1979).

2. Magnetized ion-ion instability

The magnetized ion-ion instability (Papadopoulos et al., 1971) is a counter-streaming ion-ion instability. It is a fluid-like (or hydrodynamic) instability. The turn-on condition for this instability is $V_{ii} \gtrsim 2v_i$ where V_{ii} is the relative ion-ion drift. At maximum growth one can show that $\omega_r \simeq 0$, $\gamma = \omega_{\ell h}$ and $k \simeq \omega_{\ell h}/V_{ii}$ where $\omega_{\ell h} \simeq \omega_{pi}/(1 + \omega_{pe}^2/\Omega_e^2)^{1/2}$ is the lower-hybrid frequency. However, the instability is linearly stable when $V_{ii} > V_A(1 + \beta_e)^{1/2}$ where $V_A = B/(4\pi nm_i)^{1/2}$ is the Alfvén velocity and $\beta_e = 8\pi nT_e/B^2$. The mode saturates because of ion trapping and produces strong ion heating as well as a reduction in the relative ion-ion drift velocity. The anomalous ion-ion collision frequency associated with this instability is $\nu_{an} \lesssim 10^{-1}\omega_{\ell h}$ (Lampe et al., 1975).

3. Lower-hybrid-drift instability

The lower-hybrid-drift instability (Davidson et al., 1977) is a resonant (or kinetic) instability which is excited via an ion-drift wave resonance when $V_{di} \lesssim v_i$ (here, $V_{di} = (v_i^2/\Omega_i)\partial \ln n/\partial x$ is the ion diagmagnetic velocity). The turn-on condition for the instability is $V_{di} > v_i(m_e/m_i)^{1/4}$. The instability is characterized at maximum growth by $\omega_r \simeq kV_{di} \lesssim \omega_{\ell h}$, $\gamma \simeq (V_{di}/v_i)\omega_r$ and $k\rho_e \sim (T_e/T_i)^{1/2}$ where ρ_e is the mean electron Larmor radius. This instability is relatively insensitive to the temperature ratio T_e/T_i. However, the mode is suppressed in high β plasmas because of an electron ∇B drift-wave resonance. A variety of nonlinear theories have been suggested for the lower-hybrid-drift instabilities (e.g., quasilinear relaxation, resonance broadening, ion trapping, mode coupling). Again, we will not discuss these in detail but note that the most likely saturation mechanism is mode coupling (Drake et al., 1983). The anomalous collision frequency associated with the turbulence is $\nu_{an} \simeq (V_{di}/v_i)^2 \omega_{\ell h}$ and a steady state resistivity can result from this turbulence.

IV. APPLICATION TO RECONNECTION

Prior to discussing the relevance of each instability discussed in Section III to reconnection, it is important to note a major difference between the magnetized and unmagnetized instabilities. Namely, the spatial region where these instabilities can exist. As noted in Section II the unmagnetized instabilities are limited to $|x| < x_e$, i.e., essentially the null region where $B \simeq 0$. This is precisely where one would like microturbulence to exist in order to "decouple" the plasma from the magnetic field. On the other hand, the magnetized instabilities are restricted to $|x| > x_e$, away from the null field region. Thus, these instabilities do not <u>directly</u> produce an anomalous resistivity in the null region. However, the dynamic evolution of the plasma and field in a reconnection process may allow penetration of the magnetized modes to the region $|x| < x_e$ (e.g., current steepening, convection).

A. Unmagnetized instabilities

1. Buneman instability

The Buneman instability requires a strong relative electron-ion drift to be excited (i.e., $V_d \gtrsim 2v_e$). By using Ampere's law to relate the width of the current sheet (λ) to the relative drift (V_d), one can show that $\lambda < c/\omega_{pe}$ for this instability to be excited in the diffusion region. Because of the extremely thin current sheet needed, it seems unlikely that the Buneman instability can be of any importance to collisionless reconnection processes.

2. Ion acoustic instability

A theory of reconnection incorporating the ion acoustic instability as a source of anomalous resistivity has been developed by Coroniti and Eviatar (1977). For a detailed discussion, we refer the interested reader to this paper. However, several comments on this work are in order. First, the model developed by Coroniti and Eviatar (1977) is reasonably self-consistent although a number of simplifying assumptions were required for the analysis. Second, they found that steady state reconnection could occur based upon ion acoustic wave turbulence for certain parameter regimes. Third, even though the turn-on condition for the ion acoustic instability is less stringent than that for the Buneman instability, a thin current sheet is still required to excite this mode, i.e., $\lambda \lesssim$ few (c/ω_{pe}), especially for plasmas such that $T_e \ll T_i$. Finally, although ion acoustic turbulence has been observed in laboratory reconnection experiments (Bratenahl and Yeates, 1970) its exact role is unclear. Moreover, in space plasmas, it is unlikely that current sheets develop as thin as required for this instability (e.g., the earth's magnetotail). Thus, the ion acoustic instability is probably not important for reconnection processes in collisionless space plasmas.

B. Magnetized plasmas

1. Beam cyclotron instability

The beam cyclotron instability has been discussed in regard to reconnection by Coroniti and Eviatar (1977) and by Haerendel (1978). As noted by Papadopoulos (1979), thin current sheets ($\lambda \lesssim$ few (c/ω_{pe})) are needed to produce a significant anomalous resistivity. Also, it has been shown that a magnetic field gradient (∇B) substantially reduces the growth rate of this instability (Gary, 1972; Sanderson and Priest, 1972). Thus, we conclude that the beam cyclotron instability is not important to reconnection processes.

2. Magnetized ion-ion instability

The magnetized ion-ion instability has recently been proposed as a source of anomalous resistivity for magnetotail reconnection by Lee (1982). However, the plasma configuration required is somewhat more complicated than shown in Fig. 2a. That is, a second electron and ion species is also considered as shown in Fig. 3. This second plasma is labelled untrapped. At the position $x = x_0$ in Fig. 3, the diagmagnetic drifts of the two ion species are in opposite directions so that ion counter-streaming occurs. Based on this type of plasma configuration, Lee (1982) finds that he magnetized ion-ion instability can be unstable. It should be noted that (1) the scale lengths of the density gradients need to be relatively sharp ($L_n < \rho_i$ where $L_n \simeq (\partial \ln n/\partial x)^{-1}$) in order that the instability turn-on $v_{ii} > 2v_i$; (2) the mode is stable in high β plasmas; and (3) the important effect of electron ∇B damping has been ignored in Lee (1982).

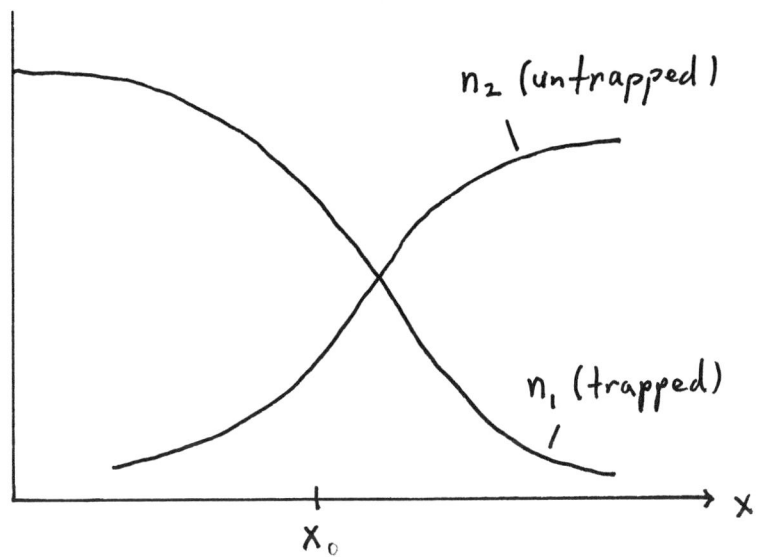

Figure 3: Equilibrium for ion-ion instability.

3. Lower-hybrid-drift instability

The lower-hybrid-drift instability was first proposed by Huba et al. (1977) as a source of anomalous resistivity for reconnection in the earth's magnetotail. Two factors in favor of this instability are (1) the mode can be excited in relatively broad current sheets ($\lambda \lesssim (m_i/m_e)^{1/4} \rho_i$), and (2) the mode is insensitive to the temperature ratio T_e/T_i. Both of these factors should be contrasted to, say, the requirements for the ion acoustic instability. A subsequent study determined that turbulence observed in the distant magnetotail was consistent with the occurrence of the lower-hybrid-drift instability (Huba et al., 1978). However, a problem with this instability (as it applies to a reconnection process) is that the mode is damped in a high β plasma ($\beta \gg 1$) because of an electron ∇B drift-wave resonance. Thus, based upon both a local and nonlocal linear analysis (Huba et al., 1980), the instability is stable in the near vicinity of the null point.

Although this result is unfavorable in directly providing an anomalous resistivity in the diffusion region, the evolution of the magnetic field in the presence of a resistivity based upon the nonlocal mode structure of the lower-hybrid-drift instability has been investigated (Drake et al., 1981). In this regard, a 1D transport equation for the magnetic field has been developed for an arbitrary resistivity profile in a field-reversed plasma. The equation is given by

$$\frac{\partial B}{\partial t} + \frac{cE}{B}\frac{\partial B}{\partial x} - \frac{2B}{B^2+B_\ell^2}\frac{\partial}{\partial x}\nu_{an}\rho_{es}^2 B\frac{\partial B}{\partial x} = \frac{2B\,B_\ell}{B^2+B_\ell^2}\frac{\partial B_\ell}{\partial t} \qquad (1)$$

where $B_\ell = B$ (outer boundary) and $\rho_{es}^2 = \rho_e^2(T_i/T_e)$. On the LHS of Eq. (1) the first term represents the time rate of change of the magnetic field, the second term represents convection because of the inductive electric field E, and the third term represents diffusion based upon an arbitrary collision frequency ν_{an}. The RHS side of Eq. (1) contains the effect of a time-varying boundary field.

We have solved Eq. (1) numerically (Drake et al., 1981). A resistivity model such that $\eta \propto B^2$ was chosen; this model has the feature that $\eta = 0$ at the neutral line, but $\eta \neq 0$ away from the neutral line. The results of this work are illustrated in Fig. 4. The initial magnetic field (Fig. 4a) and current density (Fig. 4b) profiles are labeled $\tau = 0$; the profiles at a later time are labeled $\tau = 0.2$. It is found that magnetic flux is transported towards the neutral line and that the current density increases at the neutral line which is due to a diffusion process. This leads to the possibility that waves can subsequently penetrate to the neutral region during the nonlinear evolution of the field-reversed plasma. Such an evolution has been observed in particle simulations of field-reversed plasmas (Winske, 1981; Tanaka and Sato, 1981). However, these simulations used unrealistic mass ratios and it is unclear at this time whether or not wave penetration occurs using realistic mass ratios (Quest, private communication).

Finally, recently a 2D mode coupling nonlinear theory of the lower-hybrid-drift instability has been developed (Drake et al., 1983). This theory is consistent with both laboratory measurements of the instability as well as with computer simulations. An important result from this new theory is an estimate of the anomalous resistivity associated with the turbulence: $\nu_{an} \simeq 2.4(\rho_i/\lambda)^2\omega_{\ell h}$. This value of ν_{an} corresponds to a magnetic Reynolds number of $R_m \simeq 0.5 (m_i/m_e)^{1/2}(\lambda/\rho_i)^3$. Thus, it is found that the lower-hybrid-drift instability only provides significant anomalous transport for current sheets such that $\lambda \simeq \rho_i$. Also, a discussion of this instability as it applies to substorm dynamics is given in Huba et al. (1981).

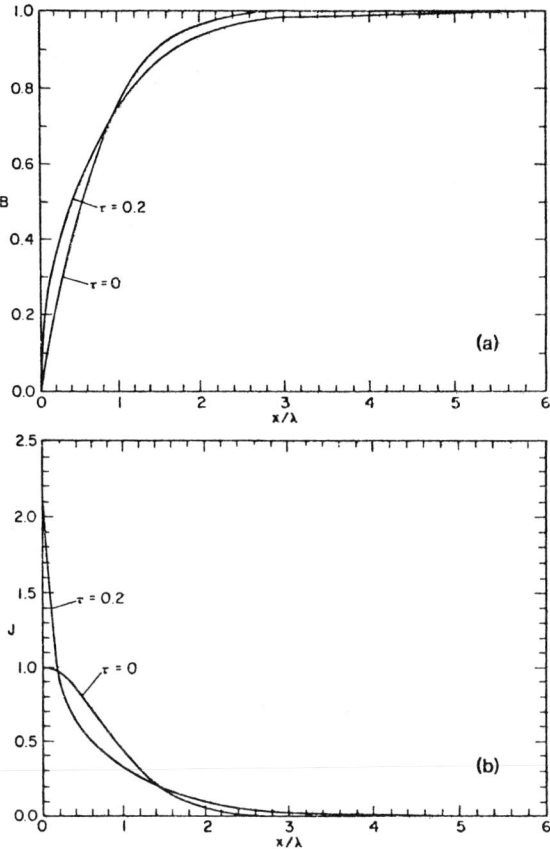

Figure 4: Time evolution of B and J based upon Eq. (1).

V. CONCLUDING REMARKS

It is well known that microinstabilities can affect the dynamic evolution of plasmas through wave-particle interactions (i.e., scattering) and cause anomalous diffusion, momentum transfer and energy exchange. The purpose of this review is to briefly discuss several instabilities that have been proposed as anomalous transport mechanisms in current sheets. The focus has been on reversed magnetic field configurations (Fig. 2a), as they relate to collisionless reconnection processes, since the presence of microturbulence in the diffusion region can influence the hydrodynamic flows. However, the stability analysis of waves in the diffusion region is difficult and simplifying assumptions are made, as noted in Section II.

The two "favored" instabilities are the ion acoustic instability and lower-hybrid-drift instability. The ion acoustic instability can be excited in the null field region but requires quite thin current sheets ($\lambda \lesssim$ few (c/ω_{pe})) and is more easily excited in hot electron plasmas ($T_e \gg T_i$). Although it has been observed in laboratory reconnection experiments where these conditions can be met, its occurrence in relevant space plasmas is rather unlikely (e.g., the earth's magnetotail). On the other hand, the lower-hybrid-drift instability has received considerable attention since it can be excited in broader current sheets ($\lambda \sim \rho_i$) and is relatively insensitive to T_e/T_i. However, the waves are strongly damped close to the null region. In a dynamic situation (e.g., forced reconnection), lower-hybrid-drift wave turbulence may penetrate the null region, but this result is tentative at this time. Nonetheless, even if this turbulence does not penetrate the null region, it is likely to exist over a substantial portion of the current sheet and can strongly affect plasma flow in the regions where the mode is unstable. One possibility is that this instability may limit the width of current sheets to $\lambda \sim \rho_i$ and inertial effects may be dominant in the null region (Coroniti, private communication).

We emphasize that a simplified plasma and field configuration has been used. It is possible that other instabilities may be excited which depend upon non-Maxwellian distribution functions which contain, say, beams and anisotropies. In this regard, laboratory experiments and in situ space observations may indicate more appropriate distribution functions.

Finally, as noted in the introduction, it is crucial to <u>self-consistently</u> incorporate plasma turbulence in the dynamic evolution of collisionless reconnection. This is an exceedingly difficult problem which, perhaps, may only be answered by 3D particle or hybrid simulations, which in themselves are also enormously difficult and beyond present day computational facilities. Maybe our grandchildren will finally solve the problem.

ACKNOWLEDGMENTS

I wish to thank Drs. J. Drake, A. Hassam, F. Coroniti and K. Quest for helpful conversations regarding this problem. This work has been supported by NASA and ONR.

REFERENCES

Biskamp, D., and Schindler, K.: 1973, Phys. Fluids $\underline{16}$, 888.
Bratenahl, A., and Yeates, C.M.: 1970, Phys. Fluids $\underline{13}$, 2696.
Buneman, O.: 1959, Phys. Rev. $\underline{115}$, 503.
Coroniti, F.V., and Eviatar, A.: 1977, Ap. J. Supp. $\underline{33}$, 189.
Davidson, R., Krall, N.A., Papadopoulos, K., and Shanny, R.: 1970, Phys. Rev. Lett. $\underline{24}$, 579.
Davidson, R.C., Gladd, N.T., Huba, J.D., and Wu, C.S.: 1977, Phys. Fluids $\underline{20}$, 301.
Drake, J.F., Gladd, N.T., and Huba, J.D.: 1981, Phys. Fluids $\underline{24}$, 78.
Drake, J.F., Guzdar, P.N., and Huba, J.D.: 1983, Phys. Fluids $\underline{26}$, 601.
Gary, S.P.: 1971, J. Plasma Phys. $\underline{6}$, 561.
Gekelman, W., Stenzel, R.L., and Wild, N.: 1982, J. Geophys. Res. $\underline{87}$, 101.
Haerendel, G.: 1978, J. of Atm. and Terr. Phys. $\underline{40}$, 343.
Hoh, F.C.: 1966, Phys. Fluids $\underline{9}$, 277.
Huba, J.D., Gladd, N.T., and Papadopoulos, K.: Geophys. Res. Lett. $\underline{4}$, 125.
Huba, J.D., Gladd, N.T., and Papadopoulos, K.: 1978, J. Geophys. Res. $\underline{83}$, 5217.
Huba, J.D., Drake, J.F., and Gladd, N.T.: 1980, Phys. Fluids $\underline{23}$, 552.
Huba, J.D., Gladd, N.T., and Drake, J.F.: 1981, J. Geophys. Res. $\underline{86}$, 5881.
Krall, N.A., and Trivelpiece, A.W.: 1973, "Principles of Plasma Physics", McGraw-Hill Co., NY.
Lampe, M., Manheimer, W.M., McBride, J.B., Orens, J., Shanny, R., and Sudan, R.N.: 1971, Phys. Rev. Lett. $\underline{26}$, 1221.
Lampe, M., Manheimer, W.M., McBride, J.B., Orens, J., Papadopoulos, K., Shanny, R., and Sudan, R.: 1972, Phys. Fluids $\underline{15}$, 662.
Lampe, M., Manheimer, W., and Papadopoulos, K.: 1975, NRL Memo Rept. 3076.
Lee, L.C.: 1982, Geophys. Res. Lett. $\underline{9}$, 1159.
Papadopoulos, K., Davidson, R.C., Dawson, J.M., Haber, I., Hammer, D.A., Krall, N.A., and Shanny, R.: 1971, Phys. Fluids $\underline{14}$, 849.
Papadopoulos, K.: 1979, Dynamics of the Magnetosphere, Ed. Akasofu, S.-I., D. Reidel, 289.
Sanderson, J.J., and Priest, E.R.: 1972, Plasma Phys. $\underline{14}$, 959.
Sato, T., and Hyashi, T.: 1979: Phys. Fluids $\underline{22}$, 1189.
Sato, T.: 1985, this volume.
Stenzel, R.L., Gekelman, W., and Wild, N.: 1982, J. Geophys. Res. $\underline{87}$, 111.
Tanaka, M., and Sato, T.: 1981, J. Geophys. Res. $\underline{86}$, 5541.
Ugai, M.: 1983, to be published in Phys. Fluids.
Vasiliunas, V.M.: 1975, Rev. Geophys. Space Phys. $\underline{13}$, 303.
Winske, D.: 1979, Phys. Fluids $\underline{24}$, 1069.
Wong, H.V.: 1970, Phys. Fluids $\underline{13}$, 757.

DISCUSSION

Van Hoven: Would you comment on the effects of the addition of B_y to your model, so that it has field shear? Does one have different spatial structure or different instabilities?

Huba: Finite B_y introduces magnetic shear into the equilibrium model. A study of this situation is given in Huba et al., J. Geophys. Res. **87**, 1697 (1982). Basically, magnetic shear has a stabilizing influence on the lower-hybrid-drift instability (and instabilities in general), and acts to inhibit penetration of the mode toward the neutral line.

Migliuolo: The addition of a y-component of the equilibrium B-field keeps β low in the central region. This might make some of the aforementioned modes (e.g. ion-acoustic) more effective in producing η_{an}.

Huba: Although a y-component of $\underset{\sim}{B}$ may keep β low in the central region, it introduces magnetic shear into the equilibrium, which is a stabilizing influence. (See reply to Van Hoven.)

Vasyliunas: Nearly all of the instabilities discussed, with the exception of lower-hybrid-drift, occur only when the thickness of the current sheet is comparable to or less than c/ω_{pe}, but on this scale the inertial terms in the generalized Ohm's law dominant and their neglect is rather questionable.

Huba: This point is discussed in Coroniti and Eviatar (1977). It is not clear that inertial terms would be dominant over the anomalous resistivity provided by, say, the ion acoustic instability. However, I agree they should be included for self-consistency.

D. Smith: Could you go into more detail on how magnetic field is transported toward the neutral line?

Huba: The transport of magnetic flux towards the neutral line in the presence of an anomalous collision model based upon the lower-hybrid-drift instability is simply a diffusion process. It is evident in the 3rd term on the LHS of Eq.(1) which is the 1D transport equation for B.

Hasegawa: MHD equilibrium requires the presence of ∇p rather than $\nabla n = 0$?

Huba: The analysis of the lower-hybrid-drift instability is based upon p = nT. If $\nabla n = 0$, then for $\nabla T \neq 0$. The lower-hybrid-drift instability can be excited when $\nabla n = 0$ and $\nabla T \neq 0$.

Coppi: In the theory of magnetic reconnection it is necessary to couple the consideration of microscopic (kinetic) effects with those of the macroscopic magnetic configuration. In fact, the theory of tearing modes in collisionless regimes for sheared magnetic field configurations shows that these tend to become stable, and this could not be foreseen without carrying out a complete analysis. Therefore it is probably premature, on the basis of the state of the theory you presented, to pass a judgement on what is adequate to explain the reconnection processes that appar to occur in the Earth's magnetotail.

Huba: I agree that a self-consistent theory which couples microturbulence to the macroscopic evolution of the plasma is needed. I have tried to emphasize this point in my paper, although the purpose of the article is to simply review the linear and nonlinear properties of various instabilities possibly relevant to reconnection processes.

ANOMALOUS TRANSPORT INDUCED BY FIELD ALIGNED CURRENTS AND ITS RELATION TO ELECTROMAGNETIC COUPLING

Christian T. Dum
Max-Planck-Institut für Physik und Astrophysik
Institut für extraterrestrische Physik, D-8046 Garching
Federal Republic of Germany

The wealth of detailed observations on transport in turbulent plasmas that has become available over the last decade from laboratory experiments, in situ space craft observations and computer simulation demonstrates that in contrast to classical transport, the various steps in the analysis of anomalous transport, involving microscopic processes and the global dynamics, are closely coupled. It also points to many new, exciting, and truly anomalous phenomena. These statements apply especially to the highly dynamic processes connected with field aligned currents. The usually vast difference in time or length scales must allow for considerable simplications in the analysis, depending on the system at hand, but requires careful consideration of the microscopic or macroscopic physics that of necessity is to be treated in a simplified manner. This point is demonstrated by using a marginal stability approach in a numerical model for field aligned currents and electromagnetic coupling in an extended system. Its aim is to see how energy is supplied to a localized dissipation region from a distant generator and how this dissipation in turn affects the global electro-dynamic structure. In addition to the earth's auroral flux tubes, for which this model was primarily designed, these questions are of importance to other extended current systems in astrophysics. Microscopic processes supporting enhanced dissipation and leading to other truly anomalous processes such as acceleration of selected particle groups are then discussed in relation to the global problem.

The aim of this paper is primarily to describe some of the main issues to be adressed in the talk and to provide some key references. In order to keep the list of references manageable, preference is given to the most recent papers that also contain extensive references to important earlier work in which many ideas may have originated.

APPROACHES TO ANOMALOUS TRANSPORT

The traditional approach to anomalous transport consists of five steps: 1) Identification of a free energy source for wave growth, i.e. the relative drift $u = -J/ne$ between electrons and ions in our case. 2) Selection of a plausible instability, usually the one with the lowest threshold $u_c < u$ for a given set of plasma parameters, using linear instability theory. 3) Determination of a quasi-steady fluctuation level for the ensuing turbulence, usually from some nonlinear theory for saturation of wave growth. 4) Relation of this fluctuation level to some effective collision frequency ν_{eff} which is supposed to replace the frequency ν_{cl} of Coulomb collisions in the transport relations for the fluid variables. 5) Solution, of the electrodynamic problem for the system at hand in the usual manner, except that ν_{cl} is replaced by ν_{eff}. In spirit this approach exactly follows classical transport theory, and thus may be termed "Enhanced classical transport theory". The effective collision frequency, however, is usually estimated only for the transport process considered to be dominant, anomalous resistivity in our case, and the resulting transport relation corresponds more closely to elementary kinetic theory than the complete transport theory that is available for Coulomb collisions (e.g. Braginskii, 1967).

This traditional approach may give some indication of the possible effects of anomalous transport. The detailed observations that are now available show, however, that the potential of anomalous transport is far greater but also that the five steps listed above, traditionally treated independently, often by different people, are in fact strongly coupled. The dominant instability often switches to other wave modes as the system evolves, depending on how the free energy sources for wave growth are maintained. The fluctuation level, as a result, is more often determined simply by the macroscopic dynamics and the evolution of the microscopic distribution functions, rather than any fancy nonlinear saturation mechanism. Theories of the latter generally assume fixed plasma parameters and (Maxwellian) distribution functions, as do the linear stability calculations that are carried out by so many people in such detail. Scattering of particles by turbulent fluctuations has a speed and angle dependence that strongly depends on the underlying wave mode but in any case is quite different from the velocity dependence of scattering by Coulomb collisions. A major reason for this fact is that in contrast to the very broad and isotropic collisional fluctuation spectrum, turbulent spectra are much more restricted in wave number and nearly always strongly anisotropic. Interaction with these spectra is in essence restricted to particles that satify certain resonance conditions (e.g. Galeev and Sagdeev, 1979). There is no known mechanism in a turbulent, collisionless plasma that would maintain for any length of time the Maxwellian distributions that are generally assumed for computational convenience. Although it is difficult, distribution functions must be determined in order not only to carry out step 4, but also for steps 2 and 3. For example, scattering of electrons by ion acoustic turbulence makes distribution functions nearly isotropic, similar to electron-ion collisions, but does not maintain Maxwellians. An anomalous transport theory for the electrons was developed which in completeness corresponds to classical transport theory (Dum, 1978 a, b). Even in this case, however, transport relations differ in structure and not just in magnitude of the effective collisions frequency, from the corresponding classical relations. Different approaches will be necessary for transport theories connected with other wave spectra.

Laboratory experiments and especially particle simulation experiments show a strong dependence of microscopic processes on macroscopic boundary conditions, e.g. whether, in current driven turbulence, the current or the applied electric field is kept constant (Dum et al., 1974; Dum and Chodura, 1979). They also show invariably that turbulence is of a transient nature. Computer experiments of course have grossly simplified boundary conditions and they cannot handle the vast difference in time or length scales that exists in astrophysical systems. This difference in scales must allow for considerable simplifications in the analyis. Microscopic processes should enter the global dynamics (step 5) only in some averaged sense. All that may matter to lowest order for the global evolution is, for example, a very rapid increase in anomalous transport once some threshold is exceeded. The system will then stay close to marginal stability, often with relaxation oscillations around this state, thus providing on the average the required dissipation, as can be demonstrated explicitly for anomalous resistivity due to ion sound turbulence (Dum, 1981). Thus, in contrast to step 3, which explicitly assumes a quasi-steady dependence of the fluctuation level on the evolving plasma parameters, no steady state is reached, and wave growth will not even enter into the nonlinear regime.

The marginal stability approach to anomalous transport has been used since the early days of plasma physics with various degrees of sophistication and success. It can be applied not only to quasi-steady states with a balance between sources and sinks but also to rapidly evolving systems such as compression by a turbulently resistive shock. The analysis proceeds essentially in the opposite direction from the traditional approach listed above, i.e. it starts with the global dynamics and ends with the microscopic physics (Manheimer and Boris, 1977). The shock wave experiments provide us with other important points for this approach. It has been shown that the unknown coefficient in one of the best known anomalous resistivity formulas would have to be varied by orders of magnitude for agreement with measured shock widths. Agreement with marginal stability for ion sound is reasonable, provided one takes the critical drift velocity for a plasma with high energy ion tails and not the much lower threshold for Maxwellian ions (Eselevich et al., 1971). The instability criterion for Maxwellian distributions in fact becomes irrelevant already very early in the growth phase of the instability, because electrons evolve to a strongly non-Maxwellian distribution for which instability is no longer sustained solely by the current, but largely by effective drifts related to density and temperature gradients (Dum, 1978b), as demonstrated by the experiments of Söldner et al. (1977). Excellent agreement between detailed measurements and a numerical code can be found by taking into account the interrelationship between wave growth and the various anomalous transport processes.

A much simpler model using the marginal stability approach is presented in the following for the global problem of field aligned currents and electromagnetic coupling in extended systems. It uses a nonlinear Ohm's law in which coefficients, although derived from microscopic theory, can be varied in order to check the marginal stability assumption. Other microscopic processes can be included step by step. It should be noted, however, that the total size of the largest particle simulation systems is at most comparable to the numerical grid spacing in our fluid code (Lysak and Dum, 1983). Owing to the high electron mobility along magnetic field lines some intrinsically microscopic features arise for field aligned currents, such as runaway acceleration and formation of nonuniform potential drops, which cannot directly be described by this approach, but for which the macroscopic model provides the proper boundary conditions. An ad hoc model of anomalous resistivity has recently been used by Sato and Hayashi (1979) in a numerical study of externally driven reconnection. The current flows perpendicular to the magnetic field in this case. Although anomalous resistivity is the primary cause for abrupt onset of reconnection, the key features of the process are ultimately controlled by the boundary conditions. This can be verified by varying the resistivity law, see also Ugai (1982).

SYSTEMS WITH FIELD ALIGNED CURRENTS

Strictly speaking, a system with field aligned currents corresponds to non-uniform magnetic fields

$$\nabla \times \underline{B} = (4\pi/c)\underline{J}_\parallel = \alpha \underline{B}$$

so called force free configurations, that are considered to be of great importance in astrophysics for two reasons. The equation of motion

$$\rho(d\underline{u}/dt) + \nabla \cdot \underline{\underline{P}} = (\underline{J}/c) \times \underline{B} = \nabla \cdot \left[(B^2/8\pi)\underline{\underline{I}} - (\underline{BB}/4\pi) \right]$$

implies that near equilibrium configurations with small ratios $\beta = 8\pi n(T_e + T_i)/B^2$ of plasma and magnetic pressure must be largely force free. Also, such a magnetic field pattern contains an excess energy over a system with current free, potential magnetic field. By a change in the configuration some of this energy may be released to produce solar flares (Gold and Hoyle, 1960; Barnes and Sturrock, 1972). More generally, we consider currents that are essentially along an ambient magnetic field B_o, such that the selfconsistent magnetic field produced by the current can be neglected to lowest order. In the laboratory B_o is produced by external field coils in order to confine the plasma. The internal magnetic field of planets provides the ambient magnetic field for the surrounding plasma, e.g. the earth's ionosphere and lower magnetosphere. The interplanetary solar magnetic field ($\nabla \times B \approx 0$) carried by the solar wind, plays of course an increasingly important role in the more distant magnetosphere. Interaction with the solar wind not only determines the shape of the magnetosphere but also leads to exciting events, magnetospheric substorms, in which this shape undergoes sudden transitions (Akasofu, 1977). The solar wind, as modulated by solar activity, provides the primary energy input, with secondary dynamos related to reconnection of magnetic field lines in the distant magnetotail. Geomagnetic disturbances and the spectacular auroral displays are manifestations of these impulsive events in the earth's polar regions (Kan, 1982; Sato, 1982). High latitude magnetic field lines are thus transmission lines for an extended electromagnetic system with a magnetospheric generator, field aligned currents and an ionospheric load, where closure of the current is through Pedersen and Hall currents perpendicular to the ambient magnetic field. According to ideal MHD, the electric field satisfies $\underline{E}+(\underline{u}/c)\times \underline{B} = 0$, implying that $E_\| = 0$ and that convection is perfectly mapped along the magnetic flux tubes, if it is also assumed that the electric field is potential, $E = -\nabla \phi$. These assumptions provide a reasonable description of the global current system, during quiet periods. Some potential drop is expected to be present also in this case. It is connected with the injection of various particle populations at the magnetospheric and ionospheric ends of the flux tubes, similar to sheath effects in laboratory devices, and the electric field that is required to overcome the magnetic mirror force in converging flux tubes (Chiu and Schulz, 1978; Kan, 1982).

The observation of energetic particles with energies much above thermal level, e.g. 10 keV as compared to thermal energies of less than 1 eV for electrons causing aurora, and especially the more recent, detailed in situ measurements of particles and electromagnetic fields by satellites with polar orbits and of apogees ranging from about 8000 km (Mozer et al., 1981) to 1500 km (Burke et al., 1980) demonstrate, however, that both assumptions fail for active periods. The fields are not static, but correspond to electromagnetic low frequency pulsations that propagate along the magnetic flux tubes, i.e. shear Alfvén waves. There are also regions of intense parallel electric fields that not only should accelerate particles but also lead to a partial decoupling of magnetospheric and ionospheric convection. In the same hightly localized regions it can be shown that field aligned currents are sufficiently intense to excite microinstabilities. Indeed, intense turbulence is observed, that correlates reasonably with the intensity of field aligned currents. Since waves, such as electrostatic ion cyclotron waves are destabilized by low energy electrons and not directly by the current, and other nonthermal features such as ion beams are often present which could also have a cause or effect relationship to these waves, identification of the free energy sources is not unambiguous however (Cattell, 1981). Intense escaping radio emission, auroral kilometric radiation is also closely related to the auroral particle acceleration process (Lee et al., 1980; Gurnett and Anderson, 1981). From what has been said already it is clear that these microscopic processes must be seen in conjunction with the global problem.

Some of the same basic microscopic and electrodynamic processes that operate in the earth-solar wind interaction for auroral regions, also operate in the interaction of Jupiter's moon Io with the Jovian magnetosphere-ionosphere. This interaction plays an important role in Jovian decametric radiation. The recent Voyager 1 and 2 encounters with Jupiter have shed new light on these processes and significant advances have been made in our understanding. Radiation is due to accelerated particles, similar to the processes along auroral flux tubes. Field aligned currents and electric fields are generated, but in this case by Io's rapid motion through the Jovian magnetic field which at Io's orbit is embedded in a dense plasma torus. The rapid motion of Io in this environment invalidates earlier, static current circuit models. Instead, a standing pattern of Alfvén waves is excited which provides the electromagnetic coupling and carries the field aligned currents (Neubauer, 1980; Goertz, 1980; Gurnett and Goertz, 1981).

It is not intended to enter into a discussion of the merits of the many solar flare models. For an excellent recent review, see the articles edited by Priest (1981). There is general agreement that the flare energy is derived from largely force free current carrying magnetic field configurations. Coronal currents are related to photospheric motions that twist the flux tubes (e.g. Heyvaerts, 1974). Whether this process is directly reponsible for the sudden energy release (Piddington, 1974) or whether this energy release is related to newly emerging flux that triggers impulsive reconnection in narrow current sheets (Heyvaerts et al., 1977) does not matter for our purpose. In both cases, the sudden changes that occur in the configuration must be communicated between photosphere and low corona by shear Alfvén waves propagating along the flux tubes. In the first type of model, Alfvén waves play a fundamental role, but they are also expected to be excited by the sudden expansion of the current sheet in the second type of model and perhaps trigger an additional release of energy in other parts of the system . Direct disruption of field aligned currents as envisioned by Alfvén and Carlquist (1967) also would produce Alfvén waves. Finally, pulsars are assumed to be rotating neutron stars which provide us with a very different range of physical parameters and fascinating new phenomena. Field aligned currents and two stream instabilities are also thought to be of importance to the electrodynamics and the radiation from these objects (Sturrock, 1971).

A MODEL FOR ELECTRODYNAMIC COUPLING IN EXTENDED FIELD ALIGNED CURRENT SYSTEMS

Our model for electromagnetic coupling (Lysak and Dum, 1983) derives from the extensive observations made along auroral magnetic field lines and the approach to anomalous transport outlined here and in a previous paper (Dum, 1981). In its present simple form it is, however, designed mainly for an answer to some fundamental questions concerning current driven turbulence in extended system. The aim is to see how energy is supplied from a distant generator to a localized dissipation region and how this dissipation in turn affects the global electromagnetic structure. As demonstrated above, these questions are not only of importance to the global aspects in a number of extended current systems, but also are a neccessary input for a study of the microscopic processes connected with field aligned currents.

The information on transient phenomena in an extended current system such as the earth's magnetosphere-ionosphere must be carried by low frequency hydromagnetic waves. Shear Alfvén waves are the only credible candidate for this purpose. Even for wave vectors with large angles to be magnetic field, the group velocity for energy transport is essentially along the magnetic field. These oblique wave modes carry the field aligned current. It is agreed that Pi 2 pulsations with periods between 40 and 150 sec are transient Alfvén waves that are directly associated with the magnetospheric substorm onset and auroras as they travel along magnetic flux tubes (Southwood and Stuart, 1979; Baumjohann and Glassmeier, 1983). A number of models for electromagnetic magnetosphere-ionosphere coupling have seen constructed on this basis (e.g. Mallinckrodt and Carlson, 1978;

Nishida 1979). More generally, any disturbance along the field lines, wether it is the release of an artificial ion cloud (Scholer, 1970), modifications of ionospheric conductivity (Maltsev et al., 1977; Sato, 1982) or sudden changes of resistivity along the field line due to some anomalous process (Lysak and Dum 1983; Arykov and Maltsev, 1983) must create shear Alfvén waves, bouncing between the origin of the disturbance and the northern and southern ionospheres. For extended structures, the electromagnetic field is across the magnetic field, according to ideal MHD. It has been pointed out by Fejer and Kan (1969) that highly oblique waves must also produce a field aligned electric field, that is related to kinetic effects such as inertia and Landau damping (see Hasegawa and Uberoi, 1982 for a recent review). These electric fields may be of some importance in the narrow structures corresponding to auroras (Goertz and Boswell, 1979).

To produce greatly enhanced parallel electric fields in localized acceleration regions, as required by observation, other effects which are connected with an instability of the field aligned current must be considered, however. We consider electrostatic ion cyclotron waves which, as has been shown by many in situ observations, are driven to large turbulence levels in these regions. Very similar fluctuation levels are observed in ingeniuous laboratory experiments (Corell et al., 1975; Böhmer and Fornaca, 1979) in, even for a parent, amazingly detailed agreement with nonlinear theory (Dum and Dupree, 1970). For the auroral acceleration zone, boundary conditions for particles and the current are likely different in some respects from the laboratory experiments and thus should be studied in more detail.

For definiteness we choose an anomalous resistivity law derived from the nonlinear theory of ion cyclotron waves (Dum, 1981). However, any other threshold dependent sink for the Alfvén wave energy, e.g. double layers, would modify the propagation of Alfvén waves in a similar manner. This point, in agreement with our discussion of marginal stability, can be demonstrated by a series of numerical experiments in which the amplitude, wave form and perpendicular wavelength of the current generator and the parameters of the dissipation region, including the anomalous resistivity law are varied (Christiansen et al., 1982). Due to self regulatory macroscopic effects, total voltage drops and dissipation become remarkably insensitive so some of these details. The basic effect on the global electrodynamic structure is a partial decoupling of the generator from the ionospheric load with multiple wave reflections from the dissipation region and the ionosphere (Lysak and Dum, 1982). In the absence of anomalous dissipation, ionospheric reflection would be the only process that eventually can establish a steady state electrostatic potential structure. Many bounces of the waves between the generator and the ionospheres would be required, however, owing to large ionospheric conductivity. In the presence of an anomalous dissipation region this process takes place much faster. Nearly electrostatic potential structures may be established for field lines in contact with the enhanced dissipation region. Details depend, however, also on boundary conditions at the generator. The model has been extended to include a moving generator or equivalently plasma convection normal to the magnetic field lines. Feedback to the generator is naturally much weaker in this case. The electric field structure stays, however, essentially electromagnetic, with large enhancements of the parallel electric field in the localized dissipation region. The interference pattern produced by a moving generator and multiple wave bounces between the dissipation region and the ionosphere corresponds to the narrow structures associated with discrete auroral arcs, in an Alfvén wave model of auroral arc formation by Haerendel (1982). Its key features can be examined with our numerical code. We have extended the uniform magnetic field model to include dipole-like ambient magnetic fields and cross tail currents near open field lines. The code thus contains the mapping of steady state structures along converging magnetic flux tubes. The variation of Alfvén speed and critical drift velocity for instability with height arises now naturally from the ambient non-uniform density and magnetic field profiles.

The model still leaves room for many improvements. The inclusion of heating in the dissipation region should lead to a self-consistent evolution of the critical drift velocity for instability. This is likely to cause repeated excitation and self-quenching of current driven instabilities over small temporal or spatial periods, if heat conduction is also included. For larger ratios of magnetic to plasma pressure, pressure effects should also be included for wave propagation. The need for an improved treatment of the generator is indicated by the observed feedback from the dissipation region. Details depend, however, on the specific system to be considered.

MICROSCOPIC PROCESSES AND CONCLUDING REMARKS

Microscopic processes should be discussed in relation to the global problem of field aligned currents. Fairly recent reviews of microscopic processes supporting anomalous resistivity have been given by Papadopoulos (1977) and Dum (1981). Heyvaerts (1981) has given an extensive review of microscopic processes related to particle acceleration in solar flares. We thus confine ourselves to mentioning a few very recent developments and concluding remarks relating to the topics that have been discussed in this paper.

We discussed the propagation of electromagnetic signals and currents along magnetic flux tubes. Propagation by hydromagnetic shear Alfvén waves bears great resemblance to wave propagation on metallic transmission lines, the reflection processes included. It is also assumed that the speed of the current carrying particles is far smaller than the phase speed of wave propagation. For a copper wire this is trivial, but for the plasma it means restriction to electron thermal velocities below the Alfvén speed, i.e. ratios of plasma pressures to magnetic pressure, smaller than the electron-ion mass ratio. This assumption can easily be relaxed for kinetic Alfvén waves, with minor effects on wave propagation. For the localized dissipation region with turbulence driven by field aligned currents, the particle populations supporting this current and the turbulence must be examined in detail, however.

It appears unlikely that the plasma in the dissipation region, which for the earth's auroral field lines is at roughly 6000 km altitude retains a perfect memory of boundary conditions at distant sources of particle injection, as is implied by laminar double layer models for potential drops. For computer simulations and laboratory experiments where boundaries are much closer and can be carefully tailored this is much more likely. These structures are nevertheless, even in these cases, associated with intense small scale turbulence. This is in sharp contrast to the purely laminar solutions of the Vlasov equation that are usually considered for an explanation (see e.g. the experiments of Guyot and Hollenstein, 1983; Hollenstein and Guyot, 1983). The problem of particle supply nevertheless is important in connection with low frequency current driven instabilities in a magnetized plasma, such as ion cyclotron waves. Conventional quasilinear theory would predict very rapid stabilization by plateau formation in the low energy electron distribution. This is confirmed by particle simulations in a homogeneous plasma. If fresh particles are injected at the boundary this process still occurs, but subsequent quenching of waves by interaction with the ions in the heated regions allows a step by step spread of the dissipation zone (Okuda and Ashour-Abdalla, 1983). Mechanisms for the spreading of microscopic dissipation processes in solar flares are reviewed by Heyvaerts (1981).

A conflict between laminar and turbulent theories for electric fields associated with field aligned currents arises only, if by analogy with a metallic conductor, one assumes enhanced steady state resistivity for current driven instabilities. The large drifts required for such instabilities or double layer mechanisms represent a very strong perturbation of the system. In contrast, perturbations introduced even by sizable currents in metallic conductors are quite negligible. This fact and high electron mobility along the magnetic field make an intermittent process with periods of nearly free acceleration much more likely for low density plasmas (Dum, 1981).

We should still discuss macroscopic effects of turbulence on the current carrying plasma column such as macroscopic stability, heat flow and filamentation of current channels, in more detail. Instead, we like to end on a note of curiosity. The pinching process in a tiny laboratory plasma (Bykowskii and Lagoda, 1982) is explained by the merging of current filaments with oppositely directed axial magnetic fields, which bears a striking resemblance to the solar flare model proposed by Gold and Hoyle (1960). In the region of reconnection intense current driven ion sound turbulence is assumed to build up. Changes in the current, intense ion fluxes and microwave radiation give evidence for this process.

REFERENCES

Akasofu, S.-I.: 1977, Physics of magnetospheric substorms, Reidel, Dordrecht.

Alfvén, H., Carlquist, P.: 1967, Sol. Phys., 1, 220.

Arykov, A.A., Maltsev, Yu.P.: 1983, Planet. Space Sci. 31, 267.

Barnes, C.W., Sturrock, P.A.: 1972, Astrophys. J. 174, 659.

Baumjohann, W., Glaßmeier, K.H.: 1983, subm. Planet. Space Sci.

Böhmer, H., Fornaca, S.: 1979, J. Geophys. Res. 84, 7239.

Braginskii, S.I.: 1967, in M.A. Leontovich (ed.), Reviews of Plasma Physics, Consultants Bureau, New York, Vol. I., p. 205.

Braginskii, S.I.: 1967, in M.A. Leontovich (ed.), Reviews of Plasma Physics, Consultants Bureau, New York, Vol.I, p. 205.

Burke, W.J., Hardy, D.A., Rich, F.J., Kelley, M.C., Smiddy, M., Shuman, J.: 1980, Geophys. Res. 85, 1179.

Bykowskii, Yu.A., Lagoda, V.B.: 1982, S. Phys. JETP 56, 61 (Zh. Eksp.Teor. Fiz. 83, 114).

Cattell, C.: 1981, J. Geophys. Res. 86, 3641.

Chiu, Y.T.: Schulz, M.: 1978, J. Geosphys. Res. 83, 629.

Christiansen, P.J., Dum, C.T., Lysak, R.L.: 1982, to be publ.

Correll, D.L., Rynn., Böhmer, H.: 1975, Phys. Fluids 18, 1800.

Dum, C.T.: 1978a, Phys. Fluids 21, 945.

Dum, C.T.: 1978b, Phys. Fluids 21, 956.

Dum, C.T.: 1981, in S.I. Akasofu and J.R. Kan (ed.), Physics of Auroral Arc Formation, Geophys. Monograph Ser., AGU, Washington, p. 408.

Dum, C.T., Chodura, R.: 1979, in P.J. Palmadesso and K. Papadopoulos (ed.), Wave instabilities in Space Plasmas, Reidel, Dordrecht.

Dum, C.T., Dupree, T.H.: 1970, Phys. Fluids 13, 2064.

Dum, C.T., Chodura, R., Biskamp, D.: 1974, Phys. Rev. Letters 32, 1231.

Eselevich, V.G., Eskov, A.G., Kurtmullaev, R.Kh., Malyutin, A.I.: 1971, Sov. Phys. JETP, 33, 898 (Zh. Eksp. Theor. Fiz. 60, 1658, 1971).

Fejer, J.A., Kan, J.R.: 1969, J. Plasma Phys. 3, 331.

Goertz, C.K.: 1980, J. Geophys. Res. 85, 2949.

Goertz, C.K., Boswell, R.W.: 1979, J. Geophys. Res. 84, 7239.

Gold, T., Hoyle, F.: 1960, Mon. Not. Roy. Astr. Soc. 120, 89.

Gurnett, D.A., Anderson R.R.: 1981, in S.-I. Akasofu, J.R. Kan (ed.) Physics of Auroral Arc Formation, Geophys. Monograph Ser., AGU, Washington, p. 341.

Gurnett, D.A., Goertz, C.K.: 1981, J. Geophys. Res. 86, 717.

Guyot, M., Hollenstein, Ch.: 1983, Phys. Fluids 26, 1596.

Haerendel, G.: 1983, in B. Hultquist, T. Hagfors (ed.), High Latitude Space Plasma Physics, Plenum, London.

Hasegawa, A., Uberoi, C.: 1982, The Alfvén Wave, Tech. Inform. Center, U.S. Dep. of Energy.

Heyvaerts, J.: 1974, Sol. Phys. 38, 419

Heyvaerts, J.: 1981, in E.R. Priest (ed.), Solar Flare Magnetohydrodynamics, Gordon + Breach, New York, p. 429

Heyvaerts, J., Priest, E.R., Rust, D.M.: 1977, Astrophys. J. 216, 123.

Hollenstein, Ch., Guyot, M: 1983, Phys. Fluids 26, 1606.

Kan, J.R.: 1982, Space Sci. Rev. 31, 71.

Lee, L.C., Kan, J.R., Wu, C.S.: 1980, Planet. Space Sci. 28, 703.

Lysak, R.L., Dum, C.T.: 1983, J. Geophys. Res. 88, 365.

Mallinckrodt, A.J., Carlson, C.W.: 1978, J. Geophys. Res. 83, p. 1426.

Maltsev, Yu.P., Lyatsky, W.B., Lyatskaya, A.M.: 1977, Planet. Spcae Sci. 25, 53.

Manheimer, W., Boris, J.P.: 1977, Comments Plasma Phys. Cont. Fusion 3, 15.

Mozer, F.S., Catell, C.A., Hudson, M.K., Lysak, R., Termin, N.N., Torbert, R.B.: 1981, Space Sci. Rev. 27, 155.

Neubauer, F.M.: 1980, J. Geophys. Res. 85, 1171.

Nishida, A.: 1979, J. Geophys. Rs. 84, 3409.

Papadopoulos, K.: 1977, Rev. Geophys. Space Phys. 15, 113.

Piddington, J.H.: 1974, Sol. Phys. 38, 465.

Priest, E.R. (ed.): 1981, Solar Flare Magnetohydrodynamics, Gordon + Breach, New York.

Okuda, H., Ashour-Abdalla, M.: 1983, J. Geophys. Res. 88, 899.

Sato, T.: 1982 in A. Nishida (ed.) Magnetospheric Physics, Reidel Dordrecht, p. 197.

Sato, T., Hayashi, T.: 1979, Phys. Fluids 22, 1189.

Scholer, M.: 1970, Plan. Space Sci. 18, 977.

Söldner, F., Dum, C.T., Steuer, K.H.: 1977, Phys. Rev. Letters 39, 194.

Southwood, D.J., Stuart, W.F.: 1979, in S.I. Akasofu (ed) Dynamics of the Magnetosphere, Reidel, Dordrecht, p. 341.

Sturrock, P.A.: 1971, Astrophys. J. 164, 529.

Ugai, M.: 1982, Phys. Fluids 25, 1027.

DISCUSSION

Kennel: Could you tell us what physical processes set the transverse (perpendicular to B) scale of the electron acceleration region in your Alfvén wave reflection model?
Dum: Primarily, the scale is set by the generator. The large effective collision frequency in the acceleration region leads to diffusion and, thus, broadening of this region. Also, reflection from the acceleration region and the ionosphere and the resulting interference between wave trains affect the complex pattern in this region. Reflection from the acceleration region is strongest for small scales and large current densities.
Kundu: I just would like to know more details about the experiment that was made to simulate the flare model of Gold and Hoyle. Specifically, I'd like to know if they should observe radiation at microwaves alone - why not over a much broader range of wavelengths? In our VLA observations, we detected a quadrupole structure only because we were observing at 6 cm.
Dum: No details on the radiation spectrum are given. Comparing with other experiments, I assume the spectrum was fairly broad. The experiment was designed to study filamentation for intense currents. The explanation for the collapse of filaments (pinching) apparently is independent of Gold and Hoyle's model, but bears a striking resemblence.

SELF-CONFINED COSMIC RAYS

Donat G. Wentzel
Astronomy Program, University of Maryland

Cosmic rays do not stream freely through the galaxy, contrary to earlier expectations. Streaming cosmic rays are slowed down by the emission of resonant Alfven waves that scatter the cosmic rays. The theory of self-confinement explains the isotropy of the bulk of the cosmic rays but not of cosmic rays above 10^3 Gev; it has been a stimulus to the theory for cosmic-ray acceleration at supernova shocks; and, on inclusion of diffusion in a galactic wind, it may explain the uniform cosmic-ray density out to 18 kpc in our galaxy. Rapidly streaming electrons in clusters of galaxies, in supernova remnants, and near solar flares are accomodated by the theory when it is expanded to include the effects of hot plasmas and other wave modes. A "resonance gap" may prevent the turning backwards of streaming particles and thus allow streaming near the particle speed.

INTRODUCTION

We have become used to measuring electric currents in the magnetosphere of the Earth, and even elsewhere in the solar system. However, the only electrical current that we can measure from beyond the solar system is a current of cosmic rays. All other astrophysical currents are merely inferred indirectly from observations of electromagnetic radiation.

Why should one expect cosmic rays to stream past the Earth? By about 1950 astronomers were fairly confident that the magnetic field in interstellar space is somewhere between 1 and 10 microgauss. In such a magnetic field, a typical cosmic ray at an energy of a few Gev has a gyroradius of the order of 10^7 km. This scale is some five orders of magnitude smaller than any interstellar structure that could then be observed. One deduced that cosmic rays should have constant magnetic moments as they propagate through the galaxy. They should be tied to the field lines and propagate along the magnetic field. If they are accelerated at supernovae, then the cosmic rays should be streaming away from these sources at roughly the speed of light. The cosmic rays would not actually provide a net electrical current, because a return current is very easily carried by thermal electrons. Nevertheless, the cosmic rays constitute a large fraction of the interstellar energy density, and therefore their rapid streaming would have been very important.

COSMIC-RAY ISOTROPY

The observations indicate that most cosmic rays are nearly isotropic. They stream past us at mean velocities less than a few 10^2 km/sec. For a few years, one could explain the isotropy by invoking cosmic rays permanently trapped on closed magnetic field lines, so that cosmic rays would reach us equally from all directions. Since then, the age of the cosmic rays has been measured. Most cosmic rays are merely some 10^7 years old (Stephens 1981). Therefore, there must be field lines along which cosmic rays can escape from the disk of our galaxy. We must conclude that some phenomenon makes the cosmic rays escape slowly.

Today we recognize that cosmic rays are scattered by Alfven waves with wavelengths comparable to the cosmic-ray gyroradii. The fundamentally new aspect recognized in the 1960's is that these Alfven waves can be generated by the cosmic rays themselves. The strongest original impact on cosmic-ray physics was a paper by Kennel and Petschek (1966) on the scattering of protons and electrons in the van Allen belts. In the same year, Tidman (1966) pointed out that suitable Alfven waves could scatter cosmic rays and influence their streaming. Lerche (1967) showed that streaming cosmic rays would generate resonant Alfven waves in a very short time. These ideas were then combined in papers by Kulsrud and Pearce (1969) and by myself (Wentzel 1969). We estimated the damping of the Alfven waves and we asked how rapidly the cosmic rays stream if the cosmic rays generate the waves as rapidly as the waves are damped. We found that the bulk of the cosmic rays can stream only barely more rapidly than the Alfven speed in the interstellar medium. This speed is between 10 and 10^2 km/sec in much of interstellar space, and it matches satisfactorily the slowest streaming rate that has been measured for the cosmic rays. Some places in the galaxy might permit much faster streaming, but the migration of cosmic rays through the galaxy is totally controlled by those places with the slowest streaming speed. The interpretation of this work is that cosmic rays cannot stream freely out of our galaxy but they are essentially self-confined to the interstellar medium. (See review by Wentzel 1974).

Have these ideas withstood the test of time? In the following I wish to outline two applications in which self-confinement is at least qualitatively important, namely the acceleration of cosmic rays and their dispersal in a galactic wind. Then I wish to stress a failure, in that observations of high-energy cosmic rays require <u>more</u> scattering than theory provides. Finally, I review the observations which require much <u>less</u> scattering than the standard theory provides, and the extensions of the theory that accomodate the observations.

ACCELERATION OF COSMIC RAYS

One success relates to the acceleration of cosmic rays. Supernovae are the most likely source of energy for the cosmic rays. One needs roughly 1% of supernova energy transferred to cosmic rays. Until a few years ago, 1% seemed like a rather large efficiency, theoretically. Yet the observers discovered ever more places in the universe where the acceleration efficiency seemed to be well above 1%. To my knowledge, the solution was first indicated by Schatzman (1963). He showed that particles could be accelerated very efficiently at hydromagnetic shocks if these particles could be made to cross the shock a large number of times. Bell (1977, 1978) pointed out that the cosmic rays near shocks would create Alfven waves that would scatter the cosmic rays back into the shock. The result is efficient acceleration. Blandford and Ostriker (1978, 1980) worked out enough details to show that this scheme is basically correct.

The cause of acceleration can be seen rather simply. Imagine a shock that is much faster than the Alfven speed. Therefore, any Alfven waves that exist are effectively moving with the gas. The increase in density across the shock requires a convergence in the velocity field, and therefore a convergence of the Alfven waves. If these waves scatter cosmic rays, then the cosmic rays find themselves in a converging set of scattering centers. Cosmic rays reflected from these centers many times experience what is called first-order Fermi acceleration. Moreover, the accelerated cosmic rays tend to diffuse away from the shock and, therefore, they will actually produce the Alfven waves that are needed.

The rate at which any one cosmic ray is accelerated depends on the frequency with which it crosses the shock; that is, the acceleration rate is proportional to the scattering rate caused by the Alfven waves. The duration of acceleration is limited by the distance beyond the shock from which the cosmic ray can still be scattered back to the shock; this duration depends inversely on the scattering rate. The product is significant acceleration largely independent of the scattering rate. Supernova shocks can easily replenish the cosmic rays that leave the galaxy. The concept of self-confined cosmic rays has increased the theoretical efficiency of acceleration from something like 1% to the order of unity. One presumes that this kind of efficiency also works for other astrophysical energetic phenomena. In fact, one sometimes hears the view that the acceleration problem is solved. I think such optimism is not yet justified (cf. Casse 1981, Volk 1981).

In the case of the cosmic rays, we have many observations concerning energy spectra, anisotropy, and the composition and ages of secondaries at various energies. Any detailed theory of cosmic-ray acceleration can be severely tested against these observations.

The theory cannot yet be this precise because of primarily three rather basic problems.

First, the cosmic rays are compressed and accelerated in the shock, but later they participate in the re-expansion of the gas and then they are adiabatically decelerated. They lose much of the energy gained by acceleration, but not all. The difference between gain and loss depends significantly on the shock strength, on the dynamics of the gas expansion, and on the relatively reduced scattering efficiency in the surrounding plasma (see discussion of the resonance gap below). Therefore, the net energy gain from a supernova is still quite uncertain (Blandford and Ostriker 1980).

Second, the accelerated cosmic rays must themselves influence the shock structure. One must therefore solve a highly nonlinear problem that involves not only shock dynamics but also the details of the acceleration process and the (possibly nonlinear) waves that are consistent with it. We are still far from such a solution (see Axford 1981, McKenzie and Volk 1982).

Third, we cannot measure interstellar cosmic rays below roughly 100 Mev, and interstellar electrons only indirectly down to 10 Mev (Bertsch and Kniffen 1983). We must depend largely on theory to explain how thermal particles can be selected to become cosmic rays. The observations of cosmic ray abundances and isotopes have not restricted these theories very much. (See Casse 1981). It is possible that cosmic rays come from ambient thermal interstellar material overrun by the supernova shock. Eichler (1980) has emphasized that a small number of particles must be accelerated out of the thermal distribution; the mere re-acceleration of existing cosmic rays would result in a spectrum of secondaries different from that observed. Eichler (1979) suggested how particles in the ambient gas may be selected and accelerated self-consistently with their resonant waves.

Despite all these difficulties, the theory is probably on the right track. When the solar wind impinges on the Earth's bowshock with the magnetic field sufficiently parallel to the wind velocity, then observations can be compared in detail with the theory for acceleration (Eichler 1981, Lee 1982). Observed details concerning waves, particles anistropy, and dependence on magnetic direction are reproduced by theory with essentially only one free parameter, namely the number of particles accelerated. The maximum energies of 100 Kev obtained in the bowshock are due to its finite extent. Much higher energies can be obtained in the much larger astrophysical shocks.

DISTRIBUTION OF COSMIC RAYS FAR FROM THE SUN

A very recent observation provides a new challenge to the theory for self-confined cosmic rays. The observation of gamma rays can tell us the density of cosmic rays in distant portions of the galactic disk. If you believe in self-confined cosmic rays, which cannot migrate very far in the galactic disk, then you expect a higher density of cosmic rays nearer the galactic center from the Sun, because supernovae are more frequent there and the cosmic rays are more easily confined to the galactic disk. You would expect a rather low density of cosmic rays well beyond the Sun's position in our galaxy. My colleague Leo Blitz and his collaborators at Leiden have recently compared the gamma-ray observations with maps of neutral hydrogen observed in 21-cm radiation. They find that the density of cosmic rays is remarkably uniform as far as 18 kpc from the center. Beyond that distance there is not enough gas to analyze the resulting gamma rays. Why should the density be so uniform where there is relatively little gas to confine cosmic rays and where there are rather few supernovae to make them? One might argue about details such as more efficient acceleration and fewer collisional losses where there is less gas, but the many possible factors are not likely to result in a uniform density.

Probably one has to relax self-confinement and incorporate it in a theory for the convection of cosmic rays. This can be done quite naturally for a galactic wind. Perhaps the wind is driven by the self-confined cosmic rays (Ipavich 1975), perhaps by the dynamics due to supernovae (galactic fountains). Owens and Jokipii (1977, see also Jokipii and Higdon 1979) have argued that a wind naturally explains why cosmic-rays are about ten times older than the time cosmic rays spend in the galactic disk. If a wind convects the cosmic rays away from the disk, individual cosmic rays may well be scattered back into the disk, but only if they have not been convected too far away. The maximum distance from which they can return determines the age of the oldest cosmic rays and also the mean age we observe at Earth. Owens and Jokipii (1977) modeled a wind moving perpendicular to the disk. At the other extreme, Ipavich (1975) and Axford (1981, Fig. 3) modeled a spherically symmetric

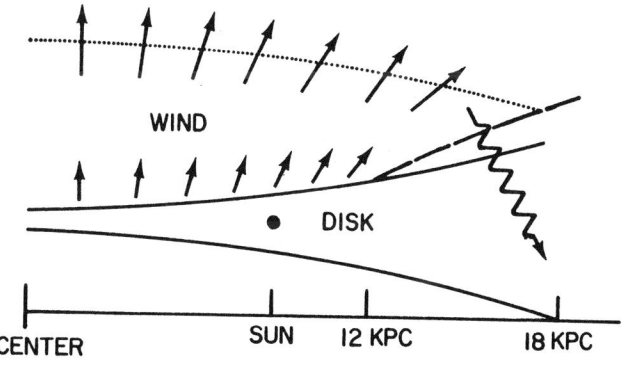

Figure 1. Cosmic rays may diffuse from the galactic wind back into the outer portions of the galactic disk.

wind. The reality is likely to be in between. In Figure 1, I assume a wind is driven from those portions of the disk where supernovae are frequent, say out to 12 kpc. Part of this wind becomes nearly radial. Because of this wind, cosmic rays can reach distant parts of the galaxy, even those rather close to the galactic disk. If one selects a fairly large but still reasonable diffusion coefficient, the cosmic rays are diffusion dominated out to the dotted line and they have a nearly uniform density out to there. If one now postulates at least a weak magnetic connection between the wind and the distant disk, perhaps caused by the equalizing of angular momenta in disk and wind, then cosmic rays can easily diffuse from the wind into the distant disk. The density of cosmic rays in the distant disk will simply mimic the density in the wind. If the density in the wind is nearly uniform, so is the density in the disk, as observed. The modeling for this kind of wind is only beginning. It presents a challenge to the concept of self-confined cosmic rays.

FAILURE: ISOTROPY OF COSMIC RAYS ABOVE 10^3 GEV

One major problem with cosmic-ray self-confinement is its energy dependence. Self-confinement is based on a balance between wave damping in the ambient medium and wave growth due to cosmic-ray streaming. The Alfven resonance requires that the Alfven wavelengths are similar to the gyroradii. The more energetic cosmic rays make longer wavelengths. The growth rate is proportional to $n(c.r.) \langle v \rangle$ where $n(c.r.)$ is the integral number of cosmic rays able to resonate with the wavelength of interest. The damping by almost any nonlinear mechanism is proportional to the wave amplitude squared, thus to the scattering rate and to $1/\langle v \rangle$. (An efficient degradation of Alfven waves applicable to the 10^6K gas in the galactic disk and halo is derived by Shukla and Dawson 1983). In balance, $\langle v \rangle$ increases with energy roughly in proportion to $n(c.r.)^{-1/2}$. Cosmic rays above 10^3 Gev would stream faster than observations allow.

Resonant waves can grow appreciably faster if they grow by a coherent process at a shock front (Wentzel 1977). One may imagine that a shock surrounds the corona of our galaxy and that the cosmic rays are scattered at this shock. The isotropy near Earth might then merely be due to our location in the plane of symmetry of such a shock. Such a model would imply a different energy dependence in the escape rate of cosmic rays below and above roughly 10^3 Gev, but no model exists to evaluate this difference.

Two external sources of scattering have been suggested. First, there might be a cascade of eddies from large-scale interstellar turbulence down to the required small scales. This is improbable in detail (Cesarsky 1980). Second, Hall (1980, 1981) has argued that interstellar supersonic turbulence causes pressure anisotropies, and these anisotropies cause the firehose and mirror hydromagnetic instabilities. These instabilities are quite attractive for explaining the pulsar scintillations, because the scintillations are due to <u>density</u> fluctuations and these are <u>not</u> due to ordinary Alfven waves. However, the fluctuations have very short wavelength. The bulk of the cosmic rays must resonate with the magnetic fluctuations at the hundredth gyroharmonic, and the energetic cosmic rays at the millionth harmonic. Whether these fluctuations can confine cosmic rays is still somewhat uncertain.

Altogether then there is a fundamental problem with high-energy cosmic rays: no existing theory adequately explains their confinement to the galaxy.

OBSERVATIONS OF RAPIDLY STREAMING PARTICLES

Non-relativistic particles from solar flares show that the scattering theory I have outlined is substantially incomplete. One knows the magnetic fluctuations and Alfven waves in the solar wind, and therefore one can compute a scattering mean free path according to scattering theory (Jokipii 1971). This mean free path should become very small at low energies. Observations of flare protons show that the mean free path decreases to about 0.3 A.U. and stays at that value for non-relativistic protons (see Figure 1 in Goldstein 1980). A similar behavior occurs for flare electrons (Ma Sung and Earl 1978). In addition, one observes scatter-free events for non-relativistic particles from flares (McDonald, Fichtel and Fisk 1974). Clearly these observations indicate some rather fundamental omission in the scattering theory.

There also exist disquieting observations in astrophysics. The Coma cluster of galaxies is one of the few clusters that has a diffuse radio emission from the entire cluster. We know that the radiating electrons lose energy to inverse-Compton interactions on the cosmic background radiation. That gives us a maximum lifetime for the electrons of about 10^8 years. If we now suppose that the electrons are accelerated at the active galaxies near the center of the cluster and also that they travel to all parts of the cluster within their maximum lifetime, then we find that these electrons must stream at about 10^4 km/sec. However, if these electrons were self-confined to the Alfven speed, they should stream only at about 10 km/sec. There is a clear contradiction. One possible resolution (Dennison 1980, Vestrand 1982) is that the electrons are secondaries produced locally throughout the cluster by collisions of primary protons. The protons have very long lifetimes. They can reach all parts of the cluster during their lifetimes even when streaming merely at the Alfven speed. The other likely resolution is to admit very rapid streaming of electrons. Then one must seek a change in scattering theory.

A third suggestion of very rapid streaming arises for all sources of cosmic rays. It has long been stressed that self-confined cosmic rays are adiabatically decelerated while they try to travel away from their sources (Wentzel 1973, Kulsrud and Zweibel 1975).

The more compact the source, the greater the adiabatic deceleration before cosmic rays reach normal interstellar space. For acceleration very near a supernova, cosmic rays escape with merely one per cent of their original energy. For acceleration at a shock surrounding a supernova remnant, they may escape with less than half the original energy. This problem has led to suggestions that self-confinement breaks down near supernova shocks, perhaps once the shock has slowed down sufficiently.

CORRECTIONS TO SIGNAL SPEED

The Alfven speed $v_A = B/(4\pi\rho)^{1/2}$ is appropriate for a cold and isotropic plasma with negligible numbers of cosmic rays. Particles can stream more rapidly if the signal speed is faster.

Hall (1981) emphasized the turbulence of the 10^6K interstellar gas. The turbulence leads to an anisotropic gas pressure. For one sense of the anisotropy, the signal speed might be as high as the sound speed. But for the other sense of the anisotropy, one obtains the firehose instability. As long as both anisotropies are equally probable, the cosmic rays are still scattered by the firehose instability and stream slowly.

Cosmic rays may also increase the signal speed. Since cosmic rays have high magnetic rigidity, they should impart extra rigidity to the medium, and this increases the signal speed. The increase can be significant where (cosmic-ray pressure)x(streaming speed/c) > $B^2/8\pi$ (Morrison 1979). This may well be important in the Coma cluster of galaxies where the field is relatively weak, especially if one assumes rapid streaming to begin with (see also Spangler 1979). For supernova remnants, the signal speed increases only by a modest factor (Axford 1981) unless the accumulation of self-confined cosmic rays is substantial. All that is needed is a signal speed that is faster than the speed of the supernova shock.

THE RESONANCE GAP

It has been known since the beginning of self-confinement theory, first, that the simple Alfven self-confinement theory breaks down when $v_\parallel < v_A$, and second, that it is crucial for standard diffusion theory that particles can cross the state $v_\parallel = 0$ (Wentzel 1969). However, it was normally assumed that this range in velocity, now called the resonance gap, was small and easily bridged. Holman, Ionson and Scott (1979) outlined the observations suggesting rapidly streaming particles and argued that the resonance gap is indeed important in these situations. Considerable work has followed from their paper. Therefore, I first review why this gap is so important. When Alfven waves cause diffusion in pitch-angle, an individual particle undergoes many <u>small</u> changes in its pitch-angle. It continues to move in the same direction along the magnetic field until many of these small changes cause the pitch-angle to change past 90° and the velocity along the field $v_\parallel$ reverses. We consider the particle to diffuse through space if it executes many such reversals in $v_\parallel$. Indeed, when we deduce a diffusion coefficient for particles observed in the solar wind, we compute the mean free path between reversals in $v_\parallel$ and we tend to ignore the many small changes in pitch-angle between the reversals. However, if for any reason the small pitch-angle changes cannot reach $v_\parallel = 0$, but they cease acting at some finite $v_\parallel = v_g$, then the particle continues to move in the same direction in which it started. Resonances can scatter the particles and make them nearly isotropic outside the gap, $v_\parallel > v_g$. If this gap is small, then the streaming speed is $0.5v$ (v = particle speed).

Nonresonant processes may "fill in" the resonance gap. The most widely considered process is mirroring from long-wavelength fluctuations in magnetic field strength. These may be caused by local gas dynamics or by the particle themselves. The minimum needed fluctuation is of order $\delta B/B \simeq (v_g/v)^2$. Alternatively (Fedorenko 1982), nonlinear processes may reverse the directions of the Alfven waves (cf. Shukla and Dawson 1983). Cosmic rays can be scattered by absorbing these re-directed Alfven waves. No resonance gap occurs. In either case, when the transfer of the particles through the resonance gap is slow, this process will determine the mean free path and the spatial diffusion rate. Meanwhile, resonant scattering may be quite rapid outside the gap. Therefore, the particle velocity distribution may frequently look as shown in Figure 2. In general, the mean free path is determined by the slower of the resonant scattering and the process causing particles to cross the resonance gap.

One must, therefore, ask two questions: how large is the resonance gap, and can any other phenomenon bridge the gap?

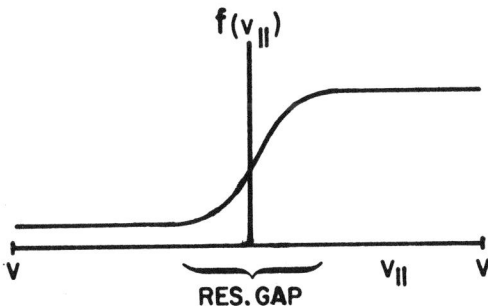

Figure 2. Velocity distribution when the transfer of particles through the resonance gap is slow compared to resonant scattering outside the gap.

We have the best chance to answer this question for solar particles in the solar wind. Goldstein (1980) suggested that the measured mean free path could be explained if the mean free path is controlled by mirroring of particles. Davila (1982) has computed the situation in considerable detail. For electrons, he finds that the gap has a width (γ = relativity factor) $v_g = 1.2$ c/γ. Electrons below 300 Kev have $v < v_g$. They are not resonant with any waves in the wind, and they form the scatter-free events. Electrons above about 2 Mev have a small enough v_g such that the known field strength fluctuations in the wind can mirror the electrons. The mean free path is then just the wavelength of the waves carrying the field fluctuations, $\lambda \simeq 0.04$ A.U. This estimate fits the observations reasonably well. At still higher energies, mirroring is more rapid than resonance scattering. Then the standard resonance scattering theory again applies.

For protons, Davila finds $v_g = 2.5$ v_A. The scattering time is λ/v_A and the mean free path is $\lambda(v_w/v_A) \simeq 0.3$ A.U., again in reasonable agreement with observations.

In many cosmic objects, the gas pressure exceeds the magnetic pressure (high-β plasma) and the ion thermal speed v_{Ti} exceeds the Alfven speed. Holman, Ionson and Scott (1979) pointed out that the resonance gap then is comparable to v_{Ti} and that such a large gap could allow streaming speeds much faster than v_A. Basically, the gap occurs because particles with $p_\parallel < m_i v_{Ti}$ could resonate only with waves that are highly damped and are effectively absent. The computation of the resonance gap is greatly simplified by the fact that all waves not traveling along the magnetic field are heavily damped; one need consider only waves along the field (Foote and Kulsrud 1979). Davila (1982) considered the gap in detail and found for supernova conditions

$$v_g(\text{protons}) = v_{Ti} (0.16 + 12/\gamma),$$

$$v_g(\text{electrons}) = v_{Ti} (0.06 + 8 \times 10^3/\gamma), \gamma < 10^5.$$

The first term in each equation is appropriate for high-β plasma and is roughly independent of $\beta \gtrsim 10$; the second term depends only logarithmically on the density of the fast particles, as long as these are not so numerous as to affect the dispersion relation.

Davila concludes that electrons with $\gamma > 10^5$ are self-confined. The confinement of the other particles depends on how they might cross the gap. In the surroundings of a supernova shock, fluctuations in field strength of order $(v_{Ti}/c)^2 \simeq 10^{-6}$ seem quite probable, but it is unknown whether they are sufficiently numerous to mirror the particles as many times as is needed for acceleration. Achterberg (1981) estimated the field fluctuations due to streaming particles when the wave energy generated by the particles is limited by nonlinear damping. Since all waves are nearly transverse, the fluctuation in field strength is proportional to the square of the wave amplitude. Achterberg found that electrons with $\gamma < 200$ would be able to escape supernova remnants, and that most electrons and non-relativistic protons would also be able to stream through and escape from any surrounding hot plasma. It is possible, however, that confinement is better if field fluctuations are higher than computed. We should remember that the field fluctuations in the solar wind are substantially larger than had been anticipated.

In any case, when particles are accelerated at a supernova shock and are once self-confined, their acceleration makes them even more thoroughly self-confined (until their gyroradius becomes too large). This is rather contrary to the hopes of people worrying about adiabatic deceleration. It appears that particle escape will occur only when the shock slows down sufficiently.

The suggestion by Holman, Ionson and Scott (1979) is attractive for clusters of galaxies because electrons streaming at a few times v_{Ti} could indeed cross the radius of the Coma cluster within their lifetimes. However, the actual streaming speed again depends on the processes that allow the electrons to turn past $v_{\parallel} = 0$. If the medium is very quiescent, then the electrons stream at $0.5c$.

If there are merely a few field fluctuations with amplitudes of at least $(v_{Ti}/c)^2 \simeq 10^{-5}$, they already slow the streaming speed down to the minimum needed for the electrons to traverse the cluster. The separations between mirrors can be no less than about 3% of the cluster radius. The galaxies moving within the cluster may well produce turbulence on such a scale.

SUMMARY

The original theory of self-confinement by Alfven waves has had to be extended to include convection, more details on the ambient plasma, and more details concerning the resonant waves, but basically the theory has fared quite well. It appears that electrical currents due to cosmic rays may alter the medium at supernova shocks, but even this is not established yet.

I acknowledge helpful discussions with D. Eichler, G. Holman and J. Davila.

REFERENCES

Achterberg, A.: 1981, Astron. Astrophys. 98, 161.
Axford, W. I.: 1981, 17th Internat'l Conf. Cosmic Rays (Paris) 12, 155.
Bell, A. R.: 1978, Mon. Not. Roy. Astr. Soc. 182, 147.
Bertsch, D. L. and Kniffen, D. A.: 1983, Astrophys. J. 270, 305.
Blandford, R. D. and Ostriker, J. P.: 1978, Astrophys. J. 221, L29.

Blandford, R. D. and Ostriker, J. P.: 1980, Astrophys. J. 237, 793.
Casse, M.: 1981, 17th Internat'l Conf. Cosmic Rays (Paris) 13, 111.
Cesarsky, C. J.: 1980, Ann. Rev. Astr. Astrophys. 18, 289.
Davila, J. M.: 1982, Ph.D. thesis, University of Arizona.
Dennison, B.: 1980, Astrophys. J. (Letters) 235, L92.
Eichler, D.: 1979, Astrophys. J. 229, 419.
Eichler, D.: 1980, Astrophys. J., 237, 809.
Eichler, D.: 1981, Astrophys. J. 244, 711.
Fedorenko, V. N.: 1982, preprint, A. F. Ioffe Physical Technical Institute, Leningrad.
Foote, E. A. and Kulsrud, R. M.: 1979, Astrophys. J. 233, 302.
Goldstein, M. L.: 1980, J. Geophys. Res. 85, 3033.
Hall, A. N.: 1980, Mon. Not. Roy. Astr. Soc. 190, 353 and 371.
Hall, A. N.: 1981, Mon. Not. Roy. Astr. Soc. 197, 977.
Holman, G. D., Ionson, J. A. and Scott, J. S.: 1979, Astrophys. J. 228, 576.
Ipavich, F. M.: 1975, Astrophys. J. 196, 107.
Jokipii, J. R. and Higdon, J. C.: 1979, Astrophys. J. 228, 293.
Kennel, C. and Petschek, H. E.: 1966, J. Geophys. Res. 71, 1.
Kulsrud, R. M. and Pearce, W. P.: 1969, Astrophys. J. 156, 445.
Kulsrud, R. and Zweibel, E.: 1975, 14th Internat'l Conf. Cosmic Rays (Munich) 2, 365.
Lee, M. A.: 1982, J. Geophys. Res. 87, 5063.
Lerche, I.: 1967, Astrophys. J. 147, 689.
Ma Sung, L. S. and Earl, J. A.: 1978, Astrophys. J. 222, 1080.
McDonald, F. B., Fichtel, C. E. and Fisk, L. A.: 1974, in "High Energy Particles and Quanta in Astrophysics", eds. F. B. McDonald and C. E. Fichtel, MIT Press, p. 212.
McKenzie, J.F. and Volk, H.J.: 1982, Astron. Astrophys. 116, 191.
Morrison, P. J.: 1979, Ph.D. thesis, University of Maryland.
Owens, A. J. and Jokipii, J. R.: 1977, Astrophys. J. 215, 677.
Schatzman, E.: 1963, Ann. d' Astrophys. 26, 234.
Shukla, P. K. and Dawson, J. M.: 1983, Astrophys. J. (Letters), in press.
Spangler, S. R.: 1979, Astrophys. J. (Letters) 232, L7.
Stephens, S. A.: 1981, 17th Internat'l Conf. Cosmic Rays (Paris) 13, 89.
Tidman, D. A.: 1966, Astrophys. J. 144, 615.
Vestrand, W. T.: 1982, Astron. J. 87, 1266.
Volk, H. J.: 1981, 17th Internat'l Conf. Cosmic Rays (Paris) 13, 131.
Wentzel, D. G.: 1969, Astrophys. J. 157, 545.
Wentzel, D. G.: 1973, Astrophys. Space. Sc. 23, 417.
Wentzel, D. G.: 1974, Ann. Rev. Astr. Astrophys. 12, 71.
Wentzel, D. G.: 1977, Astrophys. J. 216, L59.

DISCUSSION

Kennel: Aren't your two examples of non-confined cosmic rays different? In the solar wind you actually do see scatter-free propgation. But is it necessary for the Coma cluster? Why can't you have particles bound to the gas by wave-particle interactions and then have the gas expand?

Norman: The Coma cluster may not be a good example to emphasize since this phenomenon of an extended diffuse source is not at all a common phenomenon. The Coma cluster may in fact be made up of old radio sources where the transport of electrons is in energetic collimated jet flows at roughly 10^4 km/sec; these then shock and produce the acceleration at distances of 100 kpc to 1 Mpc from the source.

Bridle: There are clusters other than Coma which have diffuse steep-spectrum radio emission. An example is Abell 2256, which has a group of tailed radio sources and a steep-spectrum diffuse source which does not appear to be the integrated emission of discrete sources associated with individual galaxies.

Cowie: A third possible explanation of the electron lifetime would seem to be re-acceleration in turbulent shocks within the hot diffuse gas or in the bowshocks of galaxies. Can this be easily ruled out?

Norman: Bow shocks would have a very small filling factor, and there is no observational evidence for shocks in the general intracluster medium.

Wentzel: Certainly there are many sources where acceleration or re-acceleration throughout the source is required; some are mentioned in this Symposium. The question is: are there sites where apparent fast streaming cannot be explained by convective motions or acceleration? The arguments excluding fast convection and acceleration for the Coma cluster rest on its apparent homogeneity, including the broad, rather smooth X-ray distribution (Vestrand 1982). W. Jaffe (Astrophys. J. 212, 1, 1977) argued against convection of particles and acceleration by shocks in terms of excessive dissipation of kinetic energy, but his argument is weak if shock acceleration is efficient.

Dum: I thought the resonance gap at $v_{\shortparallel} = 0$ ($\theta = 90°$) that you mentioned arises only in quasilinear theory and that it is bridged by nonlinear orbit modification (resonance broadening). There were several papers by F. Jones, T. Birmingham and M. Goldstein that include numerical experiments demonstrating this effect.

Davila: If the scattering waves have a power at all k, as assumed in standard scattering theory, then the resonance gap is infinitessimal and resonance broadening effects can effectively cause particles to scatter across it. But when the maximum value of k is limited by ion-cyclotron damping, as assumed in resonance gap theories, the gap is too wide for resonance broadening to bridge the gap.

Wentzel: The work of Davila emphasizes the sensitive dependance of the gap on the dispersion relations of the relevant plasma modes. The nonlinear work (see M. Goldstein, Astrophys. J. 204, 900, 1976 and references therein) assumes magnetostatic turbulence.

Henriksen: Can the mirrors required for filling the resonance gap be due to the non-linear end of the resonant Alfvén wave spectrum?

Wentzel: The Alfvén waves may certainly become strong enough to interact nonlinearly, especially for k just somewhat below the resonant gap, and their degradation may fill in the gap to some degree. In fact, Achterberg shows that the mirroring is quite effective, even when the waves generated by the cosmic rays are kept fairly weak by nonlinear damping processes.

Mullan: Do particles from solar flares have time to excite Alfvén waves during their flight from the Sun to the Earth?

Wentzel: No, they are scattered by pre-existing turbulence. I used flare particles to demonstrate the importance of mirroring rather than self-confinement.

NONLINEAR EFFECTS AND THE LIMITATION OF ELECTRON STREAMING INSTABILITIES IN ASTROPHYSICS

Steven R. Spangler and James P. Sheerin
Department of Physics and Astronomy
The University of Iowa
Iowa City, IA 52242

ABSTRACT

Nonlinear effects, such as soliton collapse, will result in evolution of hydromagnetic waves excited by a field-aligned charged particle beam. If the time scale for such evolution is comparable to, or shorter than, linear time scales such as those for wave growth or pitch angle isotropization, then nonlinear effects may limit the instability. For conditions appropriate to relativistic electron streaming in a radio galaxy, the nonlinear time scale may be comparable to the linear time scales.

A topic of considerable discussion in this symposium is the stability of a relativistic, magnetic field-aligned electron beam. Linear plasma theory predicts that such an electron distribution is unstable to the growth of Alfvén waves. Quasilinear theory then predicts the electrons will resonantly interact with these waves, resulting in pitch angle diffusion, and thereby isotropization of the distribution. The net result of the linear and quasilinear processes is that the electron streaming speed is reduced to the Alfvén speed. A controversy continues as to whether the aforementioned sequence of events occurs. To date, theoretical arguments against the Alfvén streaming speed have tended to invoke linear processes, i.e., Holman, Ionson, and Scott (1979).

In this paper we consider possible nonlinear processes which might limit this instability. Nonlinear hydromagnetic waves in a finite-beta plasma are described by the Derivative Nonlinear Schrödinger equation (Spangler and Sheerin 1982),

$$i \frac{\partial \varphi}{\partial t} + i \frac{\partial}{\partial x}\left\{\frac{1}{4}\varphi\left[4 + \frac{|\varphi|^2}{(1-\beta)}\right]\right\} \pm \mu \frac{\partial^2 \varphi}{\partial x^2} = 0 \quad , \tag{1}$$

where φ is a circularly-polarized hydromagnetic wave, x is the direction of wave propagation, β is the ratio of plasma pressure to magnetic energy density, and μ a parameter determining the strength of dispersion.

The DNLS possesses a number of envelope soliton solutions (Spangler and Sheerin 1982). We now consider the consequences if the Alfvén waves excited by streaming electrons resemble these Alfvénic solitons. The most profound effects would probably occur if they undergo collapse, as do Langmuir solitons. Langmuir solitons are stable in one-dimensional systems. In two or three dimensions, however, they are unstable in the sense that they become narrower and of greater amplitude with time. At the present time we do not know if such a collapse will occur for Alfvén solitons. For the present we will assume that such a collapse will occur. In order to discuss the phenomenon in the context of the above theory, we will further assume that as the soliton contracts, it makes a transition between one-dimensional soliton states.

Three characteristics of the wave packet both insure that the soliton will collapse and furnish a means for estimating the collapse time scale. (1) A constant of the motion of the DNLS is the total energy in the soliton,

$$C_o = \int_{-\infty}^{\infty} |B|^2 \, dx \simeq 2B^2 \ell \quad , \tag{2}$$

where B is the wave amplitude and ℓ is the length scale of the soliton. (2) If the plasma beta exceeds unity, there will be an anticorrelation between the wave energy density and the plasma density, i.e.,

$$\delta n \propto \frac{B^2}{(1-\beta)} \quad , \tag{3}$$

where δn is the change in the plasma density. (3) Associated with the nonlinear wave field, there will be a ponderomotive force which tends to drive plasma out of the soliton.

We envision the evolution of the nonlinear wave packet or soliton to proceed as follows. The ponderomotive force causes a flow of plasma out of the soliton, thus lowering the plasma density and making δn more negative. The proportionality in (3) indicates that for $\beta > 1$, this entails an increase in the wave energy density. From equation (2) we see that if B^2 increases, conservation of the constant of the motion requires that ℓ become smaller, i.e., the soliton must contract. A larger energy density in the soliton and a smaller length scale mean that the ponderomotive force will be greater. The above sequence of events will continue, but at an accelerated rate. The result is, therefore, a runaway collapse.

The above arguments can be used to obtain an equation for the temporal evolution of the soliton length scale, which goes to zero on a time scale (Spangler and Sheerin 1983)

$$\tau_c = \frac{4}{3} \frac{\ell_o}{V_A} \sqrt{\frac{B_o^2}{B^2(0)}} \quad . \tag{4}$$

In equation (4) ℓ_o is the initial length scale of the soliton, and V_A is the Alfvén speed. The quantity $B^2(0)/B_o^2$ is the energy density of the initial wave packet divided by the energy density of the static field.

The time scale (4) has been derived by considering hydromagnetic waves to be solitons, but we feel that it is probably of much greater generality, and gives a good estimate of the time scale on which nonlinear effects cause a substantial change in a wave packet. In support of this statement, we note that the numerical results of Steinolfson (1981) for the time scale for steepening of waves into hydromagnetic shocks are in good agreement with (4). This time scale is also similar to that obtained by Goldstein (1978) for the decay of a large amplitude Alfvén wave by coupling to random density and magnetic fluctuations in a plasma.

The importance of soliton collapse, or related nonlinear processes, to the electron streaming instability may be determined by how the nonlinear time scale (4) compares with the familiar linear time scales, such as the time for wave growth (reciprocal of the growth rate) and pitch angle isotropization (reciprocal of the pitch angle diffusion coefficient).

Spangler and Sheerin (1983) compared the linear, quasilinear, and nonlinear time scales for conditions appropriate to an extragalactic radio source. For a plausible subrange of conditions, the nonlinear collapse time scale can be comparable to or less than the linear time scales when the wave amplitude is 1 - 10 percent of the static field. This suggests the possibility that the streaming instability would begin in the linear phase, and proceed until the excited waves reached a certain amplitude. Thereafter, nonlinear processes would determine the evolution of the waves, and might "decouple" the waves from the beam responsible for their excitation.

There remains the question of the process by which this decoupling would occur. One possibility is wave damping. Amplitude-modulated waves in an envelope of shrinking size will be spread out in wave number, possibly enhancing the mechanism discussed by Holman, Ionson, and Scott (1979). Another possible mechanism is resonance breaking, or resonance broadening. The wavenumber spreading alluded to above imposes a coherence time on the wave-particle interaction, and thereby breaks the resonance. An additional source of resonance breaking is the nonlinear phase modulation characteristic of the DNLS. If the Alfvén wave excited by an electron beam is described by the DNLS then (possibly random) phase shifts, due to this modulation, will occur in the wave train.

As pointed out by Spangler and Goertz (1981), the occurrence of such phase jumps can drastically change the magnitude and even sign of the Alfvén wave growth rate, and it seems likely that the quasilinear diffusion coefficient would be similarly effected. The net result might be an inhibition of the quasilinear relaxation.

REFERENCES

Goldstein, M. L. 1978, Astrophys. J., 219, 700.
Holman, G. D., Ionson, J. A., and Scott, J. S. 1979, Astrophys. J., 228, 576.
Spangler, S. R. and Goertz, C. K. 1981, Astrophys. J., 247, 1078.
Spangler, S. R. and Sheerin, J. P. 1982, J. Plasma Phys., 27, 193.
Spangler, S. R. and Sheerin, J. P. 1983, Astrophys. J. (in press, Sept. 1, 1983).
Steinolfson, R. S. 1981, J. Geophys. Res., 86, 535.

ACKNOWLEDGMENTS

This work is supported by grant AST-8217714 from the National Science Foundation.

DISCUSSION

Hasegawa: Alfvén waves have a very small dispersion and, hence, are unlikely to form an envelope soliton (because it can decay before the envelope soliton is formed) unless $\omega > \omega_{ci}/4$. Also, in the presence of B, collapse is unlikely.

Spangler: The dispersion may be small, but is not ignorable. Formally, this is true because the dispersive term is of a different order than the other terms in the equation, and is therefore reminiscent of the boundary layer problem in hydrodynamics. The magnitude of the dispersion sets the scale of the soliton. With regard to the remainder of your comment, I would contend that issue such as decay and collapse will be unclear until a multidimensional treatment with realistic ion dynamics is undertaken.

Benford: We have observed electromagnetic emission in laboratory relativistic beam-plasma experiments, perhaps caused by Langmuir solitons. The observed plasma line emission is suppressed or broadened by quite small magnetic fields, $\omega_c \sim \omega_p/10$. The galactic magnetic field may inhibit your Alfvén soliton collapse. Can you take this into account?

Spangler: No. As I mentioned in my talk, our treatment utilizes 1-D soliton properties and the assumption that collapse occurs. Questions regarding the details of the collapse, such as stabilization etc., and even a rigorous demonstration of its existence must wait a multidimensional treatment. We are presently working on this.

Bratenahl: I am trying to understand the structure of Alfvén solitons - they imply a pair of anti-parallel currents, do they not? I like the collapse argument (negative feed back), but I don't understand what happens to the anti-parallel currents.

Spangler: I have a somewhat different view, picturing these solitons as due to a balance between nonlinearity and dispersion due to a finite ion gyrofrequency.

Papadopoulos: Two Comments: (1) The 2-D theory of solitons, as shown by Rowland et al. Phys. Rev. Lett. 1981, shows that soliton collapse can be prevented by transverse magnetic field effects. (2) The presence of growth and damping transforms the solutions of the nonlinear Schrödinger equation to stochastic solutions, invalidating the soliton picture.

Spangler: The role of a transverse magnetic field in preventing Langmuir soliton collapse is not so clear. While an infinitely strong field would certainly prevent collapse, the work of Weatherall et al (1981, Ap. J., 246, 306) and Weatherall et al (1982, JGR, 87, 823) shows that soliton collapse does occur in the presence of a magnetic field.

With regard to your second point, I disagree that the occurrence of a transition to chaos invalidates the soliton picture. Poolen et al.(1983, Phys. Rev. Lett., 51, 335) show that chaotic solutions of the Zakharov equations possess "caviton" properties.

Finally, your comments refer to the properties of Langmuir waves and turbulence. It is not clear that they are applicable to nonlinear Alfvén waves.

Henriksen: We have looked at single relativistic particles interacting with non-linear wave packets at various scattering angles. We find that resonance persists even for very short "coherence lengths" or "phase lengths" of the packet.

Spangler: At least some of the quantities of interest, such as wave growth rates, are determined by the particle distribution function. A single particle approach may not be sufficient to reveal modifications due to large wave amplitudes, shocks, etc.

Kennel: What can you tell us about multidimensional soliton solutions that propagate at an angle to $\vec{B}_o$?

Spangler: Solitons propagating at an angle with respect to the magnetic field correspond to the kinetic Alfvén mode, and differ, for example in the nature of the dispersion, from those I have discussed.

COSMIC RAYS AND THE PARKER INSTABILITY

A.H.Nelson
Dept. of Applied Mathematics
 & Astronomy
University College
Cardiff, U.K.

  Parker (1966, 1969, 1979) has shown that the magnetic buoyancy of a uniform horizontal magnetic field will destabilize the Galactic gas layer. Perturbations of the form shown in Fig. 1 will grow in time with the magnetic loops ballooning up into the Galactic halo, and the interstellar gas draining down the field lines to collect in the midplane. Parker also showed that if the dynamical effect of the cosmic ray component of the interstellar medium is included, using an isotropic cosmic ray pressure, then the instability is enhanced.
  The instability has been regarded as significant for two reasons. Firstly it allows escape of magnetic field and cosmic rays out of the gas disc and into the Galactic halo (Parker 1967; 1969); and secondly it leads to condensations of the interstellar gas which may be the mechanism for the formation of the large molecular clouds observed in the Galaxy (Mouschovias, Shu, and Woodward 1974, Shu 1983), and hence the instability would represent an early stage in the process of star formation.
  However it is possible to argue that the assumption that the cosmic ray pressure is always isotropic may not be valid. Since the only mechanism for isotropizing the collisionless cosmic rays is scattering by magnetic field fluctuations, it is necessary that fluctuations of the correct scale length and of sufficient amplitude should be present in the interstellar plasma; but under the assumptions that the field should be dominantly uniform, and that the cosmic ray pressure is isotropic in the unperturbed state, it may be difficult for the necessary fluctuations to arise. Consequently it is of interest to allow the cosmic ray pressure to become anisotropic, in order to ascertain how crucial the assumption of isotropy is for instability.
  An appropriate approximation for the stress tensor of the cosmic rays P_{ij} is that due to Chew, Goldberger and Low (1956) in which we have

$$\underset{\sim}{P} = P_{ij} = P_{\parallel} b_i b_j + P_{\perp}(\delta_{ij} - b_i b_j)$$

where $\underset{\sim}{b}$ is a unit vector parallel to the field $\underset{\sim}{B}$, and $P_{\parallel}$ and $P_{\perp}$ are the pressures parallel and perpendicular to $\underset{\sim}{B}$ respectively. Equations governing $P_{\parallel}$ and $P_{\perp}$ can be derived under the assumptions that the Larmor radii and periods of the cosmic rays are respectively much smaller than the characteristic wavelength and timescale of the instability (Rossi and Olbert 1970). These are

$$\frac{d}{dt}\left(\frac{P_\perp}{nB}\right) = 0 \tag{1}$$

$$\frac{d}{dt}\left(\frac{P_\parallel B^2}{n^3}\right) = 0 \tag{2}$$

where n is the cosmic ray density, and d/dt is the cosmic ray comoving time derivative. These replace the equations $P_\parallel = P_\perp = P$ and $dP/dt = 0$ employed by Parker, otherwise the analysis is very similar to that of Parker, with the assumption of quasi-static cosmic rays, i.e.

$$\underline{j}_{cr} \times \underline{B} = \nabla \cdot \underline{\underline{P}} \quad ,$$

together with equations (1) and (2) determining the time evolution of the cosmic rays.

Parker's analysis showed that for instability the wavelength of the perturbation must be greater than a certain critical wavelength given by

$$\lambda_c = 2\pi h_o \left[\frac{2\alpha\gamma(1+\alpha+\beta)^2}{(1+\alpha)(1+\alpha-\gamma) - \frac{1}{2}\alpha\gamma + \beta(2+\beta+2\alpha-\gamma)}\right]^{\frac{1}{2}}$$

where γ is the polytropic exponent of the gas, α and β are respectively the unperturbed magnetic and cosmic ray pressures as a fraction of the gas pressure, and h_o is the characteristic scale height of the gas. For $\gamma = 1$ and $\alpha = 0.01$ we obtain $\lambda_c = 0.06\, h_o$ and $\lambda_c = 0.012\, h_o$ for $\beta = 0$ and $\beta = 1$ respectively, i.e. the cosmic rays decrease λ_c and hence destabilize the gas layer.

Using (1) and (2) however we obtain

$$\lambda_c = 2\pi h_o \sqrt{1+\beta/\alpha} \left[\frac{2\alpha\gamma(1+\alpha+\beta)^2}{(1+\alpha)(1+\alpha-\gamma) - \frac{1}{2}\alpha\gamma + \beta(2+\beta+2\alpha-3\gamma/2)}\right]^{\frac{1}{2}}$$

which yields $\lambda_c = 0.18\, h_o$ for $\alpha = 0.01$, $\gamma = 1$, and $\beta = 1$, i.e. in this case the cosmic rays are stabilizing since λ_c is increased.

The physical reason for this is that the cosmic ray pressure is now largest in the troughs of the magnetic field lines in Fig. 1, due to the conserved magnetic moments, and their effect is to push the gas vertically out of the troughs. Whereas under the assumption that the cosmic ray stress is isotropic and constant along the field lines, the main effect of the cosmic rays comes from the horizontal component of the gradient of this stress which acts to push the gas into the troughs. The assumption of isotropic stress for the cosmic rays is therefore crucial for the enhancement of the Parker instability. Unless convincing arguments are found to justify this assumption it would seem likely that

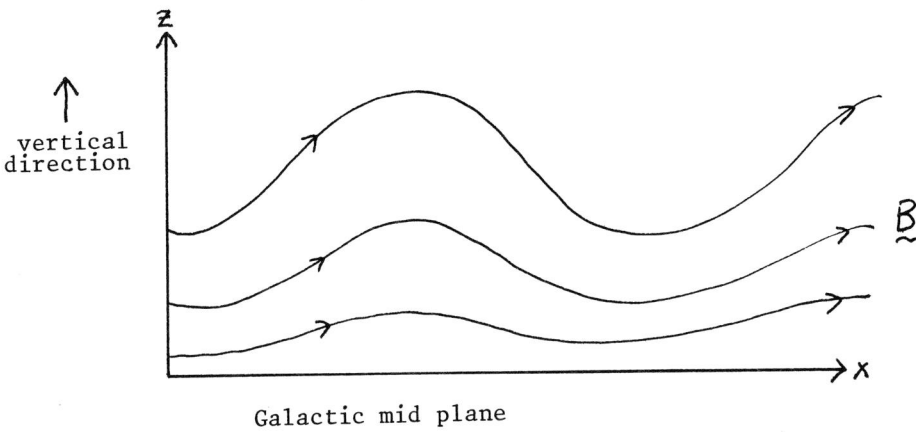

Fig. 1 Perturbation of the horizontal Galactic magnetic field lines.

the effect of cosmic rays helps to stabilize the Galactic gas layer, and makes the Parker instability less plausible as a mechanism for forming molecular clouds.

REFERENCES

Chew,G.F.,Goldberger,M.O.,and Low,F.E.: 1956, Proc. Roy. Soc. A236, 112.
Mouschovias,T.Ch.,Shu,F.H., and Woodward,P.R.:1974,Astron. Astrophys. 33, 73.
Parker,E.N.:1966,Ap. J. 145, 811.
Parker,E.N.:1967,Ap. J. 149, 535.
Parker,E.N.:1969,Space Sci. Rev. 9, 651.
Parker,E.N.:1979, *Cosmical Magnetic Fields*, Clarendon Press, Oxford.
Rossi,B. and Olbert,S.:1970, *Introduction to the Physics of Space*, pp.354-358, McGraw-Hill.
Shu,F.H.:1983, I.A.U. Symp 106.

PHASE MIXING OF PROPAGATING ALFVÉN WAVES

L. Nocera, B. Leroy, E.R. Priest
Appl. Maths. Dept., University of St Andrews, KY16 9SS (U.K.)

Among MHD waves, Alfvén waves have been proved to be the best candidates to reach the solar corona and, eventually, to be responsible for the heating of this outer part of the solar atmosphere. The problem arises, however, about the mechanism able to transform the energy stored in the waves into heat.

Recently (Heyvaerts and Priest 1983, or HP), a simple and very appealing idea has been proposed for the dissipation of Alfvén waves in an inhomogeneous medium: Let the inhomogeneity be described by a space dependent Alfvén speed $v_A(x)$ and let the oscillations, whose velocity we denote by $v = v(x,z,t)\hat{y}$, be polarized in the y-direction and propagating along the mean magnetic field in the z-direction (Fig. 1).

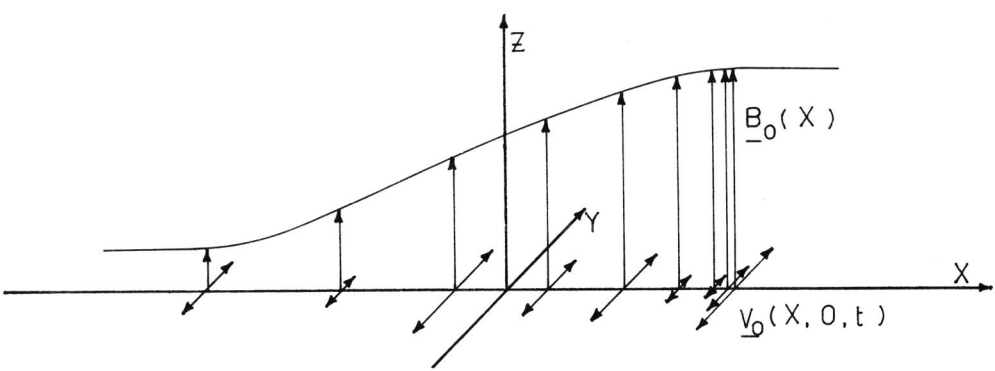

Fig. 1   The background magnetic field profile. Each field line is excited at the boundary $z = 0$ by an inhomogeneous disturbance $\underline{v}_0$

Due to the gradient in the phase speed $v_A(x)$, the oscillations of neighbouring field lines, that are in phase at some altitude, will become increasingly out of phase as the wave propagates outward, what

Heyvaerts and Priest denoted by 'phase mixing'. In other words, at high altitudes, strong lateral gradients are expected to develop in the velocity field.

If viscosity and resistivity are taken into account, now, this mechanism leads to a considerable dissipation of energy. The heating rate is indeed conspicuous provided the magnetic field is not very strong (which would lower the effective viscosity); hence the above mechanism seems to fail in active regions where, on the other hand, the heating demands are highest (see Priest, 1983).

In these regions, the magnetic field shows mainly a closed geometry, in which we expect to find <u>standing</u> waves. These waves (for which phase mixing occurs as time elapses) can, now, be shown to suffer Kelvin-Helmholtz and tearing-mode instability (see HP and Browning and Priest, 1983). If, on one hand, the appearance of instabilities increases the values of the viscosity and resistivity, it can possibly lead to the onset of a turbulent cascade, thus enhancing the heating rate and making of the phase mixing a good candidate for the heating of active regions too.

Generally speaking, both propagating and standing waves are governed by a modified wave equation taking into account wave propagation and energy diffusion as well

$$\left(\frac{\partial^2}{\partial t^2} - v_A^2(x) \frac{\partial^2}{\partial z^2}\right) v = \nu \left(\frac{\partial^2}{\partial x^2} + \frac{\partial^2}{\partial z^2}\right) v. \tag{1}$$

Here ν represents the sum of the kinematic viscosity and of the magnetic diffusivity, both assumed to be small (i.e. large Reynolds and magnetic Reynolds numbers). The lateral diffusion of energy (x-directed) can be thought of as the excitation of new oscillations on neighbouring field lines. These new oscillations, of course, can propagate both outward <u>and</u> inward. We can get rid of this complication provided the phase mixing is not much developed (see Nocera <u>et al</u>. 1983, for a detailed discussion of the various orderings quoted here).

Let the oscillations be generated at $z = 0$ at a frequency ω and let $k(x) = \omega/v_A(x)$ be their wavenumber. If we define a transverse scale length

$$a = |d \ln k/dx|^{-1} \tag{2}$$

and a total Reynolds number

$$R_{TOT} = \omega a^2/\nu, \tag{3}$$

we find, in the long wave ($k \ll R_{TOT}/a$), large altitude ($z \gg a$) limit (Nocera et al 1983)

$$v(x,z,t) \sim \exp[-(z/\Lambda)^2] \exp i(\omega t - kz) \cdot z^{-(R_{TOT}/8)^{\frac{1}{2}}} \quad (4)$$

$$\Lambda = 2^{3/4} (R_{TOT})^{1/4}/k. \quad (5)$$

This result is different from HP's one

$$v_{HP} \sim \exp[-(z/\Lambda_{HP})^3] \exp i(\omega t - kz), \quad (6)$$

$$\Lambda_{HP} = (6 R_{TOT})^{1/3}/k, \quad (7)$$

the difference arising from the relaxation of some oversimplifications introduced by HP.

The solution (6) seems to hold up to several solar radii where steep x-gradients unavoidably develop, thus contradicting the weak phase mixing hypotesis. In this circumstance the Alfven wave spreads out laterally, just like a light beam defocuses in an inhomogeneous medium.

A similar effect is found at small heights ($z \le a$), where the main energy flux is due to outgoing waves from the nearby source and, again, we can neglect the ingoing waves. Thus, if we assume that the perturbation acts on a single line of force located at $x = 0$, say, and that the plasma is strongly magnetized, we find (Nocera et al. 1983)

$$v \sim \exp(i\omega t)[2\pi k(0)z]^{-\frac{1}{2}} \exp[-w^2(x)/z], \quad (8)$$

where

$$w(x) = (2\nu)^{-\frac{1}{2}} \int_0^x v_A^{\frac{1}{2}}(s) \, ds \quad (9)$$

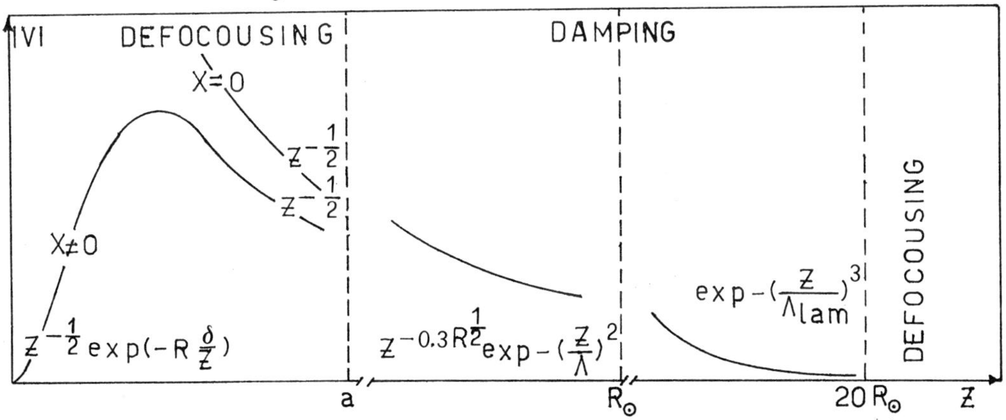

Fig. 2 Damping laws (schematic) for the wave amplitude v at different altitudes z, for a pointlike disturbance (according to Nocera et al. 1983)

So far, we dealt with propagating waves. As concerns standing waves in closed structures of length ℓ, they vibrate with eigenfrequencies $\Omega_n = 2\pi n v_A(x)/\ell$, and their main properties can be described by the Green function of the fundamental (n = 1) oscillator (see HP)

$$G(x,t) = \frac{\sin(\Omega_1 t)\exp[-(t/\tau_D)^3]}{\Omega_1} \qquad (10)$$

where

$$\tau_D = (6R_{TOT})^{1/3} / \Omega_1 \qquad (11)$$

is a damping time which, in typical coronal conditions equals about 20 wave periods.

A consistent discussion of dissipation processes cannot bypass the problem of Kelvin-Helmholtz (KH) and Tearing Mode (TM) stability, which is particularly important in a highly sheared medium. A simplified, local, stability analysis (HP) shows that propagating waves are stable whereas a standing wave, whose velocity amplitudes is u and whose transverse inhomogeneity scale length is d, shows instability with growing times

$$\tau_{KH} = d/u, \quad \tau_{TM} \simeq \left(\frac{ud}{\nu_m}\right)^{3/7} \frac{d}{u} . \qquad (12)$$

Recently, the KH stability analysis has been performed by Browning and Priest (1983) in more realistic conditions, leading to the conclusion that the instability develops well before the phase mixing reaches its strong stage; the Alfvén wave may be significantly disrupted within a very few wave periods and it reaches a turbulent state well before the laminar damping becomes significant.

The heating rate E_H can be calculated according to some guess on the turbulent spectrum (Hollweg, 1983, uses a Kolmogorov spectrum). It turns out to be of the same order (or even greater) of the input required for the heating of a coronal loop, $E_H \simeq 4.10^{-5} \div 1.8.10^{-3}$ erg cm^{-3} s^{-1}, suggested by Hollweg (1983) (see tab. 1)

	Wave period(s)	E_H(erg cm^{-3} s^{-1})
Wave amplitude = 60 km/s	3	7.2.10-2
inhomogeneity scale = 1000 km	30	1.2.10-2
Table 1	300	1.7.10-3

Turbulent heating rate for different periods of the laminar oscillations, in the solar corona (After Browning and Priest 1983).

REFERENCES

Browning, P.K., Priest, E.R. : accepted by Astron. Astrophys. (1983)
Heyvaerts, Priest, E.R. : Astron. Astrophys., 117, 220 (1983)
Hollweg, J.V. : to appear in Solar wind V (1983)
Nocera, L., Leroy, B., Priest, E.R. : submitted to Astron. Astrophys. (1983).
Priest, E.R. : Plasma Physics, 25, 161 (1983).

DISCUSSION

Ionson: In actual plasma systems, e.g. coronal loops, the field lines communicate phase primarily by $\nabla_\perp (B_\parallel B)$ rather than through the dissipative terms.

Heyvaerts: In the working hypothesis used here, there is no compressional effect. But it is true that if we go to cylindrical or toroidal geometry such effects will show up. Their effect should be to broaden the resonances, on the one hand, and also to introduce new collective modes, on the other. I expect the general effect to remain for the continuum of modes.

Migliuolo: The transverse structure B(x) dictates the form of eigenmodes of the systems. What is the relationship to the model presented here?

Nocera: The eigenvalue problem in the x-direction was not solved in the general form, since it is difficult to manage with three (x,z,t) coordinates.

At low altitudes, however, this has been done. The x- dominion is actually unbounded and the solution (equation 8) seems to vanish at $x \to \pm \infty$ for any model for $v_A(x)$ in equation (9) and for any value of the frequency ω: then, we have a continuous spectrum there.

VELOCITY SHEAR INSTABILITIES IN THE ANISOTROPIC SOLAR WIND

Stefano Migliuolo
High Altitude Observatory
National Center for Atmospheric Research
Boulder, Colorado 80307   USA

The linear and quasilinear theory of perturbations in finite-β (β is the ratio of plasma pressure to magnetic energy density), collisionless plasmas, that have sheared (velocity) flows, is developed. A simple, one-dimensional magnetic field geometry is assumed to adequately represent solar wind conditions near the sun (i.e., at $R \simeq 0.3$ AU). Two modes are examined in detail: an ion-acoustic mode (finite-β stabilized) and a compressional Alfven mode (finite-β threshold, high-β stabilization). The role played by equilibrium temperature anisotropies, in the linear stability of these modes, is also presented. From the quasilinear theory, two results are obtained. First, the feedback of these waves on the state of the wind is such as to heat (cool) the ions in the direction perpendicular (parallel) to the equilibrium magnetic field. The opposite effect is found for the electrons. This is in qualitative agreement with the observed anisotropies of ions and electrons, in fast solar wind streams. Second, these quasilinear temperature changes are shown to result in a quasilinear growth rate that is lower than the linear growth rate, suggesting saturation of these instabilities.

A. INTRODUCTION

The observations of Alfvenic waves, in the solar wind (Belcher and Davis, 1971; Bavassano, Dobrowolny and Moreno, 1978) have motivated, in part, the analysis of modes that can be driven unstable by velocity shear (Gary and Schwartz, 1980; Mikhailovskii and Klimenko, 1980; Melander and Parks, 1981; Huba, 1981a,b). The question still remains, however, whether the observed state of the solar wind (amount of velocity shear, plasma β, temperature anisotropies, etc.) is compatible with the picture of the shear, encountered at the interfaces between fast and slow solar wind streams, being the source of free energy for these Alfvenic fluctuations. Furthermore, the cause and effect relationship between the observed temperature anisotropies and fluctuations still needs to be addressed.

A second question revolves around the equation of state of the solar wind ions. As pointed out by Marsch et al. (1983) and Schwartz and Marsch (1983), it is difficult to explain the observed violation of the constancy of the first adiabatic invariant, $\mu = T_\perp/B$, without invoking some sort of wave heating. In this paper, we follow Migliuolo (1983) and answer these questions. The details of the method, the mathematics, as well as a more complete list of references can be found in Migliuolo (1983).

B. LINEAR THEORY

We shall use kinetic theory (Vlasov's and Maxwell's equations, retaining all finite-β effects), to analyse the linear stability of electromagnetic perturbations. The only restrictions are that we consider low frequency (compared to ion cyclotron), long wavelength (compared to an ion gyral radius) modes, as well as using the local approximation (wavelengths short compared to equilibrium scale lengths). A fully ionized hydrogen plasma is assumed throughout. The equilibrium includes a one-dimensional magnetic field (in the z-direction) and inhomogeneities in the x-direction (density and magnetic field gradients), as well as a sheared velocity flow, $\vec{v} = \hat{e}_z v(x)$. The computation of the dispersion relation is straightforward (Huba, 1981b; Migliuolo, 1983), and yields:

$$0 = (O_{11} O_{22} + O_{12}^2) O_{33} + \tfrac{1}{2}\beta_i (O_{13}^2 O_{22} - O_{23}^2 O_{11} - 2 O_{12} O_{23} O_{13}) \quad (1)$$

which is the determinant of a nine element matrix equation, obtained from Poisson's equation and the x and z components of Ampere's law (see Migliuolo (1983), for a definition of the elements O_{ij}). The vector, on which the matrix O acts is

$$\chi_1 = \phi_1 - A_{1z}(\omega - k_\parallel v_0)/k_\parallel c, \quad \chi_2 = (\omega/k_\parallel c) A_{1z}, \quad \chi_3 = (k_\perp T_{\perp i}/eB) A_{1x},$$

where ϕ_1, and $\vec{A}_1$ are the perturbed potentials.

Two modes are found to be unstable; first, an ion-acoustic mode that has been extensively discussed by Huba (1981); second, a compressional Alfven mode. The detailed discussion of their linear properties can be found in Migliuolo (1983). Here, we give a brief summary of their properties.

OUTLINE OF RESULTS.

(1) Effect of varying velocity shear: the ion acoustic mode has a threshold in this parameter (Mikhailovskii and

Klimenko, 1980; Melander and Parks, 1981), and is unstable for values higher than the threshold. It however has a second threshold: the mode is stabilized by Landau damping above the second threshold. The compressional Alfven mode also has a threshold for instability (sensitive on the value of β), but we found no second threshold. Furthermore, this compressional Alfven mode has a (real part of the) frequency that can be larger than a tenth of the ion cyclotron frequency, making this a candidate for Helium heating.

(2) Effect of β: the ion acoustic mode is stabilized by finite-β effects (Huba, 1981b), while the compressional Alfven mode has a threshold in β (Mikhailovskii and Klimenko, 1980; Melander and Parks, 1981). The compressional Alfven mode is stabilized for high values of β (the exact value for second threshold depends on the other plasma parameters).

(3) Both modes propogate at a large angle to the magnetic field, typically $86°$ for the ion-acoustic mode and $87°$ for the compressional Alfven mode. This last property makes these longitudinal modes: $\vec{B}_1 \cdot \vec{B}_0 >> |\vec{B}_1 \times \vec{B}_0|$. Thus they are not candidates for the fluctuations observed by Belcher and Davis (1971).

(4) Both modes are stabilized by positive temperature anisotropies ($T_\perp > T_\parallel$), with the electron anisotropy being the major agent of stabilization. The observed state of anisotropy (of the solar wind) yields growth rates that are lower that those for an isotropic solar wind: the anisotropy cannot be considered a cause for these fluctuations.

(5) The observed values for the shear parameter (Marsch et al., 1982), $A_i = (1/\Omega_i)(\Delta V/R\Delta\phi) \simeq 3\ 10^{-5}$ at $R = 0.3$ AU, yield modes that propagate nearly perpendicularly to the B-field, with small growth rates. Here, Ω_i is the ion cyclotron frequency, ΔV is the total (observed) jump in velocity, $\Delta\phi$ is the (observed) angle over which ΔV is measured. Thus, we conclude that, in order for these modes to play an important role in the physics of the solar wind (i.e., they need to have substantial growth rates, to reach finite amplitudes by $R \simeq 0.3$ AU), they need to be generated in regions of much higher velocity shear, e.g., at the edges of streamers in the solar corona.

C. QUASILINEAR THEORY

We look at the quasilinear theory of these modes, in order to examine their feedback (in a space and time average sense) on the plasma. A tedious calculation is carried out (see Migliuolo, 1983, for details), that determines the quasilinear temperature changes for the electrons and ions (the quasilinear density changes are null for both species). Two results ensue:

(1) For all cases considered, the quasilinear temperature changes were as follows (for a plasma that is initially isotropic, for both species): $T_{\perp i} > T_{\parallel i}$, and $T_{\perp e} < T_{\parallel e}$. This is exactly the trend observed in fast solar wind streams. We conclude that we have identified a wave heating mechanism (for the ions), that acts in a self-consistent manner, as was postulated by Marsch et al. (1983) and Schwartz and Marsch (1983) the growth and feedback of modes driven by velocity shear.

(2) When we compute the quasilinear growth rate (we solve the dipersion relation, incorporating the quasilinear temperature changes calculated in (1), assuming some small amplitude for these fluctuations), we find that it is always smaller than the linear growth rate (computed for the initially isotropic plasma). This suggest that these instabilities saturate, via the gentle quasilinear feedback on the temperature state of the plasma.

D. CONCLUSIONS

The linear and quasilinear theory of velocity shear instbilities (kinetic theory) is developed for the solar wind. Two modes are found to be unstable: an ion-acoustic and a compressional Alfven mode. Both modes propagate at large angles to the equilibrium magnetic field, and are primarily longitudinal in nature. The quasilinear feedback of these modes yields temperature anisotropies that are consistent with the observed anisotropy of the solar wind. The feedback also suggests that gentle quasilinear temperature temperature changes could lead to saturation of the instabilities.

REFERENCES

Bavassano, A., Dobrowolny, M, and Moreno, G.: 1978, Solar Phys. 57, pp. 445-465.
Belcher, J.W., and Davis, L.J.: 1971, J. Geophys. Res. 76, pp. 3534-3563.
Gary, S.P., and Schwartz, S.J.: 1980, J. Geophys. Res. 85, pp. 2978-2980.
Huba, J.D.: 1981a, J. Geophys. Res. 86, pp. 3653-3656.
Huba, J.D.: 1981b, J. Geophys. Res. 86, pp. 8991-9000.
Marsch, E., Mulhauser, K.H, Schwenn, R., Rosenbauer, H., Pilipp, W., and Neubauer, F.M.: 1982, J. Geophys. Res. 87, pp. 52-72.
Marsch, E., Mulhauser, K.H., Rosenbauer, H., and Schwenn, R. 1983, J. Geophys. Res. 88, pp. 2982-2992.
Melander, B.G. and Parks, G.K.: 1981, J. Geophys. Res. 86, p 4697-4707
Migliuolo, S.: 1983, J. Geophys. Res. 88 (submitted)
Mikhailovskii, A.B., and Klimenko, V.A.: 1980, J. Plasma Phy 24, pp. 385-407.
Schwartz, S.J., and Marsch, E.: 1983, J. Geophys. Res. 88 (submitted).

KELVIN-HELMHOLTZ INSTABILITIES IN A MAGNETISED COMPRESSIBLE PLASMA (SHEARED FLOW).

T.P. Ray
Physics Dept., University College Dublin, Ireland
A.I. Ershkovich,
Dept. of Geophysics and Planetary Sciences,
Tel Aviv University, Israel.

ABSTRACT.

We have investigated Kelvin-Helmholtz (K-H) instabilities for a homogeneous compressible plasma containing a uniform magnetic field and a linear velocity shear. A derivation of the relevant K-H dispersion equation and details regarding method of solution are given elsewhere (submitted to Mon. Not. R. Astr. Soc.). We present here an outline of our results.

INTRODUCTION.

It is usual, when calculating K-H instability growth rates and phase velocities, to apply the vortex sheet approximation i.e. to assume the velocity shear is of zero thickness. In practice, viscosity (we use the word in its most general sense) introduces a velocity transition region of finite thickness. The effects of this finite shear layer only become apparent for K-H wavelengths comparable to the layer's width. Early results for incompressible fluids (Chandrasekhar, 1961; Michalke, 1964) showed that such a shear layer stabilizes all wavelengths shorter than its width. Blumen (1970), Blumen, Drazin & Billings (1975) and Drazin & Davey (1977) investigated a compressible fluid containing a hyperbolic tangent velocity profile. They showed that instability occurs at all Mach numbers. Ray (1982) examined the case of a compressible fluid with a linear profile. He found results which are analogous to those found in the hyperbolic tangent case. Ferrari, Trussoni & Zaninetti (1980) investigated the effects of magnetic fields on K-H modes but using the vortex sheet approximation. Our results, an outline of which follows, concern a MHD treatment of a layer of finite thickness. Let us assume that the equilibrium flow is in the x direction, the velocity shear is in the z direction and the uniform magnetic field is in the (x, y) plane. We use a linear profile:

$$V = \frac{U_o z}{2d} \quad -d < z < d$$

$$V = U_o/2, \quad z > d \qquad (1)$$

$$V = U_o/2, \quad z < -d$$

where U_o is a constant. Thus the jump in velocity across the shear layer is U_o and 2d is its width.

We can conveniently divide up our results as follows:
(i) Incompressible Case.

It is found that, irrespective of wavelength, K-H modes do not grow if $U_o \cos \theta < 2V_A \cos \beta$, where V_A, θ and β are the Alfven speed and the angles the modes makes with the flow and magnetic field, respectively. This condition shows that the shear layer is unstable unless the flow is exactly along the field lines ($\theta = \beta$) and the velocity jump is less than twice the Alfven speed. All solutions are found to be stationary i.e. $\omega_R/k = 0$

where R denotes 'real part'. The complex frequency and real wavelength are denoted by ω and k respectively.
(ii) The Subsonic Regime. $M < 1$, $V_A > 0$.

Here M, the Mach number, is defined in terms of half the velocity jump across the layer

$$M = \frac{U_o}{2a} \qquad (2)$$

where a is the sound speed). Because of the large number of parameters involved, we shall only consider one magnetic field orientation, viz $B \parallel U_o$, in this and subsequent sections. The case of $B_o \perp U_o$ is easily dealt with as the dispersion equation becomes identical in form to its $V_A = 0$ counterpart but with the sound speed replaced by $(a^2 + V_A^2)^{\frac{1}{2}}$. Consequently the results for $B_o \perp V_o$ are the same as the non magnetic results but with M replaced by the magnetosonic Mach number

$$M_m = \frac{U_o}{2(a^2 + V_A^2)^{\frac{1}{2}}} \qquad (3)$$

The interesting thing we find for $B \parallel U_o \parallel k$ is that for $V_A/U_o \sim 0.5$, the solutions cease to be stationary. Furthermore, one requires V_A/U_o to be somewhat greater than 0.5 for stability. This compares with the incompressible case where

$$\frac{V_A}{U_0} = 0.5 \tag{4}$$

is found to be sufficient.
(iii) The Supersonic Regime, $M > 1$, $V_A > 0$

For $B \parallel U_0 \parallel k$ it is found that growth ceases ($\omega_I/k = 0$) for $V_A > a$, the sound speed. This represents a generalization of the vortex sheet result due to Sen (1964). He found that for $B_0 \parallel V_0$, modes parallel to the flow are stable for all values of V_A/V_0. We find this result applies, even when a shear layer is present, if $V_A \geq a$ ($a = 0$ in the work of Sen, 1964).

REFERENCES

Chandrasekhar, S.: 1961, Hydrodynamic and Hydromagnetic Stability, Oxford University Press.

Blumen, W.: 1970, J. Fluid. Mech. 40, pp.769

Blumen, W., Drazin, P.G., and Billings, D.F.: 1975, J. Fluid Mech. 71, pp.305

Drazin, P.G., and Davey, A.: 1977, J. Fluid Mech. 82, pp.255

Ray, T.P.: 1982, Mon. Not. R. Astr. Soc. 198, pp.617

Ferrari, A., Trussoni, E., and Zaninetti, L.: 1980, Mon. Not. R. Astr. Soc. 193, pp.469

Michalke, A.: 1964, J. Fluid Mech. 19, pp.543

Sen, A.K.: 1964, Phys. Fluids. 7, pp.1293

SESSION VI

PLASMA HEATING BY ALFVÉN WAVES — KINETIC PROPERTIES OF MAGNETOHYDRODYNAMIC DISTURBANCES

Akira Hasegawa
Bell Laboratories, Murray Hill, NJ 07974

Mechanisms of Alfvén wave heating in space-astrophysical plasmas are presented with particular emphasis on the parallel electric field generated in the magnetohydrodynamic perturbations due to the finite Larmor radius effects.

I. INTRODUCTION

Collisionless heating of plasmas by Alfven waves which has been proposed by Grossmann and Tataronis (1973) and Hasegawa and Chen (1974) is now recognized as a major mechanism of plasma heating both in laboratory, in space and in astrophysical plasmas. The original idea of the heating was based on the resonant absorption of the Alfvén wave in a inhomogeneous plasma. The resonant absorption occurs due to the fact that only a local field line at $x = x_0$ in an inhomogeneous plasma can be in resonance with the Alfvén wave with a given frequency ω and parallel wave number $k_{\parallel}$ such that $\omega = k_{\parallel} v_A(x_0)$ where $v_A(x)$ $[= B_0(x)/(\mu_0 \rho(x))^{1/2}]$ is the spatially varying Alfvén speed.

It was later recognized by Hasegawa and Chen (1976) that the resonant absorption is a manifestation of the linear mode conversion from the MHD Alfvén wave to the kinetic Alfvén wave and that the physical mechanism of the heating depends on the collisionless absorption of the kinetic Alfvén wave. The kinetic Alfvén wave is the Alfvén wave with perpendicular wavelength comparable to the ion gyroradius, ρ_i.

The important recognition here is the fact that the Alfvén wave accompanies an electric field parallel to the ambient magnetic field if $k_{\perp} \rho_i \simeq O(1)$ and the fact this electric field can produce collisionless wave-particle interactions. In space plasmas this electric field has been proposed as a mechanism of accelerating auroral electrons (Hasegawa, 1976; Goertz and Boswell, 1979; Goertz, 1981), as solar (Ionson, 1978) and Stellar (Ionson, 1982) coronal heatings, and acceleration of plasmas in Io torus (Goldstein and Goertz, 1983).

In most cases of plasma heating in space and astrophysical plasmas, the transfer of energy is from the mechanical to the wave and to the particle. Hence the Alfvén wave should first be excited by a mechanical mean. This requires an existence of well defined eigenmodes for an efficient transfer of mechanical energy to the wave energy. Usually the surface Alfvén wave (sometimes called the "kink" mode in laboratory plasmas) is considered for this purpose.

The talks will consist of a brief review of the resonant absorption and resonant mode conversion of the Alfvén wave in an inhomogeneous plasma (II), introduction of the Alfvén surface wave and its modification in the presence of a plasma flow and current (III) and derivation of the local (kinetic) Alfvén wave dispersion relation in a general geometry (IV). Much of the content of the talk can be found in the monograph by Hasegawa and Uberoi (1982).

II. RESONANT ABSORPTION AND RESONANT MODE CONVERSION

In an inhomogeneous plasma with the inhomogeneity in the x direction, the x component of the linearized plasma displacement satisfies the eigen value equation given by (Hasegawa and Chen 1976)

$$\frac{d}{dx}\left[\frac{\epsilon \alpha B_0^2}{\alpha k_\perp^2 B_0^2 - \epsilon}\frac{d\xi_x}{dx}\right] - \epsilon \xi_x = 0 \tag{2.1}$$

where

$$\epsilon(x) = \omega^2 \mu_0 \rho - k_\parallel^2 B_0^2 \tag{2.2}$$

and

$$\alpha(x) = 1 + \frac{\omega^2 v_S^2}{v_A(\omega^2 - k_\parallel^2 v_S^2)}. \tag{2.3}$$

Here $B_0(x)$ is the ambient magnetic field, $v_A(x)$ is the Alfven speed, $v_S(x)$ is the MHD sound speed and $\rho(x)$ is the mass density of the plasma. The equation (2.1) shows that for a given frequency $\omega = \omega_0$, the equation becomes singular at $x = x_0$ where $\epsilon(x_0) = 0$. Near $x = x_0$, the solutions of Eq. (2.1) becomes logarithmic,

$$\xi_x = C \ln(x - x_0 + i\delta). \tag{2.4}$$

The energy absorption rate dW/dt is obtained from the jump in the power flow in the x direction in the cross sectional area $S = L_y L_z$,

$$p = \frac{L_y L_z}{2} \text{Re}(\mathbf{E} \times \mathbf{H}^* + v p^*)$$

$$= -\frac{L_y L_z \omega_0}{2} \text{Im } \xi_x^* \frac{\epsilon}{k_\perp^2 \mu_0} \frac{\partial \xi_x}{\partial x}$$

Thus

$$\frac{dW}{dt} = -\frac{L_y L_z \omega_0}{2} \frac{|C|^2}{k_\perp^2 \mu_0} \left.\frac{d\epsilon_r}{dx}\right|_{x=x_0} \text{Im}[\ln(x - x_0 + i\delta)]_{x_0^-}^{x_0^+}$$

$$= \frac{\omega_0}{2} \pi L_y L_z \frac{|C|^2}{k_\perp^2 \mu_0} \left.\frac{d\epsilon_r}{dx}\right|_{x=x_0} \tag{2.5}$$

The singularity in Eq. (2.1) originates from the "fluid" approximations of a MHD plasma, hence it can be eliminated by taking into account the plasma "kinetic" property. In fact if the Vlasov equation is used, the singularity of Eq. (2.1) disappears due to the appearance of terms with higher spatial derivatives (Hasegawa and Chen, 1976). Then the absorbed energy can be identified to be mode converted to the kinetic Alfvén wave whose dispersion relations in a uniform plasma (local dispersion relation) is given by

$$\omega^2 = k_\parallel^2 \, v_A (1 + k_\perp^2 \, \bar{\rho}^2) \tag{2.6}$$

where

$$\bar{\rho}^2 = \left(\frac{3}{4} + \frac{T_e}{T_i}\right) \rho_i^2 \tag{2.7}$$

with

$$\rho_i^2 = \frac{T_i}{m_i} \frac{1}{\omega_{ci}^2}. \tag{2.8}$$

The kinetic Alfvén wave has the electric field in the direction parallel to the ambient magnetic field whose magnitude is

$$E_z \equiv E_\parallel \simeq E_\perp \frac{k_\parallel}{k_\perp} \tag{2.9}$$

This electric field produces the wave-particle interaction. For example, the linear response of the resonant electrons becomes (Hasegawa and Mima, 1978).

$$f_k = \pi \delta(k_z v_Z - \omega) \left[\frac{e}{m} \frac{\partial f_o}{\partial v_Z} - \frac{k_y v_Z}{\omega B_o} \frac{\partial f_o}{\partial x} \right] E_{zk}. \tag{2.10}$$

The first term produces the velocity space diffusion (momentum and energy transfer) while the second term produces the coordinate space diffusion. The heating rate for example is readily obtained by taking the quadratic velocity moment,

$$n_o \frac{dT_e}{dt} = \text{Re} \, \frac{1}{2} \, (J_{zk} E_{zk}^*)$$

$$= n_o T_e \omega \left[\frac{\pi}{8}\right]^{1/2} \frac{m_i}{m_e} \sum_k \frac{|B_{xk}|^2}{B_o^2} \frac{\lambda_S}{(1 + \lambda_S)^{3/2}} \tag{2.11}$$

$$\times F\left[\frac{v_A}{v_{Te}}\right]$$

where

$$F(x) = x^3 \exp(-x^2/2)$$

and

$$\lambda_S = (k_\perp \rho_S^2)$$

$$= \frac{T_i}{T_e} (k_\perp \rho_i)^2 \tag{2.12}$$

Note that the heating rate vanishes in the MHD limit, $k_\perp \rho_i = 0$.

III. ALFVÉN SURFACE WAVES

When there exists a sharp discontinuity in the plasma density and the magnetic field in the direction perpendicular to the magnetic field, the surface Alfvén wave appears in addition to the bulk Alfvén waves. Using ϵ defined in Eq. (2.2), the dispersion relation of the surface Alfvén wave becomes,

$$\epsilon_I + \epsilon_{II} = 0 \tag{3.1}$$

where ϵ_I and ϵ_{II} are values of ϵ at the two sides of the discontinuity. If one side is vacuum, Eq. (3.1) gives

$$\omega = \omega_S = \sqrt{2} k_\parallel v_A . \tag{3.2}$$

Since ω_S lies between the Alfvén frequencies in the vacuum (∞) and the plasma ($k_\parallel v_A$) regions, there exists a location $x = x_0$ between the vacuum and bulk region where "local" Alfvén frequency $\omega = k_\parallel v_A(x_0)$ becomes equal to ω_S. From the result of Section II, we note that the surface wave is resonantly mode converted to the kinetic Alfvén wave at $x \equiv x_0$ and the wave energy is absorbed through the wave-particle interactions.

The surface wave dispersion relation is modified when there exists a flow or a current in the plasma. In the presence of a flow $\mathbf{v} = \mathbf{v}_0$ in region I, it is given by (Chandrasekhar, 1961)

$$\frac{1}{n_{0I}[(\omega - \mathbf{k} \cdot \mathbf{v}_0)^2 - (\mathbf{k} \cdot \mathbf{v}_A)_I^2]} + \frac{1}{n_{0II}[\omega^2 - (\mathbf{k} \cdot \mathbf{v}_A)_{II}^2]} = 0 . \tag{3.3}$$

Here n_0 is the plasma number density and the subscripts I and II show quantities in the region I and II respectively and the direction of the vector Alfvén velocity is that of the ambient magnetic field. When v_0 exceeds the threshold given by

$$(\mathbf{k} \cdot \mathbf{v}_0)^2 > \left[\frac{1}{n_{0I}} + \frac{1}{n_{0II}}\right] [n_{0I}(\mathbf{k} \cdot \mathbf{v}_A)_I^2 + n_{0II}(\mathbf{k} \cdot \mathbf{v}_A)_{II}^2] . \tag{3.4}$$

the surface wave becomes unstable (Kelvin-Helmholtz instability). the mechanical energy of the flow can be converted to the heat through the excitation of the Alfvén surface wave and subsequent mode conversion to the kinetic Alfvén wave (Osawa, et al., 1976).

In the presence of a current, the dispersion relation becomes geometry dependent. For example, in a cylindrical plasma with radius a surrounded by vacuum, a current which flows on the plasma surface modifies the dispersion relation (Kadomtsev 1966) for the surface wave with a structure $f(r) e^{i(n\theta + kz - \omega t)}$,

$$\mu_0 \rho_0 \omega^2 = k^2 B_{0z}^2 + \left(k B_{0z}^e + \frac{n}{a} B_{0\theta}^e\right)^2 - \frac{n(B_{0\theta}^e)^2}{a^2} . \tag{3.5}$$

Here the subscript zero indicates the ambient quantities, z and θ are axial and azimuthal components and the superscript e indicates the value external to the plasma. In the absence of the current, $B_{0\theta}^e = 0$ and the dispersion relation gives that of the surface Alfvén wave, $\omega = \sqrt{2} k_\parallel v_A$. In the presence of the current, the kink instability sets in for $n = 1$ mode when

$$\frac{B_{0\theta}^e}{B_{0z}^e} > \frac{2\pi a}{L},\qquad(3.6)$$

where $L = 2\pi/k$ is the length of the plasma. For $n \neq 1$, the plasma is stable but the surface wave frequency becomes n-dependent. In particular a mode with frequency *lower* than the bulk Alfvén frequency appears. This mode does not suffer the resonant absorption hence remains to be an undamped high-Q mode. When the current is distributed in the cylinder, the dispersion relation does not have a simple form as shown in Eq. (3.5) and the eigen frequency should be obtained numerically (Appert et al. 1982).

IV. FINITE LARMOR RADIUS MHD EQUATIONS

In an inhomogeneous plasma, the kinetic Alfvén wave couples with drift waves, ballooning modes and other electrostatic modes through the presence of the parallel electric field. The dispersion relation including these effects as well as the wave-particle interactions can be obtained by the use of the gyrokinetic equation (Hasegawa, 1979; Freeman and Chen, 1982). However, the gyrokinetic equation is difficult to use for a nonlinear problem in particular due to the basic two components (electron-ion) property. Recently Hasegawa and Wakatani (1983) derived a new set of *fluid* equations which is capable of treating MHD problems with the finite ion gyroradius correction. The appropriate equations are listed in the following. The first equation is $\nabla \cdot \mathbf{J} = 0$ (which is equivalent to the equation for vorticity) with $J_\perp$ given by guiding center currents. In a low beta plasma,

$$\hat{\mathbf{b}} \cdot \nabla (J_{zi} + J_{ze}) = \frac{m_i n_o}{B_0^2} \frac{d}{dt} \nabla_\perp^2 \phi$$

$$+ \frac{p_{io}}{B_0^2 \omega_{ci}} (\nabla \nabla_\perp^2 \phi \times \hat{z}) \cdot \nabla \ln(p_{io} + p_i)$$

$$+ \sum_{j=i,e} \left[\nabla p_{\perp j} \cdot (\nabla B_o \times \hat{z})/B_0^2 \right.$$

$$\left. + \nabla p_{\parallel j} \cdot (\mathbf{B}_o \times \mathbf{R})/B_0^2 R^2 \right]$$

$$\equiv -\nabla_\perp \cdot \mathbf{J}_\perp,\qquad(4.1)$$

where $\mathbf{R}/R^2 = -(\hat{\mathbf{b}}_o \cdot \nabla)\hat{\mathbf{b}}_o$ is the curvature of the unperturbed magnetic field, p_j and p_{jo} are perturbed and unperturbed pressure of the jth species. The second term on the right hand side originates from the difference of $\mathbf{E} \times \mathbf{B}$ drift between electrons and ions, due to the fact that the ion sees the electrostatic field which is reduced by $\rho_i^2 \nabla^2$. $\hat{\mathbf{b}} \cdot \nabla$ and d/dt are given by,

$$\nabla_\parallel = \hat{\mathbf{b}}(\hat{\mathbf{b}} \cdot \nabla)$$

$$= \hat{z}\left(\frac{\partial}{\partial z} + \frac{B_{oy}}{B_o}\frac{\partial}{\partial y} + \frac{\mathbf{B}_\perp}{B_o} \cdot \nabla_\perp\right),\qquad(4.2)$$

where $\hat{\mathbf{b}}$ is the unit vector in the direction of the total magnetic field, B_{oy} represents the shear field,

$$\mathbf{B}_\perp(\mathbf{x},t) = \nabla_\perp A_z \times \hat{z}\qquad(4.3)$$

is the perturbed magnetic field,

$$\frac{d}{dt} = \frac{\partial}{\partial t} + v_E \cdot \nabla$$

$$= \frac{\partial}{\partial t} - \frac{\nabla\phi \times \hat{z}}{B_o} \cdot \nabla, \qquad (4.4)$$

and ϕ is the scalar potential for the electric field. Maxwell equations become

$$\nabla^2 A_Z = -\mu_o J_z,$$

Parallel component of the generalized Ohm's law which may be obtained from the first moment of the electron drift kinetic equation is given by

$$\eta J_{ze} = -\frac{\partial A_z}{\partial t} - \hat{\mathbf{b}} \cdot \nabla\phi + \frac{1}{en} \hat{\mathbf{b}} \cdot \nabla(p_{\parallel_e} + p_{oe}) \qquad (4.6)$$

Here η is the resistivity, and J_{ze} is the z component of the electron current. If $k_\parallel$ is small enough such that the parallel ion current J_{zi} is ignorable, appropriate equations of state which relate p_i and p_e to the other field variables complete the equations. One possible choice of the equations of state is incompressive ions,

$$\frac{dp_{\parallel i}}{dt} = \frac{dp_{\perp i}}{dt} \equiv \frac{dp_i}{dt} = 0$$

and isotropic, isothermal electrons, $\qquad (4.7)$

$$\cdot \frac{dp_{\parallel e}}{dt} = \frac{dp_{\perp e}}{dt} = \frac{T_e}{e} \hat{\mathbf{b}} \cdot \nabla J_{ze}.$$

The local dispersion relation for a case $J_o = \eta = 0$ may be obtained from Eqs. (4.1), (4.2), (4.3), (4.6), and (4.7). For a simple curved field line,

$$\nabla p \cdot (R \times B_o)/R^2 B_o^2 \simeq \nabla p \cdot z \times \nabla B_o/B_o^2 = -\frac{\partial p}{\partial y} \frac{1}{RB_o} \qquad (4.8)$$

where x axis is taken in the direction of the radius of curvature and that of the pressure gradient. If we define the electron and the ion drift wave frequency,

$$\omega_e^* = \frac{k_y T_e}{eB_o} \frac{\partial}{\partial x} \ln p_{eo} \qquad (4.9)$$

and

$$\omega_i^* = \frac{k_y T_i}{eB_o} \frac{\partial}{\partial x} \ln p_{io} \qquad (4.10)$$

the local dispersion relation becomes,

$$(\omega + \omega_e^*) [\omega^2 - \omega_i^* \omega - k_z^2 v_A^2$$

$$- \frac{2k_y^2}{m_i R k_\perp^2} (T_i \frac{\partial}{\partial x} \ln p_{io} + T_e \frac{\partial}{\partial x} \ln p_{eo})]$$

$$= \frac{k_z^2 v_A^2}{\omega} (\omega - \omega_i^*) [k_\perp^2 \rho_s^2 \omega + \omega_i^* (R \frac{\partial}{\partial x} \ln p_{io})^{-1}] \qquad (4.11)$$

Equation (4.11) shows that the kinetic Alfvén wave dispersion relation is modified significantly when $k_y \rho_i \simeq 0(1)$ through the couplings to the drift waves. In addition the curvature of the field line can produce an instability of the field line to balloon in the region where $2p_o/\partial x < 0$ (ballooning instability).

V. CONCLUDING REMARKS

Magnetohydrodynamic perturbations can couple to kinetic properties of a plasma through the finite Larmor radius effects. This originates from the generation of electric fields in the direction of the ambient magnetic field and subsequent wave-particle interactions. In case of the Alfvén wave, this interaction heats the bulk region of the electron phase space while accelerates ions to the Alfvén speed if $m_e/m_i < \beta < 1$. Through the mode conversion in an inhomogeneous plasma and through the wave steeping (ultraviolet catastrophy) MHD perturbations with $k_\perp \rho_i \simeq 0(1)$ can easily be excited. Thus the process discussed here is quite universal and is expected to play an important role in the exchange of energy between MHD modes and particles.

REFERENCES

Appert, K., Balet, B., Gurber, R., Troyon, F., Tsunematsu, T. and Vaclavik, J., 1982, Plasma Phys. 24, 1147.

Chandrasekhar, S., 1961, *Hydrodynamic and Hydromagnetic Stability*, Clarendon Press, Oxford, Chapt. 13.

Freeman, E. A. and Chen, L., 1982, Phys. Fluids 25, 502.

Goldstein, M. L. and Goertz, C. K., 1983, *Physics of the Jovian Magnetosphere*, Ed. by A. J. Dessler, Cambridge Univ. Press. NY, 317.

Goertz, C. K., and Boswell, R. W., 1979, J. Geophys. Res. 84, 7239.

Goertz, C. K., 1981 AGU Monography 25, *Amoral Arc Formation*, 451.

Grossmann, W. and Tataronis, J., 1973, Z. Phys. 261, 217.

Hasegawa, A. and Chen, L., 1974, Phys. Rev. Lett., 32, 454.

Hasegawa, A. and Chen, L., 1976, Phys. Fluids 19, 1924.

Hasegawa, A., 1976, J. Geophys. Res. 81, 5083.

Hasegawa, A. and Uberoi, G., 1982. *The Alfvén Wave*, Tech. Inf. Center, US Dept. of Energy, Oak Ridge, Tenn.

Hasegawa, A. and Mima, K., 1978, J. Geophys. Res. 83, 1117.

Hasegawa, A., 1975, *Plasma Instabilities and Nonlinear Effects*, Springer Verlag, NY, 128.

Hasegawa, A., 1979, Phys. Fluids 22, 1988.

Hasegawa, A., and Wakatani, M., 1983, Phys. Fluids to be published.

Ionson, J. A., 1978, Ap. J., 226, 650.

Ionson, J. A., 1982, Ap. J., 254, 318.

Kadomtsev, B. B., 1966, *Review of Plasma Physics*, Ed. by M. A. Leontovich, Consultant Bureau, NY, 174.

Osawa, Y., Nozaki, K. and Hasegawa, A., 1976, Phys. Fluids 19, 1139.

DISCUSSION

Davila: You indicate that the normal and inverse cascade processes result in a concentration of turbulent power at scales on the order of the ion gyroradius. This, however, is also where wave damping by the thermal protons is maximized. It would seem that one would expect turbulent wave power to be a minimum at scales on the order of the ion gyroradius.

Hasegawa: The spectral cascade is purely a consequence of a "fluid" picture of the magnetohydrodynamic turbulence. When the kinetic picture is taken into account one should consider the detailed balance. In an inhomogeneous plasma many different types of micro-instability emit waves in this regime leading to a quasi-equilibrium state of a high level of turbulence at $k_\perp \rho_i \sim O(1)$, a situation somewhat similar to the consequence of the fluctuation dissipation theory. In fact recent measurements of density and magnetic field fluctuations in Tokamak type plasmas have revealed that super-thermal fluctuations are concenterated at $k_\perp \sim \rho_i^{-1}$.

Lotko: Does the inverse cascade depend on the dimensionality of the turbulence and is there coupling to the compressional Alfvén wave?

Migliuolo: Is the inverse cascade a correct picture for high-β plasmas, where the shear Alfvén wave couples to the compressional Alfvén wave?

Hasegawa: "Two dimensionality" is essential in the inverse cascade. This requires a low β situation in which the magnetic field controls the plasma dynamics.

Vasyliunas: Your kinetic MHD theory holds for $(k_\parallel/k_\perp)^2 \leq \rho_i^2/R^2$, but in astrophysical and space applications one expects a minimum value $(k/k_\perp)^2 \sim (v/v_A)^2$ where v is, e.g., the speed of magnetospheric disturbance sources relative to the earth's field, and this ratio can be much larger than $(\rho_i/R)^2$. What happens to the theory in that case?

Hasegawa: Generation of perturbation with a very large value of $k_\perp(\rho_i^{-1})$ originates through 1. singularity of MHD perturbations in an inhomogeneous plasma, 2. through the so called ultraviolet catastrophy; the continuous generation of shorter wavelengths due to nonlinear mode couplings, and 3. micro-instabilities such as drift wave modes.

Norman: It would be very nice to apply the physics to jets, but here β may be at least of order unity. Do you have any ideas of how this may be done?

Hasegawa: In a high β case (β $\sim$ 1), the present expansion scheme breaks down. This could mean that the microturbulence may not be important in a high β case except for specific cases (such as the drift mirror instability).

Vlahos: In your talk you mention only the Landau resonance for obliquely propagating Alfvén waves. Are the higher order cyclotron interactions $\omega - k_\parallel v_\parallel - n\Omega_e$ important?

Hasegawa: The Landau resonance is important for "heating" because it affects the bulk of the distribution of plasma particles. The cyclotron resonance is relevant to the high energy tail, such as for cosmic rays.

Sturrock: What role do these processes play in heating the solar corona?

Hasegawa: Microscopic perturbations produce direct wave-particle interactions through the parallel electric field that they accompany. This indicates that there is a coupling between MHD perturbations and kinetic effects in plasmas. A number of papers seem to have been published on coronal heating through a process described in the talk.

Ionson: In solar loops the dissipational scale length is $\sim 10^5$ cm compared to an ion gyroradius of 10^2 cm. Therefore, since the resonance layer thickness is 10^3 times larger than ρ_{ci} (i.e. $k_\perp \rho_{ci} \sim 10^{-3}$) it appears that finite Larmor radius corrections are not necessary.

Hasegawa: You are right. Finite Larmor radius corrections and related kinetic effects are important for relatively collision free plasmas such that the ion Larmor radius is smaller than the dissipation scale length.

KINETIC THEORY OF ALFVÉN WAVE HEATING

S. M. Mahajan
Fusion Research Center, Dept. of Physics
University of Texas
Austin, TX 78712

Abstract

In a magnetohydrodynamic description of a plasma, the shear Alfvén wave is characterized by a continuous spectrum (the MHD continuum) which results from a singularity at $\omega \simeq k_{\|}(r)V_A(r)$, where ω is the frequency, $k_{\|}$ is the wave number along the direction of the magnetic field and $V_A = B_0/(4\rho\pi)^{1/2}$ is the Alfvén speed, and r is the direction of inhomogeniety. The associated electromagnetic fields become large at this resonance layer [in fact, in ideal MHD they become singular], and if some dissipation is allowed, the transfer of energy from the waves to the electrons can take place. These continuum modes are localized in space, and can be effective in heating plasmas in narrow regions around the resonanct surface. Recently,[1-3] a new class of nonsingular global eigenmodes of the plasma have been discovered which are wide spread in the plasma, and could effectively transfer energy to the bulk of the plasma. These modes arise because of a strong coupling between the shear and the evanescent compressional mode. The coupling is provided by gradients of density, equilibrium current and nonideal effects like finite ω/ω_{ci}, where ω_{ci} is the ion cyclotron frequency. In order to understand the effectiveness of these modes to heat plasmas, i.e., to determine the effective impedance, the energy deposition profiles, parametric dependence of the heating efficiency, etc., a kinetic theory with the electron parallel dynamics is needed. We have developed a detailed kinetic theory to study the structure of the continuum as well as global eigenmodes in an inhomogenous current carrying plasma. The global modes are comparatively much less damped than the continuum modes (which become discrete in the kinetic theory) and hence are easier to excite. The implication, of course, is that the large part of energy transfer in a Alfvén Wave experiment would be mediated through these global modes. This theoretical conjecture is already confirmed in laboratory experiments on PRETEXT Tokamak.

I will present the basic theoretical analysis as well as discuss the implications of these findings.

References

1. D.W. Ross, G.L., Chen, and S.M. Mahajan, "Kinetic description of Alfven wave heating", Phys. Fluids $\underline{25}$, 652 (1982).

2. S.M. Mahajan, D.W. Ross and G-L. Chen, "Discrete Alfven Eigenmode Spectrum in Magnetohydrodynamics", The University of Texas, Fusion Research Center Report #249.

3. S.M. Mahajan, "Kinetic Theory of Alfven Waves", The University of Texas, Institute for Fusion Studies Report #84.

MHD EQUILIBRIUM AND INSTABILITIES IN EXTRAGALACTIC JETS

Attilio Ferrari

Istituto di Fisica Generale, Universita' di Torino, Italy
Harvard-Smithsonian Center for Astrophysics, Cambridge, USA

ABSTRACT

Morphologies, energetics and nonthermal radiation emission of extragalactic radio sources can be explained in the framework of models based on the physics of supersonic jets interacting with the surrounding intergalactic medium. In this review current physical interpretation of acceleration, collimation, modulation and nonthermal radiation of these jets are discussed. We begin with an analysis of the equilibrium configurations of collimated flows; the propagation outside the galactic nuclei is shown to be modulated by fluid, MHD and resistive instabilities, which, at the same time, support charged particle acceleration by turbulent stochastic processes, and consequently nonthermal radiation emission from the radio to the X-ray range. Large scale morphologies (knots, wiggles and bendings) are considered from the point of view of their interpretation as global perturbations of the flow produced by fluid instabilities.

I. INTRODUCTION

The general scheme of our present knowledge about extragalactic jets has been summarized by Colin Norman in a previous review at this Symposium. Their most striking characteristic is that they emerge already collimated from the parent galactic nuclei (<< 0.1 pc) and afterwards maintain their perfect collimation over large distances until they reach the final lobes (up to a few Mpcs from the core, in some cases); all of this even though expansions, recollimations, wiggles, sharp bendings, etc., which clearly indicate a continuous interaction with the external medium. In Fig. 1 a few typical examples of this behavior are shown, namely NGC 6251 (perfect collimation and wiggles, Bridle and Perley 1983), 3C 129 (large bending, Burns 1983), 3C 418 (sharp helical structure in the nucleus, Browne et al. 1983). Fig. 2 gives the X-ray emissivity along the jet of Cyg A as measured by the Einstein Observatory (Feigelson 1982), which, when interpreted as bremsstrahlung emission, provides the density distribution of the gas around the jet (emissivity $\propto n^2$). More detailed observational data and theoretical arguments can be found in the Proceedings of the IAU Albuquerque Symposium on **Extragalactic Radio Sources** (Heeschen and Wade eds., 1982) and of the Torino International Workshop on **Astrophysical Jets** (Ferrari and Pacholczyk eds., 1983).

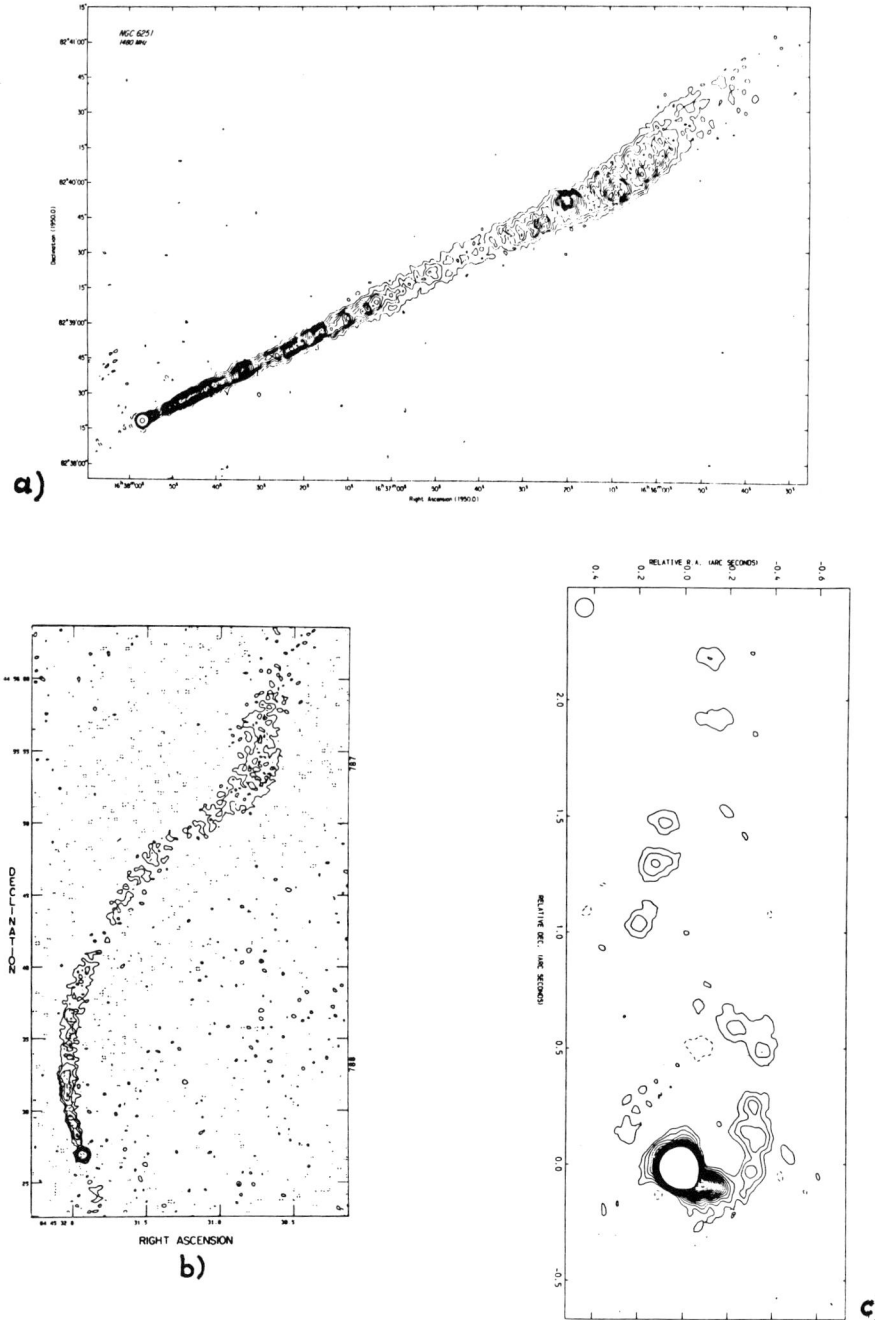

Fig. 1 - (a) VLA map of NGC 6251; (b) VLA map of 3C 129; (c) MERLIN map of 3C 418.

In the framework of the existing scenario I present in this review some of the arguments which have been recently debated about the physical processes involved in the propagation and stability of extragalactic jets; these are typical plasma physics phenomena, which have been applied in similar contexts, as it appears from the papers presented at this Symposium: I must however warn the reader that, models for jets are at present underconstrained, and many of them can actually fit the data. We have still to find out which effects are actually more important.

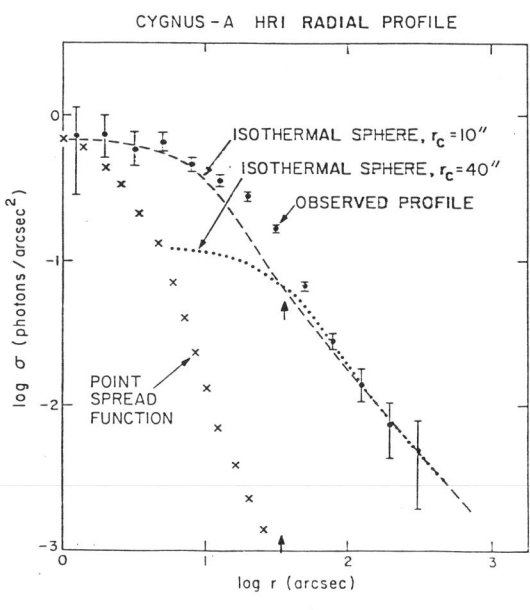

Fig. 2.
X-ray emissivity from regions surrounding the jet of Cyg A; distances are indicated from the core; arrows indicate the position of radio lobes (courtesy of E. Feigelson).

I shall devote the three following Sections to discuss collimation, stability and radiation of jets respectively; namely I shall try to emphasize which physical processes appear to be more relevant, and therefore deserve more detailed studies in the future.

In Table I the standard set of numerical values for the physical parameters of radio sources is presented: these values are derived using the incoherent synchrotron model for the radio emission from these sources, and the argument of energy equipartition between magnetic field and relativistic particles (Pacholczyk 1970). Table I illustrates the values for the four regions in which a radio source can be divided: core, jet, hot spot, extended lobe. In this respect one should keep in mind that our estimates are obviously obtained from the relativistic component of the plasma in the jet, as this is responsible for the observed nonthermal emission; the following discussion will be concerned instead with the bulk flow, and therefore with the thermal component of jets. How these different components are actually related is not clear at present, but the argument of equipartition appears to be valid also with thermal particles, at least as a constraint to estimates.

TABLE I
Average Physical Parameters in Jets

	Core	Jet	Hot-spot	Lobe-Halo
r(kpc)	<0.01	2÷1000	5	50÷1000
B_{eq}(G)	10^{-3}	10^{-5}	10^{-5}	$10^{-5.5}$
$n_{rel}(cm^{-3})$	$10^{-5.5}$	$10^{-10.5}$	$10^{-10.5}$	10^{-11}
$n_{th}(cm^{-3})$	-	$10^{-2} \div 10^{-5}$	$< 10^{-2}$	$> 10^{-5}$
% polariz.	< 2	0 ÷ 60	15	0 ÷ 60
V_b/c	< 1	$10^{-2} \div 10^{-1}$	10^{-3}	10^{-3}

Finally Table II shows a list of the measured lengths of small amplitude oscillatory perturbations (*wiggles*) in units of the observed jet radius (Bridle, private communication); they will be later confronted with the characteristics of the instabilities acting upon the flow. Note that all the jets in this list flare up after 2 - 3 oscillations of measurable amplitude.

TABLE II
Wiggle Wavelengths in Units of Jet Radius R
(R is taken as the mean HWHM of the jet
over the region of wiggling)

NGC 315 Base	1.3 kpc	6 R
M 87 Knots	0.6	22
3C 31 North	5	4
3C 449 South	17	14
NGC 6251 Outer	32	8
0326+396 West	36	12
4CT 74.17.1	56	7

II. EQUILIBRIUM CONFIGURATIONS OF JETS

The idea that extended extragalactic radio sources might indicate a continuous outflow of matter from active galactive nuclei was first proposed by Rees in 1971; until then, extended lobes had been actually considered the remnants of episodic ejections of single blobs of matter. The immediate advantage of the continuous outflow model was that the energy required from the associated galactic nuclei to support these structures could be diluted in time; but the idea was soon substantiated by high sensitivity observations which displayed bridges of emission connecting cores and lobes, and showed that the brightness distribution of these jets is actually continuous. Nevertheless direct measurements of jet velocity do not exist, if one excludes the detection of superluminal motions inside cores. Perhaps direct results will be soon available through optical line emission data, which are now being collected (Miley 1983). At present however we still rely solely upon the self-consistency of the idea.

II.1 Basic physical parameters and approximations

Originally, because of the low density suggested by observations, jets were studied in terms of the motion of test particles inside cavities plowed through the intergalactic medium by an original ejection of denser blobs of high-velocity matter (Longair et al. 1973); Rees (1971) also considered the possibility of jets powered by a low-frequency electromagnetic wave, as in pulsar models (with the appropriate scalings). Later it was realized that in fact collisionless beams of ejected particles might show collective aspects, due to the underlying action of magnetic fields, whose presence is implied by synchrotron models (Blandford and Rees 1974). I start therefore with a brief analysis of the arguments which allow to apply fluid considerations to a jet.

From the data of Table I it is easy to calculate that the collision mean-free-path is actually larger than the typical radius of a jet, while the thermal (T > 10^7 K) ion gyroradius is much smaller:

$$\ell_{coll} = \frac{(kT)^2}{n e^4 \ln \Lambda} \gtrsim R \gg r_g = \frac{m v c \gamma}{e B}. \tag{1}$$

Typical values for ℓ_{coll} are larger than 10^4 pc and for r_g smaller than 10^{12} cm. This means that the plasma in a jet is essentially collisionless in terms of short-range interactions, but actually behaves collectively due to long-range interactions. Therefore we can assume, at least as a first approximation, that fluid equations can be used to describe the jet propagation. The plasma can be considered as quasi-neutral, since its Debye length λ_{De} is smaller than 10^7 cm; nevertheless a charge separation $|[n^- - n^+]/[n^- + n^+]| \sim 10^{-15}$ can be easily present, which is sufficient to generate magnetic fields of the order of equipartition fields.

The situation is more uncertain with viscosity and pressure anisotropy, as these are related with microscopic effects (undetectable directly). Deviations from pressure isotropy are likely to be negligible in the fluid approximation, as fast microscopic instabilities are ready to restore it; in addition we know that they do not modify substantially MHD stability calculations (Sen 1963). Much more delicate is the influence of viscosity, as we shall also discuss later in relation with numerical simulations. In accordance with some results valid for the case of the solar wind interacting with the Earth's magnetosphere (Eviatar and Wolf 1968), we may estimate that the value of the Reynolds number, Re, in terms of an *effective magnetic turbulent viscosity* is:

$$Re = \frac{R}{r_g} \gtrsim 10^{10}. \tag{2}$$

As the formation of boundary layers and shears depends critically on the Reynolds number, their direct measurement would allow to put constraints to many of the models that we shall discuss in the following. However at present we can only say that the thickness of shear layers is certainly larger than r_g and probably of the order of R; this upper value is suggested by the study of instability (Sec. III).

II.2 Flow confinement

The next important question to answer before using a fluid model for jets, and for studying their stability, refers to the interaction with the external medium. A specific discussion on this point for the case of M87 jet is presented by Hardee (these Proceedings). Here I restrict myself to the general arguments.

The basic question is: are jets confined or free? Which is the evidence from observations? The internal and external pressures are both composite:

$$p_{int} = p_{th} + p_{rel} + \frac{B_{\parallel}^2 + B_{\perp}^2}{8\pi} \tag{3}$$

$$p_{ext} = p_{IGM} + \frac{B_{IGM}^2}{8\pi} + \frac{B_{\theta}^2}{8\pi}. \tag{4}$$

Namely the internal pressure is the sum of the pressures of thermal plasma, relativistic electrons and internal magnetic fields; the external pressure is due to the intergalactic medium thermal and magnetic pressures plus the confining pressure of azimuthal magnetic fields consistently generated by currents flowing along with the jet. Equipartition arguments on radio luminosities and spectra give lower limits for the value of the internal pressure (the relativistic and magnetic components only are included); X-ray emissivity of halos around jets allows to determine p_{IGM}, while the magnetic pressure of the IGM is considered negligible. No direct measurement of the azimuthal fields can be made, consistently with the fact that it has to be external to the jets, in regions of low effective emissivity.

The general trend of observational data is the following (see Miley 1980). Large scale structures and extended jets appear to be thermally confined; bright knots and hot spots are probably non-equilibrium, evolving features (e.g. shocks). Ram pressure confinement can also be important at the jet leading surface. However in at least three well-studied jets, M87 (Schreier et al. 1982), Cen A (Feigelson et al. 1981) and 4C 32.69 (Potash and Wardle 1980), p_{IGM} appear to be one or two order of magnitude below the lower limit for p_{int}. On the other hand, these jets appear to be very far from what one would expect in a free jet: namely, no regular opening angle is measured, but the flow in all cases goes through various phases of expansion; in addition the momentum flux that one derives for 4C 32.69 is so large, that it becomes impossible to explain the sharp bendings in terms of freely expanding flows. The obvious conclusion is that the azimuthal magnetic field is required for confinement (whether the above three cases are in fact exceptions cannot be decided at present). However one must keep in mind that magnetic confinement can be globally unstable if the external field required is substantially larger than the internal equipartition field (Parker 1979).

There are also indirect arguments which support the idea that jets are in some way confined. For instance Bridle (1982) has studied in some sources the extension of the beam radius against the distance from the nucleus, $R \propto z^{\alpha}$, finding that α is generally variable, in some cases even negative (recollimation), rarely close to 1, as it should be for free jets. This variable behaviour (from expansion to recollimation in a rather irregular way) indicates that jets do not simply expand at the diffusion speed (i.e. the speed of sound); on the contrary, they may go through different regimes in which they manage to reach equilibrium against some external confining agent, magnetic or thermal pressure. Clearly in these phases one expects the formation of boundary layers, with magnetic field amplification and development of viscous stresses. It will be interesting to detect observational evidences of the presence of these regions by using radio techniques with high angular resolution.

II.3 Hydrodynamic equilibrium models

Several authors have investigated the dynamics of jets, elaborating steady equilibrium models, based on different physical *ingredients*. I shall divide models in two main classes, depending on whether they include the effects of confining magnetic fields.

The pioneering paper of the first class is by Blandford and Rees (1974), who however included some pre-existent embrionic ideas (Gull and Northover 1973, Longair et al. 1973). An active galactic nucleus contains hot relativistic plasma buoyant with respect to the surrounding galactic matter; hot bubbles can then escape along the direction of least resistance, namely the rotation axis of the galaxy. The balance of hydrostatic and gravitational pressures external to the flow provides the conditions for the formation of two opposite nozzles, where the bulk flow velocity becomes supersonic at the expenses of the bubbles internal energy. Afterwards the flow remains supersonic, and the subsequent morphology is determined by the interaction with a light extragalactic medium; the extended radio lobes represent the head of this supersonic flow where a *working surface* dissipates most of the bulk kinetic energy.

In practice this model explain the acceleration by assuming a suitable shape for the confining plasma. Correspondingly, nozzles occur on galactic scale, and therefore relatively far from the core, as their position is essentially determined by the overall gravitational field; observations show instead that jets form well inside the cores. Numerical computations performed by Wiita (1978) have tried to overcome the difficulty, but still the position of the nozzle is an external input.

A new approach in the framework of fluid models has been recently devised by Ferrari et al. (1983a) and is discussed in more detail by Tsinganos in these Proceedings. The idea is to follow the acceleration and collimation of the jet from the accretion disk itself using the quasi 2-D theory of stellar winds (see for instance Kopp and Holzer 1976). The power of the method is based on reducing the problem to a single coordinate along the flow and transforming the lateral stresses in the form of geometrical effects. Obviously this means that one must assume independently the physical conditions for confinement, but from then on the flow is completely determined. It is then possible to write an equation for the flow velocity along the jet, where the forces acting on the plasma are the gravitational pull, the thermal pressure gradient and other forms of momentum deposition. In this scheme the acceleration can be produced either by the geometry of the confining medium, which determine the effective pressure gradients, or by the deposition of momentum through electromagnetic or plasma waves formed at the boundaries of the confining medium (the disk walls inside accretion funnels, or the interstellar medium outside the core). In particular one can follow the transition to supersonic velocity and define the conditions that this takes place well inside the disk funnel; also one can predict directly the formation of shocks when more critical solutions are allowed by the the parameters of the configuration. The momentum deposition needed to accelerate high-velocity beams is found to correspond to the Eddington limit; one can also predict morphologies created by the shocks induced by lateral perturbations, as for instance kink modes from Kelvin-Helmholtz instabilities.

The propagation of free jets has been calculated in some detail by Sanders (1983, preprint). What would be the appearance of a jet coming out from the galactic core into an underpressured region? Actually it will not be allowed to expand freely and smoothly, because its sudden expansion will produce lateral shocks, when the expansion velocity reaches the sonic velocity. Ultimately these shocks will provide a sort of self-collimation (see also Courant and Friedrichs 1948). This result means that, after all, a free jet cannot be very different from a confined jet because it must always be surrounded by a layer where pressure balances thermal expansion. Begelman (1982) has discussed how confined jets can simulate free jet expansion in terms of the effect of viscous stresses at the flow boundary.

II.4 MHD equilibrium models

The prototype of the class of models in which magnetic fields are taken into account consistently, is by Chan and Henriksen (1980), who have derived a self-similar solution for the dynamical MHD equations of an axially symmetric 2-D flow, including rotation around the symmetry axis. They were able to reproduce the observed morphologies and their various regimes of expansion, assuming reasonable distributions for the external pressure p_{ext} (z) along the flow and different values of the ratio

$$\epsilon_B = \frac{B_\theta^2}{4\pi\rho v_{z\infty}^2} , \qquad (5)$$

which measures the pinching effect of the azimuthal magnetic field (at the nozzle point) over the flow (asymptotic) longitudinal velocity; i.e. high-velocity jets require some large B_θ at the nozzle . Recently Krautter et al. (1983) have pushed the implications of this model further, taking into account specific equations of state for the jet plasma. They find good fit to observations for plasma actually dominated by the relativistic electron component, which is a rather interesting result. The magnetic fields required for pinching the flow in recollimation regions are always close to the equipartition fields.

In a similar framework Benford (1978) has discussed the origin of the azimuthal field in terms of a longitudinal current carried by the flow itself; to produce such a current a charge separation of the order of 10^{-15} is already sufficient. In support to this model one may hope to detect the azimuthal field. Observations so far are not favourable: strong jets have longitudinal fields, and jets with perpendicular fields have longitudinal structures at the edges. However in this model B_θ has to be external to the jet; then relativistic electrons are very few in regions where they could emit in the presence of B_θ. Therefore observations of weak background fields are necessary. A problem in this model is how to produce well ordered currents inside the jet, and the corresponding return currents outside it.

II.5 Numerical simulations

Sophisticated numerical simulations are now under study by several authors. The most detailed calculations published so far have been obtained by Norman et al. (1983) using a 2-D axially symmetric fluid code; this code allows the analysis of the rich morphology arising from the interaction of a supersonic flow with the surrounding medium *in conditions of pressure equilibrium*. The acceleration process is assumed to take place following the twin-jets model proposed by Blandford and Rees; collimation is provided by the external pressure at the nozzle and afterwards. The flow (see Fig. 3) develops perturbations at the contact discontinuity with the external medium, forming ripples and corrugations, and shedding its kinetic energy into a cocoon flowing backwards with respect to the leading edge. The cocoon is thicker for jets lighter than the external medium, corresponding to the formation of larger ripples. Internal shocks form along the jet, with spacing typically fixed by the fastest growing modes of the Kelvin-Helmholtz instability (see next Section). However the flow as a whole appears to be dynamically stable, since all perturbations moving backwards from the leading surface are progressively damped. No magnetic field effects are contemplated in this code and thus one cannot test other forms of confinement.

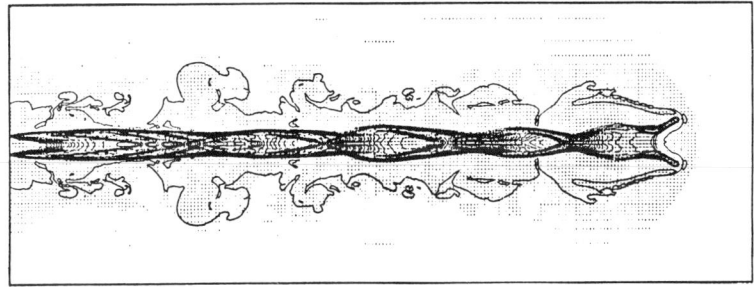

Fig. 3 - Numerical simulation of hydrodynamic jets (after M. Norman).

The numerical approach, though impressive, suffers from some fundamental inconsistencies (in addition to the fact that we are not able yet to describe the physics of the phenomena it displays).

1) The code does not allow non-axisymmetric perturbations, which on the other hand appear to be the most dangerous for the stability, if one refers to the data of laboratory experiments with collisional fluids.

2) In order to treat the formation of shocks the code has to include a numerical viscosity, which corresponds to Reynolds numbers much lower than those predicted by Eq. (2): this may excessively damp instabilities on short scales and thence stop the formation of turbulent diffusion.

Much more work is required on these points, which are in fact very critical for understanding the physics of these flows and to relate their morphology with the observed emission (I shall comment more on the second aspect in Sec. IV). However I note that several numerical works of interest in this context are already in progress. Wiita (1978) and Yokosawa et al. (1983) have addressed to the relativistic effects in the Rees and Blandford's scheme. Nepveu (1980) has directly discussed the effects of viscosity, finding results very similar to those of Norman et al., but acknowledging the inconsistency of assuming high effective viscosity. Woodward (1983) is working on a 3-D code that will permit to represent non-axisymmetric effects. Finally I mention the numerical study of MHD flows presented at this Symposium by Uchida, which, if applicable to extragalactic jets, will enable us to evaluate some aspects of the physics of magnetically confined flows.

III. STABILITY OF DIRECTED FLOWS

A simple look at standard textbooxs on fluid dynamics (e.g. Goldstein 1965, Van Dyke 1982) must make anybody wonder how extragalactic jets manage to propagate over long distances without loosing collimation, while laboratory jets become soon turbulent and spread with large opening angles. The original idea to circumvent this objection was proposing that extragalactic jets are highly supersonic and free, and therefore do not loose their momentum to a surrounding medium, but expand at the speed of sound, which is much lower than the flow velocity. However I have discussed in the previous Section that observational data on morphologies do not support that idea: jets appear to *feel* the presence of an external confinement. Whether this is a thermal or magnetic pressure does not make much difference: jets seem to be very vulnerable to instabilities of all kinds.

III.1 Kelvin-Helmholtz instabilities

A large number of papers have been written on the linear stability analysis of collimated fluid jets in extragalactic radio sources (Blake 1972, Scheuer and Turland 1976, Blandford and Pringle 1976, Ferrari et al. 1978, Ferrari and Trussoni 1978, Hardee 1979, Benford 1981, Ray 1981, Cohn 1983).

The most obvious instability from which confined fluid flows are likely to suffer is Kelvin-Helmholtz instability, arising between fluids in relative motion. The classical literature, referring to the case of an infinite plane contact surface separating (incompressible or compressible) fluids (Chandrasekhar 1961, Gerwin 1968), shows that instability sets in for all perturbation scale lengths provided their wave vector is larger than the gradient scale of the physical quantities across the contact surface. At the same time, when considering compressible flows, an upper limit on unstable Mach numbers, M_{uc}, is found (for perfect gas is $M_{uc} = \sqrt{8}$). This limit arises from the obvious fact that, when the flow is largely supersonic, a perturbation cannot satisfy causality conditions across the contact surface, as it travels typically at the sonic speed. As a matter of fact the limit actually refers to the Mach number component along the wave vector of the mode considered. In fact, Kelvin-Helmholtz modes perpendicular to the flow, but still along the contact surface, can always grow fast enough to affect any flow.

Recent works (Ferrari et al. 1981) have explored the relativistic regime in the presumption that fast enough flows would lead consistently to a progressive decrease of the growth rates. This effect is however not so large to modify the fact that jets are unstable; the region of unstable Mach numbers presents in this case a lower cutoff also, $M_{lc} \gtrsim \sqrt{3}$, due to the finite ratio between the flow speed, $V = \beta c$, and the thermal speed, $V_s \lesssim c/\sqrt{3}$, in relativistic plasmas.

Quite essential in the astrophysical context is the presence of magnetic fields. Ferrari et al. (1981) have shown that a strong, aligned magnetic field, such that the Alfven velocity is larger than the sonic velocity by about an order of magnitude, $V_a \gtrsim 10\, V_s$, can stabilize modes parallel to the flow, while perpendicular modes always tend to remain unstable (at the same time the system becomes dynamically unstable). Similarly, fields perpendicular to the flow, but parallel to the surface discontinuity, fail to slow down the growth of the perpendicular modes. Neither an external magnetic field, aligned (Cohn 1983) or wrapped (Benford 1981) around the flow, of strength of the order of the internal pressure, can provide stability, although it may regulate the equilibrium structure (as discussed in the previous Section). Finally the growth rate of instability is typically proportional to the wavenumber component parallel to the flow; short wavelengths have faster growth rates, and parallel modes grow faster than perpendicular modes.

All these results apply to wavelength perturbations much smaller than the beam transverse size, so that one can consider the contact surface infinite and plane. On the other hand, if we study large scale perturbations of beams, we must look for wavelengths of the order of the beam radius. This motivation has led to study Kelvin-Helmholtz instability in cylindrical jets. Ferrari et al. (1978, 1981), Hardee (1979), Ray (1981) and Cohn (1983) have analyzed a large set of physical situation (compressible fluids with and without magnetic fields, thermally or magnetically confined, relativistic in flow velocity and/or temperature, contact discontinuity and shear layer for the surface boundary), using the following expression for the perturbations:

$$f(r, \theta, z, t) \propto g(r)\, e^{[i(kz + m\theta - \omega t)]}. \tag{6}$$

Depending on the value of m, the azimuthal number, one can represent azimuthally symmetric perturbations or pinches (m = 0), kinks or helical modes (m = 1), fluting modes (m ≥ 2). The general outcome of these analyses is again that instability is always present for all cases, with even larger growth rates; more specifically:

1) The finite transverse dimension of the flow introduces a typical scale in the problem; thus one can estimate that the maximum growth rates correspond to modes with kR ~ 1/M (k is the wavenumber, R the beam radius) (see typical examples in Fig. 4). Both subsonic and supersonic flows are unstable; however in subsonic flows pinching modes have growth rates lower than helical modes, while in supersonic flows they are similar, due to the intervention of a new branch of unstable modes, the so-called *reflected modes*. This branch arises from a destabilization of marginally stable modes in the case of plane geometry (Gill 1965).

2) Strong magnetic fields along the flow have a stabilizing effect; for V_a/V_s >> M ordinary mode instability is actually suppressed. Reflected modes are stabilized only at large wavelengths.

3) In shear layers all modes with wavelengths smaller than the thickness of the boundary layer are stable or have very small growth rates; this fact helps to concentrate the instability to modes with kR ~ 1/M.

4) Relativistic flows behave substantially like supersonic flows.

5) Very dense and very light beams have small growth rates.
6) Numerical values for the fastest growing modes are:

$$\text{oscillation frequency} \quad \omega_r \cong kR\frac{M}{\tau_c}, \quad \tau_c = \frac{2R}{V_s},$$

$$\text{growth rate} \quad \omega_i \cong kR\frac{\Phi}{\tau_c},$$

(7)

where τ_c is the beam crossing time for sonic perturbations and Φ is a factor obtained by integrating numerically the dispersion relation, typically $\approx 0.1 \div 2$. The typical growth time for the parameters of jets is $\tau_{inst} \ll \tau_{rs} \sim 10^6 \div 10^8$ years, the typical lifetime of radio sources.

7) The analysis of the radial structures of the perturbations indicates that they are not confined to the boundaries of the beam, but actually must be considered as affecting the whole flow pattern. In particular this means that they can actually be correlated with morphological patterns. A confirmation of this result can be found in the numerical simulations referenced above. It is rather obvious that the typical perturbations which grow along the flow satisfy the condition $kR \sim 1/M$; shocks also have a repetitive pattern which agrees with the specific case of pinching modes.

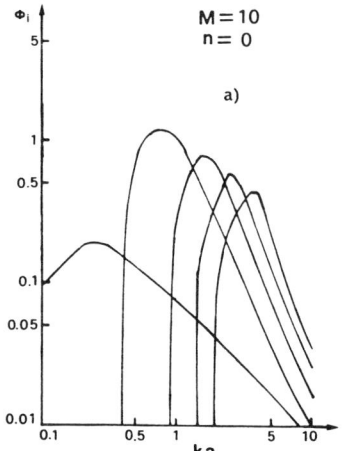
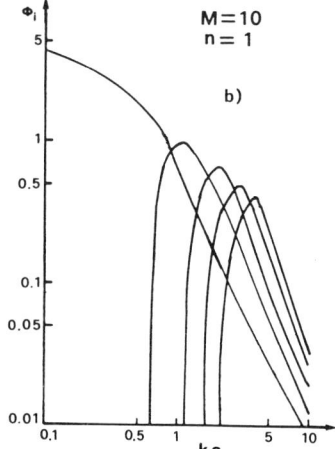

Fig. 4 - Typical growth rates, $\Phi_i = \omega_i/kV_b$, for Kelvin-Helmholtz instabilities in supersonic, cylindrical jets: a) pinching modes $n = 0$; b) helical modes $n = 1$.

There are other aspects of the study of linear instability that are now being studied. For instance Hardee (1983) has analyzed the case of conical jets for simulating a regular expansion in a medium with decreasing external pressure. The typical effect he finds is that growth rates tend to decrease in agreement with Eq. (7), which predicts $\omega_i \propto R$. It is also necessary to proceed in the study of cylindrical flows with spin around their symmetry axes; this should allow to apply the study of instability to jets in the deep core of accretion funnels, where rotation is likely to be an important effect.

Finally the effect of viscosity should also be taken into account: obviously not collisional viscosity, but what we may call an effective *magnetic turbulent viscosity*. Baan (1980) has addressed this problem for what concerns the general dynamical equations, and Nepveu (1980) has taken viscosity into account in his numerical simulations. However the effective viscosities used are too large in general for computational reasons and no specific attempt to derive a dispersion relation for Kelvin-Helmholtz modes has been completed. Furthermore we must proceed to the study of nonlinear effects, some of which are discussed below.

III.2 Resistive instabilities

Other instabilities relevant to magnetically confined jets are resistive instabilities. They are excited in the presence of electrical currents and neutral (or sheared) magnetic sheets. These conditions are verified in the magnetically-confined models, with an helicoidal field wrapped around the beam. The typical equilibrium field of such a configuration requires that the pitch-angle of the helicoidal field decreases from the axis outwards (Parker 1979). When a non-axisymmetric perturbation is created along the beam, its pitch-angle will match the equilibrium field pitch-angle at a well defined radius R_{crit}. The corresponding cylindrical surface will be a neutral sheet for the component of the magnetic field perpendicular to the perturbation wavevector, and therefore one envisage the formation of a helicoidal distribution of growing magnetic islands along this surface (the helix specifically chosen will depend on the boundary conditions). Perturbations with different pitch-angles will have different critical surfaces, according to the relation:

$$R_{crit} = \frac{m}{k} \frac{B_\theta}{B_\parallel}, \qquad (8)$$

where m is the same azymuthal number of Eq. (5). The surface also depends from the longitudinal wavelength, 1/k, which is determined by the boundary conditions at the origin of the jet. The growth time of reconnection in magnetic islands is typically related to the resistive time which is very large for the parameters of extragalactic jets ($\tau_{res} \sim 10^{10}$ years). However different islands grow out of the linear regime to reach a stage in which their transverse amplitude covers the gap between the surfaces. Merging regions become eventually stochastic (Rutherford 1980) and this phase can be explosive, with typical time scales not much longer than the MHD time $\tau_{MHD} \sim R/V_a$. In this sense reconnecting modes can modulate the magnetic field in beams on scales comparable with those of fluid MHD instabilties. The result of merging, as far as it can be predicted in terms of laboratory experiments, is to create regularly spaced pinches along helices wrapped around the flow pattern. This can be related with the jet morphological structures.

III.3 Instabilities and morphologies

In conclusion of this Section I want to justify in some quantitative way the argument that fluid instabilities can actually be related to observed morphologies. As recently discussed by Ferrari, Trussoni and Zaninetti (1983), one can derive the set of physical parameters for which the above comparison is acceptable. The idea is to compare the calculated most unstable wavelengths (see Eq.(7)) with the scales of observed features, namely knots and wiggles (see Table II); and correspondingly derive a theoretical value for the extension of the jet before disruption, as the distance after which the perturbation of pressure is so large to exceed p_{ext} by a large factor. This last part of the comparison is somewhat uncertain because it involves necessarily considerations on nonlinear stages of the instability. The theory shows that the wavelength and the growth rate of the most unstable modes depend on three parameters: the flow Mach number, M, the density contrast between internal and external plasma, n_{int}/n_{ext}, and the magnetic field strength, namely V_a/V_s. Actually the dependence on this last parameter is rather weak. Therefore we have substantially a two-parameter model, to be tested on two observational data. Applying this argument to the sources of Table II, one derives that the model can fit the data in almost all cases (M 87 is again an exception, as in several other respects); in particular wiggles should be connected with reflection modes m = 1, knots with reflections modes m = 0. The predicted flow velocities are supersonic M $\gtrsim$ 5, and the density contrast is larger than unity; thus the model suggest that jets are heavier than the external medium. This last result, however, is very sensitive to the choice of the typical length on which nonlinear effects become large; in particular the density ratio can be easily reduced to $\approx$ 1. Furthermore the equivalence of the growth rates of different instabilities allows to extend these conclusions, although the study refers to the case of Kelvin-Helmholtz modes.

In conclusion the stability analysis confirms that jets are unstable in all possible situations, and actually the perturbed global modes appear to be responsible for some aspects of the morphology. These two statements are, in a sense, contradictory, because one would expect unstable jets to break and not simply wiggle; one has to envisage the possibility of saturation mechanisms, to keep perturbations to low amplitude. This problem is rather more complex, as nonlinear model of instabilities are mathematically awkward and cumbersome.

In addition I should mention that the instability analyses so far performed did not include the dissipation of modes into the external medium, where actually they can be coupled to local modes. This point will be partly discussed in the next Section, but still require more work.

IV. INSTABILITIES AND NONTHERMAL RADIATION

The problem of instability saturation is strictly connected with interpretation of observed morphological structures and radiation emission. We have discussed in the previous Section how the typical scales of unstable perturbations are equal to the measured wiggle and knot wavelengths. On the other hand nonthermal emission coming from jets (and more generally from extended radiosources) requires continuous generation of relativistic electrons *in situ*, as their synchrotron radiation lifetime is much smaller than the travel time from the central core. As the instability is dragging energy from the ordered flow, it appears straightforward to utilize unstable MHD modes to accelerate electrons. In fact several works have aimed to establish this connection and to generate a unified picture for radiation and morphology.

Henriksen has discussed in this Symposium the specific case of turbulent jets, in which fluid MHD modes, excited by internal stresses in sheared mixing layers, can diffuse particle (specifically electrons) in energy space to very high Lorentz factors. This process can take place all the way along jets and hence represents a local acceleration process.

Actually turbulence in jets can be directly connected with fluid MHD instabilities, as discussed by Benford et al. (1980). In fact, from the previous Section, we know that in a cylindrical jet the fastest growing modes are those with wavelength ~ λ_0 = $2\pi R$. These modes have typically the right scales for modulating the flow, but can also accelerate accelerate electrons trapped in their field. A Fermi-like acceleration process sets in, with acceleration time scale typically shorter than the confinement and/or crossing time in the region of acceleration (see also Bicknell and Melrose 1983).

Furthermore nonlinear mode interactions can start, in the limit of weak magnetic fields $V_a \lesssim V_s$, a turbulent cascade from long ($\lambda \sim \lambda_0 \sim 10 \div 100$ pc) towards short ($\lambda \ll 10^{15}$ cm) wavelengths (Lacombe 1977, Eilek 1979, Benford et al. 1980), and provide the same type of physical situation invoked by Henriksen. The spectrum of turbulent MHD modes corresponds to the Kraichnan or Kolmogorov classical results, i.e. is represented by a power law, $W(k) \propto k^{-\nu}$, with $\nu \approx 1.5 \div 1.7$. Electron acceleration by small wavelength modes is now a resonant cyclotron process; electrons resonate with modes $k \approx 1/r_g$, where r_g is their (energy-dependent) gyroradius. In a stationary distribution we expect an electron spectrum with a power law $n(E) \propto E^{-\delta}$, where $\delta \approx 2.0 \div 3.0$ (typically 2.5 as consistent with emission spectra in the radio region).

It can be shown that a stochastic acceleration process can actually stop the cascade before the so-called dissipation limit; in fact, in the present situation, dissipation occurs at very large k only ($\gtrsim 10^{-13}$ cm^{-1}), when resonant interaction with thermal particles takes place. In fact, absorption of turbulent energy by relativistic electrons becomes dominant over nonlinear mode interactions at $k \sim 10^{-15}$ cm^{-1}, i.e. closer to the source of the cascade, $\sim 1/\lambda_0$. It is then possible to establish a stationary situation in which the energy released by the fluid instabilities at large wavelengths is transported to short wavelengths by nonlinear processes, where it is dissipated by particle acceleration. In particular the sequence of time scales of the different processes is such that they ultimately represent a way to saturate the instability (Benford et al. 1980).

If this saturation mechanism works, we can conclude that large amplitude modes are saturated close to the linear regime and, while non disrupting the beam, can provide both basic morphology and energetics of radiation emission. On the other hand it is also possible that the instability actually evolves beyond the linear regime, either because of different phyisical conditions or because the stationary energetic balance cannot be exactly satisfied. In this case we can easily predict that shocks will be formed, because the transverse perturbation of the collimated flow pattern cannot exceed an amplitude $\xi > \lambda/M$ (Ferrari et al. 1981): in fact, at this point the lateral velocity would exceed the velocity of sound in the external medium. Therefore shocks are likely to be, in such a context, the nonlinear mechanism which saturate instabilities; this point seems to be confirmed by the numerical simulations previously discussed. Shocks dissipate energy both heating thermal particles and accelerating superthermal electrons locally (Bell 1978, Blandford and Ostriker 1978). The problem of shock acceleration at highly relativistic energies has not been completely solved yet; Ferrari and Trussoni (1981) have shown that this process does not seem able to accelerate electrons emitting synchrotron X-rays, which are instead detected in M 87, Cen A and 3C 273. In this respect stochastic acceleration by MHD modes appears more promising.

A schematic summary of this unifying scenario is illustrated in Fig. 5 by means of a flow diagram. I have also indicated the other interesting fact that the small scale MHD turbulence can also affect the morphology directly by scattering the jet particles in real space. As shown by Benford et al. (1980), in such a model jet emissivity and morphological structures must be correlated: statistically significant observations to test this point would be interesting.

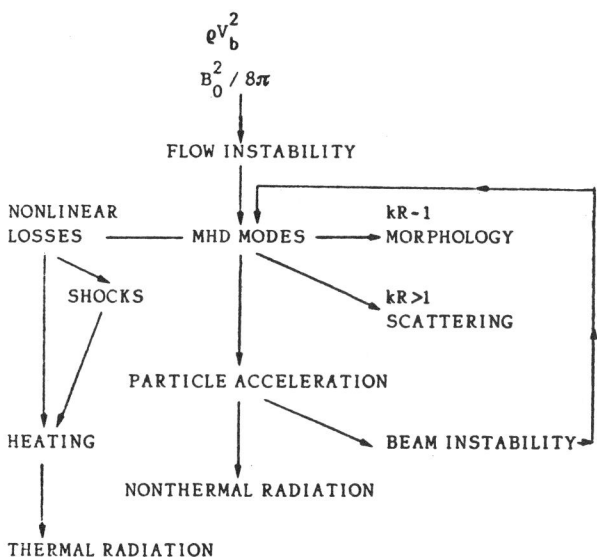

Fig. 5 - Flow diagram of the unified scheme for instability, morphology and radiation in jets.

Another possible instability saturation mechanisms in magnetically dominated flows ($V_a \gtrsim V_s$) is the coupling of surface modes with proper modes inside the jet plasma (the same idea could be applied to study the coupling of jet instabilities with modes in the surrounding medium) (Hasegawa and Chen 1976, Ionson 1978, Bodo and Ferrari 1982). In fact surface Kelvin-Helmholtz modes can be coupled resonantly with kinetic MHD waves inside the flowing plasma, whenever an inhomogeneous boundary layer is present in which density and velocity are increasing. The resonance takes place where the local frequency of Alfven waves, corrected by the Doppler effect of the flow, matches the frequency of the surface modes. These kinetic modes propagate transversally to the magnetic field towards the jet axis, where they provide local plasma heating and relativistic electron acceleration. An interesting aspect of this mechanism is that it leads to generate short wavelength MHD modes in the whole body of the jet, indipendently from any nonlinear cascade process.

V. CONCLUDING REMARKS

It is an obvious characteristic of all theoretical reviews to be a sort of personal interpretation of the achievements and failures of the models examined. I am sure that others would have approached this discussion in a different way; and I have probably left over aspects that others find crucial. Nevertheless I am convinced that the general scheme summarized in Fig. 5 can be accepted by everybody, although with different emphasis, as a starting point for future work.

I believe that the challenge put to us by the physics of jets is very stimulating. We must, in the next few years, reach a reasonable understanding of the following issues:

a) Acceleration of the flow in the inner cores of galactic nuclei: fluid acceleration or electromagnetic acceleration?

b) Collimation: does it occur at the same place where acceleration occurs, i.e. in the inner throat of accretion disks, or is it produced farther out, in nozzles shaped by the surrounding medium?

c) Confinement: is it due to the pressure of an external hot medium, or to the pressure of a self-consistent magnetic field?

d) Morphologies: are they actually related to the development of large scale instabilities, and therefore to the interaction with the external medium, or are they produced by the dynamics of the central powerhouse (precessing beams)?

e) Stability: how are large scale instabilities saturated? How important are the effects of boundary layers, dissipations, nonlinear mode interactions?

f) Nonthermal radiation: what accelerates electrons *in situ*? Turbulent modes or shocks?

Many of these questions have been already dealt with in other astrophysical situations. Some of them have also been solved, although in different ranges of parameters. But, above all, I stress that the new real element we have here to consider is the presence of supersonic, collimated flows which are able to provide all the energy needed. Of course, also in this case the ultimate energy source, deep in the core, is somewhat mysterious. However the real challenge is to explain how the ordered kinetic energy can be directly transformed into radiation, and create these incredible structures.

In the framework of this Symposium, the problem of jets is one of the latest subjects to which plasma astrophysicists have been asked to contribute. The main drawback is that the amount of data on these objects is still rather small, and also their reach is still instrumentally limited. After all we must keep in mind that the energy from extragalactic objects collected by all radiotelescopes, since Karl Jansky's first antenna, amounts to only a tiny fraction of the kinetic energy of a falling snow flake. For a real progress of our understanding of these jets we must urge for the definition of a statistically significant set of parameters of a *standard jet*. This will be achieved by the continuing operation of powerful radiotelescopes, as the Very Large Array and the Very Long Baseline Interferometry intercontinental network, and of orbiting observatories, as the Space Telescope and the Advanced X-ray Astronomy Facility.

Acknowledgements. I wish to thank E. Trussoni and K. Tsinganos for helpful discussions and comments. This work was partially supported by C.N.R. grant PSN82-015 and N.S.F. grant AST83-03522.

REFERENCES

Baan, W.A. 1980, Astrophys.J., **239**, 433
Begelman,M.C. 1982, **Extragalactic Radio Sources**, Heeschen and Wade
   eds., D.Reidel, Dordrecht, p. 392
Bell, A.R. 1978, Mont. Not. Roy. astron. Soc., **182**, 147
Benford, G. 1978, Mon. Not. Roy. astron. Soc., **183**, 29
Benford, G. 1981, Astrophys.J., **247**, 458
Benford, G., A. Ferrari and E. Trussoni 1980, Astrophys.J., **241**, 98
Bicknell, G.V. and D.B. Melrose 1983, Astrophys.J.,**262**, 511
Blake, C.M. 1972, Mon. Not. Roy. astron. Soc., **156**,67
Blandford, R.D. and J.P. Ostriker 1978, Astrophys.J.Lett., **221**, L29
Blandford, R.D. and J.E. Pringle 1976, Mon. Not. Roy. astron. Soc., **176**, 443
Blandford, R.D. and M.J. Rees 1974, Mon. Not. Roy. astron. Soc., **169**, 395
Bodo, G. and A. Ferrari 1982, Astron. Astrophys., **114**, 394
Bridle, A.H. 1982, **Extragalactic Radio Sources**, Heeschen and Wade
   eds., D. Reidel, Dordrecht, p. 121
Bridle, A.H. and R.A. Perley 1983, **Astrophysical Jets**, Ferrari and
   Pacholczyk eds., D. Reidel, Dordrecht, p. 57
Browne, I.W.A., et al. 1983, **Astrophsysical Jets**, Ferrari and
   Pacholczyk eds., D. Reidel Dordrecht, p. 27
Burns, J.O. 1983, **Astrophysical Jets**, Ferrari and Pacholczyk eds.,
   D. Reidel, Dordrecht, p. 67
Chan, K.L. and R.N. Henriksen 1980, Astrophys.J.,**241**,534
Chandrasekhar, S. 1961, **Hydrodynamic and Hydromagnetic Stability**,
   Oxford Univ. Press
Cohn H. 1983, Astrophys.J., **269**, 500
Courant, R. and K.O. Friedrichs 1948, **Supersonic Flows and Shock Waves**,
   Wiley, New York
Eviatar, J. and A.K. Wolf 1968, J. Geophys. Res., **73**, 5561
Eilek, J.A. 1979, Astrophys.J., **230**, 373
Feigelson, E.D. et al. 1981, Astrophys.J., **251**, 31
Feigelson, E.D. 1982, private communication
Ferrari, A., E. Trussoni and L. Zaninetti 1978, Astron. Astrophys., **64**, 43
Ferrari,A. and E. Trussoni 1981, **Plasma Astrophysics**, Varenna Intl.
   Workshop, Proceedings, ESA SP-161, p. 353
Ferrari, A. and E. Trussoni 1981, Space Sci. Rev., **30**, 135
Ferrari, A. and A.G. Pacholczyk (eds.) 1983, **Astrophysical Jets**,
   D. Reidel, Dordrecht
Ferrari, A., E. Trussoni and L. Zaninetti 1983, Astron. Astrophys., in press
Ferrari, A., S. Habbal, R. Rosner and K. Tsinganos 1983, Astrophys.J.Lett., in press
Gerwin, R.A. 1968, Rev. Mod. Phys., **40**, 652
Goldstein, S. 1965, **Modern Developments in Fluid Dynamics**, Dover, New York
Gill, A.E. 1965, Phys. Fluids, **8**, 1428
Gull, S.F. and K.J.E. Northover 1973, Nature, **244**, 80
Hardee, P.E. 1979, Astrophys.J., **234**, 47
Hardee, P.E. 1983, private communication
Hasegawa, A. and L. Chen 1975, Phys. Fluids, **19**, 1924
Heeschen,D.S. and C.M. Wade (eds.) 1982, **Extragalactic Radio Sources**,
   D. Reidel, Dordrecht
Ionson, J.A. 1978, Astrophys.J., **226**, 650
Kopp, R. and T. Holzer 1976, Solar Phys., **49**,43

Krautter, A., R.N. Henriksen and K. Lake 1983, Astrophys.J., **269**, 81
Lacombe,C. 1977, Astron. Astrophys., **54**, 1
Longair, M.S., M. Ryle and P.A.S. Scheuer 1973, Mon. Not. Roy. astron. Soc., **164**,243
Norman, M.L., K.H.Winkler and L. Smarr 1983, **Astrophysical Jets**, Ferrari and Pacholczyk eds., D. Reidel, Dordrecht, p. 227
Miley, G. 1983, **Astrophysical Jets**, D. Reidel, Dordrecht, p. 99
Nepveu, M. 1980, Astron. Astrophys., **84**, 14
Pacholczyk, A.G. 1970, **Radio Astrophysics**, Freeman, San Francisco, Chap. 7
Parker, E.N. 1979, **Cosmical Magnetic Fields**, Oxford Univ. Press
Potash, R.I. and J.F.C. Wardle 1980, Astrophys.J., **239**, 42
Ray, T.P. 1981, Mon. Not. Roy. Astron. Soc., **196**, 195
Rees, M.J. 1971, Nature **223**, 312
Rutherford, P.H. 1980, **Physics of Plasmas Close to Thermonuclear Conditions** Proceedings, Varenna Workshop, p. 129
Sanders, R.H. 1983, preprint
Schreier, E.J., P. Gorenstein and E.D. Feigelson 1982, Astrophys. J., **261**, 42
Sen, A.K. 1963, Phys. Fluids, 6,1154
Turland, B.D. and P.A.S. Scheuer 1976, Mon. Not. Roy. astron. Soc., **176**, 421
Van Dyke, M. 1982, **An Album of Fluid Motion**, The Parabolic Press, Stanford
Wiita, P.J. 1978, Astrophys.J., **221**, 42 & 432
Woodward, P.R. 1983, BAAS, **14(4)**, 870
Yokosawa, M., S. Ikeuchi and S. Sakashita 1983, preprint

DISCUSSION

IONSON: Could you give me the following numbers for the physical parameters along jets: Alfven transit time, electron temperature, electron density, ion gyroradius ?

FERRARI: I can quote some typical values, which are all model dependent, as I have discussed before. The Alfven transit time is typically $\cong (10 \div 100)R/c$, i.e. 100 to 100,000 years. Gyroradii for thermal particles are of the order of astronomical units; temperatures can be estimated below 10^8 K from X-ray data; electron density is $\lesssim 10^{-3}$, according to depolarization measurements (se also Table I).

STURROCK: Is it possible that a jet is really a stream of plasmoids, each of which could be confined by ram pressure ?

FERRARI: At present data cannot tell us the difference between a continuous fluid jet and a sequence of plasmoids, if these are really very small. If you mean by plasmoid something as large as the knots in M87, I am inclined to say that it cannot be the case, as we definitely see emission from the flow in between. From the theoretical point of view, ram pressure confinement requires large external densities (see Miley 1980).

BENFORD: When a jet moves to the side at an angle greater than M^{-1}, with M the Mach number, sidewise shocks develop. This means by the time you can see a wiggle it is in the nonlinear regime. Also, loading of external or confining B_θ field lines can alter the linear theory. How can you take this into account ? Won't this lead to lighter jets and lower M ?

FERRARI: I agree with you about these general statements. If one applies the linear theory of instabilities to interpret morphologies in jets, large densities and large Mach numbers are derived (Ferrari et al. 1983). The effects you indicate may change these results, if the perturbations grow to the nonlinear level. There are however processes that could saturate the growth of instabilities still in the linear regime, as I have discussed in Fig. 5.

RAY: Do you feel the observed wiggles are reflected modes ? You mentioned the coupling of short wavelength Kelvin-Helmholtz modes with kinetic Alfven waves as a source of transporting energy towards the center of the jet. Would the K-H waves not be damped by any reasonable velocity profile ?

FERRARI: As I mentioned in the answer to Greg Benford, I see the possibility that reflected modes generate wiggles, provided the concurrence of a sequence of physical processes. Concerning the coupling of surface K-H waves with Alfven waves, this actually represents a way to discuss how short wavelength modes are damped by physical gradients; in fact this coupling is studied deriving a dispersion relation for modes in the presence of density and velocity gradients. However the mechanism requires $V_a \gtrsim V_s$.

BIRN: Have you checked the possibility of a Jeans instability, i.e. a self-gravitational effect ?

FERRARI: Yes, but it can be excluded by the low densities and relatively high pressures of relativistic particles.

UCHIDA: Supplementing the question of numbers, could you say anything about the ratio of kinetic energy to magnetic energy ? In other words, whether the material stretches the field or the field guides the flow ?

FERRARI: The flow kinetic energy cannot be directly estimated, because we have no access to thermal matter. However the energy released in the whole radiosource extended lobes is several order of magnitudes larger than that required to support the emission of jets, and therefore its magnetic energy. I am inclined to believe that the kinetic energy in the collimated flow is actually dominating.

COPPI: I think that of all the assumptions that have to be made in order to develop a tractable theory the weakest is probably that the ions would have a Maxwellian distribution.

FERRARI: I agree that the fluid treatment is very approximate, as it neglects microscopic plasma effects, which are known to be very important in laboratory and space physics. However we are still at the initial stages of the theory, and look for large scale morphological and energetic effects. The sequence of processes indicated in Fig. 5 will certainly require appropriate consideration of those microscopic effects.

BURSTING PARTICLE ACCELERATION IN RADIO JETS

R.N. Henriksen
Department of Physics, Queen's University at Kingston
Kingston, Ontario K7L 3N6 Canada

I. INTRODUCTION

This work follows on the papers by Henriksen, Bridle and Chan (1982) (HB) and by Eilek and Henriksen (1983). These papers introduced a comprehensive model of (hydrodynamic) turbulence driven Alfven wave, resonant, particle acceleration. Very similar ideas were introduced independently by Fedorenko (1980), but the details were sketchy and the universality of the resulting spectral index was not realized. Eichler (1979) has come closest to these ideas previously (again independently) in a very original paper directed at ion acceleration in solar flares. He includes a back reaction on the turbulent eddies however which requires an adhoc treatment and is probably unnecessary (we replace this by the Lighthill noise analysis: HBC) and he omits synchrotron and adiabatic losses. Moreover, the quasi-linear theory is not used in calculating the distribution function. Nevertheless, it is reassuring that the same intuitive physics, even to the self-similar universality, should arise independently.

This mechanism establishes in a volume the Alfven wave spectral density necessary for the particle acceleration, as a balance between MHD wave emission from the turbulent vortices throughout the volume (Lighthill noise) and the damping losses to relativistic and thermal particles. The wave number dependence of the wave spectral density $W_A(k)$, is fixed both by the spectrum of the fluid turbulence (e.g. 'young' that is, steep; Kolmogorov or Kraichnan) and by the wave number dependence of the damping (e.g. Eilek 1979). The stochastic resonant interaction of this spectrum with relativistic electrons (second order Fermi process) can, assuming isotropy and quasi linear theory, be described as a diffusion in energy ($\varepsilon = cp$ for relativistic particles) space with coefficient $D = Gp^{\tilde{\nu}}$ (e.g. Lacombe, 1979). The index $\tilde{\nu}$ is that of the Alfven wave spectral density $W_A(k) = W_o k^{-\tilde{\nu}}$. The quanty G depends on the normalizing factors of the particle and wave spectrum, but not on energy or wave number. Eilek and Henriksen (EH) have argued that there are feed-back mechanisms operating alternately on the timescales of the synchrotron lifetime and the acceleration time which yield $\tilde{v} = 3$ and $S/G = 6-s_t$ asymptotically. Here S is the synchrotron (Compton losses may be included by adding the effective B of the radiation field to B) loss coefficient (B is magnetic field strength and other symbols are standard)

$$S = \left(\frac{4e^4}{9m_e^4 c^6}\right) B^2, \qquad (1)$$

and s_t gives the wave number dependence of the Alfven wave driving power/unit wave number (per unit volume) due to the turbulence as (κ is the eddy wave number)

$$I_a(k) = I_o(k/\kappa_T)^{-s_t}. \qquad (2)$$

The quantity I_o depends only on the turbulent flow properties, and κ_T is the Taylor scale of the turbulence. The index $s_t = 1.5$ for Kolmogorov turbulence, 1 for Kraichnan (i.e. 'old', equipartition turbulence) and 2 for young turbulence. They have also shown that under these conditions, the solution for the electron distribution function is

$$f = f_o p^{-s}, \qquad (3)$$

where s tends asymptotically to $S/G = 6-s_t$ over substantial portions of energy space. This converts to a synchrotron spectrum $I_\nu \propto \nu^{(s_t-3)/2}$ (e.g. $\propto \nu^{-0.75}$ for Kolmogorov).

Thus far however, the theory has been discussed purely locally, without reference to convection or to the expected positional dependence of the synchrotron brightness. In the extragalactic jet sources, such variation with position is the defining property (both parallel and perpendicular to the jet axis). In this paper, therefore, we study how the EH theory applies in the presence of convection and of axial variations in the local properties of the turbulence. The well studied sources NGC315 and NGC6251, are referred to for illustration.

II. THEORETICAL FRAMEWORK

We may write the equation for the electron distribution function, $f(\underline{r}',t')$, in the local proper frame of the jet in the coordinates of the background (galaxy) frame ($f(\underline{r},t) = f(\underline{r}',t')$) as

$$\frac{\partial f}{\partial t} + \underline{u} \cdot \underline{\nabla} f - \left(\frac{\underline{\nabla} \cdot \underline{u}}{3}\right) p \frac{\partial f}{\partial p} = \frac{S}{p^2} \frac{\partial}{\partial p}(p^4 f)$$

$$+ \frac{G}{p^2} \frac{\partial}{\partial p}\left[p^{2+\tilde{v}} \frac{\partial f}{\partial p}\right] \cdots \qquad (4)$$

For definiteness, we proceed by substituting the CH velocity field

(Chan and Henriksen, 1980) for the jet velocity $\underset{\sim}{u}$ so that

$$u = u_z \left(\frac{r}{R_j} \frac{dR_j}{dz}, \frac{u_\phi}{u_z} \frac{r}{R_j}, 1 \right) \tag{5}$$

where (r,ϕ,z) are cylindrical coordinates relative to the jet axis, $R_j(z)$ is a fiducial boundary stream line usually identified with a radius determined from the transverse brightness profile and $u_z = u_z(z)$, $u_\phi = u_\phi(z)$. With this ansatz the operative equation becomes

$$\frac{\partial f}{\partial t} + u_z \frac{\partial f}{\partial z} - \frac{1}{3R_j^2} \frac{d}{dz}(R_j^2 u_z) \, p \, \frac{\partial f}{\partial p}$$

$$= \frac{S}{p^2} \frac{\partial}{\partial p}(p^4 f) + \frac{G}{p^2} \left[p^{2+\tilde{v}} \frac{\partial f}{\partial p} \right] \tag{6}$$

This equation becomes complete when $u_z(z)$, $S(z)$, $G(z)$ and $\tilde{v}$ are given.

In the turbulence driven model of (EH)

$$G = \frac{(s-2)}{2\pi\tilde{v}(\tilde{v}+2)} \left(\frac{\Omega m_e}{c} \right)^{(s-2-\tilde{v})} \frac{I_o}{f_o} \kappa_T^s t,$$

$$I_o = \eta_A \left| \frac{6-4m}{3-m} \right| \rho v_a^3 (E_t/\rho v_a^2 R)^{\frac{3}{3-m}}, \tag{7}$$

$$\tilde{v} = s + s_t - 3,$$

where Ω is the classical gyrofrequency $(eB/m_e c)$, E_t is the turbulent kinetic energy density, $s_t = 3(m-1)/(3-m)$, and m is the spectral index of the hydrodynamic turbulence (e.g. $m = 5/3$ for Kolmogorov), v_a is the appropriate Alfven speed, η_A is a number of order unity and $R \gg 1$ is a factor which allows for the fraction of turbulent energy in the resonant region (below the Taylor scale) $\underset{\sim}{R} \geq (\kappa_T/\kappa_o)^{(m-1)}$. EH show $\tilde{v} = 3$ for a form stable spectrum when synchrotron losses and resonant acceleration dominate equation (6), which determines s as $6-s_t$. However this need not be so in other regimes. We also note that if the explicit model for the jet turbulent energy density given by HBC is adopted, then

$$I_o = \eta_A \left| \frac{6-4m}{3-m} \right| R^{\frac{3}{3-m}} \rho v_a^{-s_t} (u_z \, dR_j/dz)^{6/(3-m)}. \tag{8}$$

The dependence on f_o in G reflects the coupled integro-differential nature of the full problem, which we have dealt with by replacing f with $f_o p^{-s}$ in the integrals for G. We must always check for self-consistency a posteriori however. Moreover, this dependence can lead to self-limiting strong acceleration wherein G decreases as f_o increases, and hence to a relaxation oscillation.

We proceed now to investigate various separate physical regimes of equation (6).

a) Adiabatic Expansion

In this case $Sp\, z/u_z \ll 1$ and $Gp^{(\tilde{v}-2)} z/u_z \ll 1$ so that (6) becomes

$$\frac{\partial f}{\partial t} + u_z \frac{\partial f}{\partial z} - \frac{1}{3R_j^2} \frac{d}{dz}(R_j^2 u_z)\, p\, \frac{\partial f}{\partial p} = 0.$$

This has the general solution

$$f = F(t - \int \frac{dz}{u_z},\; R_j^2 u_z p^3) \tag{9}$$

where F is formally arbitrary but is usually determined by the initial conditions. If, for example, at one point in the jet a burst of relativistic electrons is produced with a spectrum of the form $f_o(t_o) p^{-s}$, then subsequently

$$f = f_o(t - \int \frac{dz}{u_z})(R_j^2 u_z p^3)^{-\frac{s}{3}} \tag{10}$$

where $t_o \equiv t - \int dz/u_z$ is the (convective) 'retarded time'. We normally observe at some fixed t.

b) Bursting Acceleration

In this case $Gp^{(\tilde{v}-2)} \gg u_z/z,\, Sp$, so that equation (6) becomes in the appropriate range of p

$$\frac{\partial f}{\partial t} = \frac{G}{p^2} \frac{\partial}{\partial p}[p^{2+\tilde{v}} \frac{\partial f}{\partial p}],$$

and the electron spectrum is dominated by the local acceleration.

The general solution for the Laplace transform of f, $\bar{f}(\omega,p) \equiv \int_0^\infty e^{-\omega t} f(t,p)\, dt$ is (when $\tilde{v} = 3$), $\bar{f}(\omega,p) = \frac{1}{G} \int_{p_{min}}^{p_{max}} p'^2 dp'\, G_\omega(p,p') f_o(p'),$

$$G_\omega(p,p') = \frac{2I_4(2\sqrt{\omega/G}p_>)K_4(2\sqrt{\omega/G}p_<)}{(pp')^2},$$

$p_>,p_<$ are respectively the greater or lesser of p,p', f_o is the initial distribution and I_4, K_4 are Bessel functions. Asymptotes may be found by using $f(\infty,p) = \lim_{\omega \to 0} \omega f$, but it is normally more convenient to find these as limits to the self-similar behaviour. When however, $f_o(p) = (N_o/4\pi p_o^2)\delta(p-p_o)$, then for $p > p_o$

$$f(\infty,p) = \lim_{\omega \to 0} \{\frac{\omega}{4G} \frac{N_o}{4\pi}\} p^{-4}.$$

This supports our interpretation of the asymptote (13b) below. For $p < p_o$, the result is

$$f(\infty,p) = \lim_{\omega \to 0} \{\frac{\omega}{4Gp_o} \frac{N_o}{4\pi p_o^3}\}.$$

We look for the asymptotic behaviour by setting

$$x \equiv Gp^{(\tilde{\nu}-2)}t$$
$$g \equiv 4\pi p^3 f \tag{11}$$

so that the self-similar equation to be solved is

$$\frac{dg}{dx} = (\tilde{\nu}-2) \frac{d}{dx} [(\tilde{\nu}-2)x^2 \frac{dg}{dx} - 3gx].$$

This has the general solution (see also EH and Lacombe, 1979)

$$g = x^{3/(\tilde{\nu}-2)} \exp(-x^{-1}/(\tilde{\nu}-2)^2) [A_1 + A_2 \int \frac{dx \exp(x^{-1}/(\tilde{\nu}-2)^2)}{x^{(2\tilde{\nu}-1)/(\tilde{\nu}-2)}}] \tag{12}$$

and the asymptotic region is found with $x \gg 1$. In this latter limit the dominant terms of the two modes are

$$f_1 \sim \frac{A_1}{4\pi} (Gt)^{\frac{3}{\tilde{\nu}-2}} \quad \ldots \text{(a)}$$

$$f_2 \sim \frac{A_2}{4\pi} (Gt)^{-1} p^{-(\tilde{\nu}+1)} \quad \ldots \text{(b)}. \tag{13}$$

The physical mode is (13b) as it gives an upward flux of particles in momentum space and corresponds to the non-trivial steady state $f \propto p^{-(\tilde{\nu}+1)}$. The solution is consistent with $s = \tilde{\nu}+3 - s_t$ for all $\tilde{\nu}$

if $s_t = 2$ (young turbulence), so that the initial spectrum can be expected to be maintained ($\tilde{\nu} \simeq 3$).

The mode (13a) gives a downward flux of particles in momentum space and corresponds to the trivial steady state f = const. It requires an initial supply of high energy particles to be realized ($p < p_o$ in the Laplace Transform example above).

We note finally that the inverse dependence on G in (13b), is due to $f(p)$ varying inversely with the rate at which particles are accelerated beyond p.

c) Synchrotron Loss Dominated

Here we require that $Sp \gg u_z/z$, $Gp^{(\tilde{\nu}-2)}$, so that (6) is simply

$$\frac{\partial f}{\partial t} = \frac{S}{p^2} \frac{\partial}{\partial p} (p^4 f).$$

This equation has the general solution

$$f = (F(t - 1/Sp)p^{-4},$$

where F is arbitrary. However the initial value problem is not well behaved past $t = 1/Sp$ for strict initial power laws ($f \propto (1/Sp-t)^{s-4} p^{-4}$ if $f_o \propto p^{-s}$), so that it is generally more useful (EH) to use the asympote

$$f = \frac{1}{S} \frac{A_1}{4\pi p^4} \frac{1}{(t - 1/Sp)} \cdot \qquad (14)$$

for $t \gg 1/Sp$. The particles stream downward in momentum space.

d) Form Stability

This case has been discussed at length in EH who argue for $\tilde{\nu} = 3$ and $S/G = s$. We require further that $Sp\, z/u_z$ and $Gp\, z/u_z \gg 1$, so that (6) becomes

$$\frac{\partial f}{\partial t} = \frac{S}{p^2} \frac{\partial}{\partial p} (p^4 f) + \frac{G}{p^2} \frac{\partial}{\partial p} [p^5 \frac{\partial f}{\partial p}].$$

A true steady state may be achieved locally in two modes;

$$f_1 = A_1\, p^{-S/G}$$

and (15)

$$f_2 = F_o/(4\pi G(S/G - 4))p^{-4},$$

if $S/G \neq 4$. The self-similar asymptotes have two corresponding modes which are at large $x \equiv Gpt$

$$f_1 \sim A_1(Gt)^{3-S/G} p^{-S/G} \tag{16}$$

$$f_2 \sim A_2(Gt)^{-1} p^{-4}$$

with $S/G = 6 - s_t = s = 3 + 2\alpha$.

We expect this form stable condition to apply in the mean in a region where adiabatic losses are not important. Turbulent bursting acceleration followed by synchrotron decay may well be superimposed on this mean behaviour in a form of relaxation oscillation. The remarkable conclusion to be drawn from equations (10), (13b), (14) and (16) is that the spectral index is not expected to vary greatly in any of these processes, and may be expected to be close to $\alpha = 0.5$ in the bursting mode (young turbulence and synchrotron loss).

III. PREDICTED SYNCHROTRON BRIGHTNESS VARIATIONS

We may calculate the surface brightness along a pathlength $d(\theta)$ (distance d in the source at an angle θ to the magnetic field) produced by optically thin synchrotron radiation of relativistic electrons as

$$I_\nu = c_5(\alpha) \, 4\pi c^{2\alpha} \left(\frac{\nu}{2c_1} \right)^{-\alpha} (B \sin\theta)^{\alpha+1} f_0 \, d(z), \tag{17}$$

with

$$d_\parallel (z) = 2R(z) \, \mathrm{cosec} \, \theta$$

$$d_\perp (z) = 2R(z) \, \mathrm{sec} \, \theta,$$

where θ is the angle between the line of sight and the magnetic field (pitch angle), and $\parallel$ and $\perp$ refer to the direction of the field relative to the axis of the jet. There are various cases. We express our results in terms of the observed index $\alpha \equiv (s-3)/2 = (3-s_t)/2$.

a) Adiabatic Expansion

If at any point (the 'source point') in the jet, a steady distribution function becomes adiabatic then equation (10) gives $f \propto (R_j^2 u_z)^{-(1+ \frac{2\alpha}{3})}$ subsequently. If moreover $B \propto z^{-1}$ and u_z, dR_j/dz are constant, then (17) gives $I_\nu \propto z^{-(2+7\alpha/3)}$, which agrees with a well known result. Note however, that if $u_z \propto z^{-1}$ (Landau-Squires jet) then $I_\nu \propto z^{-(1+5\alpha/3)}$. But such a velocity decline is probably excessively wasteful of energy in the large scale jets.

In the general case of equation (10) we find

$$I_\nu = f_o (t - \int \frac{dz}{u_z})(R_j^2 u_z)^{\frac{-(1+2\alpha)}{3}} B^{(1+\alpha)} d(z). \tag{18}$$

The time dependent factor will not change rapidly until $z = 0(u_z t)$ where t is the age of the outburst. This may be significantly close to the source point (in terms of real brightenings) but it is unlikely to negate the steady adiabatic decline over many scale in z.

b) Bursting Mode

Suppose that G increases suddenly at a point in the jet due to a sudden increase in the turbulent input power I_o through the burst factor $(\kappa_T/\kappa_o)^{1/2}$ (see e.g. EH) or otherwise (e.g. rapid variation in ρ - a cloud encounter - or rapid increase in R_j). One expects a phase in which particles are pumped up in energy space according to (13b). We see readily from equation (19) that $I_\nu \propto (z/G) B^{1+\alpha}$ in such a phase, where G might be calculated initially from equation (7) and (8) keeping f_o constant. However with G large the time scale of this phase is very short, and G will vary rapidly as the number of relativistic particles (f_o) increases. Hence we proceed directly to the 'trickle down' phase of the outburst when synchrotron losses are bringing the particles back down in energy space. Either equation (14) (if S >> G for a time) or equation (16) should apply in this phase. From the first of the form stable modes (16), and (17) we find $f \propto G^{-2\alpha}$ and,

$$I_\nu \propto (d(z)/G^{2\alpha}) B^{1+\alpha}$$

or, since $G \propto S \propto B^2$ in this mode,

$$I_\nu \propto d(z) B^{(1-3\alpha)}. \tag{19}$$

Consequently, if $d(z) \propto z$, $I_\nu \propto z$, $z^{1.5}$ or z^2 according as B is constant, $\propto z^{-1}$ or $\propto z^{-2}$. We note that the actual magnetic field is unlikely to be the equipartition magnetic field in these bursts because of the sudden production of relativistic particles.

This brightening phase must end at p when

$$Sp\, z/u_z \geq 1 \tag{20}$$

at which point an adiabatic decline will begin, in the absence of additional bursting. Identifying such behaviour at sufficiently high frequency can lead to a useful lower limit for u_z.

Should the strictly form stable mode (see 15) be achieved in a region, we see that f_o = constant there. This yields from (17) that

$$I_\nu \propto d(z) B^{1+\alpha} . \tag{21}$$

Therefore if $d(z) \propto z$, $I_\nu \propto z$, $z^{-\alpha}$, $z^{-(1+2\alpha)}$ according as B = constant, $\propto z^{-1}$ or $\propto z^{-2}$. This gives brightness declines from the burst peak which are less steep than adiabatic and quite symmetrical in form relative to the rising edge. In a region of relaxation ascillation where first G dominates and then S dominates, this probably corresponds to the mean behaviour.

IV. NGC6251 AND NGC315

The large source NGC315 (Willis et al. 1981; HBC and unpublished data of Bridle, Fomalont, Palimaka and Henriksen) has a main jet which is remarkably smooth in its brightness variation. There is a rapid rise to a peak at $z \sim 10'$, followed by $I_\nu \sim z^{-0.6}$ out to $z \sim 20"$. This is followed by $I_\nu \propto z^{-1.6}$ out to $z \sim 100"$, and after $z \sim 120"$ there is a nearly adiabatic decline. This latter decline is associated with $dR_j/dz \to 0$. HBC have interpreted this brightness behaviour simply as proportional to the turbulent driving associated with a dynamical model, which corresponds to the strict steady state of the form-stable mode (EH). It is tempting to interpret NGC315 with our more detailed theory here, as a single slow burst. In the steady form-stable mode interpretation we must have $B \simeq$ constant at the base, then $B \propto z^{-1}$ from 10" to 20" and $B \propto z^{-1.6}$ from $z = 20"$ to $\sim 100"$ (we have used $\alpha = 0.6$). This latter variation does not correspond well to the equipartition field variation (Willis et al. 1981) which tends to be flatter (as one might expect if relativistic particles are being produced).

If we interpret the adiabatic decline at $z \sim 120"$ (1" ≡ .314 kpc) according to equation (20) and use $B = B_{eq} \simeq 3~\mu G$, then $u_z \gtrsim 4.4 \times 10^2~\nu_{max}^{1/2}~cms^{-1}$, where ν_{max} is strictly the maximum frequency at which the decline is seen. This gives $u_z \gtrsim 4.4 \times 10^7~cms^{-1}$ if $\nu_{max} \sim 10^{10}$ Hz.

The spectacular one-sided source NGC6251 has been thoroughly studied by Perley, Bridle and Willis (1983). It has a complex bursting structure which we suggest is due to many rapid events of the same general type as the single slow burst in NGC315. We will leave a detailed discussion to a later work. Here we observe that the various $I_\nu(z)$ observed in the different segments of the jet can generally be understood in terms of our ideas above. Thus, the inner jet has $I_\nu \propto z$ for $9" < z < 13"$, so that $B \simeq$ constant based on the form stable mode. The mean behaviour out to $z \sim 100"$ is $I_\nu \propto z^{-1.5}$, which, as in NGC315, requires a form stable steady state and $B \propto z^{-1.6}$. The central, faint region of the jet has $I_\nu \propto z^{-2.2}$ (120" < z < 180") which suggests a form stable steady state with $B \propto z^{-2}$.

When $dR_j/dz \neq$ constant, it is better to use the steady form stable brightness variation in the form $I_\nu \propto B(R_j)^{(1+\alpha)} R_j$. Thus a $B \propto R_j^{-1.6}$ gives symmetrical $I_\nu \propto R_j^{-1.5}$ (I_ν rising with decreasing R and vice versa) as is observed. The same is true with $B \propto R_j^{-2}$ (or a little steeper) to give symmetrical $I_\nu \propto R_j^{-2.2}$. The form stable brightening mode (eq. (19)) may be distinguished from this steady behaviour by brightness peaks that coincide with maximum radii, rather than the converse as above, and as is mostly observed. Thus these variations can be mainly dynamically driven (through R_j), with steady turbulence.

There are two anomalous regions at $z = 50'$ and at $z = 210''$ which have the adiabatic slope ($\alpha = 0.6$) both rising and falling. That is, approximately, $I_\nu \propto R_j^{-(2+7\alpha/3)} = R_j^{-3.4}$. Applying equation (20) is not too useful for these burst however, because $(\Delta z)_{burst} \ll z$ and only a very low, lower limit is obtained at radio frequencies.

The oscillations following the burst at $z = 20''$ out to $z \simeq 110''$ appear to be almost dynamical 'ringing' of R_j while the turbulence is relatively undisturbed from the form stable steady state except at $z \simeq 50''$ where the compression and expansion is so rapid as to be adiabatic. The brightness oscillations in the outer jet region appear to be slipping in phase relative to R_j. This suggests that the turbulence is becoming intermittent, with new particles produced again in bursts.

Acknowledgements:

The author acknowledges the support of the National Sciences and Engineering Research Council of Canada (NSERC A3160), and Mrs. Janie Barr for her assistance in preparing this manuscript.

References:

1. Chan, K.L., and Henriksen, R.N., 1980, Ap. J. 241, 534.
2. Eichler, David 1979, Ap. J., 229, 413.
3. Eilek, Jean, A. 1979, Ap. J., 230, 373.
4. Eilek, Jean, A., and Henriksen, R.N., 1984, Ap. J., Feb., 15 - in press.
5. Fedorenko, V.N., 1980, Sov. Astron., 24, 294.
6. Henriksen, R.N., Bridle, A.H., and Chan, K.L., 1982, Ap. J. 257, 63.
7. Lacombe, C., 1979, Astr. Ap., 71, 1969.
8. Perley, R.A., Bridle, A.H., and Willis, A.G., 1983, NRAO preprint.
9. Willis, A.G., Strom, R.G., Bridle, A.H., and Fomalont, E.B., 1981, Astr. Ap., 95, 250.

DISCUSSION

Norman: How does this generalize to the important particle acceleration processes generated in shocks?

Henriksen: This is a volume mechanism, whereas the shock wave acceleration is more discrete. Moreover, our mechanism is second order Fermi acceleration rather than first order. Shock waves in laboratory jets (not necessarily the same regime) tend to fade into volume turbulence. However, the shock wave mechanism can also produce a universal spectrum and it may well operate in exceptionally dissipative regions.

Vlahos: Coronal streamers are jets that are closer to us. Have you scaled down your process of particle acceleration to such objects, and what are the results?

Henriksen: No. I have not attempted to apply this to the sun. It should be done. The mechanism becomes inefficient, however, if the thermal particle density is 'too high'. However, the scaling is not obvious, as it depends on magnetic field strength, spatial scales, turbulent intensity, and so on.

Sturrock: What damps the Alfvén waves? Do the effects discussed by Hasegawa play an important role?

Henriksen: The Alfvén waves are damped by the resonant interaction with the relativistic particles primarily, and also by interaction with thermal particles at high wave numbers. Dr. Hasegawa's effects seem to be relevant to the high wave number regime.

Sturrock: What determines which turbulence spectrum (Young, Kolmogorov, Kraichnan) is appropriate?

Henriksen: Kolmogorov appears to be the fully developed initial range of turbulence before equiparition with the magnetic field has occurred on all scales of interest. We know (e.g. De Young, Ap. J. $\underline{241}$, 81, 1980) that this equiparition develops first at small scales and only later at larger scales. Thus Kraichnan turbulence is 'old'. The young turbulence represents a state where relatively more energy is in the larger spatial scales.

Kennel: Can you tell anything about the variety of impulsive acceleration mechanisms by observing the spectral index, and its variation, downstream of the acceleration region?

Henriksen: Yes. We predict the range $0.5 \leq \alpha \leq 1.0$, with the normal condition being $\alpha \simeq 0.75$. This is frankly the main reason for studying the mechanism. We don't expect significant downstream variation in α (except for localized anomalies) until synchrotron losses dominate and have depleted the store of high energy particles. The 'trickle down' of the reservoir is the 'relaxation' mode that can give $\alpha = 0.5$ in the absence of strong turbulence.

INSTABILITIES IN ASTROPHYSICAL JETS: DISEASE AND CURE

David Eichler
Astronomy Program, University of Maryland

I. Introduction

A striking feature of extragalactic radio jets is that they are so narrow. Theories of collimation should account for opening angles of order $1°$ or less (as opposed to those of laboratory jets which become turbulent and typically open up to more than $10°$). This makes instabilities particularly troublesome for theories of collimation in which they are present. While instabilities might not entirely destroy the general bipolar nature of the flow, they decollimate it by definition.

There have been many papers concerning instabilities in jets and they are reviewed by Norman and Ferrari at this conference. The hydrodynamic (Kelvin-Helmholtz type) instabilities are basically "centrifugal", that is, if a small bend in the flow develops, the centrifugal force of the flow as it shoots around the bend pushes it out further. For subsonic flow, the work done by the centrifugal force exceeds the energy in the acoustic deformation of the fluid that is associated with it, and the remaining energy goes into the growth of the instability.

Helical magnetic fields have been invoked to contain jet material by magnetic tension. However, as Parker (1977) has pointed out, balancing the pressure in the field with its tension requires just the right pitch, but in a steady-state, field-dominated configuration in a column of varying cross section, the pitch necessarily varies along the axis of the column. Too much pitch ($B_\phi/B_z \gtrsim 2$) causes the coiled magnetic field to be under compression, in which case it is unstable to buckling.

The general criterion for the various instabilities is that internal stress, thermal or magnetic, supports the disruption of the flow. Following an argument (section II) that instabilities are a serious problem for jet models that invoke dynamically significant internal stresses, a model is presented in section III for the hydrodynamic collimation of jets having no significant internal stresses.

II. Are the Instabilities Critical?

It is often suggested that instabilities do occur in astrophysical jets but that they are not critical. Some of these suggestions and their difficulties are reviewed here.

Blandford and Rees (1974) suggested that the turbulence at the edge of the jet is constantly cleansed by the convective action of the moving jet material. But this does not address global instabilities (i.e. those having a scale as large as the jet itself). Several authors (e.g. Hardee 1979, Cohn 1983) have suggested that the instabilities help the knots form that are sometimes observed. But a Kelvin-Helmholtz type instability that has grown to large amplitude destroys the collimation of the flow and there is no obvious reason why the jet, knotty or not, should maintain its collimated form.

Axisymmetric 3-D numerical simulations (Norman et al 1981) demonstrate a fluid nozzle operating, without being disrupted by axisymmetric perturbations, over a limited range of jet power that corresponds to "fat" jets, where the jet thickness is comparable to the ambient scale height. However, this simultaneously implies poor collimation near the nozzle even for a stable jet. Moreover, 2-D simulations (Woodward, in preparation) suggest that kink-mode instabilities are even more dangerous than axisymmetric ones; thus, a true 3-D jet would have an operating range that is more narrow than that given by the axisymmetric simulations and possibly vanishing.

It is sometimes conjectured that a sharp density contrast between the jet material and the stationary ambient medium can solve the stability problems, because the growth rate decreases with decreasing jet density ρ_j. However, the velocity at which ripples at the contact surface propagate in the direction of the jet is proportional to ρ_j whereas the imaginary part of the frequency varies only as $\rho_j^{1/2}$; thus, decreasing the jet density only serves to increase the number of e-folding times over which a perturbation remains within a given scale height.

The basic difficulty underlying all these suggestions can perhaps be stated in a general way by noting that collimation, like instabilities, is just the bending of flow lines. In the former case, the flow lines are bent systematically towards the axis, whereas in, say, a kink instability, they are all bent in the same direction. There is no obvious reason, given that the instabilities do exist, why flow lines should be bent preferentially towards the axis. While the instabilities may be "weak" in some sense for some parameter regimes (Blandford and Pringle 1976), neither is there any obvious reason why the flow should resemble the symmetric pattern that the theorists perturb around. The contention of the author is that the stability problems for jet confinement (when there is internal pressure equilibrium) are serious and motivate alternative models specifically designed to avoid them.

III. Jets Without Internal Stress

An alternative confinement scheme for radio jets that is purely hydrodynamic has been proposed by myself (Eichler 1982, 1983) for radio jets. Canto and co-workers independently proposed a scheme that is basically the same at the general level in the context of bipolar outflows from young star systems in the galaxy (Canto et al 1981) which recently have been associated with the problem of radio jets. The key assumption of the model is that there is no significant internal pressure or magnetic stress; the ejected material is assumed to be a supersonic, high Alfven Mach number wind by the time it expands to the collimation scale. Acoustic and magnetic signals cannot propagate through the jet over the dynamical timescale as would be necessary to support global instabilities. The collimation occurs close to the wall of the channel opened up by the outflow. Shocks would occur near the channel wall and, in order to maintain a high Mach number, the internal energy of the post-shock material would have to be dissipated somehow. Possible dissipation mechanisms include cyclotron emission or other photon radiation, escape of high energy particles or their collisional by-products, or heat conduction into the channel wall, depending on the field strength, optical depth, and other parameters of the flow. Many of these possibilities obviously would predict that much of the beam power escapes as some sort of radiation from the scale of collimation.

The axisymmetry in this model is enforced by the axisymmetry or rapid rotation of the central object emitting the supersonic wind; the interior of the jet is mechanically decoupled from the confining cloud so it can neither be destabilized nor symmetrized by it. There are possibly very small scale instabilities within the sheath of shocked material at the channel wall, but they do not seem nearly as dangerous as the global instabilities that have been eliminated by the assumptions of the model.

The analysis of the channel wall shape and the extent of collimation, which neglects the jet's internal pressure and equates its inertial forces with the ambient pressure, is in a sense the reverse of Blandford and Rees (1974), who keep the internal pressure and neglect its inertial forces. It can be shown that for a pressure profile proportional to r^{-4} and an axisymmetric wind, the channel shape is a perfect cylinder, so for $P \propto r^{-\alpha}$; $\alpha < 4$, the ambient pressure focuses the jet to a point on the axis.

The equation that described the channel shape can be shown to be (Eichler 1982)

$$P_a = v\frac{d\dot{M}}{d\Omega}(r\frac{d\theta}{dr})^2/r^2[1+(r\frac{d\theta}{dr})^2] +$$
$$(r\cos\theta\ R_c)^{-1}\int_0^\theta v\cos\psi(\theta\,\check{}\,)\frac{d\dot{M}}{d\Omega}\cos\theta\,\check{}\ d\theta\,\check{} \qquad (1)$$

Here P_a is the ambient pressure, v is the velocity, the channel shape is expressed as a function $\theta(r)$ in polar coordinates ($\theta \equiv 0$ at the equator), ψ is the angle of impact with the channel wall, R_c is the radius of curvature of a meridianal cross-section of the channel, and $d\dot{M}/d\Omega$ is the ejected mass flux per unit solid angle. The first term on the right hand side is the ram pressure of the material hitting the channel wall and the second is the centrifugal pressure of the material as it shoots around the channel wall subsequent to its impact.

The cylindrical channel shape
$$\cos\theta = \frac{1}{r} \qquad (2)$$
for which $R_c^{-1} = 0$, is a solution when $P_a = 1/r^4$, and v and $dM/d\Omega$ are spherically symmetric (unit for all Ω). Thus an ambient pressure that decreases more slowly than r^{-4} focuses the flow inward.

Given dissipation of the transverse motion by shocks at the focal point, the final collimation is arbitrarily good. That is, the "small number" in the theory that can be associated with the very small opening angles of many radio jets is the fraction of internal energy that is retained by the post shock material.

This confinement scheme does not require a carefully tailored pressure profile to draw the jet material into a slender form over many pressure scale heights as do the schemes invoking internal pressure equilibrium within the jet (leaving aside stability questions). Given effective dissipation, good collimation results from any pressure profile that falls off less rapidly than r^{-4} along the channel wall as long as the confining cloud is oblate enough to allow the fluid to escape along the axis.

A second analytic solution to (1), which illustrates the recollimation of a precessing beam by a uniform ambient medium, can be obtained by taking P_a to be constant, and $d\dot{M}/d\Omega$ to be proportional to $\delta(\theta-\theta_o)$, corresponding to a precession cone of angle θ_o. For $\theta > \theta_o$, the ram pressure term in (1) vanishes and the centrifugal term simplifies to a term proportional to R_c^{-1}. Writing R_c^{-1} in cylindrical coordinates z (height) and x (radius) — $R_c^{-1} = d/ds$ arctan $(\frac{dz}{dx})$, where $ds = (dz^2 + dx^2)^{1/2}$, yields the generalized solution

$$z\,\check{} = \frac{(\frac{x}{2} + \sin\theta_o)}{[1 - (\frac{x}{2} + \sin\theta_o)^2]^{1/2}} \qquad (3)$$

This is readily integrated numerically, and one derives a focal length z_f of about $3(L/\pi v P_a)^{1/2}$, where L is the jet power, at which point the jet converges back into the axis. This value for the focal length is consistent wiht observations of laboratory jets, where the focal lengths (which are expected to be slightly shorter because the jets are filled in, so that the average opening angle is only $\sim 2/3\ \theta_0$) are between 2 and 3 times $(L/\pi v P_a)^{1/2}$. The calculation is relevant to jets that appear to have been focused while supersonic, such as the one in the supernova remnant W50 (SS433) and possibly some extragalactic jets, where observational limits can be set on L, P_a, and z_f.

In conclusion, cold, pressureless jets seem to be easier to collimate than those that are in pressure equilibrium wiht their surroundings. They can in fact be focused, and are qualitatively more stable during collimation. Because shocks form, the shocks must be dissipative in order to keep the jets could over many focal lengths. But perhasp this assumption is worth making if it yields a satisfactory theory of jet collimation.

References

Blandford, R. D. and Pringle, J. E.: 1976, M.N.R.A.S., 176, 433.
Blandford, R. D. and Rees, M. J.: 1974, M.N.R.A.S., 169, 395.
Canto, J., Rodriguez, L. F., Barral, J. F. and Carrel, P.: 1981, Ap.J., 224, 102.
Cohn, J.: 1983, Ap.J., 269, 500.
Eichler, D.: 1982, Ap.J., 263, 571.
Eichler, D.: 1983, Ap.J., 272, (in press).
Hardee, P.: 1979, Ap.J., 234, 47.
Norman, M. L., Smarr, L., Wilson, J.R. and Smith, M. D.: 1981, Ap.J., 247, 52.
Parker, E. N.: 1977, Ann. Rev. Astron. and Ap., 15, 45.

DISCUSSION

Henriksen: I have several comments: i) Jets are not limb brightened in general. ii) Compressible turbulence can be recollimated by the external pressure. iii) Turbulent jets that are compressible may have a non-uniform viscosity and a core-halo structure. iv) Eddy scale dissipation is $R/(dR/dz) \sim z$. Locally in our models (Henriksen, Bridle, Chan) dissipation power is about what is required for synchrotron brightness. v) Turbulence is regenerated by entrainment (HBC).

Eichler: I agree that jets are not, in general, limb brightened on large scales. On scales smaller than 1 pc, where the intitial collimation may take place, no one knows.

The main problem that I see with turbulent jets is that, as you seem to agree, the turbulence at some point z along the jet dissipates over a scale of about z. Over a very large increase in scale size, it seems likely that most of the energy in the jet, including the flow energy, would be dissipated, and the jet would peter out. This is what happens in laboratory turbulent jets, including those that start out supersonic. The challange to proponents of truly turbulent extragalactic jets where there are global instabilities, much entrainment of surrounding material etc. - is to explain what difference about them from laboratory jets enables them to survive the development of turbulence.

Incidentally, I agree that some turbulence is present in these jets, I merely argue that global instabilities are not present.

Benford: The hidden assumption here is cylindrical symmetry, imposed by spinning the cloud, or a precessing jet. Doesn't it seem equally simple to use the instrinsic cylindrical symmetry of a current, which imposes a confining B_θ and can also ameliorate sidewise instabilities if there is a parallel magnetic field as well?

Eichler: The assumption of a cylindrically symmetric wind is there, and mentioned explicitly. On the other hand, I know of no demonstration in the literature that currents in a dynamically active $\vec{B}$ field are "intrinsically" symmetric; the kink instability seems to be a counter example. The challenge that should be met by proponents of magnetic confinement is to show that B_ϕ/B_z can be large enough to collimate the flow but not large enough to destabilize it in or near the zone of collimation. There is much skepticism that this can be achieved, even with careful planning.

Gilden: If the stability of the walls requires that the source not be dynamically coupled, how is force balance achieved? Is there not a shock propagated into the cloud and then overall expansion?

Eichler: No, the cloud is assumed to be in hydrostatic equilibrium or some other steady state. The only shocks are in the jet.

Norman: For a collimated jet propagating in an atmosphere with a density profile $\rho \propto r^{-2}$, $\alpha > 2$, the jet will break free in a few scale heights. Essentially, the transverse velocity exceeds the sound speed. Firstly, why doesn't this happen here? Secondly, what is the physical basis for your models critical exponent of $\alpha = 4$?

Eichler: Your first sentence is true only according to a particular, popular set of assumptions; ruining the assumptions doesn't necessarily ruin the jet. I agree that if $\alpha < 2$, the transverse velocity will generally exceed the sound speed and shocks will form. This does happen here. But if the post-shock fluid dissipates its energy somehow, as I hypothesize, the jet does not break free. The "critical exponent", as you call it, is just the value of α below which the thin sheath of shocked material is bent back towards the axis. It is $d + 1$, where d is the dimensionality of the flow, since, for a supersonic wind striking a cylindrical shell, $\rho v^2 \sin^2 x$ goes as $r^{-(d+1)}$; i.e. $\rho \propto r^{-d+1}$, $v = $ const, and $\sin x \propto r^{-1}$.

Coppi: Do you have any comment about the accretion funnel model that was proposed some time ago?

Eichler: I presume that the accreting material, in this model, acts as the confiruing material. So it seems to be a special case of hydrodynamic confinement, applied to a region very close to the central compact object in active galactic nuclei. I have no quarrel with the location, and am not very familiar with some of the other details.

Bridle: The jet in NGC 6251 definitely shows alternating regimes of rapid and slow expansion, on scales of tens of kiloparsecs. The 3° opening angle that has been referred to here is only an average behavior. The detailed collimation behavior shown by the VLA observations (Perley et al, to appear in Ap.J. Supplements) clearly shows there some recollimation goes on over scales comparable with that of the associated galactic atmosphere, rather than being set once and for all on the parsec scale of the VLBI jet.

Eichler: I agree (see Eichler, D., Ap.J., Sept. 1983); the jet in W50 is another example. But I still feel that a jet as striking as the one in NGC 6251 could not have suffered global instabilities at any point along its length.

CURRENT SYSTEMS IN RADIO JETS

Jean A. Eilek
Physics Department, New Mexico Tech
Socorro, N. M. 87801 U.S.A.

The structure of the magnetic field in radio jets is a topic of recent interest, especially due to the possibility that some high pressure jets are confined by a magnetic pinch. Several such jets have been found which cannot be confined by external cluster gas pressure, on which there are observational limits; nor can they be in free expansion, since they do not show evidence of adiabatic expansion losses. Recent radio interferometer observations of surface brightness and polarization allow the possibility of determining the magnetic field structure. In this paper I present some basic considerations of the current and field structure required if the observed jets are to be magnetically confined.

BASIC MAGNETIC CONFINEMENT

The simplest MHD equilibrium in a cylindrical geometry is the classic Bennett pinch, in which the tension of an azimuthal field, B_ϕ, balances the plasma pressure. The field, pressure and axial current are related by

$$B_\phi^2 \sim 8\pi p ;$$
$$j_z = cB_\phi/2\pi r ; \qquad (1)$$

if r is the jet radius. For radio jets with $p \gtrsim p_{eq} \sim 10^{-10}$ dyn cm^{-2} (note that equipartition estimates are lower limits for the internal pressure), $B_\phi \sim B_{eq} \sim 10$ to 100 µG is needed, hence a current density $j_z \sim 10^{-16}$ cgs. The net current is $I = \pi r^2 j_z \sim 10^{18}$ A. If this current were to arise from a charge imbalance in the streaming plasma, one extra electron in $\sim 10^{12}$ particles would be required. This small imbalance nonetheless would lead to a charge accumulated in the end of the jet or the radio lobes which would be large enough to stop the jet (which is carrying $\rho v_j^2 \pi r^2$ of kinetic energy) in only 10 to 10^3 years. Thus, a complete circuit must exist.

Very little can be said about the path of the return current. The observations discussed below do not admit solutions in which the return current lies on the jet surface. Alternative possibilities may be on the boundary layer between a "cocoon" of the radio source and the cluster gas (Benford, 1978), or an even more diffuse (field aligned?) current in the cluster gas itself. The former would be akin to, say, magnetopause currents driven by the solar wind dynamo. The latter would require a cluster field with the right configuration; very few observational constraints have yet been put on any such field.

THE NATURE OF JET FIELDS AND CURRENTS

At least two separate lines of argument suggest that the jet field has a flux rope-like structure. Near the axis the field must be mostly longitudinal (pitch angle $\to 0$), and away from the axis the field must become mostly azimuthal (pitch angle $\to \pi/2$). The radio luminous plasma must lie mostly in the region where B_z dominates; the B_ϕ region must be lacking in luminous plasma, and is probably nearly force-free. The arguments for this structure are as follows:

(a) This structure is consistent with observations, which often show a projected field parallel to the jet in luminous (high energy density) cases (Bridle, 1982), and yet it can provide the magnetic confinement through the external B_ϕ. (Note, however, that some small amount of B_ϕ in the luminous plasma will still produce a projected parallel field; Laing, 1981).

(b) Calculations of static cylindrical magnetic pinches suggest that this type of configuration may be more stable than the pure theta pinch, which is notoriously unstable to pinching and twisting modes (e.g., Goedblod, 1971; Cohn, 1983).

Of course the MHD equilibrium condition,

$$\nabla p + \frac{\bar{j}}{c} \times \bar{B} = 0 ,$$

$$\bar{j} = \frac{c}{4\pi} \nabla \times \bar{B} , \qquad (2)$$

admits many solutions satisfying these general conditions, with boundary conditions p, B_z = finite as $r \to 0$; p, B_z, $B_\phi \to 0$ as $r \to \infty$. Typical choices of solutions are shown in Figure 1; more detailed calculations of field and current structure and integrated surface brightness will be presented elsewhere (Eilek, 1984).

ENERGETIC AND DYNAMO CONSIDERATIONS

One basic problem is "what drives the current?" One can address this problem by considering the energy budget of the jet flow, and by looking at general dynamo physics, with hopes of restricting the set of allowed models to less than all of parameter space.

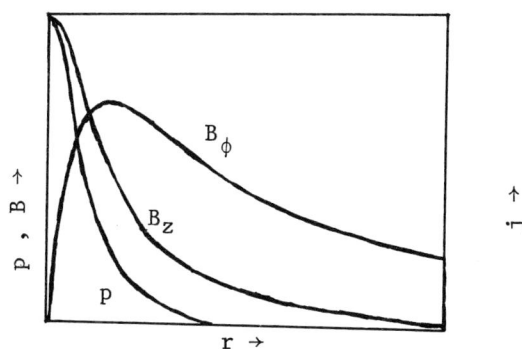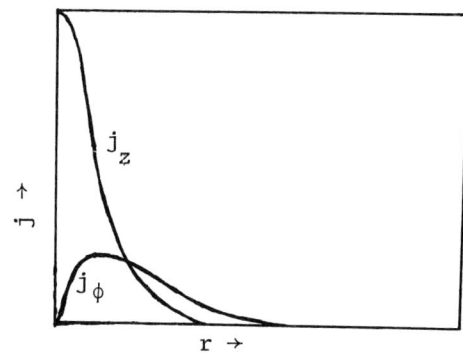

Figure 1. Behavior of p(r), B(r) and j(r) in typical equilibrium.

The size of the voltage drop and local dissipation associated with the confining current are important quantities. Both depend on the plasma conductivity, σ, which is of course hard to say anything about. Anomalous effects can lower the conductivity by several orders of magnitude over the traditional Lorentz gas value (which relies on electron-ion collisions to regulate the transport), due to higher collision rates in the presence of microturbulence. It turns out, however, that in radio sources even the most extreme estimate of the anomalous conductivity (when the collision rate is taken equal to the electron plasma frequency) still results in quite small voltages ($\Delta V \propto \sigma^{-1} \int j \cdot dl \sim 3 \times 10^6$ volts over a 10 kpc jet) and local ohmic losses ($j^2 \sigma^{-1} L \ll \rho_j v_j^3$, where the latter is the conventional estimate of local energy source in a jet flow). In the Lorentz gas limit, $\sigma \to \infty$ and both the voltage drop and the local dissipation are negligible.

Another important consideration is the nature of the battery or dynamo which drives the current. Two possibilities arise.

One model would attribute the dynamo to the "engine" in the galactic nucleus. This unspecified machine is also the source of the jet flow. Blandford (1976) and Lovelace (1976) have considered rotating compact objects as electromagnetic dynamos, and they estimate central electric potentials from objects which are well above that needed for the confining current. Application of such models to radio jets leads to the standard problem of coupling nuclear models (involving scales $\lesssim 1$ pc) to the extended structure (scales $\sim 10^3$ to 10^6 pc); in particular, in this case the question of how the current path is determined must be addressed. This area deserves further work.

An alternative model would use the jet itself to provide the dynamo. Since the kinetic energy flux in the jet, $\rho_j v_j^3$, is generally thought to be the local source of radiation and heat (through MHD/plasma turbulence generated in the jet and local particle reacceleration processes), it may be that this turbulent energy conversion can also supply an *in situ* kinematic dynamo. It is well established that fluid turbulence with non-zero helicity ($\overline{v} \cdot \nabla \times \overline{v} \neq 0$) can generate a net EMF and large scale magnetic field (e.g., Steenbeck, Krause and

Rädler, 1966; De Young, 1980). It may be that turbulence which develops in a jet which has an initial azimuthal velocity field could supply a local magnetic field with the proper configuration; specific calculations are required to address the dynamo problem in this context, and this author is not aware of any such work to date.

Energetic arguments, however, may indicate the limits of applicabiity of this idea, as follows. The best, "maximal helicity", case of turbulent generation of fields reaches a limit of rough equipartition, $B^2/8\pi \sim \rho_j v_t^2$, between the large scale fields and the turbulent energy density (De Young, 1980). But the MHD equilibrium requires $B^2/8\pi \sim p$. The turbulent velocity $v_t < v_j$, as generation of turbulence tends to be inefficient. Thus, this model may work only if $\rho_j v_t^2 \sim p < \rho_j v_j^2$, that is, for "lukewarm" jets. Some jets observed may satisfy this condition at present; but the inefficiency of turbulent field generation (much of the energy input goes into dissipation and plasma heating) may make it hard to maintain the $B^2/8\pi \sim p$ balance over the source lifetime.

OBSERVABILITY

The flux rope models above clearly can satisfy the general observational constraints, both in terms of surface brightness and projected field. It may be possible to discriminate between transverse profiles from the rate at which the surface brightness falls off, using well-resolved sources. The azimuthal confining field is harder to test, as by assumption it lies outside of the luminous plasma. It may be marginally detectable as a foreground Faraday screen, if the jet lies in a sufficiently dense ionized cluster gas. The rotation measure ($\chi \propto \int n_{ICM} B_\phi \cdot ds$) should vary systematically across the jet, as $B_\phi \cdot ds$ changes sign, and may be just large enough to be detectable.

It is a pleasure to acknowledge useful conversations with Dave De Young, Phil Hardee and Frazer Owen.

REFERENCES

Benford, G., 1978, M.N.R.A.S., 183, p. 29.
Blandford, R. D., 1976, M.N.R.A.S., 176, p. 465.
Bridle, A. H., 1982, in "Extragalactic Radio Sources", I.A.U. Symposium 97 ed. D.S. Heeschen and C. W. Wade, p. 121.
Cohn, H., 1983, Ap. J., 269, p. 500.
De Young, D. S., 1980, Ap. J., 241, p. 81.
Eilek, J. A., 1984, in preparation.
Goedbold, J. P., 1971, Phys. Rev. Letters, 24, p. 253.
Laing, R. A., 1981, Ap. J., 248, p. 87.
Lovelace, R. V. E., 1976, Nature, 262, p. 649.
Steenbeck, M., Krause, F. and Rädler, K. H., 1966, Z. Naturforsch., 21a, p. 369.

DISCUSSION

Sturrock: I do not see how a jet can generate the B_ϕ magnetic field in the surrounding intergalactic medium, since it has virtually infinite conductivity.

Benford: The return currents are necessarily set up by the electric field E_z induced by penetration of the jet head, which carries an increasing current. This field is spread over several jet radii, and probably even more, considering that E_z is made at the hot spot working surface, which is wider than the inner jet.

Vasyliunas: To make a flux rope, you have to twist at one end and hold it (or twist differently) at the other end. One end can be anchored in the central object, but what holds the other end, to twist the jet rather than have it simply rotating with the central object?

Eilek: If the jets are indeed flux ropes, perhaps the "working surface" when the jet hits the external plasma provides the anchor.

Coppi: Can you estimate, roughly, what the gross stability of the magnetic configuration may be?

Eilek: One can, of course, estimate stability to first order by anology with static pinches; but specific calculations which address the instability growth rates (in time and space) of jet flows in cluster environments are needed.

Elphic: A lesson from the Venus flux ropes: The return current you are worrying about is very diffuse in this case. It is distributed over a much larger area than the intense central axial current. In your situation the return current may be very tough to observe.

Guillory: If you could see the return current, you could probably estimate the age of the structure, since the field must penetrate the conducting plasma surrounding the jet.

IS THE JET IN M87 MAGNETICALLY CONFINED?

Philip E. Hardee
Department of Physics and Astronomy
University of Alabama
Tuscaloosa, Alabama 35486

Abstract: Calculations of the minimum pressure in the M87 jet show that the jet pressure is an order of magnitude in excess of the thermal pressure in the interstellar medium calculated from the X-ray emission. The continuous energy input required to produce the jet emission is incompatible with a freely expanding jet. On the other hand, the uniform jet expansion, the jet emission and the jet polarization can be explained if the jet is self-confined by j x B forces.

There are now several jets in which the minimum jet pressure appears to exceed the thermal pressure of the surrounding medium. One of these jets is associated with the quasar 4C32.69 (Potash and Wardle 1980). Other jets associated with extragalactic radio sources may also have minimum jet pressures that exceed the pressure in the surrounding medium (Burns et al. 1983). Calculations of the minimum jet pressure in the M87 jet have been made by Biretta et al. (1983). The pressure of the surrounding medium calculated from the X-ray data combined with a model of the cooling accretion flow in M87 decreases with distance from the nucleus with $P \propto z^{-1.5}$ (Lea et al. 1982) and is as much as an order of magnitude lower than the minimum jet pressure interior to Knot A. Static pressure confinement is not ruled out interior to Knot D. Only if this jet were extremely relativistic, $\gamma > 50$, could the jet be statically pressure confined. In part, this is because the relativistic jet must lie near the line of sight and the jet extends further outwards from the nucleus where pressures in the external medium are lower. The observations also show that the jet expands at a constant rate with opening angle $\sim$ 5 degrees at least to Knot A, that the polarized emission at 2 cm is somewhat limb brightened with the electric vector oriented perpendicular to the jet axis and that the radio brightness falls very slowly along the jet and the jet is nearly isothermal.

The constant opening angle of the jet could be produce by a freely expanding jet with initial Mach number of about 25. With no energy input such a jet must expand adiabatically. Energy input into a jet will be the result of dissipation of turbulence in the jet fluid. Turbulence is generated at the jet external medium interface and may be produced in the jet near the origin. In the free jet turbulence produced at the interface is restricted to a thin surface layer because jet expansion is more rapid than the internal sound speed. Ordering of the magnetic field in the jet prevents relativistic particle diffusion into the jet interior (Eilek 1982) and only a limb brightened configuration is produced. Turbulence in the interior can be produced at the origin of the jet but turbulent velocities decay and the fundamental eddy size increases as $\Delta u = \Delta u_o (t_o/t)^{5/7}$ and $1 = 1_o (t/t_o)^{2/7}$, respectively, with result that the energy dissipation rate per unit volume $\dot{E} \sim \rho \Delta u^3/l$ decreases rapidly as (Landau and Lifschitz 1959)

$$\dot{E} \sim \rho_o \Delta u_o^3 (z_o/z)^{31/7}/l_o$$

where $z = v_z t$, v_z is the jet velocity and ρ_o, Δu_o and l_o are the initial density, turbulent velocity and initial eddy size ($l_o \sim$ jet radius), respectively. Thus the requirement that there be approximately constant energy input into the M87 jet along the jet cannot be met if the jet is free. On the other hand, confinement of the M87 jet allows for continuous energy input throughout the jet through dissipation of turbulence generated at the jet boundary (Hardee 1983) which can keep the jet nearly isothermal (Begelman 1981).

Although the M87 jet cannot be confined by pressure of the external medium except perhaps interior to Knot D, the jet can be self-confined by $j \times B$ forces. If the jet velocity v_z remains nearly constant, the steady state dynamics of the magnetically confined jet in the radial direction provided the expansion is slow is given by (Chan and Henriksen 1980)

$$dv_r/dz = 2 [P_j + B_z^2/8\pi - B_\phi^2/2n\pi]/\rho R v_z$$

where P_j is the internal thermal pressure at the jet axis, B_z is the component of the magnetic field parallel to the jet axis (here assumed constant across the jet) and B_ϕ is the strength of the confining magnetic field at the jet surface and $n = 2$ or 4 if the current is uniformly distributed across the jet or is a surface current, respectively. Note that $B_z^2 \propto R^{-4}$, $B_\phi^2 \propto R^{-2}$ and the $P_j \propto R^{-2}$ if the jet is isothermal. Thus, jet expansion will remain nearly constant if $P_j \sim B_\phi^2/2n\pi$. We must assume that pressure balance is established somewhere near the jet origin. Such a result might be produced by a hydromagnetic flow from an accretion disk (Blandford and Payne 1982).

The polarization observations provide a clue as to the nature of the current and orientation of the jet if it is magnetically confined. For a Faraday thin jet and helical magnetic field we can use the results obtained by Laing (1981) which describe the polarization properties of a cylindrical jet. We find that the polarization at jet center, the mild polarization limb brightening and the position angle of the electric vectors at 2 cm imply that the jet is oriented within 30 degrees of the plane of the sky and that at the jet surface the field helicity is less than 45 degrees, i.e., $B_\phi < B_z$ at the jet surface. The fact that B_ϕ should increase rapidly along the jet relative to B_z as the jet expands may be taken to imply that B_ϕ inside the jet must be restricted to a surface layer. If the conductivity of the material is high near the jet axis but decreased near the jet surface, perhaps because of increased turbulence, a self-consistent picture of the magnetic field emerges. Current flows are near the surface and the magnetic field is frozen into the fluid in the jet interior parallel to the flow. If the fluid carries angular momentum we expect small B_ϕ in the jet interior with B_ϕ increasing rapidly near the jet surface. The jet can be thought of as confined by the field that is outside the jet which is produced by a surface current. This configuration reproduces both the position angles of the electric vectors perpendicular to the jet axis and the modest limb brightening in the polarized intensity.

Finally we need to address the stability of this magnetically confined configuration. A magnetically confined jet will be sufficiently stable to pinching if the jet Mach number, M, is greater than about 3 (Cohn 1983), i.e., the rate of growth is too slow to affect the jet dynamics unless the jet is very long. This result is similar to that obtained for a jet that is thermally confined. An estimate of the minimum M87 jet luminosity when combined with an estimate of the jet's sound speed implies that M > 3 for the M87 jet and it is clear from the observations that the knots are not pinches in the flow. In general, the stability properties of a supersonic and superalfvenic hydromagnetic flow confined by j x B forces will be similar to the thermally confined case. The reason for the similarity is that the energy in such a system resides in the fluid flow and not in the magnetic fields and associated currents. Thus, results obtained for thermally confined jets (see the review by Ferrari) may be applied to magnetically confined jets. Some allowance must be made for the effects which result because the external magnetic field ties the external medium to the jet. This increases the jet's inertia. For example, this slows the growth of jet helicity (Benford 1981) and a supersonic flow with M > 5 will be sufficiently stable to helical twisting which will not affect the dynamics significantly. The fact that the M87 jet exhibits significant helicity downstream of Knot A implies different jet conditions in the inner and outer portions of the jet. Jets are also unstable to fluting wave modes which are harmonics of the helical wave mode. The fluting wave modes grow rapidly at wavelengths less than the jet radius. Short wavelengths can be suppressed by velocity shear (Ray 1982) but at least a few fluting wave modes with characteristic wavelength less than the jet radius will be sufficiently unstable to drive jet turbulence (Hardee 1983) and provide the heating and particle acceleration needed in the jet in M87.

REFERENCES

Begelman, M.C. 1982, in IAU Symposium 97, Extragalactic Radio Sources, ed. D.S. Heeschen and C.M. Wade (Dordrecht: Reidel).
Benford, G. 1981, Ap.J., 247, 792.
Biretta, J.A., Owen, F.N. and Hardee, P.E. 1983, Ap.J. Letters, in press.
Blandford, R.D. and Payne, D.G. 1982, MNRAS, 199, 883.
Burns, J.O., Basart, J.P. and DeYoung, D. 1983, preprint.
Chan, K.L. and Henriksen, R.N. 1980, Ap.J., 241, 534.
Cohn, J. 1983, Ap.J., 269, 500.
Eilek, J.A. 1982, Ap.J., 254, 472.
Ferrari, A. 1985, in IAU Symposium 107, Unstable Current Systems and Plasma Instabilities in Astrophysics, ed. M. Kundu and G. Holman (Dordrecht: Reidel)(this volume).
Hardee, P.E. 1983, Ap.J., 269, 94.
Laing, R.A. 1981, Ap.J., 248, 87.
Landau, L.D. and Lifschitz, E.M. 1959, Fluid Mechanics (New York: Pergamon).
Lea, S.M., Mushotzky, R.F. and Holt, S.S. 1982, Ap.J., 261, 42.
Potash, R.I. and Wardle, J.F.C. 1980, Ap.J., 239, 42.

DISCUSSION

Henriksen: I have two comments: 1. Large scale magnetic fields are not inconsistent with turbulent jets. Both the velocity spectrum and the magnetic field spectrum can have large scale components. 2. If $\beta \propto R^{-2}$, then $p \simeq \rho v_t^2 \propto R^{-2}$.

Hardee: Particularly if turbulence in subsonic is a supersonic flow the magnetic fields will appear relatively well ordered in this observer' reference frame.

Uchida: Shibata and myself have done a calculation very much related to this model by Dr. Hardee and that by Dr. Eilek. A large difference is that ours deals with the transient process in which the B_ϕ produced by rotating motion in a $\beta \gg 1$ region relaxes into the $\beta \ll 1$ region, and in this dynamical process the front of the relaxing packet of B_ϕ carries the material with it.

Hardee: At least approximately this is how I imagine that the jet becomes self-confined.

Bratenahl: Is there, within observational constraints, a more or less uniform B-field in space through which the jet passes? If so, is the strength $\sim$ microgauss? I am excited about this -- at UC Riverside we are trying to understand the mechanism by which our plasma jet penetrates the external field by polarization currents at the front.

Hardee: Yes, $\sim$ microgauss. This compares to a jet magnetic field of several hundred microgauss.

T. Ray: Could the known presence of the observed Hα filaments near the M87 jet give you any handle on the extent of the B_ϕ fields? (My reasoning being that if the Hα filaments where inside the B_ϕ field it would exert pressure on them).

Hardee: The B_ϕ field will fall off rapidly as $1/r$ outside the jet and may be screened by return currents.

Wilson: I wonder how confident you are about ruling out thermal confinement. The minimum internal jet pressure (magnetic and cosmic ray) is subject to the usual uncertainties, such as the exact geometric configuration of the field and particles, and the X-ray measurements of the external density refer to only a certain range of gas temperature. In such terms, a factor of 10 difference between the pressures may not be insurmountable.

Hardee: I am confident that the computation of the minimum jet pressure is accurate and is a firm lower limit. I am less certain about the X-ray data, but two separate computations using different assumptions arrive at estimates of the pressure in the external medium that differ by only a factor of 2 at 2 kpc from the nucleus. It is difficult to imagine any assumptions that could increase the external pressure by an order of magnitude.

Sturrock: What is your interpretation of what is happening at the knots?

Hardee: Pinching of the jet can be ruled out as a means of producing the knots. Aside from knot A, which is almost certainly a shock, I like the idea of turbulent bursts in the jet material as discussed by Henriksen.

SESSION VII

LABORATORY PLASMA PROCESSES OF ASTROPHYSICAL INTEREST

C. S. Liu
University of Maryland, College Park, Maryland 20742

and

GA Technologies, San Diego, California 92138

In recent years with the advance of large fusion devices, very high temperature plasmas (>1 keV) with long confinement times (> tens of msecs) are produced in the laboratory, offering the possibilities of studying many basic plasma processes of astrophysical interest in laboratory plasmas. In the following, I wish to review some of the recent progress in fusion plasma physics bearing on astrophysical and space plasma processes:

(i) turbulent relaxation towards the state of minimum magnetic energy;

(ii) resonant radial diffusion of energetic ions;

(iii) wave acceleration of electrons to sustain a current in a plasma.

I. TURBULENT RELAXATION OF UNSTABLE PLASMAS TOWARDS THE STATE OF MINIMUM MAGNETIC ENERGY IN PINCHES

In toroidal confinement schemes, there are basically two types of discharges characterized by the degree of twist of magnetic field lines, termed "rotational transform," $\iota/2\pi$, which is the average change of the poloidal angle following the field line once around the torus, $\overline{\Delta\theta}/2\pi$. In a pinch experiment a toroidal magnetic field B_0 is produced by external coils and then a toroidal current I is induced. For high currents, the magnetic field line is tightly twisted, $\iota/2\pi > 1$ and the plasma column is unstable to kink modes, much like a tightly twisted rubber band to form loops. After an initial phase of violent instability, however, the plasma remarkably relaxes to a quiescent state which depends only on the pinch ratio $\Theta = 2I/aB_0$, where a is the plasma radius. For Θ exceeds a certain value, the toroidal magnetic field plasma at the plasma edge is reversed in the quiescent state.

Recently, J. B. Taylor proposed an elegant theory,[1] based on the theory of minimum energy state first developed by Woltjer[2] and successfully explained many of the observations. The essence of the theory is that the plasma seeks a state of minimum magnetic energy

$$W_B = \int_{V_0} dV B^2/8\pi$$

subject to the constraint that the toroidal and poloidal flux linkage or magnetic helicity over the total volume of the plasma V_0:

$$K_0 = \int_{V_0} dv \, \vec{A} \cdot \vec{B}$$

be invariant, where $\vec{A}$ is the vector potential. This is greatly relaxed over the ideal MHD constraint which requires $K = \int dv \, \vec{A} \cdot \vec{B}$ to be conserved on each line of force, because of the effect of finite resistivity. By varying W subject to the constancy of K: $\delta(W_B - \lambda K_0) = 0$, one finds the minimum energy state is a force-free state governed by $\nabla \times \vec{B} = \lambda \vec{B}$ depending on only one parameter λ. A particular solution is

$$B_r = 0, \quad B_\theta = \alpha J_1(\lambda r), \quad B_z = \alpha J_0(\lambda r) ,$$

for which $\lambda a/2 = \Theta = 2I/aB_0$. If $\Theta < 1.2$, then the toroidal field is reversed in good agreement with experimental observation. This field-reversed quiescent state is maintained over a period of several milliseconds, much longer than the classical skin time for magnetic field diffusion: $\tau_s = 4\pi a^2/\eta c^2$, where $\eta = m\nu_{ei}/ne^2$ is the plasma resistivity and ν_{ei} is the electron-ion collision frequency, suggesting that perhaps some kind of turbulent dynamo action is at play, counterbalancing the resistive diffusion. This is presently an active area of theoretical and experimental investigation and the understanding of these processes will undoubtedly help us to understand better the phenomenon associated with solar flare and magnetic stars.

II. RESONANT RADIAL DIFFUSION OF ENERGETIC-IONS FROM TOKAMAKS BY TRAPPED ION-INDUCED INTERNAL KINKS

Tokamaks, unlike pinches, have strong toroidal magnetic fields so that the field line is only weakly twisted: $\iota/2\pi < 1$. However, due to a thermal instability of ohmic heating during which the current tends to peak where the temperature is high resulting in a current distribution peaked at the center, and a local region near the center where $\iota/2\pi > 1$. The kink instability can then develop in this localized region and is called "internal kink." Nonlinearly this instability causes the electron temperature and current in the unstable central region to be flattened so that $\iota/2\pi \sim 1$ at center. This unstable buildup and relaxation of the electron temperature is observed in the "sawtooth"-like variation of x-rays emitted from a tokamak plasma.[3]

Ohmic heating alone, however, is insufficient to heat a tokamak plasma to the temperatures required for thermonuclear ignition because of the decreasing plasma resistivity with increasing temperature and the amount of current limited by $\iota/2\pi < 1$. Thus, auxiliary heating schemes employing neutral beam or radio frequency waves are needed. In neutral beam heating, energetic neutral atoms ($\geqslant 40$ keV) were injected into a tokamak plasma which, upon ionization and charge exchange, become energetic ions in the plasma. These energetic ions are then slowed down by Coulomb collisions with background plasma which in turn are heated by the energetic ions. The neutral beam injection has been quite successful as a way of plasma heating and a record in temperature of 7 keV was attained in Princeton Large Torus (PLT) several years ago with neutral beam injection of several megawatt of power into the plasma.[4]

Recently up to 7 MW of neutral beam power is injected almost perpendicular to the magnetic field into a plasma in the Princeton Divertor Experiment (PDX). Above certain injection power, the efficiency of plasma heating is reduced. One reason for the decreased efficiency is the observed ejection of energetic ions from the tokamak, associated with the observed oscillation of the poloidal magnetic field and electron temperature, at the ∇B drift frequency of the energetic ions around the torus (≈ 20 kHz), so-called "fish-bones."[5] This oscillation has been interpreted as an induced internal kink mode destabilized by the resonance with the drift of trapped energetic ions,

$$\omega \stackrel{\sim}{=} \omega_d = cE/eBR ,$$

where E is the energy of the ions, R is the major radius. The resonance between the wave and the drift then causes the energetic ions to diffuse radially outward, successively interacting with different poloidal modes coupled by toroidicity, each peaked at different radii, eventually ejected out of the tokamak. White et al.[6] have obtained very good agreement between theory and experiment. This is similar to the mechanism for the building up of the ring current belt of energetic protons ($\geqslant 40$ kW) in the earth's magnetosphere.[7] During magnetic storms the energetic ions from solar wind come into the magnetosphere when the particle drift frequency once around the earth is in resonance with the frequency of external magnetic perturbation associated with the magnetic storm. Such interaction would preserve the invariants of the magnetic moment $\mu = mv_\perp^2/2B$ and $J = \int mv_\parallel d\ell$ while diffusing the ions radially inward. Such μ J-conserving drift-resonant diffusion process can simply be described with the drift kinetic equation for the guiding centers in a magnetic field $\vec{B} = \nabla\psi \times \nabla\theta$.[8] For total energy $H = (1/2)mv_\parallel^2 + \mu B + e\phi$ implicitly expressed as function of μ, J, ψ, θ, through

$$J(\mu,H,\psi,\theta) = \int d\ell [2m(H - \mu B - e\phi)]^{1/2} .$$

The drift equations of motion in ψ (radial) and θ (azimuthal) are Hamiltonian equations

$$\dot{\psi} = -\frac{c}{e}\frac{\partial H}{\partial \theta}\Big|_{\mu,J,\psi} \qquad \dot{\theta} = \frac{c}{e}\frac{\partial H}{\partial \psi}\Big|_{\mu,J,\theta}.$$

The Vlasov equation in the drift approximation for the guiding center distribution $F(\mu,J,\psi,\theta,t)$ is

$$\frac{\partial F}{\partial t} + \frac{c}{e}\left[\frac{\partial H}{\partial \psi}\frac{\partial F}{\partial \theta} - \frac{\partial H}{\partial \theta}\frac{\partial F}{\partial \psi}\right] = 0.$$

For a perturbed magnetic and electric field δB and $\delta\phi$ with conservation of J: $\delta J = 0$, there is a corresponding

$$\delta H = \int \frac{d\ell}{v_\parallel}(\mu\delta B + e\delta\phi)/\int d\ell/v_\parallel = \langle \mu\delta B + e\delta\phi \rangle.$$

For perturbations periodic in θ, we may Fourier analyze

$$\delta H = \sum_n \delta H_n \exp(in\theta - i\omega t)$$

and

$$\delta F = \sum_n \delta F_n \exp i(n\theta - \omega t),$$

and linearized Vlasov equation gives

$$\delta F_n = \left(\frac{c}{e} H_n \frac{\partial F_o}{\partial \psi}\right)/(n\omega_d - \omega).$$

The ensembled averaged distribution $F_o(\mu,J,\psi)$ then evolves according to quasilinear equation

$$\frac{\partial F_o}{\partial t} = \frac{\partial}{\partial \psi} D \frac{\partial F_o}{\partial \psi},$$

where

$$D(\mu,J,\psi) = \frac{c^2}{e^2}\sum_n |H_n|^2 \pi \delta(\omega - n\omega_d).$$

Note that the diffusion coefficient is nonvanishing only for resonant particles with $n\omega_d(\mu,E,\psi) = \omega$. For spatially overlapping modes with the same frequency, large scale radial excursion is possible only for particles with ω_d weakly depending on ψ, which is indeed the case in PDX tokamak.

III. ELECTRON ACCELERATION BY LOWER HYBRID WAVES

Present tokamak operation is pulsed in nature because the toroidal current is driven by an inductive electric field, which is undesirable for an electricity-generating power plant requiring more or less continuous operation. In an attempt to operate a steady-state tokamak, it has been proposed that radio frequency waves can be used to drive a current by imparting wave momentum directly to electrons. Recent experiments on PLT and other devices showed that a plasma current of several hundred kiloamperes, carried mostly by energetic electrons can be sustained for several seconds by lower hybrid (LH) waves of a few hundred kilowatts of rf power.[9] There are several interesting features of the PLT experiment:

1. The launched LH waves have very high phase-velocity, $\omega/k_\parallel$, along the magnetic field. Typically the energy spectrum of the wave peaks at $\omega/k_\parallel \approx c/2$, where c is the speed of light.

2. Hard x-ray emission (E > 30 keV) shows marked increase with RF on, with maximum energy comparable to $m/2\,(\omega/k_\parallel)^2$ and mean energy about 150 keV. The mean energy as well as the maximum energy of electrons increase with increasing phase velocity.

3. Concurrent with the appearance of the energetic electrons, there is enhanced plasma wave radiation near the plasma frequency, which is much higher than the launched LH frequency.

The acceleration of the electrons to high energy is accomplished by the combined action of the initial ohmic electric field which can accelerate electrons to energies about 10 keV before the LH wave is launched and the LH waves with a minimum phase velocity in resonance with these accelerated electrons can further accelerate them to the maximum phase velocity $\sim c/2$. This process can be described by solving the Fokker-Planck, quasilinear equations

$$\frac{\partial f(v_\parallel,t)}{\partial t} = \frac{\partial}{\partial v_\parallel} D \frac{\partial f}{\partial v_\parallel} + \frac{e}{m} E_\parallel \frac{\partial f}{\partial v_\parallel} + C(f) ,$$

where

$$D = \frac{\pi e^2}{m^2} \sum |\tilde{E}_\parallel|^2 \delta(\omega - k_\parallel v_\parallel)$$

is the quasilinear diffusion coefficient due to waves, $E_\parallel$ is the ohmic electric field, and C(f) is the Coulomb collision term. Eventually a steady-state is established when the collisional drag balances the wave acceleration. The resulting distribution, however, has very large anisotropy with average parallel energy $T_\parallel$ much greater than perpendicular $T_\perp$: $T_\parallel \gg T_\perp$, because the electrons are accelerated by the LH waves primarily along the magnetic field. This anisotropic distribution of energetic electrons can destablize plasma waves $\omega = \omega_p k_\parallel/k$ via the anomalous Doppler resonance $\omega + \Omega = k_\parallel v_\parallel$,[9] where $\Omega = eB/mc$ is the electron gyrofrequency, typically larger than the plasma frequency ω_p.

As a result of this instability, lower phase velocity but higher-frequency plasma waves are excited resulting in copious plasma radiation observed during the experiment. The unstable plasma waves in turn scatter the energetic electrons in pitch angle to increase their perpendiuclar energies, leading to enhanced cyclotron radiation in x-mode. When this process becomes sufficiently strong, relaxation oscillation in cyclotron radiation at the fundamental frequency, (ordinary mode measuring $T_\parallel$) out of phase with that at the 2nd harmonic (x-mode measuring $T_\perp$) along the burst of radiation at plasma frequency, qualitatively described by the nonlinear theory using the moment equations of the quasilinear equation for $T_\parallel$, $T_\perp$, and $W = |\tilde{E}|^2/8\pi$ the energy density of the plasma waves.[10] Similar processes may also occur in the Bow shock of the earth's magnetosphere.

REFERENCES

1. J. B. Taylor, Phys. Rev. Lett. 33, 1139 (1974).
2. L. Woltjer, Proc. Nat. Acad. Sci. 44, 489 (1958).
3. H. Eubank et al., Plasma Physics and Controlled Nuclear Fusion Research, 7th Conference Proc. 1, 167 (1980).
4. S. von Goeler, W. Stodiek, and N. R. Santhoff, Phys. Rev. Lett. 33, 1201 (1974).
5. K. McGuire et al., Princeton Plasma Physics Lab. Report PPPL-1946 (1982).
6. R. B. White et al., PPPL-1992 (1983); H. Strauss et al. PPPL-2023 (1983).
7. M. P. Nakata, J. W. Dungey, and W. N. Hess, J. Geophys. Res. 70, 4777 (1965).
8. C. S. Liu, Phys. Rev. Lett. 27, 399 (1971).
9. S. Bernabei et al., Phys. Rev. Lett. 49, 1255 (1982).
10. C. S. Liu, V. S. Chan, D. K. Bhodra, and R. W. Harvey, Phys. Rev. Lett. 48, 1479 (1982).
11. M. Rosenberg, V. S. Chan, C. S. Liu, and F. McLain, to be published; V. V. Parail and O. P. Pogutse, Nucl. Fusion 18, 303 (1978).

ACCRETION DISK ELECTRODYNAMICS

F. V. Coroniti
Departments of Astronomy and Physics
University of California
Los Angeles, California 90024

Accretion disk electrodynamic phenomenae are separable into two classes: 1) disks and coronae with turbulent magnetic fields; 2) disks and black holes which are connected to a large-scale external magnetic field. Turbulent fields may originate in an $\alpha - \omega$ dynamo, provide anomalous viscous transport, and sustain an active corona by magnetic buoyancy. The large-scale field can extract energy and angular momentum from the disk and black hole, and be dynamically configured into a collimated relativistic jet.

1.0 INTRODUCTION

This review emphasizes some recent developments on accretion disk and black hole electrodynamics. The broader subject of accretion disk structure has been excellently reviewed by Pringle (1981). Regrettably, we cannot discuss accretion onto magnetized neutron stars or white dwarfs.

Accretion disk electrodynamics can be divided into two broad points of view, which are not necessarily exclusive. The first view argues that the strong Keplerian differential rotation and convective turbulence within the disk produces a highly disordered magnetic field which is susceptible to Parker's (1972; 1979) topological dissipation - the rapid reconnection relaxation of localized regions of turbulently induced magnetic shear. Although it may be primordial, the disk field is probably maintained against buoyancy losses by a regenerative $\alpha - \omega$ dynamo. The alternative view is that the disk remains connected to a large-scale external magnetic field. The disk and the black hole possess a magnetosphere which can inertially extract angular momentum from the Keplerian flow and the hole's rotational energy, and establish outflowing jets aligned along the rotation axis.

In Section 2.0 we briefly review the standard accretion disk model. Section 3.0 discusses the turbulent view of disk electrodynamics.

Section 4.0 discusses hydromagnetic disk winds, and in Section 5.0 we review black hole electrodynamics.

2.0 ACCRETION DISK HYDRODYNAMICS - THE STANDARD MODEL

Although earlier progenitors exist, the (so-called) standard model of a thin accretion disk surrounding a central black hole was developed by Shakura and Sunyaev (1973) and Novikov and Thorne (1973) who solved the conservation laws of steady state hydrodynamics in conjunction with radiation transport. The strong central gravity requires the disk material to be in circular Keplerian orbits. The vertical component of gravity is balanced by the gradient of the total pressure; in the inner region near the black hole, radiation pressure (P_r) usually greatly exceeds the thermal gas (P_g) pressure. A radially inward accretion flow results from the viscous dissipation of orbital angular momentum; viscosity diffuses angular momentum to larger radii thus allowing the gas to diffuse inward. For rapid accretion the viscous stress ($t_{r\phi}$) must be anomalous, arising from either hydrodynamic and/or hydromagnetic turbulence; a common ansatz is to scale $t_{r\phi}$ to one of the basic pressures - P_r, P_g, or the magnetic pressure $B^2/8\pi$.

Accretion liberates gravitational binding energy which must be radiated or transported vertically away from the disk. For a mass accretion rate $\dot{M}$, the total luminosity is $L = \varepsilon \dot{M} c^2$ where ε is determined by the radius of the last bound orbit; $\varepsilon = 0.06 (0.42)$ for a Schwarzschild (maximally rotating Kerr) black hole. An accretion rate of 10 $M_\odot$/yr ($10^{-8} M_\odot$/yr) will produce a luminosity of $\sim 10^{47}$ (10^{38}) ergs/sec which is comparable to the energy output of quasars and active galactic nuclei (galactic x-ray sources). This very efficient conversion of rest mass energy to luminosity remains the strongest argument that energetic astrophysical objects are powered by accretion onto black holes.

In the standard model, the energy output is radiative, and occurs as either black body radiation if the disk is optically thick, or as various forms of Comptonized radiation if electron scattering dominates free-free absorption. The effective radiation temperature of stellar mass ($\sim M_\odot$) scale disks corresponds to x-ray energies, whereas for supermassive ($\sim 10^9 M_\odot$) disks the effective temperature drops 10^5 to 10^6 °K.

Critiques of the standard model abound in the literature (see Pringle, 1981). For our purposes, 3 points are of particular significance. The anomalous viscosity mechanism is highly uncertain. The vertical structure of a radiation-supported disk is quite likely to be thermally unstable (Bisnovatyi-Kogan and Blinnikov, 1977). A thermally convective disk might have an active magnetic dynamo, a mechanically or hydromagnetically excited corona, and an escaping wind or jet. If a significant fraction of the disk's energy output is deposited in a corona or a wind, the disk structure and predictions of the standard model would be fundamentally altered.

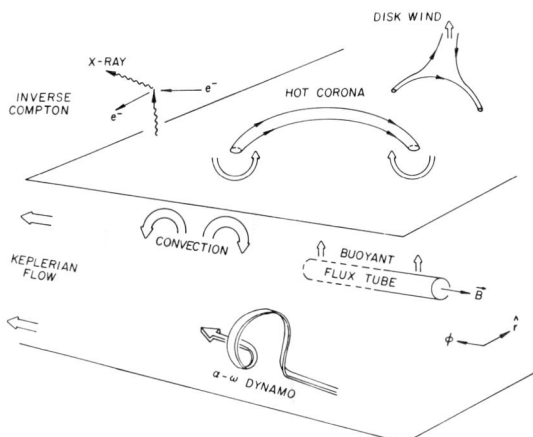

Figure 1. Hydromagnetic phenomenae in thin accretion disks.

3.0 ACCRETION DISK HYDROMAGNETICS

Since accretion disks crudely resemble flat stars, their hydromagnetic structure and dynamics possess the full panoply of difficulties associated with solar and stellar magnetism (Figure 1). For vertically thin disks, a basic question is whether magnetic buoyancy (Parker, 1979) will rapidly remove magnetic flux from the disk (Section 3a). We summarize the attempts to incorporate magnetic dynamos into the theory of disks in Section 3b. A model for the anomalous viscosity due to turbulent magnetic Maxwell stresses is presented in Section 3c. Magnetized disk coronae are discussed in Section 3d.

3.a. Magnetic Buoyancy

In a gravitationally bound atmosphere, a magnetic flux tube experiences a buoyancy force since pressure equilibrium requires that the tube's internal density be less than that in the surrounding atmosphere (Parker, 1979). For an isothermal, gas pressure-dominated atmosphere a flux tube with $B^2/8\pi \ll P_g$ is weakly buoyant, and rises (against gravity) slowly until it reaches density equilibrium. If $B^2/8\pi \gg P_g$, the buoyant rise is rapid, reaching a drag-limited velocity comparable to the Alfven speed. In an adiabatic atmosphere, the density contrast is preserved as the tube rises so that the atmospheric pressure gradient expels even weakly magnetized flux tubes (Moreno-Insertis, 1983).

For an accretion disk we have two additional considerations: 1) in the inner region ($P_r \gg P_g$) does strong buoyancy set in when $B^2/8\pi \sim P_g$ or P_r; 2) differential rotation can shear the flux tube thereby attempting to change the flux tube's volume and field strength. Coroniti (1981) and Stella and Rosner (1983) have argued that flux tubes become strongly buoyant when $B^2/8\pi \sim P_g$, independent of the ratio P_r/P_g. Recently Sakimoto (1983) has carried out extensive numerical calculations on buoyant flux tubes including the effects of shear, heat flow into the tube, inertia, and Newtonian drag. For $P_r \gg P_g$, the general conclusions are: 1) heat flow maintains the tubes near temperature equilibrium; 2) for an adiabatic disk atmosphere, even $B^2/8\pi \ll P_g$ tubes are strongly buoyant, although shear can increase the magnetic pressure to above P_g; 3) for an isothermal atmosphere shear drives even $B^2/8\pi \ll P_g$ tubes out of the disk; in the absence of shear, $B^2/8\pi \ll P_g$ tubes rise until density equilibrium is reached, while $B^2/8\pi \gg P_g$ tubes are rapidly lost.

Most flux tubes reach a drag-limited velocity $V \sim C_s (R_T \Delta\rho/H\rho)^{1/2}$ where C_s is the sound speed based on the total pressure ($C_s^2 \sim P_r/\rho$), R_T is the tube radius, H is the disk scale height (H $\approx C_s/\Omega$ where Ω is the Keplerian angular velocity) and $\Delta\rho/\rho$ is the density contrast. The buoyant flux tubes rise much faster than the Alfven speed ($\ll C_s$), and escape from the disk in 1 to 10 Keplerian rotation periods. Hence buoyancy places the stringent limit $B^2/8\pi < P_g$ on the magnetic field strength which can be retained in an accretion disk.

3.b. Accretion Disk Dynamos

The strong differential rotation and likely thermal convective turbulence have tempted several authors to develop an $\alpha - \omega$ dynamo model (Parker, 1955; 1979; Moffatt, 1978) for accretion disks. As a maximal model Takahara (1979) assumed that all of the accretion energy was invested in the dynamo generation of a strong toroidal magnetic field. Since the disk is cold and thus magnetically supported against gravity, the flux is buoyantly lost on the Alfven time scale, and accumulates in a low density corona. A similar model was developed by Galeev, et al. (1979) who argued that meridional convective turbulence could lead to the exponential growth of the toroidal field.

In these models the dissipation of accretion energy into heat and radiation occurs in the corona where the high Alfven speed permits rapid reconnection of the field. Although a magnetically active corona is attractive (see next section), the underlying dynamo disk models remain incomplete since the source of the convective turbulence required for the regenerative α-effect is uncertain. If the disk has a finite temperature so that the convection is thermally driven, it may be difficult to avoid $P_r \gg P_g$ in the disk's inner region; if so, the buoyancy limit $B^2/8\pi < P_g$ would severely reduce the strength of the coronal magnetic fields and reconnection dissipation. As an alternative, Stella and Rosner (1981) suggest that the buoyancy instability may generate the necessary convective turbulence in the disk; although promising, this "bootstrap" dynamo model requires further elaboration.

In a different approach, Pudritz (1981a,b) and Pudritz and Fahlman (1982) have applied the Mean Field Dynamo Theory (Steenbeck and Krause, 1969; Moffatt, 1978) to accretion disks. Pudritz (1981a) attempts to

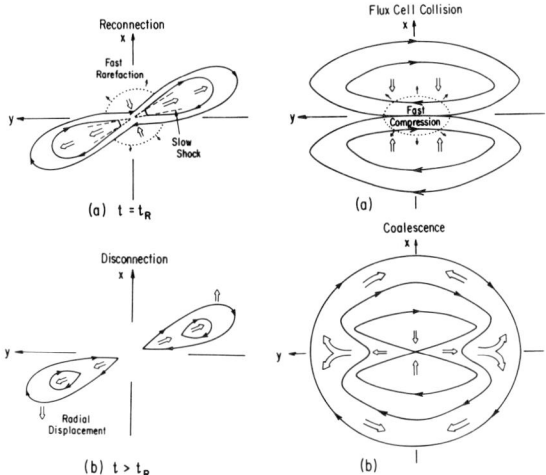

Figure 2. Flux cell dynamics and magnetic viscosity.

self-consistently determine the turbulent magnetic stresses which produce the required anomalous viscosity and to calculate the dynamo-generated large-scale magnetic field (Pudritz, 1981b). A potential difficulty is that the turbulent magnetic field is much larger than the spatially-averaged mean field, which is comparable to the radiation pressure in the inner disk (Pudritz and Fahlman, 1982). Buoyant losses will greatly reduce the strength of the turbulent field on much shorter time scales than the Ohmic diffusion time.

In conclusion, buoyancy severely limits the strength of the turbulent magnetic fields in accretion disks. Even the presence of a relatively weak magnetic field probably requires some form of dynamo activity to replace the buoyancy flux losses. However, as for the solar dynamo problem, a self-consistent accretion disk dynamo model and its coupling to a magnetically active corona still resides in the future.

3.c. Magnetic Viscosity

Early disk models (Lynden-Bell, 1969; Shakara and Sunyaev, 1973; Novikov and Thorne, 1973) recognized that a turbulent Maxwell stress ($B_r B_\phi/4\pi$) would exert local torques, which could diffuse angular momentum. Eardley and Lightman (1975) suggested a model for this magnetic viscosity in which topological dissipation breaks up the initial magnetic field into localized flux cells; for simplicity, we assume that the cells are oriented parallel to the disk plane and extend roughly over the disk height. Figure 2 outlines the "life of a flux cell" in a later version of this model (Coroniti, 1981).

We assume that the initial magnetic field is weak ($B^2/8\pi \ll P_g$) in order to minimize buoyancy losses. The Keplerian differential rotation (y-direction) stretches the flux cell until the Maxwell stress builds up to stop the shear flow within the cell; the plasma within the cell shear increases the field strength (at the expense of the flow kinetic energy) and decreases the radial (x-direction) extent of the cell; both effects greatly reduce the time scale for the oppositely-directed magnetic fields to reconnect at the center. Fast reconnection bifurcates the cell, and dissipates the stored magnetic energy into heat across the slow shocks which stand in the incident flow (Petschek, 1964).

After disconnection, the plasma in the inner (outer) cell has too little (too much) angular momentum to be in local Keplerian equilibrium; the cells move radially until a new equilibrium is achieved. If the cells are closely packed, collisions between cells result in their reconnection coalescence; during coalescence the internal plasma mixes within the new cell. Averaging over many disconnection-coalescence events, the plasma undergoes a random walk in radial location.

The radial diffusion coefficient (ν) depends on the size of the flux cells (L) and the reconnection-coalescence time scale. If $L > C_g/\Omega$ where $C_g = (P_g/\rho)^{1/2}$ is the gas sound speed, the shear flow raises $B^2/8\pi$ above P_g, and buoyancy rapidly removes the cell from the disk; thus $L < C_g/\Omega$. The reconnection time depends on the shear distortion and compression of the flux cell, and is estimated to be $\lambda t_R \approx M_A^{-1}(C_A/\lambda L)^{1/3}$ where $\lambda = 3/2 \, \Omega$, C_A is the Alfven speed, and M_A is reconnection Alfven Mach number which is of order $M_A \sim 1/10$ to $1/20$ (Petschek, 1964; Parker, 1979). We then obtain $\nu \sim 9/4 \, C_g^2 M_A^{2/3}/\lambda$, which yields an anomalous viscous stress of $t_{r\phi} = \rho\nu\lambda \approx 9/4 \, M_A^{2/3} P_g$.

Although greatly over-simplified, the flux cell model demonstrates that a dynamic disk magnetic field can provide an anomalous viscous transport. The scaling of $t_{r\phi}$ with P_g is consistent with the conclusion of Stella and Rosner (1983) and leads to accretion disks which are much denser and somewhat hotter in their inner regions than predicted by the standard model with $t_{r\phi} \propto P_r$ (Sakimoto and Coroniti, 1981).

3.d. Accretion Disk Coronae

An inevitable consequence of convective turbulence and magnetic buoyancy will be to form a hot, magnetically active corona (Stella and Rosner, 1983). The significance of coronal dissipation is that the hard x-rays observed form Cygnus X-1 and the softer x-rays from quasars cannot originate in the disk, but must be emitted in the corona. The soft disk photons are Compton up-scattered to x-rays by the hot or mildly relativistic coronal electrons (Price and Thorne, 1975; Liang and Price, 1977; Eardley, et al., 1978).

A difficult problem is to determine what fraction of the disk's energy output goes into heating the corona, which, in turn, will bound the system's x-ray luminosity. Drawing on recent developments in solar physics, Galeev, et al. (1979) suggested that the corona is confined and heated in strong magnetic loops which have buoyantly emerged from the disk. The twisted magnetic fields of the loops are dissipated by rapid reconnection in the high Alfven speed coronal plasma. Instead of being cooled by thermal conduction or bremstrahlung, the hot electrons in the loops lose energy by Compton scattering the soft disk photons. Along similar lines, Ionson (1983) has scaled the solar coronal heating function to disks, and concluded that a large fraction of the disk's energy can be dissipated in the corona. He argues that the corona may be optically thick to Compton scattering so that the emergent coronal spectrum will have the power law form of unsaturated Compton scattering (Shapiro, et al., 1976; Rybicki and Lightman, 1979); a power law spectrum is observed for the hard x-ray component of Cygnus X-1 and in quasars.

4.0 HYDROMAGNETIC DISK WINDS

The existence of an active corona suggests that disks should have winds (Piran, 1977; Liang and Thompson, 1979). Since the gravitational binding energy is of order 450 R_s/r MeV ($R_s = 2GM/c^2$ is Schwarzschild radius), a thermally driven wind is unlikely. However, if the disk's luminosity exceeds the Eddington limit ($\sim 10^{38}$ $M/M_\odot$ ergs/sec), the disk's inner region becomes almost spherical and a radiatively driven wind will develop (Shakura and Sunyaev, 1973; Meir, 1979, 1982). In addition, if the outer disk flares appreciably, the absorption of x-rays emitted from the inner region can stimulate a thermal wind (Begelman, et al., 1983). Although thermal and radiative winds undoubtedly occur, we will limit discussion to electrodynamically driven winds.

Despite the cogent arguments that disk magnetic fields should be turbulent, the disk may remain connected to the large-scale field which originally threaded the infalling plasma. Blandford (1976) realized that the external field and plasma are coupled to the disk by a system of field-aligned currents which close by flowing across the disk's magnetic field. The closure currents exert a $\underline{J} \times \underline{B}$ spin-down torque on the disk, thus extracting angular momentum from the Keplerian flow and permitting the disk material to flow radially inwards. Disk angular momentum is carried away either by Alfven (or electromagnetic) waves, or, if the disk also loses mass, as part of a hydromagnetic wind.

Blandford and Payne (1982) have proposed a rotationally driven hydromagnetic wind which takes the form of a magnetically collimated jet aligned along the disk's rotation axis (Figure 3). Extragalactic radio sources have large-scale jets (10kpc to 2Mpc) which are often aligned with the VLBI jets ($\sim$ 1pc) emanating from the nuclei of galaxies and quasars (compact radio sources)(Miley, 1980). Their high luminosity implies that the jets are powered by gravitational accretion or by the central black hole (Blandford and Znajek, 1977; Macdonald and Thorne, 1982; next section).

Blandford and Payne (1982) assume that the magnetic pressure of the large-scale field dominates the coronal gas pressure. The coronal

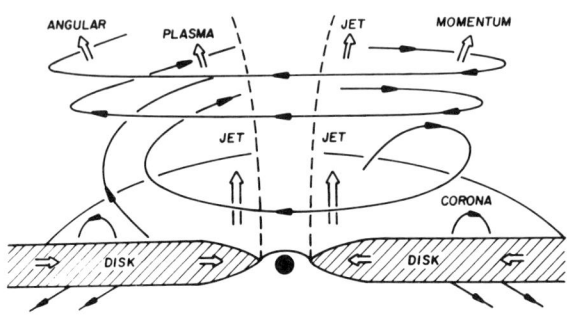

Figure 3. Centrifugally driven hydromagnetic disk wind.

magnetic field is assumed to rigidly connect to the disk so that the field lines rotate with the Keplerian velocity at the radial distance where they enter the disk; this requires that the Alfven speed in the corona must exceed the Keplerian velocity. Since the disk's field convects with the plasma, the magnetic pressure in the disk must be much less than the dynamic pressure of the Keplerian flow. Hence in order to obtain the required high coronal Alfven speed, the coronal density must be low so that the hydromagnetic wind transports only a relatively low mass and particle angular momentum flux.

If the coronal field lines make an angle of less that 60° to the disk plane, the cold coronal plasma is centrifugally slung outwards along the rigid lines of force. Near the wind's Alfven singular point, the plasma's inertia bends the lines of force forming a toroidally wound field which carries off the disk's angular momentum. As is typical with cold magnetic sling solutions, the wind does not pass through a fast critical point (which formally resides at infinity)(Mestel, et al. 1979; Michel, 1973). Since the asymptotic wind remains dominated by the magnetic Poynting flux, the hoop stress of the toroidal field collimates the flow into a jet. (The wind equations also have a solution in which the fast Mach number exceeds unity at infinity, but the jet is pinched-off by the magnetic field at a finite distance above the disk. These solutions probably correspond to the unphysical solutions of radial stellar wind theory (see Kennel, et al., 1983). The physical solutions with a fast Mach number of unity at infinity are not pinched; these will remain the correct physical solutions when finite pressure is included.)

Since the rate of gravitational energy liberation maximizes in the disk's inner region, the jet's velocity is largest near the rotational axis, and decreases with radius; consequently the jet has a high speed core which can approach relativistic velocities. The transport of angular momentum maximizes in the outer region of the jet. The efficiency with which the wind extracts energy and angular momentum from the disk depends on the strength of the large scale field. In the Blandford-Payne model, the critical field strength at which the entire disk luminosity (L_D) is carried off by the jet is roughly $B_c \simeq (L_D/\sqrt{GM}r_0^3)^{1/2}$ where r_0 corresponds to the inner disk radius; for $M = M_\odot(10^9 M_\odot)$, $B_c \sim 10^8(10^4)$ Gauss which is comparable to the turbulent disk field strengths in the standard model if Maxwell stresses supply the dissipation torques. A field of this strength could easily be achieved by the accretion compression of a large scale stellar ($M \sim M_\odot$) or interstellar ($M \sim 10^9 M_\odot$) magnetic field.

The large-scale field is convected inward with the accretion flow, and must ultimately fall in toward the black hole. The electrodynamic interaction of this magnetic field with the black hole can also power an outflowing jet.

5.0 BLACK HOLE ELECTRODYNAMICS

In the late 1960's two distinct power sources were suggested to explain high energy astrophysical objects - accretion onto a compact gravitational object, and the extraction of the rotational kinetic energy of a compact object by electromagnetic torques. The accretion process was associated with galactic x-ray sources (Prendergast and Burbidge, 1968) and quasars (Lynden-Bell, 1969). Gold's (1969) hypothesis that pulsars are rapidly rotating, highly magnetized neutron stars remains unchallenged. Morrison (1969) generalized the pulsar concept to any object - the spinar - whose evolution, including slow gravitational collapse, is driven by the electromagnetic loss of rotational kinetic energy; supermassive spinars can have quasar-scale luminosities.

Blandford (1976) and Lovelace (1976) realized that an accretion disk which possessed a large-scale magnetic field would be a flat spinar. In direct analogy with the Goldreich-Julian (1969) theory of pulsars, Blandford (1976) developed a disk model in which the electromagnetic torque was exerted by a nearly charge-separated, force-free magnetosphere which extends beyond the light cylinder. The Blandford and Payne (1982) jet model is just the hydromagnetic version of Blandford's (1976) force-free model; when the Alfven speed is much less than the speed of light, plasma inertia influences the field and flow structure well inside the light cylinder.

In a fundamental conceptual advance, Blandford and Znajek (1977) demonstrated that the rotational energy or the reducible mass of a Kerr black hole could be extracted by electromagnetic torques if the hole was threaded by a large-scale magnetic field; black holes could also be spinars. Recently, MacDonald and Thorne (1982) reformulated the

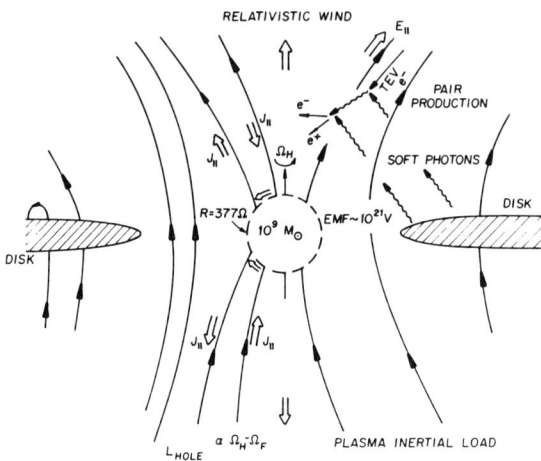

Figure 4. Coupling of black holes and accretion disks.

Blandford-Znajek concept into an absolute space – universal time description of General Relativity which, by using the familiar concepts of nonrelativistic hydromagnetics, completely illuminates the analogy between black hole and pulsar electrodynamics. In this section we present the MacDonald and Thorne (1982) arguments which lead to the conclusion that relativistic jets can be powered by the rotational energy stored in Kerr black holes.

5.a. Magnetized Black Holes

Suppose that the plasma in the accretion disk is connected and frozen-in to a large-scale magnetic field. As the plasma falls into the black hole, the magnetic field becomes causally disconnected from and slips relative to the plasma. Due to the gravitational redshift, the magnetic field appears to concentrate just outside the event horizon (Figure 4). Price's (1972) "no hair" theorem states that the black hole cannot have an intrinsic field unless it carries a net charge, an unlikely possibility in the presence of external plasma. However, to an external observer, the flux concentration at the horizon will make the black hole appear to be threaded by a magnetic field.

Blandford and Znajek (1977), Znajek (1978), and Damour (1978) showed that the magnetized black hole could be mathematically represented as if it were a magnetized conducting body which satisfied conventional hydromagnetic boundary conditions at the event horizon:
1) electric field lines terminate on surface charges; 2) the tangential component of the magnetic field is shielded from the hole's interior by surface currents; 3) the resistivity of the horizon is $R_H = 4\pi/c =$
$= 377$ ohms; and 4) the tangential electric field (E_H) is related to the

surface current density (j_H) by Ohm's law $E_H = R_H j_H$. In the absence of the highly conducting disk, the finite resistance of the horizon would cause the magnetic field to "diffuse" away at the speed of light. Thus the disk and its magnetic field are essential to maintaining a magnetized black hole.

5.b. Pulsar - Black Hole Electrodynamics

Suppose that the black hole's magnetosphere is initially a vacuum, is symmetric about the rotation axis, and rotates with the hole's angular frequency Ω_H; of course this aligned-rotator configuration does not emit magnetic dipole radiation. As in the Goldreich-Julian (1969) analysis for pulsars, the external solution to the vacuum Maxwell equations has a large electric field ($E_\parallel$) component which is parallel to the magnetic field; the "source" of $E_\parallel$ is the Goldreich-Julian space charge which polarizes the vacuum in the corotating frame. For pulsars, $E_\parallel$ causes field emission from the neutron star's surface. The extracted particles are accelerated to relativistic energies and radiate curvature photons by the synchrotron process. As the photons cross field lines, their perpendicular momentum is converted into electron-positron pairs in a quantum electrodynamic cascade. The dense pair plasma shorts-out $E_\parallel$, limiting the acceleration to a narrow vacuum gap (Sturrock, 1971; Ruderman and Southerland, 1975; Arons, 1979).

Blandford and Znajek (1977) suggested that $E_\parallel$ might also lead to pair creation on black hole field lines; for a 10^9 M, maximally rotating Kerr hole, a total emf of $\cdot 10^{21}$ volts is possible, making acceleration to relativistic energies very likely. If a "seed" electron enters the vacuum $E_\parallel$ region, after acceleration it will Compton scatter soft disk photons to γ-ray energies. The γ-rays can then collide with soft photons to produce pairs provided that the product of the two photon energies exceeds $(mc^2)^2$.

Blandford and Znajek (1977) assumed that the pair plasma density would increase until the force-free condition $\eta \underline{E} + 1/c \underline{J} \times \underline{B} = 0$ was satisfied throughout most of the magnetosphere; η is the charge density. The parallel electric field would occur in a narrow gap somewhere between the horizon and the light cylinder. Some of the pair plasma falls into the black hole, while some flows outward along the expanding magnetospheric field lines. Near the light cylinder, the plasma inertially loads the magnetic field, causing the force-free condition to be violated and driving a system of field-aligned currents. The currents close by flow across the magnetic field at the event horizon and in the load region. The $\underline{J} \times \underline{B}$ stress in the load exerts a torque which attempts, but of course fails, to maintain the angular frequency of the plasma and magnetic field (Ω_F) equal to that of the hole. The plasma's inertia bends the poloidal lines of force into the toroidal direction, so that angular momentum flows outward. The integrated $\underline{J} \times \underline{B}$ stress in the horizon exerts a spin-down torque, thus extracting the rotational energy of the black hole.

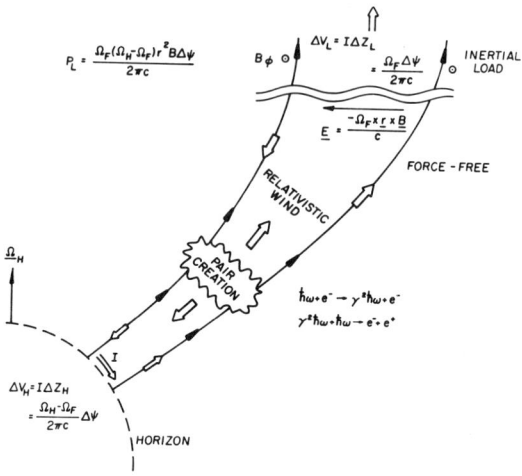

Figure 5. Equivalent circuit for black hole-magnetospheric coupling.

5.c. Equivalent Electrodynamic Circuit

MacDonald and Thorne (1982) developed a circuit analogy (first suggested by Blandford, 1979) which summarizes the magnetosphere-black hole interaction, and permits a simple estimate of the power extracted from the hole. The great advantage of the absolute space – universal time formulation is that we can forget about General Relativity, curved space-time, and the dragging of inertial frames, and treat the electrodynamics exactly as in Goldreich-Julian (1969).

Consider a ring of magnetic flux $\Delta\psi = 2\pi r B_H \Delta\theta$ which threads the black hole at a perpendicular distance r from the rotation axis and over a latitude $\Delta\theta$; ignore the fact that r is really the magnitude of a Killing vector. Just outside the horizon the magnetic field rotates at $\Omega_F \neq \Omega_H$ so that the electric field satisfies $c\underline{E} = -\Omega_F \times \underline{r} \times \underline{B}_H$. In the horizon, frame dragging forces the magnetic field to corotate with the hole; hence in the "rest frame" of the hole, there is an electric field $c\underline{E}_H = (\Omega_H - \Omega_F) \times \underline{r} \times \underline{B}_H$ which is directed opposite to $\underline{E}$ if $\Omega_H > \Omega_F$ as expected from the inertial load. Across the latitude strip $\Delta\theta$, E_H provides a voltage drop $\Delta V_H = (\Omega_H - \Omega_F)\Delta\psi/2\pi c$ (Figure 5). The impedance across the strip is $\Delta Z_H = R_H \Delta\theta/2\pi r = \Delta\psi/\pi c r^2 B_H$. By Ohm's law, the total current (I) through the horizon satisfies $I\Delta Z_H = \Delta V_H$, yielding $I = 1/2(\Omega_H - \Omega_F)r^2 B_H$.

As a simple model of the load region, MacDonald and Thorne (1982) assume that the angular velocity of the plasma and field lines drops rapidly from Ω_F to zero. Hence the electric field in the magnetosphere $c\underline{E} = -\Omega_F \times \underline{r} \times \underline{B}$ appears across the load, and corresponds to a voltage drop $\Delta V_L = \Omega_F \Delta\psi/2\pi c$. If the impedance of the load region is ΔZ_L, the current through the load satisfies $I \Delta Z_L = \Delta V_L$, and I must equal the current through the horizon; ΔZ_L is unspecified, but is presumably related to the plasma's inertia.

From the above circuit analysis, we can determine the power which is dissipated in the load as $\Delta P_L = I^2 \Delta Z_L$ or

$$\Delta P_L = \frac{(\Omega_H - \Omega_F)\Omega_F \, r^2 B_H \Delta\psi}{4\pi c}$$

Properly interpreted, the expression for ΔP_L corresponds to the exact General Relativistic result. The power vanishes if $\Omega_H = \Omega_F$ so that the magnetosphere exactly corotates with the black hole, and maximizes for $\Omega_F = 1/2\,\Omega_H$, which corresponds to matched impedances ($\Delta Z_H = Z_L$) between the hole and the load.

As a rough estimate for the total power extracted from the black hole, we take $\Delta\psi \sim \pi r^2 B_H$ and set $r \sim GM/c^2$ amd $r\Omega_H \sim c$ (maximally rotating Kerr hole) to obtain

$$L_H \sim \frac{(\Omega_H - \Omega_F)}{\Omega_F} \left(\frac{\Omega_F}{\Omega_H}\right)^2 \frac{B^2 G^2 M^2}{4c^3} \sim 4 \times 10^{45} B_4^2 M_9^2 \text{ ergs/sec}$$

with $\Omega_F = 1/2\,\Omega_H$, $B_4 = B/10^4$ Gauss and $M_9 = M/10^9 M$. Hence the luminosity of the black hole is comparable to the total power observed in the jets of extragalactic radio sources and even approaches the luminosities of quasars and active galactic nuclei. The partitioning of L_H between plasma thermal and flow energy, radiation, and even hydromagnetic waves depends on the (unknown) dissipation precesses in the load region. Presumably the plasma will be accelerated to relativistic energies, and at least part of L_H will flow outward in a jet along the rotation axis.

Although the black hole's luminosity does not depend on the structure of the accretion disk, except through its high conductivity, the infalling plasma can resupply the angular momentum lost by the hole and can add new magnetic flux to the hole. An interesting possibility would be if the disk possessed regions of opposite magnetic polarity to that of the black hole. Since all velocities near the hole are comparable to the light speed, upon falling in reconnection would rapidly dissipate the anti-parallel fields. The tension of the newly reconnected field lines would accelerate twin relativistic jets parallel to the rotation axis. These reconnection events would probably be sporadic, and may contribute to the optical, x-ray, or radio outbursts observed in quasars.

6.0 Discussion

In such a brief review we cannot do justice to the extensive efforts which have established accretion disk and black hole electrodynamics as a fundamental problem in high energy astrophysics. Although, in the astrophysical lexicon, the present models are all reasonable, they only represent a hydromagnetic skeleton which outlines the possible dynamics occurring in galactic x-ray sources, quasars, and active galactic nuclei. In order for those hydromagnetic models to acquire explicative and predictive powers, local plasma physical dissipation processes must be incorporated into the hydromagnetic flow solutions. As accretion disk and black hole models evolve, plasma physics advances in solar physics, planetary magnetospheres, and pulsars will continue to provide guidance and inspiration.

It is a pleasure to acknowledge many illuminating discussions with J. Arons, R. Blandford, C. Kennel, R. Rosner, and P. Sakimoto. This work was partially supported by NASA Grant NSG-7341.

References

1. Arons, J.: 1979, Space Sci. Rev., 24, p. 437.
2. Begelman, M. C., McKee, C. F., and Shields, G. A.: 1983, Ap. J., submitted.
3. Bisnovatyi-Kogan, G. S., and Blinnikov, S. J.: 1977, Astron. Astrophys., 59, p. 111.
4. Blandford, R. D.: 1976, Monthly Notices Roy. Astron. Soc., 176, p. 465.
5. Blandford, R. D., and Znajek, R. L.: 1977, Monthly Notices Roy. Astron. Soc., 179, p. 433.
6. Blandford, R. D.: 1979, in Active Galactic Nuclei, eds. C. Hazard and S. Mitton, Cambridge University Press, p. 241.
7. Blandford, R. D., and Payne, D. G.: 1982, Monthly Notices Roy. Astron. Soc., 199, p. 883.
8. Coroniti, F. V.: 1981, Ap. J., 244, p. 587.
9. Damour, J.: 1978, Phys. Rev. D., 18, P. 3598.
10. Eardley, D. M., and Lightman, A. P.: 1975, Ap. J., 200, p. 187.
11. Eardley, D. M., Lightman, A. P., Payne, D. G., and Shapiro, S. L.: 1978, Ap. J., 224, p. 53.
12. Galeev, A. A., Rosner, R., and Vaiana, G. S.: 1979, Ap. J., 229, p. 318.
13. Gold, T.: 1969, Nature, 22, p.25.
14. Goldreich, P., and Julian, W. M.: 1969, Ap. J., 157, 869.
15. Ionson, J. A.: 1984, Ap. J. (in press).
16. Kennel, C. F., Fujimura, F. S., and Okamoto, I.: 1983, Geophys. Astrophys. Fluid Dynamics, in press.
17. Liang, E. P. T., and Price, R. H.: 1977, Ap. J., 218, p. 247.
18. Liang, E. P. T., and Thompson, K. A.: 1979, Monthly Notices Roy. Astron. Soc., 188, p. 421.
19. Lovelace, R. V. E.: 1976, Nature, 262, p. 649.
20. Lynden-Bell, D.: 1969, Nature, 223, p. 690.
21. MacDonald, D., and Thorne, K. S.: 1982, Monthly Notices Roy. Astron. Soc., 198, p. 345.
22. Meir, D.: 1979, Ap. J., 223, p. 664.

23. Meir, D.: 1982, Ap. J., 256, p. 681.
24. Mestel, L., Phillips, P., and Wang, Y.-M.: 1979, Monthly Notices Roy. Astron. Soc., 188, p. 385.
25. Michel, F. C.: 1973, Ap. J., 180, p.207.
26. Miley, G.: 1980, Annual Reviews Astronomy and Astrophysics, 18, p.165.
27. Moffatt, H. K.: 1978, Magnetic Field Generation in Electrically Conducting Fluids, Cambridge University Press.
28. Moreno-Insertis, F.: 1983, Astron. Astrophys., 122, p. 241.
29. Morrison, P.: 1969, Ap. J. Letters, P.173.
30. Novikov, I. D., and Thorne, K. S.: 1973, in Black Holes, ed. by C. DeWitt and B. DeWitt, Gordon and Breach.
31. Parker, E. N.: 1955, Ap. J., 122, p. 293.
32. Parker, E. N.: 1972, Ap. J., 174, p. 499.
33. Parker, E. N.: 1979, Cosmical Magnetic Fields, Oxford: Clarendon Press.
34. Petschek, H. E.: 1964, AAS - NASA Symposium on the Physics of Solar Flares, ed. W. N. Hess, (NASA SP-50), p. 409.
35. Piran, T.: 1977, Monthly Notices Roy. Astron. Soc., 180, p. 45.
36. Prendergast, K. H., and Burbidge, G. R.: 1968, Ap. J. Letters, 151, p. 183.
37. Price, R. H.: 1972, Phys. Rev. D., 5, p. 2439.
38. Pringle, J. E.: 1981, Annual Reviews Astronomy and Astrophysics, 19, p. 137.
39. Pudritz, R. E.: 1981a, Monthly Notices Roy. Astron. Soc., 195, p. 881.
40. Pudritz, R. E.: 1981b, Monthly Notices Roy. Astron. Soc., 195, p. 897.
41. Pudritz, R. E., and Fahlman, G. G.: 1982, Monthly Notices Roy. Astron. Soc., 198, p. 689.
42. Ruderman, M. A., and Sutherland, P. G.: 1975, Ap. J., 196, p. 51.
43. Rybicki, G. B., and Lightman, A. P.: 1979, Radiative Processes in Astrophysics, John Wiley.
44. Sakimoto, P. J., and Coroniti, F. V.: 1981, Ap. J., 247, p. 19.
45. Sakimoto, P. J.: 1983, Ph.D. Thesis, University of California, Los Angeles.
46. Shakura, N. I., and Sunyaev, R. A.: 1973, Astron. Astrophys., 24, p. 337.
47. Shapiro, S. L., Lightman, A. P., and Eardley, D. M.: 1976, Ap. J., 204, 187.
48. Steenbeck, M., and Krause, F.: 1969, Astron. Nachr., 291, p. 49.
49. Stella, L., and Rosner, R.: 1983, Ap. J. (in press).
50. Sturrock, P. A.: 1971, Ap. J., 164, p. 599.
51. Takahara, F.: 1979, Prog. Theor. Phys., 62, p. 629.
52. Thorne, K. S., and Price, R. H.: 1975, Ap. J. Letters, 195, p. 6101.
53. Znajek, R.: 1978, Monthly Notices Roy. Astron. Soc., 182, p. 639.

DISCUSSION

Benford: Dissipation and particle production near black holes are, of course, messy problems. But can't we learn about the final wound-up field, B_o, by simply extrapolating the confining field in jets back from the extragalactic region?

Coroniti: Yes, this may be the best way.

Sturrock: As I understand it, a jet produced by a black hole will comprise an electron-positron plasma. Can this low density plasma be reconciled with the dynamics of fully developed jets?

Coroniti: The usual argument is that the positronic jet picks up hydrogenic as it propagates into the galactic and extragalactic environment, and is converted into an ordinary jet. I do not believe that this pick-up has been demonstrated, nor that the jet could survive the inertial drag.

Henriksen: The sub-parsec scale jets are the energy sources of the VLA jets, but they are probably not in the same physical state. The small scale jets will have highly ordered, high specific energy which will be degraded by mixing and entrainment as the jet expands to the VLA scale. See e.g. Henriksen, Bridle, and Chen, 1982.

D. Smith: In the case of pulsars, we have been working for years with very detailed observations, but there is no agreement on the dissipation mechanisms. Using this as a gauge, is there any hope that we could be in any better shape in the case of quasars?

Coroniti: Hope springs eternal in an astronomer's breast. In both cases, we need to construct a good numerical simulation in order to self-consistently study the basic dissipation. Even if the simulations are not realistic (parametrically), they will still be essential for sorting out which of the possible plasma dissipation processes are important.

Vasyliunas: Is the 377 ohm effective resistivity of the black hole low enough to enforce corotation of the field lines, as required for application of pulsar magnetospheric physics?

Coroniti: If the inertial load is large enough, the field lines would be decoupled from the hole. However, once decoupled, the source of emf which ultimately drives the pair production goes away, thus decreasing the inertial load. Hence the system may be self-regulating.

Uchida: You talked about the magnetic field hung up on the event horizon while the mass is absorbed into the black hole. Is the process of this defreezing of the magnetic field from the mass explored?

Coroniti: Frame-dragging at the Kerr hole's horizon causally decouples the plasma from the magnetic field. The same effect leads to the finite, free-space resistivity of the horizon.

Vlahos: Are there impulsive events observed in accretion coronae? What are the scales of the variability of the emitted energy?

Coroniti: Yes, both Cygnus X-1 and many quasars exhibit temporal variability in their x-ray and (for quasars) optical light curves; quasars also have radio outbursts. The smallest time scales are comparable to the light time across the Schwarzschild radius. Whether the observed variability is due to unsteady coronal processes or a major instability in the disk structure is not known.

A. Ray: How collimated are the jets from the black holes?

Coroniti: I do not know whether this analysis has been done yet. Sterling Phinney is attempting to calculate the jet's properties, so I suppose that the degree of collimation would come out of his relativistic wind analysis.

Davila: Is there any way to directly observe the rotation of the black holes at the center of these disks, perhaps gravity wave radiation for instance?

Coroniti: I do not think that the gravitational radiation is detectable.

Gilden: In accretion disks around black holes the efficiency of gravitational radiation is only 2%, while the photon efficiency may be $\sim 40\%$.

ULTRA-HIGH ENERGY COSMIC RAY PRODUCTION BY CURRENT DISRUPTION IN ACTIVE GALACTIC NUCLEI

George Lake
Bell Laboratories

R. E. Pudritz
Astronomy Dept., University of California

ABSTRACT

Electrodynamic models for the activity of galactic nuclei are shown to be current systems which can be examined in terms of equivalent circuits. The resulting inductive circuit which describes the coupling of the generator (black-hole and accretion disk) to the distant load (jet plasma) is prone to various instabilities. We consider the disruption of this current system and propose that ultra-energetic cosmic rays ($E \sim 10^{19} - 10^{21}$ eV) could be produced during the discharges, which occur at distances of $\sim 10^{16} - 10^{18}$ cm from the central massive hole ($M \sim 10^8$ $M_\odot$). Such discharges will also produce variable γ-ray and X-ray activity and we discuss observations of Cen A in this regard.

1. INTRODUCTION

Current observations of ultra-high energy cosmic rays ($E > 10^{18}$ eV) show an "ankle" in the spectrum at $E \gtrsim 10^{19}$ eV (Cunningham et al. 1980) with $n(E) = 3.1 \times 10^{12} E^{-2.31}$ particles m^{-2} eV^{-1} s^{-1} and a maximum detected energy of 1.6×10^{20} eV. There may be an anisotropy in arrival directions towards the Virgo cluster although this is as yet not certain. The energy density of particles of energy $> 10^{19}$ eV of $\sim 4 \times 10^{-8}$ eV cm^{-3} which needs to be replenished every 10^{16} s due to photopion production by microwave background photons. This yields a source energy density of 1.6×10^{38} ergs s^{-1} Mpc^{-3}. Type 1 Seyfert galaxies qualify as potential sources as they have a spatial density of 10^{-4} Mpc^{-3} and a 0.5–4.5 keV output of 10^{44} ergs s^{-1} or 10^{40} erg s^{-1} Mpc^{-3}. If NGC4151 (Lawrence 1980) is typical then the combined x and γ-ray energy densities are such that the efficiency of $E > 10^{19}$ eV particle production need only be 10^{-3} or 10^{-4}.

The relativistic jets of many active galactic nuclei have stimulated several electrodynamic models for the required production of highly collimated energetic flows (Lovelace 1976, Blandford and Znajek 1977, Blandford and Payne 1982). Potential drops in excess of 10^{19} eV may be produced by magnetized accretion disks around massive black-holes ($M \sim 10^8 - 10^9$ $M_\odot$) at the "heart" of the engine. The electrodynamic coupling between the black-hole plus disk (generator) with the jet (load) might produce even larger potential drops during the disruption of this current system accelerating particles to 10^{21} eV on scales of $\sim 10^{-2} - 10^{-1}$ pc from the central engine. We will describe this phenomena of current disruption using the technique of equivalent circuits, borrowed from the analogous solar and terrestrial cases (Spicer 1982, Chiu and Cornwall 1980).

2. THE COUPLING OF ACCRETION DISK AND JET PLASMA

It seems inevitable that accretion disks possess dynamically important magnetic fields ($10^3 - 10^4$ Gauss) due either to the compression of plasma containing field (eg. Lovelace 1976) or by in situ generation by a turbulent dynamo (Pudritz and Fahlman 1982). The radial electric field generated by matter in a rotating disk ($E_r = (v_\phi/c) B_p$) generates large potential drops across the disk, e.g.

$$V_{gen} \sim 10^{19} (B/10^3 \text{ Gauss}) (M/10^8 \text{ M}_\odot) \text{ Volts} \qquad (1)$$

Now magnetic field lines which are tied to the disk will connect to plasma in the jet region. Imperfect coupling of disk and jet plasmas means field lines are not equipotential surfaces, (Lee and Kan 1981), so acceleration of particles (protons in our case) can occur. If the magnetic coupling of the earth's magnetospheric plasma with the ionosphere is any guide (Chui and Cornwall 1980), the global boundary conditions necessitate a description as a current system. In the terrestrial context, the auroral zone produced as a result of such coupling yields a parallel electric field $E_\parallel$, which a magnetohydrodynamic analysis cannot consistently explain.

Field lines attached to the differentially rotating disk are twisted on a time-scale $t_{dyn} = 0.7 \times 10^4$ ($M/10^8$ M$_\odot$) $(\sigma/\sigma_i)^{3/2}$ s where $\sigma_i \sim 10^{14} (M/10^8$ M$_\odot)$ cm is the inner disk radius and σ is the position of the footpoint. This twisting drives a current through the disk magnetosphere of magnitude

$$I_o = V_{gen}/R_o \qquad (2)$$

where R_o is the load resistance (jet plasma) to which the generator is coupled. In steady state, the power generated by the (black-hole and disk) is dissipated in the jet plasma at a rate $\mathscr{L}$ ($\mathscr{L}$ is the observed jet luminosity);

$$R_o = V_{gen}^2/\mathscr{L} = f\, R_{sp.ch.}$$

$$f \equiv (B/10^3 \text{ Gauss})^2 (M/10^8 \text{ M}_\odot)^2/(\mathscr{L}/10^{44} \text{ ergs. s}^{-1}) \qquad (3)$$

and $R_{sp.ch.} = c^{-1} = 25$ohms is the space charge limited resistance so that the current is $I_o = 3 \times 10^{17}$ f[$V_{gen}/3 \times 10^{19}$V] Amps. This current system has an associated magnetic field in which energy is stored, and an inductance L_o characterizing this global circuit is introduced. Because the magnetic flux of the system is large ($L_o I_o$), perturbations of the magnetospheric current system will result in instability and "discharge" of this circuit converting stored magnetic energy into high-energy particle production (see Alfvén and Carlquist 1967).

The time-scale on which energy is stored is

$$t_{store} = L_o/R_o \qquad (4)$$

while the voltage across the discharge region is

$$V_{disch} = L_o I_o/t_{disch} = (t_{store}/t_{disch}) V_{gen} . \qquad (5)$$

The time-scale to store energy in this system ought to be roughly $t_{store} \gtrsim t_{dyn}$. The magnetic flux associated with the disk increases outwards (see Blandford and Payne 1982) so that for a disk extending to $\sigma = 10 - 10^2 \sigma_i$, $t_{store} \gtrsim 10^6 - 10^7$ sec. Observations of x-ray variability in active galactic nuclei (Lawrence 1980) indicate discharge times of order $10^4 - 10^5$ s, so that $V_{disch} \sim 10^2 V_{gen}$ is reasonably obtained.

We argue that some type of electrostatic field with $E_\parallel \neq 0$ developes in the discharge region as in the theory of double layers (e.g., Carlquist 1982) or variants (e.g., Chiu and Cornwall). A magnetically active accretion disk will have many buoyant loops of intense field visiting into the disk corona (Pudritz and Fahlman 1982) and these perturbations may be sufficient to cause the large scale magnetospheric current system to "discharge" (see Lake and Pudritz 1983 for a more detailed discussion).

3. DISCUSSION

An observable consequence of such a scheme for particle acceleration is the presence of intense, highly variable γ-ray activity. During discharge, cold electrons from the jet are accelerated towards the accretion disk and inverse scatter soft disk photons to produce hard γ-rays, softening towards x-rays at the end of the disruptive event. As an example, if the nucleus of Cen A is the primary source of the observed ultra-high energy particles; the power must be $\sim 5 \times 10^{42}$ ergs s^{-1} and the associated γ-ray luminosity would be $\sim 10^{42}$ ergs s^{-1}, with a characteristic energy of $\sim 3 \times 10^{10}$ eV. This compares favorably with the time-varying flux of 10^{41} erg s^{-1} seen at energies of $10^{12} - 10^{14}$ eV (Grindlay 1982). NGC4151 may also show such variable γ-ray activity (White et al. 1980) at lower energies (1-20 MeV) and certainly any such trend in Type I Seyferts would enhance the prospects of such a theory.

ACKNOWLEDGEMENTS

We thank J. Arons, A. Hasegawa, K. Jacobs, J. Linsley, C. McKee, J. Ostriker, and A. Watson for stimulating conversations. REP acknowledges financial support of NSF Grant AST 79-23243 to the University of California at Berkeley.

REFERENCES

[1] Alfvén, H. and Carlquist, P.: 1967, Solar Phys. 1, 220.

[2] Blandford, R. D. and Payne, D. G.: 1982, MNRAS 199, 883.

[3] Blandford, R. D. and Znajek, R. L.: 1977, MNRAS 179, 443.

[4] Carlquist, P.: 1982, Astrophys. Sp. Sc. 87, 21.

[5] Chiu, Y. T. and Cornwall, J. M.: 1980, J. Geophys. Res. 85, 543.

[6] Cunningham, G., Lloyd-Evans, J. Pollock, A., Reid, R. and Watson, A.: 1980, Ap. J. (Letters) 236, L71.

[7] Grindlay, J. E.: 1982, Proceedings of Int. Workshop on VHE Gamma Ray Astronomy (Ootacamund, India) in press.

[8] Lake, G. and Pudritz, R. E.: 1983, submitted Ap. J.

[9] Lawrence, A.: 1980, MNRAS 1982, 83.

[10] Lee, L. C. and Kan, J. R.: 1981, "Physics of Auroral Arc Formation" Eds. S-I. Akasofu and J. R. Kan, American Geophys. Union.

[11] Lovelace, R. V. E.: 1976, Nature 262, 649.

[12] Lynden-Bell, D.: 1969, Nature 223, 690.

[13] Pudritz, R. E. and Fahlman, G. G.: 1982, MNRAS 198, 689.

[14] Spicer, O.: 1982, Sp. Sc. Rev. 31, 351.

[15] White, R. S., Dayton, B., Gibbons, R., Long, J. L., Zanrosso, F. M., and Zych, A. D.: 1980, Nature 284, 608.

THEORY OF STRONGLY TURBULENT TWO-DIMENSIONAL CROSS FIELD CONVECTION OF CURRENT CARRYING SPACE PLASMAS*

M. J. Keskinen
Geophysical and Plasma Dynamics Branch
Plasma Physics Division
Naval Research Laboratory
Washington, D.C. 20375

ABSTRACT
   The "direct interaction approximation" of Kraichnan as modified by Kadomtsev is employed to develop a two-dimensional strong turbulence theory which predicts both nonlinear frequency broadening and a power law for the spectrum of a convecting plasma containing a gravitationally induced cross field current. These results are favorably compared with experimental observations, numerical simulations, and previous studies[1] of turbulent cross field convection of current-carrying plasma.

*Work supported by ONR and DNA.

1. Sudan, R.N. and M.J. Keskinen, Phys. Fluids, 22, 2305, 1979; Keskinen, M.J., Phys. Rev. Lett., 47, 344, 1981.

DISCUSSION

   Migliuolo: What is the cause of the westward drift of ionospheric bubbles (in the simulation)?
   Keskinen: The westward drift is approximately 60 m/s.
   Wentzel: In ordinary hydrodynamic turbulence there is a qualitative difference between two- and three-dimensional turbulence. Can you identify any restrictions inherent in your 2-D approximation for the solar turbulence? Is there a preferred direction to justify 2-D?
   Keskinen: In two dimensions the flux function ψ, $\underline{B} = \hat{\lambda} \times \nabla\psi$, $\hat{z}$ in the direction of the magnetic field, is conserved. In three dimensions $K \equiv \int d^3 x \, \underline{A} \cdot \underline{B}$ is conserved in an ideal fluid where $\underline{B} = \nabla \times \underline{A}$.
   Liu: You included only $\vec{V} \cdot \nabla \vec{V}$ nonlinearity in the equation of continuity, but neglected $\vec{V} \cdot \nabla \vec{V}$ nonlinearity in the equation of motion. Why?
   Keskinen: We are studying length and time scales such that inertial effects in the equation of motion can be neglected.
   Lotko: Can you say under what conditions we can apply the direct interaction approximation to MHD turbulence?
   Keskinen: These conditions are adequately discussed in Sudan and Pfirsch (Phys. Fluids, 1984, to be published).

Magnetic Field Reconnection in Differentially Rotating Accretion Disks

David Gilden
The Institute for Advanced Study

T. Tajima
Institute for Fusion Studies, University of Texas at Austin

ABSTRACT

Differentially rotating accretion disks threaded by a uniform magnetic field have been numerically simulated. Fast reconnection followed by coalescence allows the magnetic field to drive small amplitude radial oscillations in the disk. These oscillations may be observable as the viscous stresses cause the disk to brighten and fade as the disk expands and contracts. Episodes of reconnection may also be observable as hot spots produced locally at the sites of coalescence. Cataclysmic variables, and in particular dwarf novae, provide a natural interpretation for these calculations.

1. Introduction

Accretion disks, in order to transfer mass, must diffuse angular momentum outwards and this requires a viscous couple between adjacent radial zones. The dynamics of accretion disks are presently uncertain because the key element, the viscosity, is not well understood. Ordinary molecular viscosity is too small to account for the apparent radial inflow rates and luminosities of stellar disks and it is generally considered necessary that some anomalous viscosity due to turbulence or magnetic fields be present. Accretion disks are natural environments for the amplification of field fluctuations through differential rotation and radial inflow. Studies of magnetic viscosity have thus generally concentrated on chaotic field geometries (Eardley and Lightman 1975; Ichimaru 1976; Sakimoto and Coroniti 1981; Coroniti 1981). Although the evolution of chaotic fields is central to the viscosity problem, it seems worthwhile to investigate the much simpler, but still quite formidable, problem of the action of a disk upon a field with a definite geometry. The problems that are posed in this paper concern the windup of a uniform field by a differentially rotating disk. Even in this much restricted domain, there are several interesting phenomena; magnetic field reconnection and coalescence, and large scale disk oscillations. In addition, the repeated episodes of reconnection cause a net diffusion in the plasma. Although the geometry of the uniform field is contrived, many of the observations made here carry over to more complex and chaotic magnetic field configurations.

2. Simulation of a Disk Plasma Threaded by a Uniform Magnetic Field

The initial conditions in the simulation described here were those appropriate to a "cold" disk threaded by a uniform field at $t = 0$. "Cold" here refers to purely Keplerian motions. The plasma is initially confined to a torus surrounding the compact star, which is not present on the grid except as a source of gravitation. In other simulations, the disk was allowed to have a larger extent, with the inner regions filled with a rigidly rotating plasma. The continuous disks were not distinguishable from the tori in the field evolution nor in the types of oscillations that were set up in the plasma. In the simulation discussed here, the initial field strength was chosen so that the Keplerian velocity, v_ϕ was about 10 times the Alfven speed ($v_A = B/\sqrt{4\pi\rho}$). This allowed the torus to rotate without being disrupted by the magnetic field. In simulations employing much weaker fields, it was found that electrostatic effects became very important and tended to obscure the role of the field. These latter simulations will be discussed in a future publication.

Two critical moments in the evolution of the plasma and field are shown in Figs. 1A and 1B. These plots show magnetic field lines along with the non-Keplerian components of the ion velocity field.

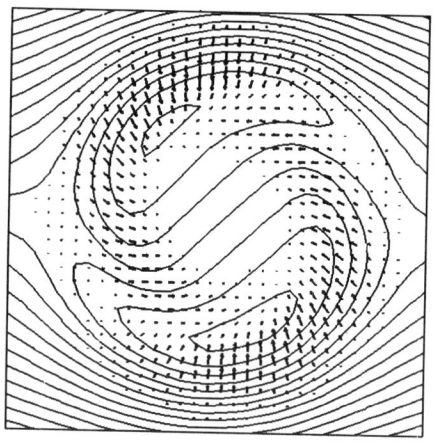 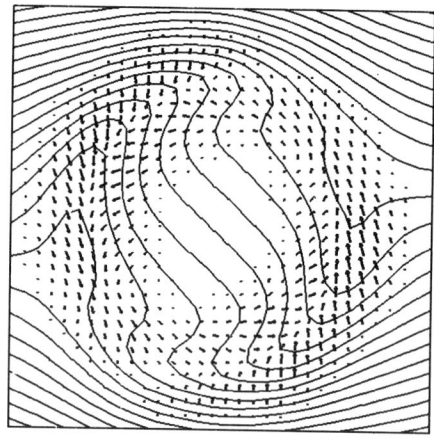

Fig. 1A   Fig. 1B

Figure 1: Simulated disk and field for the reconnection and coalescence epochs in the initial revolution of the torus. The torus in Fig. 1A has completed ½ revolution while in Fig. 1B the torus has completed ¾ revolution.

The field lines, which are frozen to the plasma, are wound up by the orbital motions in the torus. The compression of the field causes the particles to be slightly deflected from purely Keplerian orbits. During this phase, the deflection is inward since the field is retarding their motion. Once the field has rotated through π radians, X type neutral points are formed and the field undergoes fast reconnection. We emphasize here that the reconnection occurs on a timescale much shorter than the resistive timescale set by particle collisions (which have been reduced through the finite size particle effect). In Fig. 1A the formation of magnetic islands is evident. These islands rotate with the plasma and then coalesce as shown in Fig. 1B. The tendency of the magnetic field during coalescence is to snap back to a more uniform configuration. This causes the particles which were accelerated inwards during field compression to now drift outwards. This cycle is repeated in subsequent rotations of the torus. Reconnection in this geometry conserves flux; the number of field lines linking the torus is constant.

The first cycle of reconnection is special in the sense that the torus was started cold. Ion inertia causes the particles positions to lag behind the instantaneous field geometry. A steady state is eventually reached when the particles are at their maximum radial extent at peak field compression and at minimum radius when the field is returned to uniformity. This relationship is most clearly seen in Fig. 2 which depicts the energy history of the field and plasma. Time in these calculations is measured in terms of the inverse electron plasma frequency, $\omega_{pe} = \sqrt{4\pi n e^2/m_e}$.

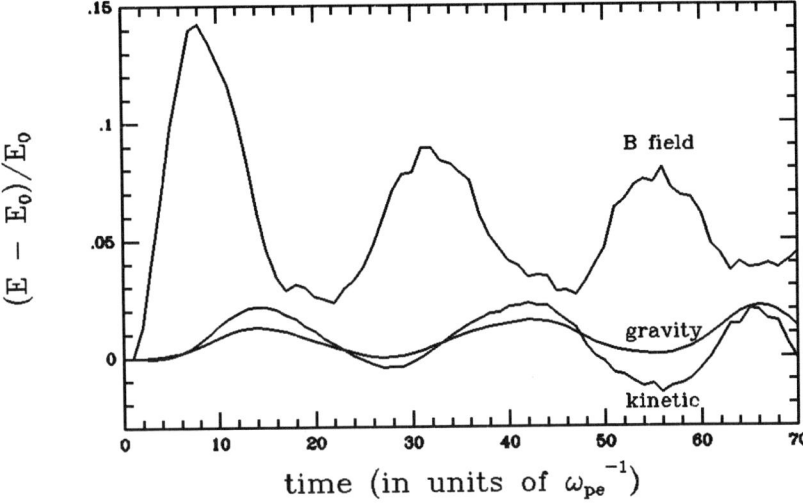

Figure 2.: Energy history of an oscillating disk plasma. Peaks in the magnetic field energy occur just prior to reconnection and at the same time as maximum disk expansion.

The particles positions can be deduced from the *kinetic* and *gravity* tracks. At maximum radius, the particles have a minimum of binding energy and consequently a minimum orbital velocity; the kinetic energy and binding energy are in phase with one another. The magnetic field energy is π out of phase, a relationship which takes several rotations to achieve. At peak compression (maximum field energy) the particles are given an inward acceleration, but it takes ½ cycle for the particles to drift inwards. Similarly, during magnetic field coalescence (return to minimum field energy) the particles are accelerated outwards with their positions lagging by ½ period.

The cycle of reconnection followed by coalescence causes the torus to radially oscillate. The maximum radial excursion is estimated by the product of the non-Keplerian drift velocity and the orbital period. The non-Keplerian ion drifts induced by coalescence are on the order of the Alfven velocity of the plasma at peak magnetic field compression. Since reconnection occurs every half cycle of revolution, the maximum field strength is roughly twice the initial field strength; there is twice as much flux per unit volume threading the torus after it has turned through π radians. The relative amplitude of the oscillation is therefore $\Delta R/R \sim 2\pi v_A/v_\vartheta$. The effect of this oscillation on the luminosity of the torus can be estimated by assuming that the torus emits the energy deposited by viscous dissipation as blackbody radiation. The viscous dissipation rate per unit area per unit time in an axisymmetric disk is (Pringle 1981) $D(R) = 2.25\, M G \nu \Sigma / R^3$, where ν is the viscosity (whatever its source), M is the mass of the compact star, and Σ is the surface mass density of the torus. The disk will brighten according to $\Delta D/D \sim 6\pi v_A/v_\vartheta$. Evidently, disks which are only mildly superalfvenic, will undergo significant changes in their brightness if this type of oscillation is driven.

3. Astrophysical Disks and Quasi-Period Oscillations

A particularly interesting application of these calculations is to the oscillations observed in a subclass of the cataclysmic variables, the dwarf novae. These oscillations are observed only during eruption, lose their coherence every few cycles (and are therefore referred to as quasi-periodic oscillations), last typically about 5 days, and have periods about 50 s and amplitudes of about 0.005 mag (Robinson and Warner 1983; Robinson and Nather 1979, Patterson 1981). In some dwarf novae, quasi-period oscillations with several different periods are present simultaneously (Robinson and Warner 1983). The range of periods, the multiplicity of oscillations, and their incoherency all suggest that the oscillations are produced in the disk. The duration of the oscillations is typical of the time it takes for the disk material to drain onto the compact star in a state of high viscosity (for a discussion of the instability leading to eruption see Bath and Pringle 1982; Meyer and Meyer-Hofmeister 1981). The period lengths themselves suggest some type of orbital phenomenon although vertical oscillations (see Cox 1981) will necessarily have periods that appear to be orbital if the disk is pressure supported and not self-gravitating.

Magnetic field reconnection provides a mechanism for driving disk oscillations. The observed brightness variations require relatively modest fields; $v_A/v_\vartheta \sim 10^{-4}$. There are two difficulties that any theory must face; that the oscillations appear only during eruption (when the disk is draining onto the compact star) and that only a few discrete periods are present and these periods are relatively stable. The simplicity of our model prevents a complete accounting but we can suggest several ways in which these problems may be resolved. Since field reconnection should always be occuring (field diffusion is very slow due to the relatively high densities in these disks) it may be that it is either masked by other oscillations (the orbiting of the hot spot) or that the oscillations are too weak in a quiet disk state to be observed. Once the disk is destabilized and begins to accrete at a fast rate, flux conservation will enhance the field ($\sim R^{1-2}$) and the oscillation amplitude will increase. The discreteness of the period may be related to the observation that increased viscosity causes the disk to simultaneously expand as the bulk of the plasma drains onto the compact star. The period of the oscillation is not set by where the plasma is concentrated, but by where the field is sufficiently twisted to form X points. In the twisting of a uniform field by plasma extending over widely disparate radii, there may be several sites where reconnection occurs with a period appropriate to the characteristic radius of each site. These sites may be fixed in space although shifting with respect to the mass profile of the disk. In future calculations we hope to investigate the possibility of multiple sites of reconnection and also to include a chaotic component in the magnetic field.

This research was supported by NSF grants ATM-82-14730 and PHY-82-17352

References

Bath, G.T., and Pringle, J.E.: 1982, Monthly Notices Roy. Astron. Soc. 199, 267.
Coroniti, F.V.: 1981, Astrophys. J. 244, 587.
Cox, J.P.: 1981, Astrophys. J. 247, 1070.
Eardey, D.M., and Lightman A.P.: 1975, Astrophys. J. 200, 187.
Ichimaru S.: 1977, Astrophys. J. 214, 840.
Meyer, F., and Meyer-Hofmeister, E.: 1981, Astron. Astrophys. 106, 34.
Patterson, J.: 1981, Astrophys. J. Suppl. 45, 517.
Pringle, J.E.: 1981, Ann. Rev. Astron. Astrophys. 19, 137.
Robinson, E.L., and Nather, R.E.: 1979, Astrophys. J. Suppl. 39, 461.
Robinson, E.L., and Warner, B.: 1983, preprint.
Sakimoto, P.J., and Coroniti, F.V.: 1981, Astrophys. J. 247, 19.

DISCUSSION

Wilson: Could you please describe the outer boundary condition on the magnetic field in your model?

Gilden: Periodic boundary conditions were used. A mesh sufficiently large to minimize the effects of this symmetry was necessitated.

Vasyliunas: In your simulation, the rotation period is about 20 inverse plasma periods. This means that the system in the simulation is relatively small in relation to the microscopic plasma length scales, so that non-MHD effects are much more important than they would be in the real, large system. This may be the reason for the rapid reconnection seen in the simulation, which may not scale up to the real case.

Gilden: Numerical limitations do not allow processes with widely disparate timescales to be simulated in a single model. Your criticism is valid, but only underscores the difficulty in interpreting numerical results, and in designing a numerical experiment. We have explored a range of ratios of the relevant time and length scales, and have not observed any sensitivity in the rate of reconnection. However, it cannot be said with certainty what the rate would be with the ratios observed in nature.

Heyvaerts: Can you guess what the difference may be between your 2-D (space) simulation, and the real flat disk behavior?

Gilden: One important difference will be the vertical motion of bouyant flux tubes out of the plane of the disk. A second important 3-D (space) effect will be the twisting of the field component normal to the disk.

Kennel: Will the exchange of gravitational energy with magnetic be so important in the real case where the Keplerian time is very much larger than the plasma period?

Gilden: Our results do not rely on the relative magnitudes of the plasma and orbital periods, as long as they are not comparable, in which case electrostatic effects become important. The exchange between magnetic and gravitational energy proceeds through windup and then reconnection of the magnetic field. Although our simulation code cannot reproduce the ratio of plasma to orbital timescales observed in nature, we have attempted to use a ratio, $\sim 10^{-2}$, which allows the energy transfer to not be corrupted by purely numerical effects.

Tajima: Because of the smaller disparity of plasma time and Keplerian time than in the accretion disk, the electrostatic energy is perhaps unrealistically large. However, there is no strong evidence that the electrostatic noise wipes out other physics. However, the two time scales are separate in the code. Once again, we should use our theoretical insight to interpret the raw data.

Krishan: Have you included radiation pressure in your simulations, since most of these accretion discs are radiation supported?

Gilden: We have not included radiation pressure. However, the accretion disks associated with the cataclysmic variables are not radiation pressure supported throughout. In their quiet state, they are cool enough so that gas pressure is dominant.

Van Hoven: What is the source of the fluctuations which the simulation produces, having started from a mixture of cylindrical and planar symmetries? Is this numerical noise?

Gilden: Fluctuations are not caused by the symmetries in the calculation, but by the use of a finite number of particles. The finite size particle technique is designed to minimize this noise.

Steinolfson: You remarked that the reconnection in your results was fast. What is it fast relative to?

Gilden: Fast relative to the collision timescale in the plasma.

CORONAL ARCADES IN THE SUN AND THEIR HYDROMAGNETIC STABILITY

A. Ray
Tata Institute of Fundamental Research
Homi Bhabha Road
Bombay 400005 India

Two ribbon flares on the sun are sometimes preceded by luminous arcade like structures in the corona with fillament activity. The arcades which are associated with the coronal magnetic field are sometimes without these fillaments when they are found to be relatively long lived. The reasons underlying the stability and instability of these structures may be relevant in understanding the basic mechanism of two ribbon flares. Here we present results on the equilibrium structure of coronal arcades and their hydromagnetic stability taking into account the effects of magnetic field line-tying on the photosphere.

The configuration of the arcade structures can be idealized as an elongated tunnel or series of magnetic arches. On the photosphere, Zeeman-splitting measurements show that the base of the active region of this structure has two distinct regions of opposite magnetic polarity separated by an inversion line. Observations of the Hα-fibrils show that the field lines lying lower towards the photosphere become progressively more parallel to the magnetic neutral line while at higher altitudes X-ray observations made from the Skylab imply the luminous arcs (and the underlying magnetic fields) to be more nearly circular with their planes perpendicular to and centered on the neutral line. The equilibrium hydromagnetic structure of the arcade was modeled (Ray and Van Hoven 1982a) in a simple analytic form in two dimensions and was considered to have a cylindrical symmetry in the upper half plane above the photosphere. As the pressure gradients in the corona are expected to be small in quiescent conditions, and the coronal magnetic fields involved are several hundreds of Gauss, the magnetic field configuration within a certain radius (about the neutral line) was modeled by a force free field. The force free field joins smoothly to a potential field outside this radius R. In this model, there are no unrealistic sheet currents in the equilibrium configuration. In cylindrical coordinates about the neutral line, the field structure inside the radius R is described by the Lundquist (1950) solution of force-free fields:

$$B_z = B_0 J_0(\alpha r); \quad B_\theta = B_0 J_1(\alpha r) \qquad (1)$$

where J_0 and J_1 are Bessel functions of order zero and one respectively. The solution has the feature that near the axis of the cylinder, the field is mostly in the z-direction, whereas near the cylinder surface defined by $\alpha R = 2.4$, the field is mostly in the azimuthal θ direction. Outside this particular radius, the force-free field is smoothly joined to a potential field:

$$B_z = 0; \quad B_\theta = B_0 \left(R\, J_1(\alpha R)/r \right) \qquad (2)$$

Having modeled the equilibrium structure of the arcades, a linear stability analysis (Bernstein et al 1958) of the solution was performed using a prescription due to Newcomb (1960).

The field lines that protrude from the photosphere into the corona satisfy certain boundary conditions on the surface. Since there is a sudden change in density at the photosphere, the field lines are essentially rigidly tied to the solar surface. The only displacement on the photosphere that is possible is a movement in the direction of the magnetic field at the footpoints (Van Hoven et al.1981). Thus, there are nodes of the radial and perpendicular (to the B_0-field) components of of $\underset{\sim}{\xi}$ but there is a complete freedom for the parallel (to B_0) motions. Each component of the displacement $\underset{\sim}{\xi}$ was expanded in a series in sine and cosine $m\theta$ and appropriate combinations of sine and cosine kz (for detailed expansions see Ray and Van Hoven 1982a) and was used in the energy integral to obtain a form (after integration over the θ and z co-ordinates):

$$\delta W_{m,r}(\underset{\sim}{\xi}) = \int_0^\infty dr \left[f\, \xi_r'^{\,2} + g \xi_r^{\,2} \right] \qquad (3)$$

where ξ_r is the radial component of $\underset{\sim}{\xi}$ and f and g are functions of r, m, k, α and R. The detailed forms of f and g inside and outside $\alpha R = 2.4$ are given Ray and Van Hoven (1982a). The Euler-Langrange equation obtained from eq. (3) can be used to provide a necessary and sufficient condition (Newcomb,1960) for hydromagnetic stability in the plasma pinch. As the obtained Euler-Lagrange differential equations do not have any singular points,(i.e. f does not vanish in the interval 0 to ∞,) to ensure that a given equilibrium mhd structure is stable,one needs to check that the solution ξ_r of the Euler-Lagrange equation which is finite at the origin does not vanish in a finite interval of r. This is a convenient global condition for stability and the Euler-Lagrange equation for $Q = r\, \xi_r$ obtained from eq.(3) was integrated numerically to test this. In order to satisfy the boundary conditions on the photosphere, the radial and perpendicular (to B_o) components of ξ were taken to be proportional to sin mθ but both sine and cosine functions of kz were allowed. The wave numbers k were unrestricted but the angular mode numbers m were taken to be either even or odd integers. The most general decomposition of ξ_r and $\xi_\perp$ in the angular interval $0<\theta<\pi$ would involve both even and odd mode numbers m. Allowing such a general perturbation would however prevent a term by term (m,k) analysis of the stability and would force one to take into account effects of mode coupling (between m and m'). With the somewhat restricted set of perturbations (of only odd or only even m) not only a term by term analysis is possible, but the surface energy integral δW_s which must vanish for the applicability of the stability theorem of Bernstein does indeed vanish.

Newcomb (1960) had argued that the m = 1 mode is the most dangerous mode as far as the possibility of instability in a plasma pinch is concerned. Higher m modes are more stable than m = 1 case. The stability of the equilibrium configuration with respect to the m= 1 mode was tested and found to be stable. The m = 0 mode is not allowed in our equilibrium model which is line-tied on the photosphere at $\theta=0,\pi$.

Thus, in a linear stability analysis of a force-free sheared field surrounded by a potential field, the equilibrium configuration was found to be stable to a reasonably general perturbation summation. With a view to handle mode couplings between even and odd modes encountered in a more general perturbation, the criterion for stability of a plasma configuration given by Newcomb (1960) was extended to find the conditions for positive definiteness of bilinear forms involving cross terms (Ray 1983; Ray and Van Hoven, 1982b). With such a generalization of the stability criterion it is possible to test equilibrium configurations with respect to more general mode mixing perturbations.

The condition of field line-tying on the perturbations impose a strong stabilizing influence. Migliuolo and Cargill (1983) have performed a normal mode analysis of the marginal stability of several force-free equilibria and have arrived at similar conclusions. They have also argued that due to line-tying at a reference boundary and the absence of mode-rational surfaces (where $(B_o \cdot \nabla)\xi_r$ could be zero) resistive modes do not contribute to an instability. Dissipation of magnetic energy due to field-line reconnection is thus ruled out unless some of the field lines are not tied to the photosphere. Hence such arcades would not be able to rapidly release a great deal of energy as is observed in X-ray flare activity. Indeed, coronal arcades that are without filaments are generally not observed to erupt. Two-ribbon flares are in general associated with some filament activation and it could be that an imbedded prominence in the coronal arcade is a necessary condition. It is also possible that some arcades may be made unstable through non-linear mhd wave coupling or due to emerging magnetic flux from the photosphere.

This work was in part supported by the Solar and Heliospheric Physics Branch of National Aeronautics and Space Administration. Hospitality of the Lewes Center for Phyiscs and discussions with the members of Solar Physics Group at the University of California at Irvine are thankfully acknowledged.

References

Bernstein, I.B., Frieman, E.A., Kruskal, M.D. and Kulsrud, R.M. 1958, Proc. Roy. Soc. A244,17.
Lundquist, S. 1951, Phys. Rev. 83,307.
Miglinolo, S. and Cargill, P. 1983 Ap. J. (to be published)
Newcomb, W.A. 1960, Ann. Phys. 10, 232.
Ray, A. 1983, J. Math. Phys. (in press).
Ray, A. and Van Hoven, G. 1982a, Solar Phys. 79, 353.
Ray, A. and Van Hoven, G. 1982b, Phys. Fluid. 25, 1355.
Van Hoven, G. Ma. S.S. and Einandi, G. 1981, Astron. Astrophys. 97, 232.

DISCUSSION

Ionson: In order for line tying to hold true, the diffusion time in the photosphere as well as the convective turnover time must be longer than the instability growth time. Are these conditions satisfied?

A. Ray: Both the resistive diffusion timescale and convective turnover timescale are longer than the relevant timescale, which is the Alfvén wave crossing time at the base of the corona.

Heyvaerts: I would like to stress that fact that stability analysis is perhaps not the best way to look at arcade evolution, because it supposes that the equilibrium under study has already been realized. Of more significance, perhaps, is the fact that a sequence of MHD equilibria may cease to exist at some catastrophy point. This is a property of non-linear force free fields, as opposed to constant α force free fields. (See Heyvaerts et al. 1982, Astron. Astrophys. 111, 104 and references therein, and Aly, these proceedings).

A. Ray: The arcades are seen to presist in the corona for long periods of time and it is not unrealistic to investigate an MHD equilibrium. The equilibrium may of course be changed at a later time by a change of conditions at the photosphere, when its subsequent development may lead through a disruptive phase.

MHD ANALYSIS OF THE EVOLUTION OF SOLAR MAGNETIC FIELDS AND CURRENTS IN AN ACTIVE REGION

S. T. Wu and J. F. Wang*
The University of Alabama in Huntsville
Huntsville, AL 35899, U. S. A.

E. Tandberg-Hanssen
NASA/Marshall Space Flight Center
Marshall Space Flight Center, AL 35812, U. S. A.

ABSTRACT

We have used a self-consistent magnetohydrodynamics model to study the evolution of solar magnetic fields in an active region. The problem has been cast as an initial boundary-value problem based on explicit mathematical formalism (i.e., method of projected characteristics), whereby a variety of horizontal photospheric motions can be treated. In this paper we deal specifically with photospheric shear motions in the active region. Our results show the evolution of the magnetic energy, the electric current systems, including the distributions of $J_\perp$ (current perpendicular to the magnetic field) $J_\parallel$ (current parallel to the magnetic field), the magnetic field configuration and a comparison between the build-up of magnetic energy in a pre-twisted field and in an initial untwisted field due to photospheric shearing motions. From these results we conclude that the energy build-up is confined within a certain region near the neutral line at the photospheric level. The possible location of the particle acceleration also can be studied.

INTRODUCTION

Recently, Wu et al. (1982) developed a non-planar magnetohydrodynamic (MHD) model to study coronal dynamics. Mathematically, this numerical model pertains to a spherical coordinate system for global studies. To study localized phenomena, Wu et al. (1983a) utilized the same theory with different techniques (Hu and Wu, 1983) to develop a MHD model in rectangular coordinates. This model has been used to study wave/mass motions in the solar atmosphere (Wu et al., 1983a) due to photospheric shear motions and to assess the energy build-up in an active region (Wu and Hu, 1981; Wu et al., 1984). However, the detailed structure of magnetic properties has never been presented.

In this paper, we use this newly developed non-planar MHD model to investigate the field and electric current structures due to shear motions in an active region. The numerical results presented are the magnetic energy intensity, electric current intensity, and magnetic field configuration. Further details are presented in Wu et al. (1983b).

ANALYSIS AND RESULTS

The governing equations and the method of solution have been reported by Wu et al. (1983a) and Hu and Wu (1983), and will not be repeated here. Since we are studying an active region, the initial magnetic field is represented by a twisted configuration (Wu et al., 1984). In particular we use a force-free magnetic field representation with a 40° twist in this investigation. The initial undisturbed atmosphere is isothermal and in hydrostatic equilibrium and is characterized by $n_o = 6.5 \times 10^{14}$ cm^{-3} and $T_o = 5 \times 10^4$ K; where n is the number density and T the temperature, and where a subscript "o" refers to the value at the lower boundary, which is the level immediately above the transition region. The initial magnetic field strength (B_o) at the lower boundary is 1,500 G leading to an initial plasma beta (β_o) of 0.1, where β is the ratio of plasma pressure to magnetic pressure (i.e., $\beta = 16$ nkT/B^2). In this initial state a shear motion of 1.0 km s^{-1} is introduced, and our MHD model reveals the ensuing responses.

Figure 1a,c,d show the two-dimensional structures for magnetic energy intensity, line and electric current intensities (i.e., $J_{||}$ and $J_\perp$) in the vertical X-Y plane. The simulation domain is 8,000 km x 8,000 km. We note that concentrations of magnetic energy (ergs km^{-3}) and of electric current (A km^{-2}) develop in similar ways. The location of these concentrations is in the neighborhood of the neutral line and in the lower parts of the atmosphere. Furthermore, the hill-valley type structures of the magnetic energy intensity and the current intensity are also observed during this development state. In general the development of the magnetic energy and electric current intensity is much like that of twisted ropes. It has been shown theoretically (Sokolov and Kosovichev, 1978) that magnetic field gradients of the order of 1 G km^{-1} will trigger instability, and we propose that the type of shear motion under study will lead to a catastrophic state of energy release which may be identified as a flare. In addition, the present self-consistent MHD model gives the physical parameters as functions of time and space and therefore can be tested by observations.

Figure 1b shows the initial magnetic field configuration (solid line and magnetic field configurations twenty minutes after introduction of th shear motion (broken lines). Some interesting physical phenomena are revealed, e.g., the lower portions of the magnetic field lines are lifted up and the higher portions of the lines are pushed downward. This may possibly be interpreted as one of the sources of atmospheric oscillations

Finally, we compare the build-up of excess magnetic energy in an initially pre-twisted (i.e., force-free) field with the build-up in an untwisted (i.e., potential) field configuration due to photospheric shearing motions, see Figure 2. The excess magnetic energy build-up in an

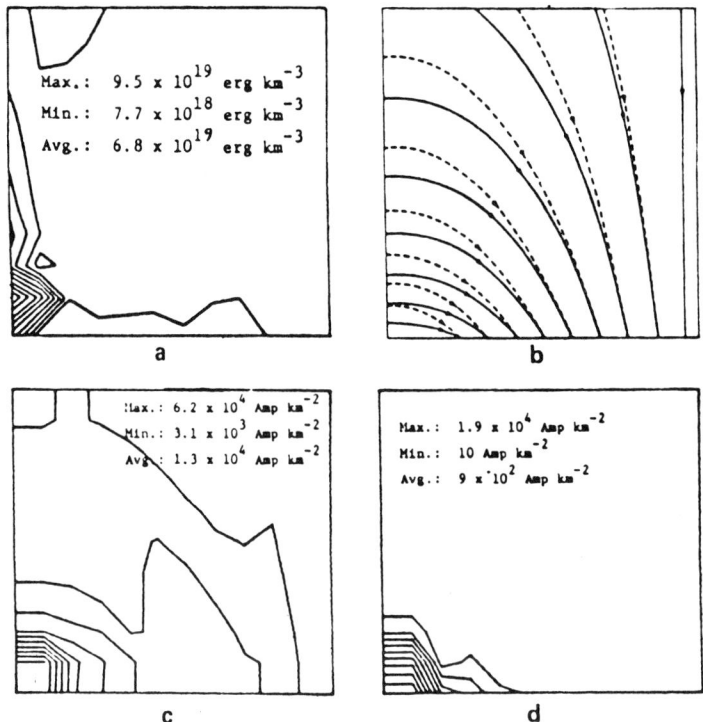

Fig. 1. Two-dimensional distribution of (a) magnetic energy intensity (b) magnetic field lines, (c) parallel current ($J_\parallel$) intensity and (d) perpendicular current intensity 1200 s after introduction of a photospheric shear motion (1 km s^{-1}) in a pre-twisted magnetic field (i.e., force-free field).

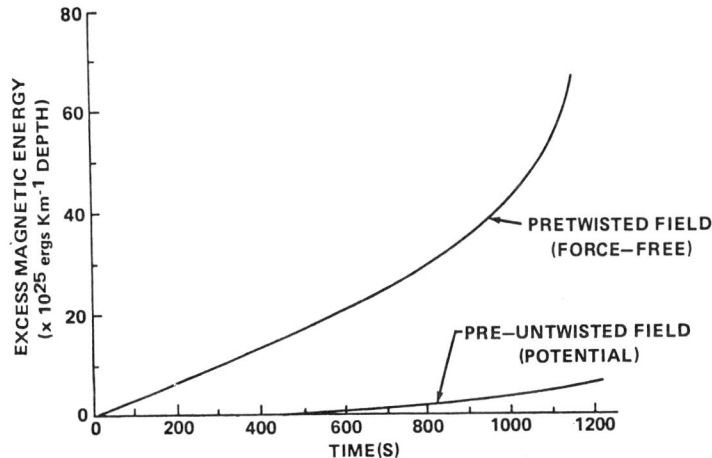

Fig. 2. Magnetic energy build-up due to photospheric shear motion (1 km s^{-1}) for both pre-twisted (force-free) and (un-twisted) (potential) magnetic field configurations.

initially pre-twisted field is almost two orders of magnitude higher than in the case of an initially un-twisted field. Specifically, the excess magnetic energy built to 8×10^{30} ergs in twenty minutes after the solar motion was introduced. This energy would be enough to produce solar activity in the active region.

CONCLUDING REMARKS

In this paper, a self-consistent model was used to investigate the evolution field and current structures due to photospheric shearing motions. It has been demonstrated that this model is capable of predicting physical parameters that may help improve our understanding of the evolution of field and current structures in an active region. We have not shown any results concerning the typical plasma parameters (i.e., density and temperature) since, it has been demonstrated earlier that the variation of these parameters in this particular case is negligibly small Wu and Hu (1981). This implies that the photospheric plasma parameters remain unchanged during this particular energy build-up process (i.e., for $\beta_o = 0.1$ and a shearing motion of 1.0 km s^{-1}). In addition, we have shown the distribution of parallel and perpendicular current systems which may help in the analysis of the evolution of particle acceleration in an active region. These results will be discussed further in Wu et al. (1983b).

ACKNOWLEDGEMENT

The work done by S. T. Wu is supported by a NASA Grant (NAGW-9) and AFGL (F19628-83-K-0019).

REFERENCES

Hu, Y. Q. and Wu, S. T.: 1983, J. of Computation Phys. (submitted).
Sokolov, V. S. and Kosovichev, A. G.: 1978, Solar Phys., 57, 73.
Wu, S. T. and Hu, Y. Q.: 1981, SMY-Proceedings of International Workshop Simferopol, March, 1981. (V. N. Obridko and E. V. Ivanov, eds). IZMIRAN, Academy of Sci. of the USSR. Moscow Vol. I, p 148.
Wu, S. T., Nakagawa, Y., Han, S. M. and Dryer, M.: 1982, Ap. J. 262, 369.
Wu, S. T., Hu, Y. Q., Nakagawa, Y. and Tandberg-Hanssen, E.: 1983a, Ap. J. 266, 866.
Wu, S. T., Wang, J. F., and Tandberg-Hanssen, E.: 1983b (in preparation).
Wu, S. T., Hu, Y. Q., Krall, K., Hagyard, M. J. and Smith, Jr., J. B.: 1984, Solar Physics 90, 117.

* Permanent Address: Dept. of Space Phys., Wu-Han University, China.

BENDING WAVES ON CURRENT SHEETS

G. Bertin and B. Coppi
Massachusetts Institute of Technology Cambridge, MA 02139, USA
and Scuola Normale Superiore, 56100 Pisa, Italy

ABSTRACT. Current sheets are found to be subject to bending waves described by a dispersion relation indicating that these are, essentially, modified surface Alfvén waves. Applications to the observed magnetic polarity sectors in the solar wind and to other astrophysical environments, such as planetary magnetospheres, are suggested.

1. INTRODUCTION

Recent in situ measurements of the solar wind plasma have demonstrated the need for a new theoretical description of both its large scale structure and of its microscopic properties. This would be in contrast to the "classical" model (Parker 1958) of the solar wind which assumes that the wind is the result of a spherically symmetric expansion of the coronal plasma, and that at the base of the corona the transport of thermal energy can be described by the collisional electron thermal conductivity.

Here we start from the well known observation of a sector structure in the magnetic field polarity and recall that, except for occasional short intervals in which the polarity is not well defined, two, four, or occasionally six major sectors are observed per solar rotation. As a result, a model where a current sheet surrounds the Sun and is inclined roughly 15° to the solar equator (Smith 1979; but see also Hoeksema et al. 1982 and Bruno et al. 1982) has been proposed. We note the following points: (1) The tilted current sheet model is only a small departure from a situation where no tilt is present, since the required inclination is small. (2) The sector structure is a large scale phenomenon where the polarity reversal is observed an even number of times (two, sometimes four, more rarely six times). (3) The sector structure is related to a current sheet, since it is not observed above a certain heliographic latitude; it is a very gentle disturbance on the current sheet. Therefore, we propose that the large scale topology of the interplanetary magnetic fields should be explained in terms of large scale bending waves of relatively small amplitude which are naturally excited on an otherwise axisymmetric current sheet disk.

Our point of view is supported by a strong analogy that the above mentioned preliminary data on the disk structure of the solar wind have with the geometry of galactic disks. In fact a promising theory for warps in self-gravitating disks has been developed and warps have been suggested to be the manifestation of large scale bending waves (Bertin and Mark 1980; Bertin and Casertano 1982; see, also, Hunter and Toomre 1969). The theory of bending waves has found recently an attractive application to the (driven) corrugations of Saturn's rings (Shu et al. 1983).

2. BENDING WAVES ON A CURRENT DISK

We consider an equilibrium axisymmetric current disk characterized by a thickness δ_z, a mass density ρ_D, and an outward streaming velocity $\underline{u}$. For mathematical purposes it may be convenient to ignore the physical thickness δ_z and work on an infinitesimally thin disk model. If we refer to the cylindrical coordinates (R,ϕ,z), this is described by $\rho_D(R,z) = \sigma(R)\delta(z)$, and $\underline{J}_D = \underline{j}(R)\delta(z)$, where $\sigma(R)$ and $\underline{j}(R)$ are surface mass and line current densities. Here $\delta(z)$ is the standard delta-function. Close to the current sheet the magnetic field $\underline{B}_D$ is perpendicular to the local current streamlines $\underline{j}$ and its strength B_D is approximately given by $B_D = (2\pi/c)j$.

We are interested in perturbations that bend the current disk from its planar equilibrium configuration. These waves, produce components of the transverse perturbed magnetic field $(\tilde{B}_z)$ that are odd in z. We recall that, in a finite thickness model configuration, even modes have different physical characteristics and produce local current enhancements without breaking the planar symmetry of the current layer. In the infinitesimally thin disk model the bent sheet can be described by the local height $h(R,\phi,t)$ of the bent disk from the equatorial plane. In particular, we look for normal modes of the form $h(R,\phi,t) = \text{Re}\{\tilde{h}(R)\exp[i(\omega t - m\phi)]\}$.

The equation of motion for the bent disk can be written as

$$\sigma \frac{D^2 h}{Dt^2} = F_z \qquad (1)$$

where the total (convective) derivative is defined by $D/Dt = i[\omega - (m/R)u_\phi + ku_R]$. We have introduced a radial wavenumber $k = k(R)$, that is appropriate for those perturbations that can be studied by a WKBJ analysis, as we assume $|Rk| > 1$.

The force F_z arises in part from the action of the current sheet on itself. The reason is that bending the current disk modifies the associated magnetic field. We have to evaluate this field $\underline{\tilde{B}}$ at $z = h(R,\phi,t)$ where the current j is displaced. The relevant integrals are found to be of the same form as the integral that describes self-gravitating disks. As a result we can make use of the asymptotic theory of bending waves in galaxies given by Bertin and Mark (1980) and find, in the WKBJ limit ($|kR| > 1$ and $m = O(1)$)

$$\tilde{F}_z = \frac{1}{c}(j_x \tilde{B}_y - j_y \tilde{B}_x) \simeq -\frac{2\pi}{c^2} j^2 |k|\tilde{h} = -\frac{B_D^2}{2\pi}|k|\tilde{h} \quad , \qquad (2)$$

This corresponds to a <u>restoring</u> force. Inserting Eq. (2) into Eq. (1) we find the relevant <u>dispersion relation</u>:

$$\left(\omega - \frac{m}{R} u_\phi + k u_R\right)^2 = \frac{B_D^2}{2\pi} |k| = \frac{v_A^2}{R \delta_z} |kR| \qquad (3)$$

where $v_A = B_D/\sqrt{4\pi\rho}$ is an appropriate Alfvén speed constructed with the typical plasma density of the sheet and the magnetic field B_D generated just outside the current layer. Here δ_z is defined by $\sigma = 2\delta_z \rho$. This dispersion relation governs the flapping of the current sheet and describes essentially a surface Alfvén wave.

We expect that the dispersion relation given above should include, on the right hand side, an additional term ω_s^2 which depends on the fields and currents in the equilibrium, and describes the effects of a restoring magnetic force on the current sheet due to the action of fields not generated by the local current J_D.

We now consider the case of low frequency bending waves, $\omega \sim m\Omega_s$, corresponding to disturbances that are almost neutral in the frame rotating at the angular velocity Ω_s of the Sun, and therefore rotate rigidly with angular velocity close to Ω_s in the inertial frame. Since we require $|kR| > 1$, we can take $|\omega - (m/R)u_\phi| \ll k u_R$. If we assume further $\omega_s^2 < v_A^2/(R\delta_z)$, we find $k \sim k_o$ with

$$|k_o(R)| = \frac{v_A^2}{u_R^2 \delta_z} \qquad (4)$$

This expression for k_o indicates the scalelength of low frequency bending waves. Our WKBJ analysis requires $|Rk_o| > 1$. Finally, from the dispersion relation we find $c_g = -\partial\omega/\partial k \sim u_R/2$. Therefore for these waves, the group velocity indicates that they naturally propagate outwards at half the radial speed of the wind. This suggests that if quasi-stationary bending structures are observed in the Ω_s-rotating frame they must have a good source of energy upstream, i.e. close to the symmetry axis.

3. EXCITATION MECHANISMS FOR BENDING WAVES

We may point to three forms of excitation that can apply to our case and may be not mutually exclusive:

3.1 Driving at the source

The Sun, and in particular the solar corona, is observed to possess irregular magnetic structures that undoubtedly are related to a slowly evolving, moderately nonaxisymmetric system of currents. Obviously such a source, at the center of the current disk, affects the dynamics of the current sheet far from the Sun. This is somewhat reminiscent of the processes that can drive spiral structures in galaxies that possess a well developed central bar (see Lin and Bertin 1981). Of course the lower multipoles (m = 1,2,3) are the first to be induced in this situation. Therefore it is not surprising to have two (a tilt), four, or six sector structures among the common cases in the solar wind.

3.2 Internal mechanisms of instability

Another possibility is that the solar wind current disk be internally un-

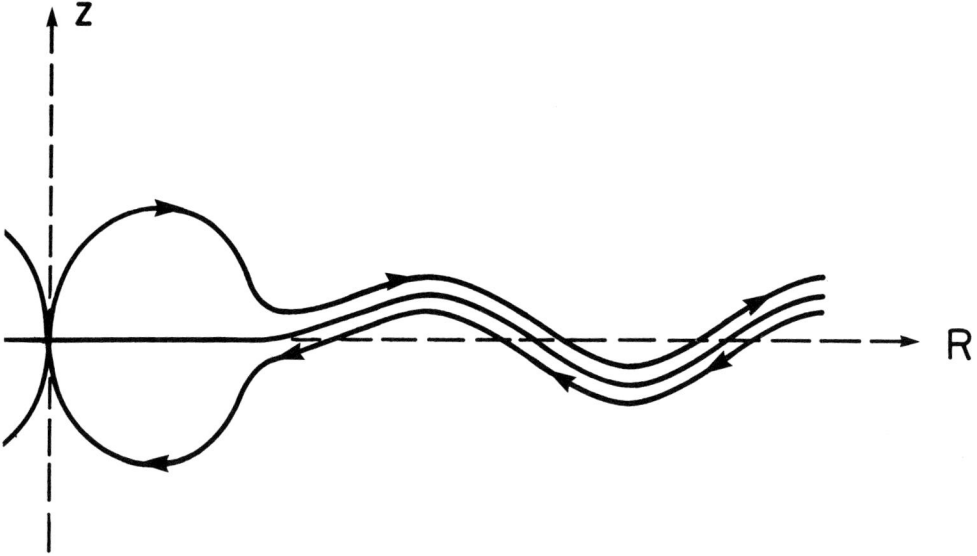

Fig. 1: Qualitative representation of bending wave in the current disk of the solar wind.

stable against bending waves. It may well be that, for the excitation of these bending waves, an important role is played by the internal structural and dynamic properties of the current disk, such as the distortion of the distribution function of electrons and ions from a Maxwellian or the gradient of the current density, that go beyond our simplified MHD analysis. Indeed a few excitation mechanisms based on "microscopic" dynamics could be identified, such as magnetic instabilities which rely on temperature anisotropy for their excitation. Our previous studies of magnetic configurations relevant for the solar wind environment suggest that the above processes are quite sensitive to detailed properties of the particle distribution function (Bertin and Coppi 1980). In particular there are modes (Coppi 1982) involving magnetic reconnection around the neutral sheet that possess a symmetry which would naturally couple them with the bending waves studied above. Quite likely bending waves excited by internal mechanisms are the essence of what Alfvén suggested (1977) and is known under the name of "ballerina effect". To our knowledge, Alfvén sketched only briefly the concept of the flapping of the current sheet but gave no quantitative explanation of his suggestion, nor did he identify the relevant restoring forces.

3.3 Wind Driving

If a current disk is embedded in a medium characterized by varying stream velocities we expect a flapping instability similar to that of Kelvin-Helmohltz to develop. This situation would apply to the solar wind current disk, if available and future data should indicate sizeable velocity gradients across the disk. A circumstance of this type is more likely to occur in planetary magnetotails, such as those that are known to associate with the Earth, Jupiter, and Saturn. In this latter case the velocity gradient is naturally provided by the relative motion between the planet magnetosphere and the solar wind. The geometry would be different from that used above but similar physical concepts should apply.

This work was sponsored in part by the Consiglio Nazionale delle Ricerche of Italy and by A.F. Geophysics Laboratory, (Bedford, Massachusetts).

REFERENCES

Alfvén, H.: 1977, Rev. Geophys. Space Sci. 15, 271.
Bertin, G., Coppi, B.: 1980 Proc. Intern. Conf. on Plasma Physics, Paper 9P-II-33, Nagoya (Japan); and MIT report PRR-79/17 (1979).
Bertin, G., Mark, J.W-K.: 1980, Astron. Astrophys. 88, 289.
Bertin, G., Casertano, S.: 1982, Astron Astrophys. 106, 274.
Bruno, R., Burlaga, L.F., Hundhausen, A.J.: 1982, J.G.R.87(A12) 10, 339.
Coppi, B.: 1982, MIT Report PTP-81/18R, Ap. J. Letters in press.
Hoeksema,J.T., Wilcox,J.M., Scherrer,P.H.: 1982, J.G.R.87 (A12) 10, 331.
Hunter, C., Toomre, A.: 1969, Astrophys. J. 155, 747.
Lin, C. C., Bertin, G.: 1981, in Plasma Astrophysics, ESA Publications SP-161, pp. 191-205.
Parker, E.N.: 1958, Astrophys. J. 128, 664.
Shu, F.H., Cuzzi, J.N., Lissauer, J.J.: 1983, Icarus, 53, 185.
Smith, E.J.: 1979, Rev. Geophys. Space Sci. 17, 61.

DISCUSSION

Migliuolo: Since the mode varies with radius (k_r), this model predicts different polarizations of B at different distances from the sun. Hence we should observe it with Helios and other spacecraft.

Coppi: You have a good point. It should be possible to verify this model by correlating simultaneous measurements of the solar wind magnetic field at widely separated distances. However I do not know how yet to resolve certain ambiguities that an experiment of this type seems to involve.

Vlahos: You have made the commment several times in this conference that a Maxwellian distribution may not be a good approximation for astrophysical calculations of microinstability analysis. My question is, how far from a Maxwellian distribution do you have to be to alter the results significantly?

Coppi: In order to excite the reconnecting type of modes I discussed, a small temperature anisotropy ($0 < T_\perp/T_\parallel - 1 < 1$) is sufficient.

Kennel: Would you repeat the arguments about the direction of energy flow in your mode? Is it a negative energy mode?

Coppi: The spiral wave that we have found has positive energy in the frame that has been considered. The direction of the energy flow is outward.

QUASI-TWO-DIMENSIONAL COSMIC JETS

K. Tsinganos [†], A. Ferrari [‡], R. Rosner [†]

[†] Harvard-Smithsonian Center for Astrophysics, Cambridge, MA 02138, USA.
[‡] Istituto di Fisica Generale, Universita di Torino, I-10125 Torino, ITALY.

INTRODUCTION

One of the major discoveries in solar physics over the past decade has been the association of coronal holes with high-speed solar wind streams (Zirker 1977 and references therein). On the other hand, advances in X-ray and radio instrumentation (e.g., *Einstein*, VLA, VLBI, etc.) in the past few years have allowed detailed observations of collimated outflows from rather more distant objects, such as young stars and active galaxies (Beer 1981, Lada 1982, Ferrari and Pacholczyk 1983 and references therein). The remarkable structural similarities between jets of magnetized gas from our Sun, other active stars, and active galactic nuclei suggest that these phenomena may be manifestations of similar hydrodynamic processes operating on both small and large scales. In this article, we shall use the experience gained by studying the nearest known astrophysical jet -- high-speed solar wind streams -- to address some of the problems of astrophysical jet acceleration and collimation associated with objects as diverse as SS 433, star-forming molecular clouds and, in particular, jets associated with galaxies and quasars.

MODEL ASSUMPTIONS AND GOVERNING EQUATIONS

The three questions that we have chosen to address in the context of the jet phenomenon are:
   (i)   the initial acceleration of the jet within the inner regions of the "power source";
   (ii) the collimation of the accelerated ionized gas;
   (iii) the physical mechanism(s) responsible for the *in situ* particle reacceleration, and corresponding multiwavelength emission from the bright knots along the jet axis.

In the case of extragalactic jets, these questions are motivated by the rich collection of observational data available in the optical and radio and, recently, at X-ray wavelengths (cf. Heeschen and Wade 1982, Ferrari and Pacholczyk 1983). These observations show that jets are highly collimated and that the nonthermal emission is peaked in bright knots along the jet axis; and imply that the jets originate deep within the "power center". From the theoretical point-of-view, several ideas have been proposed to account for jet acceleration and collimation (see review talks by A. Ferrari and C. Norman in these Proceedings). For our purposes, however, we will assume (with Rees *et al.* 1982) that a central mass is surrounded by an accretion disk (Fig. 1). Two narrow channels are formed along the rotation axis of the system (Abramowicz and Piran, 1981). This disk "throat" is not empty, however, because wave and radiation pressure acting on the plasma which diffuses inward from the disk "walls" initiate an outward flow above some stagnation distance z_0. A Parker-type wind then emerges from the throat.

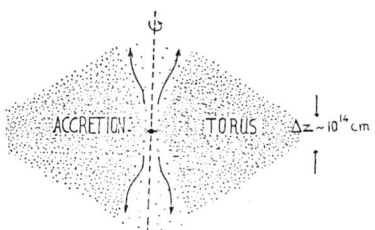

Figure 1

Along the symmetry axis of the channel, the flow speed $v(z,t)$ and density $\rho(z,t)$ are governed by the familiar hydrodynamic conservation equations for mass and momentum (Parker 1963, Kopp and Holzer 1977, Habbal and Tsinganos 1983),

$$\frac{\partial}{\partial t}(\rho A) + \frac{\partial}{\partial z}(\rho v A) = 0 , \qquad (1)$$

$$\frac{\partial v}{\partial t} + \rho v \frac{\partial v}{\partial z} = -\frac{\partial P}{\partial z} - \frac{GM}{z^2} + \rho D(z,t) , \qquad (2)$$

where $A(z,t)$ is the (macroscopic) channel cross-sectional area and $\rho D(z,t)$ represents the local momentum deposited by waves, radiation, etc., emitted by the accretion disk or the central object. The study of the energetics of the flow is temporarily postponed by assuming a polytropic equation of state $P = k\rho^\alpha$. In the following, however, we will further simplify the problem by assuming an isothermal flow, $\alpha = 1$, and hence a constant sound speed v_s. The flow $v \equiv \partial \Phi / \partial z$ is then governed by the following single equation

$$\frac{\partial^2 \Phi}{\partial t^2} + 2v \frac{\partial^2 \Phi}{\partial z \partial t} + (v^2 - v_s^2) \frac{\partial^2 \Phi}{\partial z^2} + v(\frac{GM}{z^2} - \frac{2v_s^2}{z}) = \frac{\partial \Delta}{\partial t} + v \frac{\partial \Delta}{\partial z} , \qquad (3)$$

where

$$\Delta(z,t) = v_s^2 \ln f(z,t) + \int D(z,t) dz \qquad (4)$$

represents the total effect of nonthermal momentum addition (either directly, via $D(z,t)$, or indirectly, via non-spherical expansion such that $f(z,t) = A(z,t)/z^2$).

There are two fundamental properties of the above governing equation of motion (3) which are of particular interest to our problem:

(I) The time-dependent Eq. (3) is a second order *hyperbolic* partial differential equation. As a result, perturbations in the flow might "break", and hence develop shocks. For example, suitable changes in the magnitude of Δ have been shown numerically to lead to shock formation (Habbal *et al.* 1983, Tsinganos *et al.* 1983).

(II) The time-independent form of Eq. (3) is the familiar Mach number equation (Parker 1963). For appropriate forms of Δ, however, its solution topologies are characterized by multiple critical points, possibly degenerate transonic solutions, and possible standing shocks, as it can be seen from the following results.

RESULTS

The following figures 2 and 3 are plots of the steady solutions of Eq. (3) for a gaussian form of the nonthermal momentum deposition D(s),

$$D(s) = D_o \exp[-(s-s_m)^2/\sigma^2] \tag{5}$$

Fig. 2 corresponds to the galactic case ($M=10^8$ $M_\odot$, $T=10^8$ K, $z_o=10^{15}$ cm). In 2a the parameters of nonthermal momentum addition are $D_o=10$ dyn/gm, $s_m=50$, $\sigma=10$, and in 2b $D_o=20$ dyn/gm, $s_m=50$, $\sigma=10$. In (a) the wind solution crosses the outermost critical point at s=500 while in (b) the innermost one at s=39.6. The wind solution corresponds to the branch WXW in (a) and WX'W in (b); the accretion solution to branch AXA in (a) and AX'A in (b); and the "breeze" solution to branch B. The subsonic branch XSS'H satisfies the Rankine-Hugoniot relation $M_1 M_2 = 1$ with respect to the supersonic branch X'C'CW in (b). The vertical solid lines CS and C'S' denote the position of the standing shocks. In (a) the jet becomes supersonic at the outermost critical point, as in the case of a Parker-type galactic wind. In (b) however enough nonthermal momentum has been added to the subsonic portion of the unperturbed wind such that the jet becomes supersonic much earlier, at the inner critical point X'. Note the presence of three (degenerate) solutions -- two of them being discontinuous -- for the same initial velocity: the two discontinuous solutions WX'C'S'SXW' and WX'CSXW' occur because the shock transitions C'S' and CS connect the inner transonic solution WX'W to the outer transonic branch crossing the critical point X.

Fig. 3 is the analogous jet outflow from a stellar object ($M=M_\odot$, $T=10^6$, $z_o=10^{11}$ cm). The continuous curve in Fig. 3a is the familiar Parker-type solution with D=0 and one critical point at s = 4.8. Two new critical points are added in Figure 3b; however, the flow still remains subsonic for s < 4.8. In 3c-3d the flow becomes supersonic at the innermost critical point, s =2.7. Fig 3c contains the two shock transitions C'S' and CS as in 2b. These shocks are not allowed for the values of the flow parameters of Fig 3d.

Figure 2

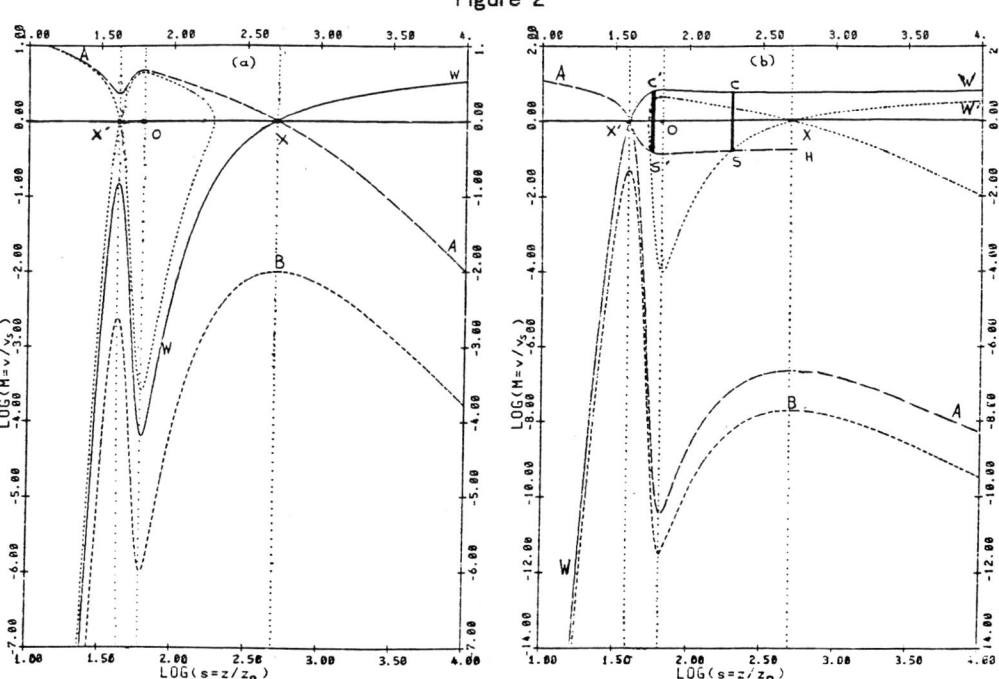

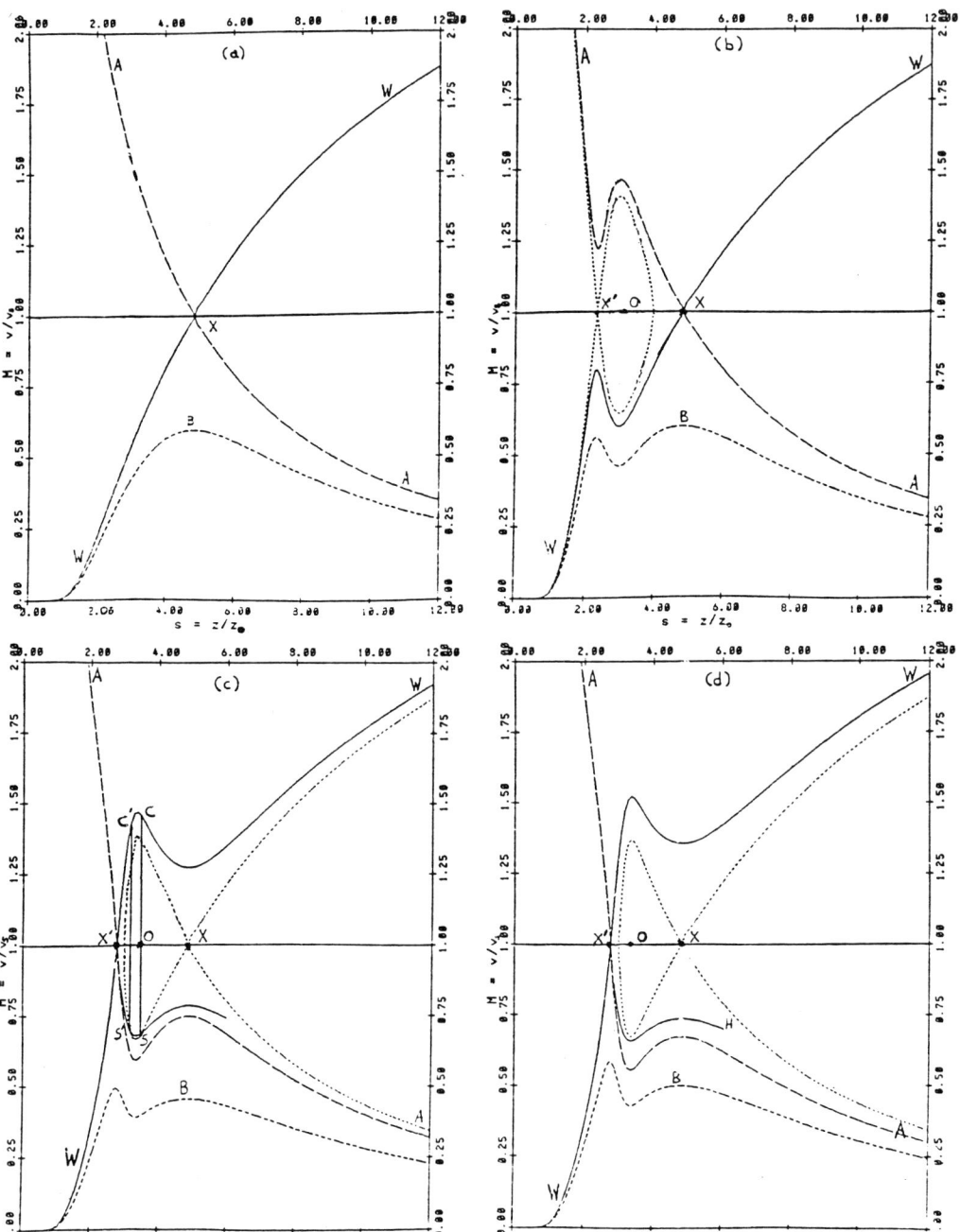

Figures 3a-3d: Mach number $M=v/v_s$ vs. distance $s=z/z_o$, (v_s = 140 km/s, $z_o=10^{11}$ cm) for Parker-type outflows from a stellar object of mass $M=M_o$ and temperature $T=1.2 \times 10^6$ K in a channel of cross-section $A(s)$ and/or with momentum addition $D(s)$. The parameters $[D_o$ dyn/gm, s_m, $\sigma]$ of the nonthermal momentum deposition $D(s)$ [cf. Eq. (4)] are [0, 3, 0.3], [3000, 2.5, 0.5], [2700, 3, 0.3] and [3000, 3, 0.3] in (a), (b), (c), and (d), respectively.

SUMMARY

There are two novel and interesting features of the above solutions of the wind equations which are introduced either by the geometry of non-radial flow tube expansion, or by direct momentum addition within the flow. The *first* is that the flow can become supersonic much closer to the central "power source" than without some effective momentum deposition; the funnel of an accretion disk offers the appropriate environment for beam collimation and non-thermal momentum addition. The *second* has to do with the propagation of the jet outside the nuclear region. As a result of hydrodynamic or hydromagnetic instabilities, the functional dependence of the cross-sectional area of the jet on the linear distance from the source may differ substantially from $A(z) \propto z^2$, opening the possibility for shocks to exist in the external region as well (as illustrated in Fig. 2b). These shocks in turn may be associated with local acceleration of particles (cf., Blanford and Ostriker 1978), resulting in the nonthermal emission associated with the bright knots aligned along the axis of, for example, M87.

REFERENCES

Beer, P. 1981, *Vistas in Astronomy*, **25(1/2)**, (Pergamon Press).
Blandford, R.D. and J.P. Ostriker, 1978, *Ap. J. Lett.*, **221**, L29.
Ferrari, A. and A.G. Pacholczyk (eds.), 1983, **Astrophysical Jets**, (Dordrecht: D. Reidel).
Habbal, S.R. and K. Tsinganos 1983, *J. Geophys. Res.*, **88(A3)**, 1965,
Habbal, S.R., R. Rosner and K. Tsinganos 1983, in **Solar Wind 5**, (in press).
Heeschen, D.S. and C.M. Wade, 1982, **Extragalactic Radio Sources**, (Dordrecht: D. Reidel).
Kopp, R. and T. Holzer 1976, *Solar Phys.*, **49**, 43.
Lada, C.J. 1982, *Scientific American*, **247(1)**, p. 82.
Parker, E.N. 1963, **Interplanetary Dynamical Processes** (New York: Interscience).
Rees, M.J. 1982, in **Extragalactic Radio Sources**, (Dordrecht: D. Reidel), p. 211.
Tsinganos, K., S.R. Habbal and R. Rosner 1983, in **Solar Wind 5**, (in press).
Zirker, J.B. 1977, **Coronal Holes and High Speed Wind Streams**, (Colorado University Press).

DISCUSSION

BENFORD : What is collimating the beam?

TSINGANOS : The geometry of the funnel and the rotation of the beam around the z-axis are expected to assist collimation.

BRATENAHL : This modeling is interesting, but have you checked the existence of shocks in the coronal holes?

TSINGANOS : Yes, we are looking into this question. The cross-sectional area $A(z,t)$ of the flow tube is a function of the time t and the height z; and in numerical simulations we indeed obtained the shocks for suitable temporal variations of the cross-section $A^*(z,t)$ and/or $D^*(z,t)$. We should expect therefore shocks to occur in the inner solar wind for $A(z,t) = A^*(z,t)$ and/or $D^*(z,t)$. It would be interesting to look at the white light or radio data from the solar corona for observational signatures of these shocks.

SESSION VIII

CREATION OF HIGH-ENERGY ELECTRON TAILS BY THE LOWER-HYBRID WAVES AND ITS RELEVANCE TO TYPE II AND III BURSTS

Motohiko Tanaka and K. Papadopoulos
University of Maryland, College Park, Maryland 20742
U.S.A.

It is commonly anticipated that high-energy electrons play an important role for the wave emission in flare bursts. For instance, electrons with >100 KeV are considered to create microwave emissions through gyro-synchrotron process and hard x-rays may be due to bremstrahlung with >25 KeV electrons. However, electron acceleration mechanism itself is still in speculations.

Recently, Holman(this symposium) proposed the runaway acceleration of electrons by existing dc electric field parallel to the magnetic field. According to his theory, the number of the accelerated electrons is determined by scattering rates of electrons into the runaway regime. He gave the acceleration time of 0.1 sec which is shorter than the observed acceleration time of 1 sec. However, the origin of the dc electric field must be explained for his mechanism to work.

Another candidate for the electron acceleration involves plasma waves which are self-consistently generated by drift current in the plasma. The drift current is associated with gradients of plasma condition (such as density) and exists at the shock front, the edge of the plasma and the magnetic neutral sheet. This drift excites the plasma instability, especially at low plasma beta condition. Most plausible among various plasma instability is the lower-hybrid or modified two-stream instability whose frequency is close to the lower-hybrid frequency and it propagates nearly perpendicular to the ambient magnetic field.

According to the closed or open geometry of the magnetic field lines, two different processes involving the waves are possible for the electron acceleration. First, when the magnetic field lines are partly open such as in Type III bursts, the waves excited in the closed magnetic loop must propagate in space into the open field line region. Otherwise, the accelerated electrons never escape into the free space. This may be called indirect process. Secondly, when the acceleration takes place only in the closed magnetic field, the acceleration can be direct and more efficient. In the latter case, the excited waves act as catalyst and accelerate electrons in the source region (for example, Type II bursts). In this case, it is possible that the final electron energy is comparable to the drift current energy, which could be by orders of magnitude larger than the wave energy itself.

Recent simulation study by Tanaka and Papadopoulos (1983) discovered selective acceleration of electrons via plasma instability even at low drift speed condition. Their mechanism is physically quite self-consistent and does not need to assume "external" electric field. Figure 1 shows formation of high-energy electron tails in the $(x, v_\perp)$ phase space. This process occurs in the time scale of $50\omega_{LH}^{-1}$. Final shape of the electron distribution function vs. $v_\parallel$ is shown in Figure 2. For the case shown in Figure 1 and 2, the simulation was done in 1-D. In 2-D run, the high-energy electron tails appear on the both side of $v=0$. The number of electrons being accelerated is approximately 10% and their energy is 50 to 100 times of the background electron temperature.

Formation of the energetic electron tails is possible when the following conditions are satisfied: (1) Plasma instability (so called modified two-stream instability - McBride et al.(1972)) must occur. This gives the lower threshold to the drift speed v_d as $v_d > c_s$ where c_s is the sound speed, (2) Ions must be trapped first and saturate the instability so that "heavy" electrons may not be confined in the wave potential field. This is possible when $v_d < 3v_i$ where $v_i = (2T_i/m_i)^{1/2}$ is the ion thermal speed, (3) Plasma beta must not be high, say $\beta \lesssim 0.1$. This is because the saturation level of the instability (hence the wave amplitude) is a rapidly decreasing function of the plasma beta around $\beta \sim 1$, and also because the most unstable wave propagates almost perpendicular to the magnetic field at low beta condition. The latter yields $v_{ph,\parallel} = \omega/k_\parallel \sim (T_i/T_e)^{1/2} v_e$ ($k_\parallel/k \sim (m_e/m_i)^{1/2}$ for $\beta \ll 1$). Combining (2) and (3), yields the condition for the selective acceleration of electrons as $c_s \lesssim v_d \lesssim 3v_i$. This criterion is new and different from the previous study by McBride et al. (1972) which showed bulk heating of electrons at high drift speed conditions.

Theoretically, the high-energy electron tails are the result of the wave-particle interaction where the wave mode changes from the slower one to the faster one with $v_{ph,\parallel} = 5 \sim 7 v_e$. Rowland et al.(1983) tried to explain this process in terms of the eigenmode shift due to ion heating. However, what is happening in the simulation is not ion heating but clealy ion trapping (cf. right column of Figure 1). The eigenmode shift is then explained in terms of ion trapping (cf. Tanaka and Papadopoulos(1983)). The dispersion equation for the ion trapping state is given by

$$D = 1 + \frac{2\omega_e^2}{k^2 v_e^2}[1 + \zeta_0 Z(\zeta_0)e^{-\lambda}] - \frac{\omega_i^2}{(\omega - k_\perp v_d)^2 - k_\perp^2 u_0^2} = 0, \quad (1)$$

where ω_e, ω_i are the electron and ion plasma frequency, respectively,

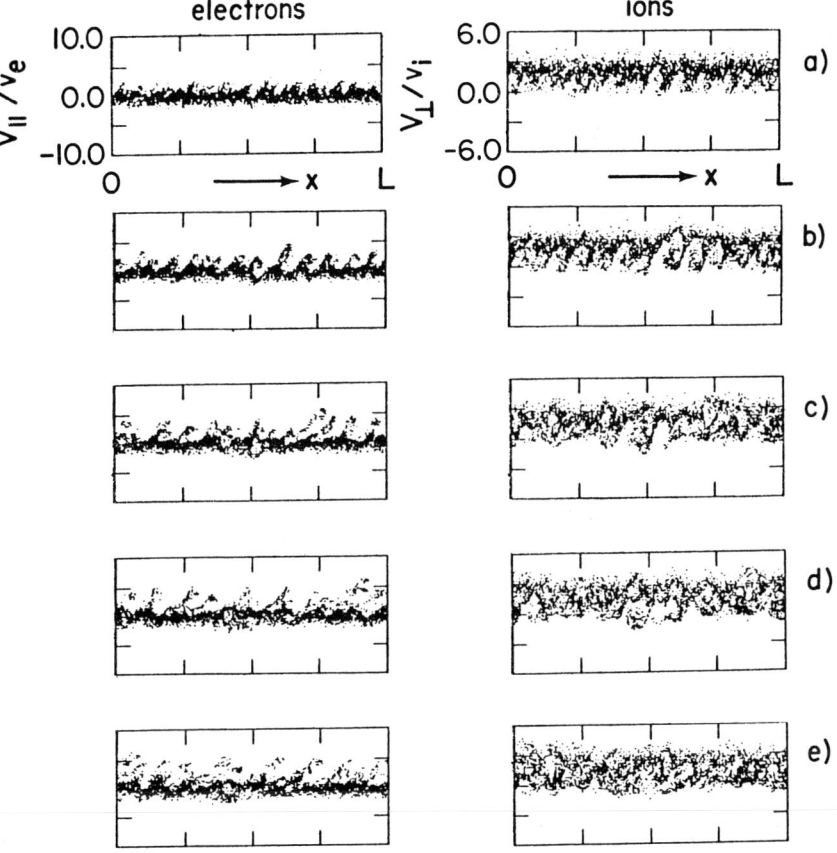

Fig.1 Phase space plots of electron and ion distributions. a)-e) correspond to time $20, 30, 40, 50, 60 \omega_{LH}^{-1}$.

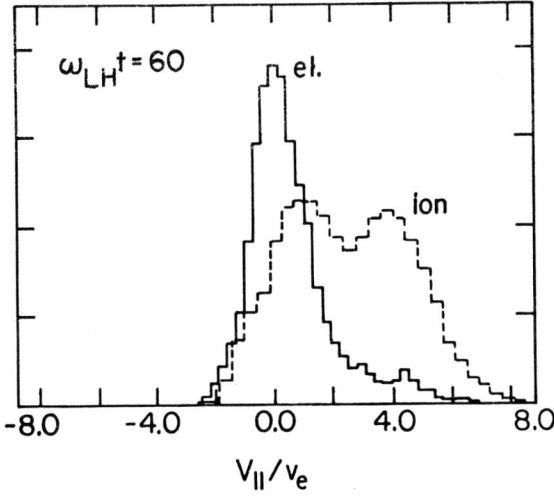

Fig.2 Distribution functions corresponding to Fig.1e).

v_e is the electron thermal speed, $\zeta = \omega/k_\| v_e$, $\lambda = k_\perp^2 v_e^2/2\Omega_e^2$, v_d is the cross-field drift and u is the "effective" temperature of the trapped (square) ion distribution function. Equation (1) gives the frequency of the faster mode as

$$\omega \gtrsim [\ 1 + (k_\|^2/k^2)(m_i/m_e)\ ]^{1/2},$$

and $v_{ph,\|} \sim 6 v_e$. This very well explains the simulation results.

Finally, some applications are shown here. One attempt was made by Smith(this symposium) to explain hard x-ray bursts. Another possible application is to the Type II bursts (closed field line case). If we assume a travelling spherical shock with the radial speed of V and the diamter L, the number of electrons picked up per unit time is estimated by $\pi(L/2)^2 Vn$ where n is the electron density at the shock front. This yields 10^{35} electrons/sec which appears to be large enough to explain x-rays and microwaves (L=10^4Km, V=10^3Km/sec, n=10^9cm^{-3} are used. To get the number of energetic electrons we have to multiply some factor to 10^{35}. However, this factor is order of 0.1). The energy of these electrons can be 50 to several hundred KeV, and this acceleration is achieved in the time of $50\omega_{LH}^{-1} \sim 10^{-6}$ sec (B=100G). This is of course short enough to explain the observed flare and emission growth.

References

McBride, J.B., Ott, E., Boris, J.P., Orens, J.H.: 1972, Phys. Fluids 15, 2367.
Rowland, H.: 1983, Astronomy Program Report No.73.
Tanaka, M., Papadopoulos, K.: 1983, Phys. Fluids 26, 1697.

DISCUSSION

D. Smith: For your model of type II bursts, even though you have a couple of orders of magnitude to spare, the condition for the angle between the ion flow and magnetic field is fairly critical and is probably only satisfied in small regions of the total shock. Shouldn't you include some factor for this in your estimates?

Tanaka: The plasma drift v_d, which is a free energy source for the instability, is actually dependent on the angle between the shock normal and the magnetic field. The drift may be largest for the perpendicular shock. However, once $v_d \gtrsim c_s$ is satisfied, the instability and high-energy electrons are resulted.

Steinolfson: How do you envision that the electric fields needed for this instability could be produced?

Tanaka: My mechanism proposed is physically self-consistent and does not need a priori assumption on the electric field.

CURRENT STATUS OF THE DISSIPATIVE THERMAL MODEL FOR SOLAR HARD X-RAY BURSTS

Dean F. Smith
Berkeley Research Associates, P. O. Box 241
Berkeley, California, U.S.A.

Up until about five years ago all models for hard X-ray bursts consisted of streaming nonthermal electrons interacting with an ambient plasma (Brown 1975). Even in its most efficient form of thick-target emission in which electrons are stopped in the ambient plasma, this type of model is very inefficient because the electrons lose about 10^5 times more energy in Coulomb collisions with the ambient plasma than in X-rays resulting from bremsstrahlung. As a result, according to the latest estimates, at least 20% of the dissipated flare energy must go into accelerated electrons at the peak of the impulsive phase (Duijveman et al. 1982). Stimulated by observations of hard X-rays with thermal spectra (Crannel et al. 1978; Elcan 1978), analysis of a thermal model in which all the electrons in a given volume are heated to a temperature $T_e \simeq 10^8$K was begun (Brown et al. 1979; Smith and Lilliequist 1979; Vlahos and Papadopoulos 1979). It was recognized from the beginning that some electrons in the tail of the distribution would escape through the conduction fronts formed and mimic nonthermal streaming electrons. This thermal model with loss of electrons or dissipation became known as the dissipative thermal model (Emslie and Vlahos 1980). If the escaping electrons are not replenished, they will cease to make a contribution after a fraction of a second and the source will become a pure thermal source. It will be shown below that collisional replenishment (Smith and Brown 1980) is too slow.

The Solar Maximum Mission (SMM) and HINOTORI results with spatially resolved X-rays up to 30 keV and millisecond time resolution to $\sim$ 500 keV have done much to resolve the nonthermal versus thermal controversy for hard X-ray models. These observations show the following pattern. In some flares two or more footpoints separated by tens of thousands of kilometers brighten simultaneously to within the time resolution of 1.5s. In some flares this phase is absent although for compact flares it is difficult to tell. This phase which we shall call the nonthermal phase lasts $\sim$ 40s and can only be explained by thick-target interactions of streaming electrons with the chromosphere. After this phase a single source appears between the original footpoints which lasts for $\sim$ 10 minutes and has a temperature $T_h \sim 4 \times 10^7$K. The energy dissipated during this phase which we shall call the thermal phase is ∼ 10 times that of the nonthermal phase. Neither phase is purely nonthermal or purely thermal. In the nonthermal phase the region between the footpoints emits X-rays with about one-third the flux of the footpoints. During the thermal phase μ-waves and hard X-rays indicating electron energies of several hundred keV continue to be emitted. Thus the names of the phases simply refer to the dominant energy loss mechanism. Since most of the energy is dissipated during the thermal phase and a dissipative thermal model with additional acceleration can explain the nonthermal phase, it appears that this model is the best candidate for explaining hard X-ray bursts.

It has been shown (Duijveman 1983) that the mean free path, λ, for electrons in the thermal phase is of order the temperature scale height. Under these conditions classical heat conduction is no longer appropriate and saturated heat conduction, the conduction of freely flowing electrons, applies (Smith and Lilliequist 1979). There is at present no observational evidence that T_e becomes sufficiently large for conduction anomolously limited by the ion-acoustic instability to apply. Although the spectra in this case agree with observations for the nonthermal phase (Smith and Auer 1980; Smith and Harmony 1982), the predicted spatial distributions clearly do not. The 1980, July 14, 08:24 UT flare consists of two spikes which show no footpoints in the 16-30 keV X-rays. There is no time delay between the 30 and 100 keV emission for the first spike and a 1.4s delay of the 100 keV emission relative to the 30 keV emission for the second spike (Orwig, private communication).

As a basis on which some results for dissipative thermal models can be discussed, we summarize the following results. When a current is flowing in a loop with a primary toroidal magnetic field, the current produces a poloidal field which can be dissipated by tearing mode instabilities (Spicer 1977). In the case of the collisional tearing mode, about 47% of the energy released will go into ion motion (Arion 1983) which is perpendicular to the primary toroidal field of the loop. When this drift velocity, v_D, is in the range of 2-3 v_i, where v_i is the ion thermal velocity, the kinetic cross-field streaming instability can create an electron beam at ∼ 6 v_e, where v_e is the electron thermal velocity (Tanaka and Papadopoulos 1983). This beam travels along the toroidal field and ∼ 50% of the ion energy goes into the electrons. For $v_D > 3 v_i$, all of the electrons are heated out to ∼ 6 v_e with ∼ 50% of the ion energy going to electrons. This instability will only work if $\Omega_e/\omega_{pe} \gtrsim 1$, where Ω_e and ω_{pe} are the electron gyro- and plasma frequencies, respectively.

We show that collisional replenishment of the tail of a Maxwellian distribution is too slow to explain the number of electrons required for the nonthermal phase. The tail of a Maxwellian is populated like an advancing wave in velocity space on a time scale $\tau = 2\lambda(v_b)/v_b$, where v_b is the velocity required for an electron to escape through the conduction front (∼ 2 v_e). During the nonthermal phase, $T_e \simeq 10^8 K$, $2v_e = 10^{10}$ cm s^{-1} and the density $n_e \simeq 10^{10}$ cm^{-3}. These lead to $\tau = 1.2$s which is much longer than the escape time of electrons ≲ 0.1s for a thermal source of scale ∼ 10^9 cm. Thus continuous acceleration is required.

Using a flux corrected transport code with implicit correction for heat conduction and radiation, we have followed a single velocity two-temperature fluid along a 47,000 km loop heated near its top. Both classical and saturated heat conduction are included and the maximum T_e was 8×10^7 K. The X-ray yields at 30 and 100 keV for the case where the heating is cut off at 8s are shown in Figure 1. Note the ∼ 1.5s time delay between the peaks of the 30 and 100 keV emissions.

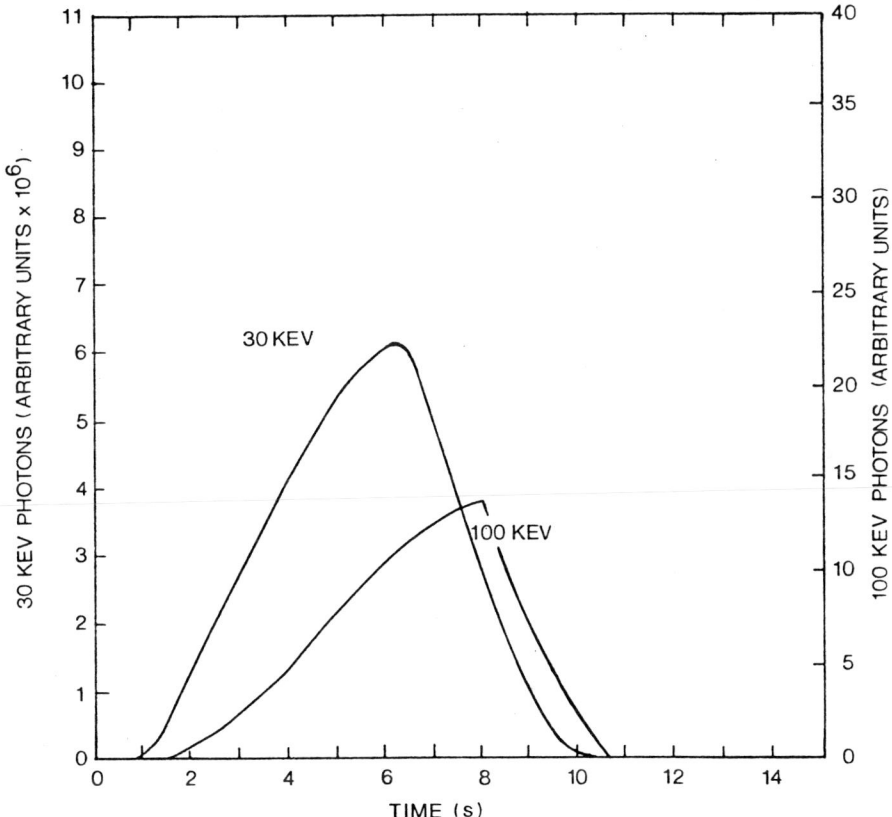

Figure 1. The 30 and 100 keV emission versus time with an energy input of 2.8×10^2 erg cm^{-3} s^{-1}.

A possible scenario for the dissipative thermal model is the following. During the nonthermal phase the loop density is comparatively low and $\Omega_e/\omega_{pe} \gtrsim 1$. Tearing mode instabilities give rise to ion flow at the Alfven speed v_A of the poloidal field across the primary toroidal field. For plausible field values ∼ 140G with $n = 10^{10}$ cm^{-3}, $v_A = 3 \times 10^8$ cm s^{-1} which is about $3v_i$ for an ion temperature T_i of 10^8K. This instability gives rise to electrons streaming along the primary toroidal field at ~ 6 v_e which leads to ~ 160 keV electrons for $T_e = 10^8$K. The maximum efficiency for conversion of flare energy into streaming electrons is $\sim 23\%$. These electrons cause plasma to be boiled off the chromosphere which flows back up the loop causing $\Omega_e/\omega_{pe} < 1$ and effective acceleration ceases. The hydrodynamics must still be followed in this stage to determine when effective acceleration stops. During the thermal phase tearing mode instabilities continue to occur in the loop, mostly in regions where $\Omega_e/\omega_{pe} < 1$, resulting in heating with a small amount of acceleration. In the nonthermal phase the efficiency is almost the same as in a nonthermal streaming model. In the thermal phase it is almost the same as in a thermal model.

The nonthermal phase would give rise to footpoints for sufficiently large loops and no time delays between 30 and 100 keV emissions. This is in agreement with observations as long as efficiencies $< 23\%$ are required. The thermal phase would give rise to time delays ~ 1.5s between 30 and 100 keV emissions and a single source. Some continuing acceleration in regions where $\Omega_e/\omega_{pe} \gtrsim 1$ or via another mechanism is required to explain the μ-wave and harder hard X-ray observations.

Thus the dissipative thermal model with additional acceleration is the most likely candidate for the whole hard X-ray burst. In both phases, but especially in the thermal phase, the plasma beta approaches unity and two-dimensional modelling of the hydrodynamics is required. Following the accelerated electrons only without taking into account the response of the bulk of the plasma is inadequate. A useful approach might be a multifluid one using ~ 15 fluids with the possibility of transfer between fluids in one- and two-dimensions. Thus, while this model contains much promise, many details remain to be worked out.

This work was supported by NASA contract NASW-3603.

REFERENCES

Arion, D.: 1983, Phys. Fluids, submitted.
Brown, J.C.: 1975, Solar Gamma-, X-, and EUV Radiation, ed. S.R. Kane (Dordrecht: Reidel), p. 245.
Brown, J.C., Melrose, D.B., and Spicer, D.S.: 1979, Astrophys. J. 228, p. 592.
Crannell, C.J., Frost, K.J., Matzler, C., Ohki, K., and Saba, J.L.: 1978, Astrophys. J. 223, p. 620.
Duijveman, A.: 1983, Solar Phys. 84, p. 189.
Duijveman, A., Hoyng, P., and Machado, M.E.: 1982, Solar Phys. 81, p. 137.
Elcan, M.J.: 1978, Astrophys. J. 226, p. L99.
Emslie, A.G., and Vlahos, L.: 1980, Astrophys. J. 242, p. 359.
Smith, D.F., and Lilliequist, C.G.: 1979, Astrophys. J. 232, p. 582.
Smith, D.F., and Auer, L.H.: 1980, Astrophys. J. 238, p. 1126.
Smith, D.F., and Brown, J.C.: 1980, Astrophys. J. 242, p. 799.
Smith, D.F., and Harmony, D.W.: 1982, Astrophys. J. 252, p. 800.
Spicer, D.S.: 1977, Solar Phys. 53, p. 305.
Tanaka, M., and Papadopoulos, K.: 1983, Phys. Fluids 26, p. 1697.
Vlahos, L., and Papadopoulos, K.: 1979, Astrophys. J. 233, p. 717.

NON-STOCHASTIC ACCELERATION OF PROTONS IN THE MAGNETIC NEUTRAL SHEET

Jun-ichi Sakai, Dept. of Applied Mathematics and Physics,
Faculty of Engineering, Toyama University, Toyama 933 Japan and
Astronomy Program, University of Maryland, College Pk. MD. USA

Ryo Sugihara, Institute of Plasma Physics, Nagoya University
Nagoya 464 Japan

A rapid non-stochastic proton acceleration mechanism by electrostatic waves during the substorm activity in the magnetospheric tail is presented to explain the origin of energetic protons (up to MeV). The protons are accelerated normal to the neutral sheet. Near a reconnection point, however, the protons are also accelerated along the sheet by a second process.

During a substorm activity, high energy protons (up to MeV) are observed in the distant tail[1]. Zeleny et al.[2] considered the proton acceleration in terms of an inductive electric field near X-point of the magnetic field during the nonlinear tearing mode instability in the tail current sheet. Gurnett et al.[3] observed electrostatic waves propagating almost perpendicular to the local magnetic field. Cattell and Mozer[4] also report that instantaneous field magnitudes rise up to 30mV/m when a substorm is active.

In this paper we present a very rapid proton acceleration mechanism by an electrostatic wave with a frequency near the lower hybrid resonance one during the reconnection phase near the tail neutral sheet. This type of charged particle acceleration in a homogeneous plasma has been predicted by Sagdeev and Shapiro[5], and Sugihara and Midzuno[6]. Dawson et al.[7] proved the possibility by a computer simulation and Yoshizumi[8] et al. revealed the process experimentally.

We examine this acceleration mechanism in the neutral sheet by solving the equation of motion for a proton. The magnetic field configurations we here consider are schematically drawn in Fig. 1, where in (a) a normal magnetic field component $B_n \vec{e}_x$ is zero and in (b) there exists a weak normal component. The configuration of the field in Fig. 1 (b) is more alike the magnetospheric tail near the X-point and the field is given by

$$\vec{B} = B_n \vec{e}_x + B_\infty \tanh(x/a) \vec{e}_y \tag{1}$$

where a is the thickness of the current sheet and B_n is constant.

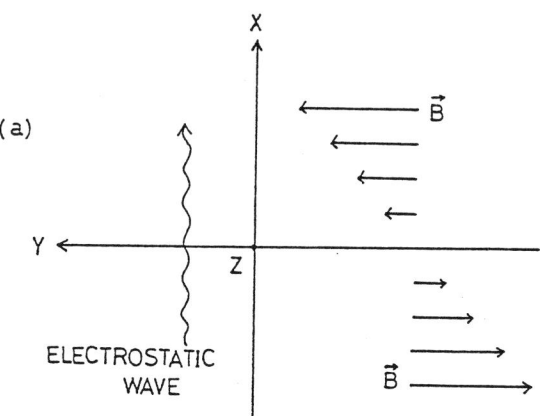

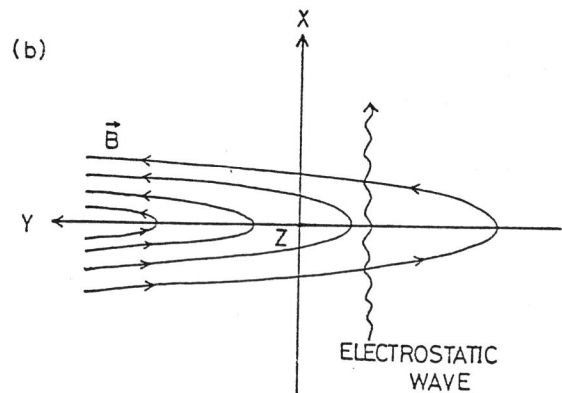

Fig. 1 Sketch of (a) $B_n = 0$ and (b) $B_n = 0$ magnetic field in the tail. The electrostatic wave propagates in the χ direction.

We assume that the driving wave is a lower hybrid one, has a form of $E_x \sin(kX - \omega t)$ and propagates across the current sheet in the positive X direction.

The equations of motion for a proton will be rewritten in the wave frame moving with the phase velocity. The non-dimensional forms are

$$\ddot{\xi} = \varepsilon \sin\xi - K\dot{Z} \tanh[(\xi + WT)/K] , \qquad (2)$$

$$\ddot{Y} = \varepsilon_n \dot{Z} , \qquad (3)$$

$$\ddot{Z} = K^{-1}(W + \dot{\xi}) \tanh[(\xi + WT)/K] - \varepsilon_n \dot{Y} , \qquad (4)$$

where lengths, the time T and velocities are scaled with a, ω_c^{-1}, and the proton thermal velocity V_t, respectively, and $K = ka$, $W = \omega/\omega_c$, $\varepsilon = KV_{max}/V_t$, $V_{max} = cE_x/B_\infty$ and $\varepsilon_n = B_n/B_\infty$.

At first we examine the case of $\varepsilon_n = 0$. Suppose a proton initially trapped in a potential well of the wave and choose $\xi = 0$ at $T = 0$. From eq. (4), the proton is accelerated as it feels $\vec{V}_{ph} \times \vec{B}(dc)$ electric field and the proton velocity in the X irection increases linearly with time. As time elapses the second term in the right hand side of eq. (2) becomes large and finally overcomes the electrostatic term, and then the proton detraps from the wave potential well. In Fig. 2 a typical example is shown for a proton which initial coordinates are $X = Z = 0$ and $V_z(0) = \dot{\xi}(0) = 0$. The parameters chosen are $\varepsilon = 150$, $K = 10$, $W = 10$, $V_{ph} = 1$.

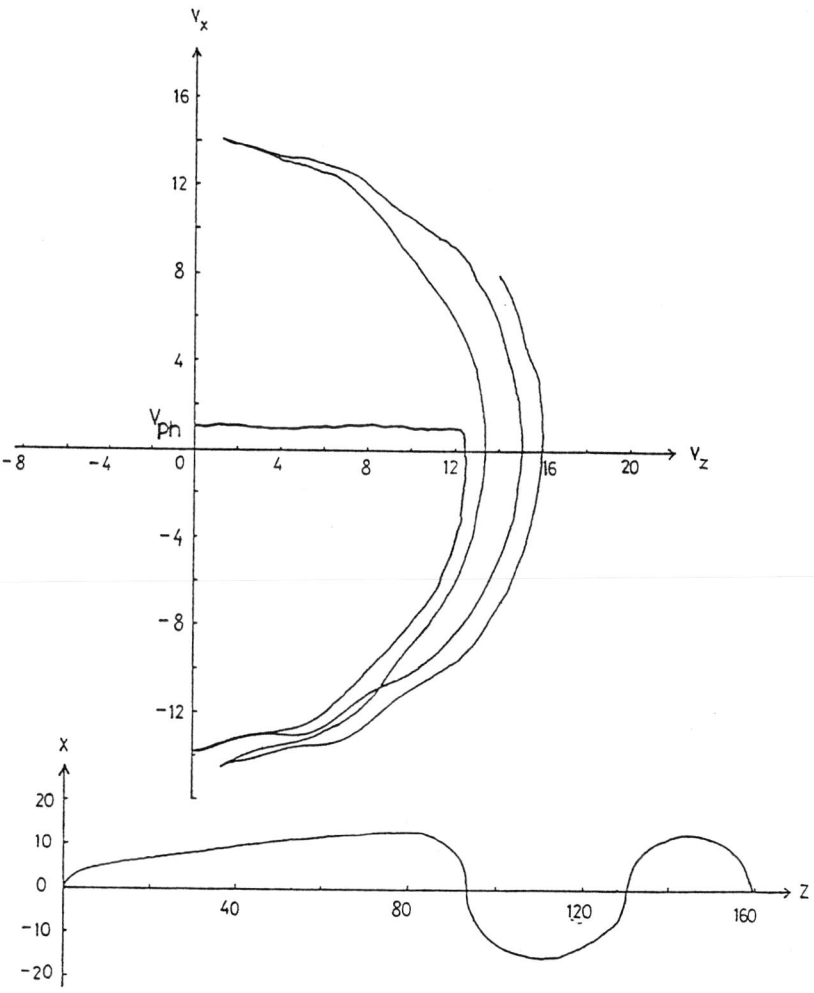

Fig. 2 Velocity-space orbit (upper) of a proton and orbit in the neutral sheet ($B_n = 0$). $V_{ph} = 1$, $V_{max} = 15$, $K = 10$, $W = 10$. The proton is initially trapped and $V_z(0) = 0$. After detrapping from the potential well, the proton moves crossing the neutral sheet.

The V_z at the point of detrapping is 12.4 which is a little bit smaller than $\tilde{V}_{max} = 15$ in the scale of V_t as shown in Fig. 2. Subsequently the acceleration time is a little shorter than the ideal τ_a. After detrapping, the proton moves crossing the neutral sheet or does the so-called meandering motion as shown in the lower portion of Fig. 2 and may be accelerated in the Z direction stochastically.

Next we consider the case where a weak normal magnetic field exists in the X direction as shown in Fig. 1(b). When ε_n is small, the linear acceleration in the Z direction can still occur as seen from eq. (4). Following this acceleration the particle is accelerated in the positive Y direction, which is seen from eq. (3). A theory[9] which treats the acceleration of a charged particle by a wave propagating obliquely to a homogeneous magnetic field predicts that $(V_z)_{max} = (\omega_c/\omega_{cn})V_{ph}$ and $(V_y)_{max} = 2(\omega_c/\omega_{cn})V_{ph}$ when the praticle does not detrap or $(V_z)_{max} <$ CE/B. This theory may be applicable to the present case because computed V_{max} and τ_a are very close to cE_x/B_∞ and $cE_x/(B_\infty \omega_c V_{ph})$. After the theory we expect that $(V_z)_{max} = 10$ and $(V_y)_{max} = 20$, which exactly agree with the computed results[10]. Then we anticipate that the maximum V_y is $2 \cdot V_{max} = 2 CE/B_\infty$ which is twice larger than that in the case of perpendicular propagation of the wave.

During the substorm activity, it is known that the plasma sheet becomes thinner and thinner to the order of proton Larmor radius[11]. This implies that the ion counter streams across the tail magnetic field is present. Now we assume the presence of such streams. If the relative velocity U of the streams exceeds V_t, the modified two stream instability (MTSI) can be excited. The MTSI may generate electrostatic waves propagating across the magnetic field with frequencies near the lower hybrid resonance one ω_{LH} and with rapid growth rates, $\gamma_{max} \simeq 0.1 \omega_{LH}$. A computer simulation[12] shows that the saturation of the electrostatic waves is estimated by the ion trapping in the wave potential. The saturated electrostatic field is given by

$$E_x = \frac{km}{4e}(U - V_{ph})^2, \quad (5)$$

where k is the wave number of the wave. In explicity estimating E_x, instead of k we use $k_{max} = 17\omega_c/V_t$ which gives the maximum growth rate of the MTSI[13], and $V_{ph} = U/2$ which corresponds to k_{max}. Now V_{max}/V_t is given by

$$V_{max}/V_t = CE_x/(V_t B_\infty) \simeq (U/V_t)^2 \quad (6)$$

We estimate quantities required when protons are accelerated up to 0.5 MeV by the above-mentioned process. Suppose that the thermal energy of the proton be 1 keV and that the perpendicular acceleration be only responsible for this acceleration. Hence V_{max}/V_t must be 22 or $U/V_t \simeq 5$ from eq. (6). The latter seems to be a reasonable value. The intensity E_x is required to be 100 mV/m for $B_\infty = 10\gamma$. The time τ_a of the acceleration is

$$\tau_a = V_{max}/(\omega_c V_{ph}) = 2V_{max}/(\omega_c U) = (2/\omega_c)(U/V_t)$$

which is about 10 second.

Thus using a plausible assumption of the presence of the ion counter streams across the neutral sheet we semi-quantitatively interpret the existence of the electrostatic wave which produces results consistent with observed data.

References

[1] Sarris, E.T., Krimingis, S.M. and Armstrong, T.P.: 1976, J. Geophys. Res. 81, 2341.
[2] Zeleny, L.M., Lipatov, A.S., Lominadze, D.G. and Taktakishuili, A.L.: 1982, Space Research Institute No. 697, Academy of Sciences, USSR.
[3] Gurnett, D.A., Frank, L.A. and Lepping, R.P.: 1976, J. Geophys. Res. 81, 6059.
[4] Cattell, C.A. and Mozer, F.S.: 1982, Geophys. Res. Lett. 9, 1041.
[5] Sagdeev, R.Z. and Shapiro, V.D.: 1973, Pis'ma Zh. Eksp. Teor. Fiz 17, 389 (JETP Lett. 17, 279, 1973).
[6] Sugihara, R. and Midzuno, Y.: 1979, J. Phys. Soc. Japan 47, 1290.
[7] Dawson, J.M., Decyk, V.K., Huff, R.W., Jechart, I., Katsouleas, T. Leboeuf, J.N., Lembege, B., Martinez, R.M., Ohsawa, Y. and Ratliff, S.T.: 1983, Phys. Rev. Lett. 50, 1455.
[8] Yoshizumi, M., Nishida, Y. and Sugihara, R. (to be published).
[9] Sugihara, R. and Sakai, J.I. (to be published).
[10] Sakai, J.I. and Sugihara, R.: 1983, I.P.P.J., Institute of Plasma Phys. Nagoya University, Japan, June.
[11] Nishida, A. and Fujii, K.: 1976, Planet, Space Sci. 24, 849.
[12] McBride, J.B., Ott, E., Boris, J.P. and Orens, J.H.: 1972, Phys. Fluids 15, 2367.
[13] Tanaka, M. and Papadoupoulos, K.: 1983, Phys. Fluids 26, 1697.

DISCUSSION

R. Smith: You showed single-particle trajectories in which the particle made many bounces before becoming untrapped. But if a wave saturates by ion trapping, there will be many trapped particles and sidebands will grow, detrapping the particle within just a few bounce periods. Thus there will be much less acceleration.

Sakai: I think that the growing time of sidebands may be very slow compared with the typical acceleration time which is of the order of the proton gyro-period.

Huba: The ion-ion modified two stream instabilities are stable in high β systems such as the neutral sheet. They cannot provide electrostatic turbulence in the neutral sheet. It is very difficult for electrostatic turbulence to exist in the magnetotail neutral sheet - the turbulence is probably electromagnetic.

Sakai: The observation by Cattell and Mozer (1982) reports that the instantaneous electrostatic field amplitudes rise up to 30 mV/m when a substorm is active. During the dynamical thinning phase of neutral sheets, the ion-ion counter stream instability may create electrostatic waves.

D. Smith: Jim Eastwood calculated electron trajectories in neutral sheets (Planet. Space Sci. '73, '74). Have you compared your results with his?

Sakai: I think that Eastwood did not calculate the trajectory of a trapped proton by electrostatic waves in the neutral sheet, which is essential in this acceleration mechanism.

BEAM-RETURN CURRENT SYSTEMS IN SOLAR FLARES

D.S. Spicer
Geophysical and Plasma Dynamics Branch
Plasma Physics Division
Naval Research Laboratory
Washington, D.C. 20375

R.N. Sudan
Laboratory for Plasma Studies
Ithaca, NY 14853

The importance of electron beams in solar flare dynamics is well known. In order to understand the dynamics produced by beams it is essential to have a clear understanding of the role beam driven return currents play and whether electrostatic or inductive electric fields maintain the return current. We show that inductive electric fields are responsible for driving return currents under solar conditions. The significant conclusions that follow from this result, as applicable to solar flares, are the following:

(a) If there is no source of anomalous resistivity, either due to a beam driven two stream instability or due to the return current becoming electrostatically unstable, our results will not alter previous results;

(b) however, if there is a source of anomalous resistivity, either due to a beam driven two stream instability or due to the return current becoming electrostatically unstable, our results modify significantly previous results;

(c) if (b) is true, we find

(1) a beam under solar conditions cannot exist for times greater than

$$\Delta t_{max} \lesssim t_D \ln\left[\frac{1}{1 - \frac{acB_z}{2I_B(0)}}\right],$$

where t_D is the resistive diffusion time, as determined by the level of resistivity and the beam radius a, B_z the axial guide field along which the beam propagates, and $I_B(0)$ is the beam current in statamps at t=0;

(2) if the beam radius is determined by the spatial scale of the acceleration region and the resistivity is anomalous we find $\Delta t_{max} \lesssim 0.5$ secs is a typical maximum beam pulse time;

(3) the plasma temperature that results from the return current being unstable is ~ 15 kev;

(4) a typical flare x-ray burst of ~ 10 secs at 15 kev must be made up of many beam pulses of ~ 0.5 secs.

(5) beam energy losses due to the larger induction electric field resulting from anomalous resistivity will exceed direct classical collisional beam energy losses; and

(6) anomalous return current Joule heating will be the dominant heating mechanism of the beam and a quasi-thermal x-ray source will result.

COLLISIONLESS EFFECTS ON BEAM-RETURN CURRENT SYSTEMS IN SOLAR FLARES

Loukas Vlahos and H.L. Rowland
Astronomy Program
University of Maryland

ABSTRACT

A large fraction of the electrons which are accelerated during the impulsive phase of solar flares stream towards the chromosphere and are unstable to the growth of plasma waves. The linear and non-linear evolution of plasma waves as a function of time is analyzed with the use of a set of rate equations that follow in time the non-linearly coupled system of plasma waves-ion fluctuations. The nonthermal tail formed during the stabilization of the precipitated electrons can stabilize the Anomalous Doppler Resonance instability and prevent the isotropization of the energetic electrons. The precipitating electrons modify the way the return current is carried by the background plasma. In particular, the return current is not carried by the bulk of the electrons but by a small number of high velocity electrons. For beam/plasma densities $\gtrsim 10^{-3}$, this can reduce the effects of collisions and heating by the return current. For higher density beams where the return current could be unstable to current driven instabilities, the effects of strong turbulence anomalous resistivity is shown to prevent the appearance of such instabilities. Our main conclusion is that the beam-return current system is interconnected and how the return current is carried is determined by the beam generated strong turbulence.

1. Introduction

The intensity of non-thermal electrons, with energies between 10-200 keV, which is necessary to explain the observed x-ray emission in these energies is relatively high e.g. $F_b \simeq 10^{36}$ electrons/s (Hoyng et al. 1976). Assuming an area for the emitting source $\simeq 10^{18}$ cm^2 and an average speed for the precipitating electron $v_b \simeq 10^{10}$ cm/sec one concludes that the density of the precipitating electrons is $n_b \simeq 10^8$ cm^{-3}. The plasma density in the low corona lies between $10^9 < n_o < 10^{11}$ cm^{-3} which leads us to conclude that $10^{-3} < n_b/n_o < 10^{-1}$. The consequences of such a large flux of non-thermal electrons streaming towards the chromosphere have up to now been discussed as two separate problems. The beam stabilization was addressed by Lifshitz and Tomozov (1974) and Vlahos and Papadopoulos (1979). The role of the return current on the other hand was addressed assuming that the beam was stable and the velocity of the "bulk" return current was estimated from the relation $v_r = (n_b/n_o) v_b$ (see Hoyng et al. 1978, Brown and Hayward (1982) and references therein). Vlahos and Rowland (1983) and Rowland and Vlahos (1983) suggest that the presence of a linearly unstable beam effects the way the return current is carried and the beam/return current must be viewed as one unified system.

2. Beam stabilization and strong turbulence effects

The dispersive characteristics of the plasma are modified dramatically when the beam generated waves W_r exceed a threshold value $(W_{th}/n_o T_e) \gtrsim (k_o \lambda_D)^2$, where T_e is the ambient temperature, k_o is the wave number of the beam driven waves ($k_o \sim \omega_e/v_b$), ω_e is the plasma frequency and λ_{De} is the electron Debye length). It is easy to show that for $n_b/n_o \gtrsim 10^{-5}$ the quasilinear saturation level of beam driven waves is above the threshold for strong turbulence. In other words, the beam generated waves will start forming solitons and reduce the wavelength of the beam driven plasma waves once $W_r > W_{th}$. The reduction of the wavelength of the unstable waves has several important consequences: (a) the waves that are non-resonant with the beam (W_{nr}) have small phase velocity ($\omega_e/k_{nr} \simeq (2-3)v_e$) and are damped in the tail of the thermal distribution. As a result of the interaction of the non-resonant, low phase velocity waves with the tail of the thermal distribution low energy non-thermal tails with energies around $5 \sim 10$ keV will be formed. (b) The formation of solitons is coupled with the ions and forms ion cavities that are strongly coupled with the soliton pulses. Cavitons play an important stabilizing role in the newly arriving electron beams. A detailed analytical and numerical discussion of these processes has now been published and we refer the reader to the original papers and the references therein for further study (see Papadopoulos 1975 and Rowland 1980). The problem of beam-plasma interactions described above differs from the related work in type III bursts in two important ways. (1) the plasma is strongly magnetized ($\omega_p/\Omega_e \lesssim 1$, where Ω_e is the gyrofrequency) and (2) the beam density is much stronger ($n_b/n_o \gtrsim 10^{-4} - 10^{-1}$. Both factors suggest that the system is one-dimensional and solitons collapsed from two-dimensional evolution will not be important (see Rowland, Lyon and Papadopoulos 1981). A system of rate equations for the regime that is described above is given below (see Vlahos and Rowland for detail discussion).

$$\frac{dW_r}{dt} = \gamma_L W_r - \gamma_{NL} W_{nr} \theta(W_r - W_{th}) - a_{NL} W_r \qquad (1)$$

$$\frac{dW_{nr}}{dt} = \gamma_{NL} W_{nr} - \gamma_d W_{nr} - a_{NL} W_{nr} \theta(W_s - W_2) \qquad (2)$$

$$\frac{dW_s}{dt} = \gamma_{NL} W_{nr} - \nu_s W_s \theta(W_s - W_2) \qquad (3)$$

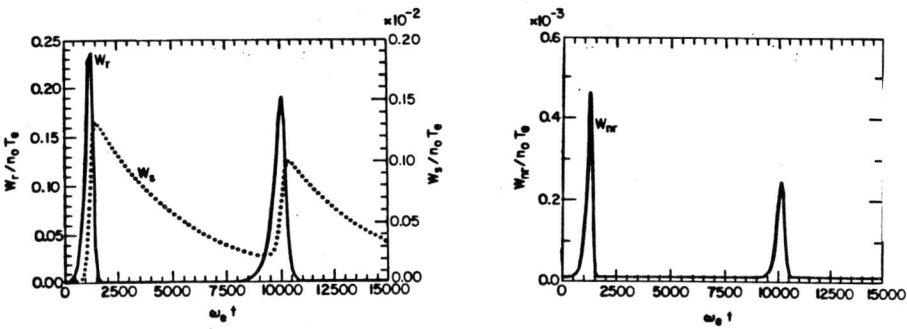

Fig. 1. Solution of the rate eq. (1)-(3)

where W_s is the ion wave energy, γ_L is the linear growth rate, γ_{NL} is the rate that the resonant waves are transferred to low phase velocities, γ_d is the Landau damping of the non-resonant waves in the tail of the thermal distribution, $\theta(x)$ is the step function, a_{NL} is the scattering of the high frequency waves from the ion density fluctuations ν_s is the damping of the ion density fluctuations. Solving the system eqs. (1)-(3) we found that the system of waves is pulsating periodically as it is shown in Fig. 1. Rowland (1980) demonstrated this behavior in fully kinetic, Vlasov simulations.

The main conclusion from our study of the beam-plasma interaction can be summaried as follows (Vlahos and Rowland, 1983):
 a. The beam losess a few percent of its initial energy to maintain the pulsations shown in Fig. 1. The beam losses are minimized because the ion acoustic waves stabilize the beam between pulses.
 b. The energy lost from the beam does not heat the bulk of the electrons but forms 5-10 keV non-thermal tails.
 c. If $n_b/n_o > 10^{-2}$, strong turbulence does not prevent the initial quasilinear relaxation. However, since $W_r \gg W_{th}$, strong turbulence effects still appear (non-thermal tails, solitons, cavitons, strong low frequency turbulence etc.). In particular, the high levels of low frequency turbulence will again decouple the beam from the plasma and later arriving beam electrons will lose little energy in the volume.
 d. By increasing the Landau damping, the non-thermal tails can prevent the pitch angle of the beam particles by the anomalous Doppler resonance. Similarly the enhanced low frequency turbulence by increasing the scattering of the high frequency waves to lower phase velocities will increase the damping and quench this instability. Thus the particles that propagate along the field lines will suffer little cross field scattering and subsequent trapping on the coronal part of the loop.

3. Collisionless effects on the return current

The presence of beam excited solitons plays an important role on the return current. Low velocity electrons in the bulk of the distribution are trapped between solitons (Rowland et al. 1981). Only particles with velocity $\gtrsim$ (2-3) v_e can freely escape. A result of the trapping of the bulk of the ambient distribution is that the return current will be carried by fewer and thus more energetic electrons that are not trapped between solitons. The exact number of the electrons that will carry the return current and their final velocity depends upon the strength of the precipitating beam and the rise time of the beam acceleration process. We found (Rowland and Vlahos 1983) that

a. For weak beams $(n_b/n_o) < 10^{-3}$ and a slow rise time (a few secs) the return current is carried by particles that suffer several collisions before escaping from the loop. In this case the return current collisional heating estimated by several authors (see Brown and Hayward 1982) is modified in two important ways; (1) only a few percent of the ambient electrons carry the return current and (2) their average velocity is $\simeq$ (2-3) v_e.

b. For strong beams $n_b/n_o \gtrsim 10^{-3}$ and a fast rise time the return current is carried by a few collisionless electrons that obtain relatively high speed (4-6) v_e before escaping from the loop.

c. Because the return current is carried by superthermal electrons and the bulk of the plasma remains stationary, current driven instabilities (ion cyclotron, ion acoustic, etc.) will not appear even with very strong beams $(n_b/n_o > 10^{-2})$.

In summary we conclude that the beam-return current is a highly interactive system that must be discussed in a unified way. A first attempt in this direction is made recently by Vlahos and Rowland (1983) and Rowland and Vlahos (1983).

Acknowledgement

This work was supported by NASA grant NAG W-81 and NSF ATM-81-17093.

References

Brown, J. and Hayward, J.: 1982, Solar Phys., 80, 129.
Hoyng, P., Brown, J.C. and van Beek, T.F.: 1976, Solar Phys., 48, 197.
Hoyng, P., Knight, K.W. and Spicer, D.S.: 1978, Solar Phys., 58, 139.
Lifshitz, M.A. and Tomozov, V.N.: 1974, Astr. Zh., 51, 560.
Papadopoulos, K.: 1975, Phys. Fluids, 18, 1769.
Rowland, H.L.: 1980, Phys. Fluids, 23, 508.
Rowland, H.L., Lyon, J.G. and Papadopoulos, K.: 1981, Phys. Rev. Lett., 46, 346.
Rowland, H.L., Palmadesso, P.J. and Papadopoulos, K.: 1981, Phys. Fluids, 24, 832.
Rowland, H.L. and Vlahos, L.: 1983, Astr. Ap., (to be submitted).
Vlahos, L. and Papadopoulos, K.: 1979, Ap.J., 223, 717.
Vlahos, L. and Rowland, H.L.: 1983, Astr. Ap., (to be submitted).

DISCUSSION

Ionson: In what way do you and Spicer disagree?

Vlahos: The analysis presented by Spicer was based on two assumptions (a) the precipitating electrons were confined in a flux tube with 10 km diameter and (b) the beam (which was assumed stable) drives inductively a return current which is unstable. Using weak turbulence theory he estimated the anomalous resistivity and the decay of the return current. As I described, above the beam goes unstable first and the presence of solitons inhibits the bulk of the electrons from carrying the current; thus we concluded that the return current driven instabilities are out of the question. The assumption that weak-turbulence theory is valid is highly questionable because of the presence of strong parallel magnetic fields. Last, but not least, I would like to point out that if the area that carries the beam is $(10 \text{ km})^2$, the beam density has to be $n_b \sim 10^{12} \text{ cm}^{-3}$ to explain the x-ray observations. Furthermore, if one assumes that the loop is filled with small tubes, still the problem is not solved, since you have to explain the interaction of the tubes, the density of beam electrons in each tube, etc. I honestly do not see why one has to assume such a filamentation of the precipitating beams. I know of no observation that supports such a claim and we know very little about the acceleration of these electrons to argue that such a filamentation really happens.

Spicer: I stated that the beam radius was determined by the acceleration region, not by the X-ray observations.

Also, how can you claim the inductive electric field associated with the decay of the return current only causes minimal beam losses if Rowland's code doesn't have Faraday's Equation in it?

Vlahos: The role of the beam driven inductive electric field was used in our analysis (see Rowland and Vlahos) to estimate the number of particles that are accelerated in the tail to neutralize the current carried by the precipitating electrons. From the simulations one can determine the strength of the inductive electric field and the amount of energy required to drive the needed return current. From this, one can find the beam energy loss.

Guillory: Similar relaxation-oscillation behavior, and effects due to cavitons and concomitant field spikes, were discussed in an astrophysical context (jets in galactic nuclei) by Rose et al. at the recent AAS meeting and are contained in a recently sumitted paper (Ap.J.).

D. Smith: When you combine this with the decay of return current a la Lee and Sudan, what will happen?

Vlahos: The decay of the current will take as much as 10^{12} secs. The reason being that we use a radius for the current carrying flux tube $R_L \sim 10^8 - 10^9$ cm and that the bulk of the plasma is stable. The return current is carried from the tail and suffers only Coulomb collisions, thus the Spitzer conductivity is a good approximation when the collison rate is calculated using only the small number of high velocity electrons. Taking these two factors into account you can show that the decay time for the return current will be $\tau \simeq \tau_c (R_L^2/\lambda_e^2) \simeq 10^{12}$ sec! (where τ_c the collision time, $\lambda_e = c/\omega_e$, c is the speed of light and ω_e is the plasma frequency).

SESSION IX

SUMMARY OF CONFERENCE

Vytenis M. Vasyliunas
Max-Planck-Institut für Aeronomie
Katlenburg-Lindau, Federal Republic of Germany

For a meeting of people from such widely different fields, this Symposium has exhibited a remarkable degree of unity. There has been one key concept running as a thread throughout the Symposium: the concept of magnetic field line reconnection, or magnetic field line merging as I prefer to call it. It was dealt with directly in many papers, and many others dealt indirectly with it and various related aspects. The concept was applied in the Symposium to an amazing variety of objects and was examined from many points of view and by many different techniques. Magnetic field line reconnection or merging is a paradoxical concept. It clearly depends upon magnetohydrodynamics (MHD); for example, constraints imposed by the MHD relation between the magnetic field and the plasma flow are essential to set it up - without these constraints (if, for example, the electric field parallel to the magnetic field could assume any desired value) the problems we discuss under the heading of magnetic reconnection would merely be moderately complicated problems of magnetostatics. At the same time, departures from ideal MHD are also an essential and unavoidable part of the concept.

It is thus appropriate, before dealing with magnetic field line merging itself, to discuss its prerequisite, the MHD coupling between the magnetic field and the plasma flow. An important example of this coupling is provided by what is sometimes called a magnetic flux rope. It is a system where a magnetic flux tube, typically with the ratio of plasma pressure to magnetic pressure $\beta \gg 1$ at both extremes, is subject to a twisting or shearing flow at one extreme and a different twisting or shearing flow at the other extreme, so that the magnetic field between the two extremes of the flux tube becomes twisted or sheared; in many cases, $\beta \ll 1$ within the in-between region and the magnetic field assumes a nearly force-free configuration. Many examples were discussed at this Symposium, ranging from flux ropes at Venus, directly observed in situ (Russell, Elphic), to coronal loops and similar structures on the sun, remotely observed (discussed by Drake, Alay, Ray, Wu, and by others in connection with coronal heating), to speculative flux ropes associated with extragalactic jets (Eilek). More generally, we may regard a situation where different plasma motions are imposed at the two ends of a magnetic flux tube and the question is what happens in between as the prototypical problem of cosmic electrodynamics. As examples, in addition to the flux ropes already mentioned, on the sun one has motions imposed by the photosphere at the two ends; in the magnetospheres of planets the motion is imposed by the solar wind at one end and by the planet's ionosphere at the other; in binary star systems the motion is imposed by the rotation of one star and by the rotation or the orbital motion of the other; in accretion disks the motion may be imposed by the disk at one end and by the rotation of the accreting object at the other, or else, analogously to the sun, by turbulent motions within the disk acting on flux tubes anchored therein. Thus we have a very general problem, and the key point, in my opinion, is that in order to construct magnetic field models one must really understand the imposed motions, what it is that is twisting up the field. In the cases of the solar wind and (to some extent at least) the sun, we can actually observe the motions; in other cases we are for the most part limited to speculation. A piece of advice particularly for the galactic jet theorists: look for the plasma flows if the twisted magnetic field model is to be placed on a firmer footing.

The electrodynamical description of this problem is identical with what is commonly called, in discussions of planetary magnetospheres, the theory of magnetosphere-ionosphere coupling. That name was mentioned during the Symposium only in connection with limited and localized aspect related to the earth's polar aurora, but in fact it is a very general theory of the electrodynamic interaction between a high-β region (the ionosphere) and the region threaded by magnetic field lines from it (the magnetosphere). The theory constitutes one of the most highly developed and extensively tested facets of magnetospheric physics (see, e.g. reviews by Vasyliunas, 1980; Wolf, 1975, 1983; Boström, 1974), and astrophysicists should take note that it has advanced well beyond the stage of simple circuit analogs.

Now the general effect of such twisting or other imposed motions is to store energy into the magnetic field, which is then available to be released if a suitable mechanism for the release can be found; also, the associated changes of the magnetic field configuration may set the stage for the magnetic field line reconnection or merging process. These concepts were discussed extensively, particularly for the case of the sun and the solar corona. The gradual release of the energy in the twisted magnetic field is a candidate for the heating of the corona, the so-called D.C. heating discussed here by Heyvaerts and by Ionson. It is an alternative to the heating of the corona by the interaction and dissipation of hydromagnetic waves, also widely discussed here, both specifically for the sun (Heyvaerts, Van Hoven, Nocera) and in a more general context (Hasegawa). It is not always clear where exactly the dividing line lies between the D.C. heating, from twisting of the field by turbulent motions, and the wave heating, since the effects of imposed motions must propagate out as waves; in practice, though, there do seem to be two distinct types of theories. The old theory of coronal heating by acoustic waves is considered to be dead; there does not seem to be a sufficient energy flux in acoustic waves to be of any importance.

Having twisted the magnetic field lines and set up the nearly antiparallel fields or whatever the required configuration might be, we now come to the discussion of magnetic field line reconnection or merging. At the Symposium there were two general reviews of magnetic merging, one slanted toward the terrestrial magnetosphere (Sonnerup) and the other toward the sun (Priest). The reported developments represent a significant step beyond what are sometimes called the "classic" models of reconnection, developed in the period from the late '50's to the early '70's (see e.g., review by Vasyliunas, 1975). The name has absolutely nothing to do with the distinction between classical and anomalous transport properties; it is used for the irrelevant reason that these models are mostly contained in a few papers regarded as "classic" - and we should recall the definition given by Hines (1974): "a 'classic' paper is one that many researchers no longer read to see what it actually says, for they think they know what it must have said." These models were developed as simplified two-dimensional steady-state treatments, not because anyone thought that was a particularly good approximation, but because they were intended to address the simple and basic questions which were then current: can the merging process occur at all, at any reasonable rate? If it does occur, what does it look like? What signatures should we look for? That was the classic phase, a sort of existence proof: it showed that the process does exist and has certain well-defined attributes (antiparallel field components, plasma streaming out, and so on). We have now gone beyond this. The emphasis now is on three-dimensional and time-dependent effects and on global aspects.

Magnetic merging in the magnetosphere of the earth was rather little discussed at the Symposium - papers by Russell and by Sonnerup and a discussion of some possibly related magnetotail observations by Lui. (Workers in this area may be saving their papers and travel budgets for the Chapman Conference on Reconnection scheduled to take place two months after this Symposium.) Most of the discussion was concerned with reconnection on the dayside of the magnetosphere, where the geometry is complicated, the field line configuration is skewed and highly variable, and rather little is known as yet, compared with reconnection on the nightside of the magnetosphere. There were brief mentions of magnetic merging in the magnetospheres of Mercury and Jupiter (Russell, Aly). In the solar wind, magnetic merging within the interplanetary current sheets was discussed by Coppi as a possible mechanism for bending the current sheet up and down (such bending can explain the sector structure of the interplanetary magnetic field.)

Most of the discussion of magnetic merging at the Symposium was concerned with the case of the sun (Priest, Van Hoven, and others). A long-standing idea is that a sudden release of magnetic energy through the reconnection process is what produces the energy dissipation associated with a solar flare. Merging of magnetic fields on two different spatial scales was proposed by Sturrock as a way of accounting for two phases of a flare, a gradual phase (lasting some tens of minutes) and a superimposed short-lived impulsive phase. There is now some direct evidence, from observations of polarization, for solar magnetic field changes of the required character occurring during flares on short time scales (Kundu).

Both the rapid magnetic merging in the corona (flare-associated effects) and the gradual release of stored magnetic energy (DC coronal heating) were scaled up to apply to flare stars by Mullan, who reported fairly reasonable agreement between the scaled-up solar-based models and the observed properties of flare stars. More speculative applications of magnetic merging to other astrophysical objects included discussions of RS CVn binaries (Uchida), dwarf novae (Gilden), and accretion disks around black holes possibly related to quasars or galactic jets (Coroniti, Gilden).

Plasma experiments in the laboratory received considerable attention, with reviews by Stenzel and Bratenahl on experiments designed specifically to study the process of magnetic field line merging; in addition, Liu discussed various plasma phenomena in tokamaks and other fusion plasma devices that may be of interest in connection with, or analogous to what happens in, astrophysical plasmas (the devices themselves, of course, were not designed for studying the problems of astrophysical plasmas as such). Laboratory experiments do not have the extreme parameter range of space and astrophysical plasma systems (such as the huge magnetic mirror ratios or the enormous spatial scales) but they do allow controlled conditions and provide detailed in situ measurements with global coverage - in contrast to space observations on the one hand, where in situ measurements are possible but only locally, at isolated spots, and to solar and other astrophysical observations on the other, where only remote sensing is possible, with global but fairly coarse coverage and by indirect methods.

The results reported from laboratory investigations of magnetic merging emphasize, again, three-dimensional effects and time variations. Of particular interest is an observed time variation known as current interruption which may perhaps be similar or analogous to flares on the sun or substorms in the earth's magnetosphere. There was also much emphasis on global aspects and on the role of the external circuit. It is not entirely clear what would be the analog of the external circuit in space and astrophysical applications, where there is no external circuit as such. There does exist an analog to walls, as pointed out by Bratenahl: the photosphere of the sun or the ionosphere of a planet act in many respects, in relation to the overlying plasma, similarly to walls in a laboratory plasma experiment (thus walls may not necessarily be the unmitigated nuisance they are usually regarded as being).

Processes in tokamaks and other fusion devices that may have analogs in space and astrophysical plasmas include particle acceleration, excitation of plasma waves, and spatial diffusion of particles, possibly driven by similar mechanisms. With all the differences between the laboratory on one side and space and astrophysical plasmas on the other, there are enough similarities to suggest that a comparison of notes between people in the two areas might prove mutually illuminating.

Computer simulations concerned with magnetic merging and related problems were discussed by Birn, Tajima, Sato, and by others in connection with anomalous transport properties. Computer simulations allow, of course, very detailed control of inputs and very extensive diagnostics - you can pull out any number you want and see what is happening. On the other hand, they must deal (because of limited computer capabilities if for no other reason) with relatively simplified configurations, often though not always two-dimensional, and with limited time development - in most cases one cannot follow the system for very long periods, nor explore very many types of possible time variations. Nevertheless, the results are interesting and instructive. There are some discrepancies between the results of the various simulations, possibly (and in some cases almost certainly) attributable to the use of different boundary conditions. What is perhaps lacking to some extent is a careful discussion and understanding of the implications of different boundary conditions. For example, a "free boundary" condition may be assumed on some edge of the simulation region, but in cosmic plasmas there are no free boundaries; this boundary condition is equivalent to some assumption about the physics of the problem, and one would like to know what the assumption is. More generally, I am not sure if we really know yet what constitutes a well-posed problem in this context, what boundary conditions one is allowed to specify. The computer code will always return an answer, whether the problem is well-posed or not, but, to really know what the answer means, a proper discussion and understanding of boundary conditions is essential.

My general impression about the problem of magnetic field line merging, on the basis of both laboratory work and computer simulation, is one of continued solid progress, starting from and on the basis of the previous "classic" models (and not in opposition to them or in a completely different direction). Many ideas which were previously guessed at or derived only intuitively have not been confirmed, refined, or extended. The results, even when they support the intuition of the early pioneers, represent a significant advance beyond their achievements.

As I mentioned before, reconnection always involves essential aspects of departure from MHD as well, and these too were extensively discussed at the Symposium. One approach is to consider the non-MHD dissipative effects in a global and time-dependent framework; to keep the problem tractable, one is then forced to adopt a rather simple model for the non-MHD terms, usually a representation simply as an effective (so-called "anomalous") resistivity. The most important example of this approach was the discussion of the tearing mode instability and its development (Drake, Steinolfson). A similar approach in a somewhat different context, development and effects of large-scale instabilities in the ionosphere, was discussed by Keskinen. The alternative approach is to consider in great detail the anomalous plasma effects due to instabilities (reviewed by Huba, Dum); then one usually has to treat a simplified geometry on a local scale, an important special case being the so-called diffusion region around the magnetic neutral line associated with magnetic field line merging. The classical approach is to first identify the instability, then compute its linear growth, and then the non-linear (or, in some cases, quasi-linear) development to obtain an effective collision frequency that determines the transport coefficients. But Dum emphasized that the system may be so complicated and the final effect (on, e.g., the distribution functions) may represent such a profound modification of the initial conditions that it may be necessary to treat all the steps of the calculation essentially simultaneously rather than in sequence; there is little point in worrying much about the instability of the initial distribution if that distribution is going to be changed into something completely different.

I have the impression that in the solar and astrophysical community there exists a general view that enhanced resistivity is necessary for magnetic reconnection. Most theorists at the Symposium who tried to make solar flares looked for ways of producing an anomalous resistivity; the one exception was Van Hoven who pointed out that the classical Coulomb-collision resistivity could be greatly enhanced, by cooling the plasma through a radiative instability, to the point where it was no longer as negligibly small as usually assumed. In the terrestrial magnetospheric physics community, on the other hand, there seems to be more reliance on inertial and/or finite gyroradius effects as a way of producing departures from ideal MHD; they were mentioned, in the context of magnetic merging, by Sonnerup, and Hasegawa presented a theory of finite gyroradius effects on MHD waves.

Electrostatic double layers constitute another example of non-MHD effects. They appear in some laboratory plasma experiments; Stenzel described a particularly dramatic development of a double layer associated with current interruption. The general theory of double layers was reviewed by R.L. Smith. It is widely thought that double layers exist in the earth's magnetosphere and are responsible for the acceleration of auroral electrons, although the direct evidence for strong double layers (as distinct from general evidence for electric fields parallel to the magnetic field) is perhaps not as complete as commonly assumed. Smith pointed out that existing computer simulations and laboratory experiments on double layers refer to parameter ranges that are very different from what is found in space and astrophysical plasmas, so that any extrapolation should be viewed with great caution. Having sounded this warning, Smith proceeded to make a far-reaching extrapolation himself. In laboratory experiments one typically finds one strong double layer, but there is now one experiment where, when the length of the system was increased, two weaker double layers appeared instead; extrapolating to the very large astrophysical scales, Smith suggested that there one may have very many double layers, each very weak. Now there are two extreme approaches to the non-MHD phenomenon of electric fields parallel to the magnetic field: one is the double layer approach, where the parallel fields are concentrated over a very narrow region, forming essentially a discontinuity, and the other is the anomalous resistivity approach, where they are distributed over a wide spatial scale. Evidently, Smith's extrapolation might point to a way of bridging the gap between the two approaches.

Plasma turbulence, whether hydromagnetic in character or associated with non-MHD effects, leads to wave-particle interactions, which in turn may lead to particle acceleration, a topic that was discussed at the Symposium in very many contexts: acceleration of ions in tokamaks and other laboratory devices (Liu), acceleration of ions above the auroral ionosphere (Lotko, André), acceleration of ions in the geomagnetic tail (Sakai). Acceleration of electrons in solar flares was discussed in a number of papers (Holman, Krishan, Tanaka, D.F. Smith), some of which were aimed at explaining observations of very short X-ray bursts; an important related problem, the return current to the electron beam, was discussed by Spicer and by Vlahos from two different points of view that seem to be roughly analogous to the previously discussed anomalous-resistivity and double-layer approaches to non-MHD effects - broadly distributed vs. localized. Going to a larger scale, there were papers on cosmic rays in the galaxy and their interaction with Alfvén waves, to account for their isotropy or otherwise, their escape from the galaxy, and possible effects of galactic winds (Wentzel, Spangler, Nelson), and, on an even larger scale, acceleration of electrons in extragalactic jets, to account for their synchroton emission and in particular for bright knots and other localized emission regions (Henriksen). A somewhat related topic was the hydromagnetic treatment of a very energetic electron-positron plasma assumed present in the Crab nebula (Kennel); the model, where such a plasma flows out as a wind from the Crab pulsar and interacts with the swept-up interstellar medium, forming two shocks (one the outward-propagating blast wave, the other propagating inward into the plasma), is able to account for some of the luminosity and other features of the Crab nebula.

The general impression on both the topics of turbulence and acceleration and of non-MHD effects is one of enormous complexity: there are very many possibilities in the theory, very many instabilities of various kinds, but at the same time there are very many different phenomena in nature. The real problem is to match the appropriate one of the many theoretical possibilities to the appropriate one of the many phenomena observed in various systems in nature and, equally important, to recognize when one has made a proper matching, given the fact that the observations are limited and the theories are simplified. There appears to be some tendency for people working in these very complex areas to split off from the rest of the community. It is important that this should not happen, that we should continue talking to each other and not withdraw each into one's own specialist shell; researchers on instabilities and turbulence should make an effort to remain intelligible to the others, and conversely, the others should make an effort to understand what is being done on instabilities and turbulence.

Finally, there is the topic of astrophysical jets, mostly extragalactic (or perhaps one should say supergalactic). General reviews were presented by Norman and by Ferrari. The practically universal assumption is that jets are produced by outflow or ejection of matter from the central galaxy. I am an outsider in this field, hearing the evidence for this assumption essentially for the first time, and my first impression is that the evidence, although certainly as good as anything on the average in astrophysics, is not as compelling as, say, that for the binary nature of X-ray pulsars; sometimes I fancy that perhaps one has adopted a fairly implausible model because anything else one can think of is even more implausible.

Various models for jets were discussed: pure hydrodynamic flow models, where the problem is to account for the observed very narrow collimation and to avoid instabilities (Eichler, Ferrari); magnetically confined models, with questions of magnetic field structure, whether flux ropes or something else (Eilek, Hardee, Uchida); turbulent acceleration models (Henriksen); nozzle models, with solar-wind-like acceleration to supersonic flow (Tsinganos). The main problem is that, as was said by Ferrari, there are too many theories that all seem to fit the data. In this respect we are at the opposite extreme from the situation in magnetospheric and solar physics, where one sometimes has the feeling that none of the theories fit the data (and therefore, presumably, there are as yet too few theories - a theory that fits the data exists, one hopes, but has not yet been found). And so this Symposium may be viewed as an encounter between people from both ends of the spectrum of the parameter (ratio of theories to observations); the ratio is very high at one end (extragalactic jets) and very low at the other (the magnetosphere, the sun), and we may hope that the encounter will help both of us to approach our common ideal - each set of observations explained by one and only one theory.

REFERENCES

Boström, R.: 1974, in B.M. McCormac (ed.), *Magnetospheric Physics*, D. Reidel, Dordrecht, pp. 45-59.
Hines, C.O.: 1974, *The Upper Atmosphere in Motion*, American Geophysical Union, Washington, D.C., p. 933.
Vasyliunas, V.M.: 1975, Rev. Geophys. Space Phys. 13, pp. 303-336.
Vasyliunas, V.M.: 1980, in C.S. Deehr, J.A. Holtet (eds.), *Exploration of the Polar Upper Atmosphere*, D. Reidel, Dordrecht, pp. 229-244.
Wolf, R.A.: 1975, Space Science Rev. 17, pp. 537-562.
Wolf, R.A.: 1983, in R.L. Carovillano, J.M. Forbes (eds.), *Solar-Terrestrial Physics*, D. Reidel, Dordrecht, pp. 303-368.

PERSPECTIVES ON SPACE AND ASTROPHYSICAL PLASMA PHYSICS

C. F. Kennel
University of California, Los Angeles, California
J. Arons
University of California, Berkeley, California
R. Blandford
California Institute of Technology, Pasadena, California
F. Coroniti
University of California, Los Angeles, California
M. Israel
Washington University, St. Louis, Missouri
L. Lanzerotti
Bell Laboratories, Murray Hill, New Jersey
A. Lightman
Harvard University, Cambridge, Massachusetts
K. Papadopoulos
University of Maryland, College Park, Maryland
R. Rosner
Harvard University, Cambridge, Massachusetts
F. Scarf
TRW Systems, Redondo Beach, California

ABSTRACT

We summarize the discussion of the current status and future prospects of space and astrophysical plasma research prepared by the Panel on Space and Astrophysical plasmas, a part of the study on Physics administered by the National Research Council of the National Academy of Sciences. The Study on Physics is chaired by W. Brinkman of Bell Laboratories and will be completed in 1984.

1. INTRODUCTION

Developments in understanding plasmas in the laboratory, in space, and in astrophysics have gone hand in hand throughout the 20th century. In the 1920's plasma oscillations were discovered in the laboratory and and radio waves were first reflected from the plasma in the earth's ionosphere -the very edge of space. From 1930-1960, the foundations of plasma physics were created as a by-product of ionospheric, solar-terrestrial, and astrophysical research, motivated by such diverse concerns as understanding how radio waves propagate in the ionosphere, how solar activity creates magnetic storms and auroral displays at earth.

By the 1950's, it was clear that fully ionized plasmas at high temperatures would be collision free -an essential property that forced us to focus on the processes that are fundamental to plasmas, as distinct from ordinary gases, and thereby to create the modern discipline of plasma physics.

Modern plasma physics began in the years 1957-1960. Two events symbolizing the deeper intellectual currents of those years were the first successful launch of an artificial earth satellite by the Soviet Union and the revelation, through declassification, that both the United States and the Soviet Union had been trying to harness the energy source of the sun -thermonuclear fusion -for peaceful purposes. Then as now, the obstacles to achieving controlled fusion lay, not in our ignorance of nuclear physics, but of plasma physics. In 1958, the terrestrial radiation belts were discovered and in 1960, the solar wind, both by spacecraft. Thus it became clear that our exploration and future understanding of the earth and sun's space environment would also be couched in terms of plasma physics.

After 1960, modern plasma physics would develop in two separate but converging directions. Fusion research seeks a source of energy accessible to human use that will last for a time comparable with the present age of the earth. Space research seeks useful comprehension of nature's processes on a global and, indeed, solar-system scale, in recognition of man's dependence on his environment. It is both symbolically and substantially significant that the same discipline of physics -plasma physics -defines the basic language used both in fusion research and in solar-system plasma physics. Moreover the plasma phenomena in the solar system have proven to be examples of general astrophysical processes. Not only do magnetohydrodynamics and plasma physics describe both solar system and astrophysical phenomena, but the solar system has become a laboratory in which astrophysical processes of great generality can be studied <u>in situ.</u>

2) RELATIONSHIP BETWEEN LABORATORY, SPACE, AND ASTROPHYSICAL PLASMA RESEARCH

2.1) <u>Definitions of Space and Astrophysical Plasma Physics</u>

Space and astrophysical plasma physics comprises many subjects with distinct historical origins. Solar system plasma physics includes solar and solar wind physics, planetary magnetopheric physics, ionospheric physics, and part of cosmic ray physics. Solar research stands at the interface between space physics and astrophysics. The sun's proximity makes possible measurements pertinent to the sun's interior structure (solar neutrinos, global oscillations) and of the plasma phenomena in its photosphere, chromosphere and corona that are obtainable for no other star. In astrophysics, our subject includes the generation of magnetic fields in planets, stars, and galaxies; the plasma phenomena occurring in stellar atmospheres, in the interstellar and intergalactic mediums, in neutron star magnetospheres, in active radio galaxies, in quasars; and, once again, part of cosmic ray physics. Each of these subjects depends upon, and contributes to, laboratory plasma physics. Each has traditionally been pursued independently. Only recently has there been a tendency to view them as one unified subject. For this reason, a discussion of the relationship between laboratory, space, and astrophysical plasma physics is timely.

2.2) Relationship between Laboratory and Space Plasma Physics

The Study Committee on Space Plasma Physics (NAS/1978) expressed this relationship as follows:

> "Space and laboratory experiments are complementary. They explore different ranges of dimensionless physical parameters. Space Plasma configurations usually contain a much larger number of gyroradio and Coulomb mean free paths than is achieved in laboratory plasma configurations. In the laboratory, special plasma configurations are set up intentionally, whereas space plasmas assume spontaneous forms that are recognized only as a result of many single-point measurements. Space plasmas are free of boundary effects, laboratory plasmas are not, and often suffer severely from surface contamination. Because of the differences in scale, probing a laboratory plasma disturbs it; diagnosing a space plasma usually does not. The pursuit of static equilibria is central to high-temperature laboratory plasma physics, whereas space physics is concerned with large-scale time-dependent flows . . .
>
> Certain problems are best studied in space . . . certain problems could be more conveniently addressed in the laboratory . . . Theory should make the results of either laboratory or space experiments available for the benefit of the whole field of plasma physics."

Upon the recommendation of the Study Committee, NASA took significant steps to strengthen theoretical space plasma physics. This, together with the increasing capability of space plasma instrumentation and the natural advantage of the space environment for certain types of measurements, means that the experimental diagnosis and theoretical interpretation of certain plasma processes now matches in precision the best of current laboratory practice. This is especially true in the field of wave-particle interactions, where non-Maxwellian particle distributions, and the plasma waves they create, were measured with such high resolution that theoretical instability models had to be increased in precision.

2.3) Relationship between Space and Astrophysical Plasma Research

The interplay between small and large scale processes is characteristic of space and astrophysical plasmas. Magnetohydrodynamics, or MHD, describes the large scale fluid systems in which can be identified the locations, scale sizes, and functions of the small scale plasma processes that regulate the global dynamics of such systems. In general, the MHD and associated plasma problems must be attacked simultaneously to achieve complete and self-consistent understanding.

Many of the MHD systems identified in solar system research have important analogs in astrophysics. These naturally give rise to similar plasma problems. We will illustrate these remarks by discussing two types of MHD systems, winds and magnetospheres, and two important plasma processes, reconnection and particle acceleration, that occur in them.

2.3a) Magnetohydrodynamic Winds

The outer layers of the sun are a convective heat engine producing both large and small-scale hydromagnetic motions. These motions create a dynamo magnetic field, in itself a poorly understood phenomenon. The solar magnetic field does not spread uniformly over the solar surface but is concentrated in intense, small-scale flux tubes. In addition, the turbulent motions in the outer convective layer of the sun heat the solar corona. Thus, activity at the solar surface sets the stage for the generation of the solar wind by providing a complex magnetic topology from which the heated solar corona must escape into interplanetary space. Since the coronal pressure greatly exceeds that of the interstellar medium, that part of the corona not strongly confined by the solar magnetic field expands in a flow that is subsonic near the sun and supersonic throughout interplanetary space. This solar wind carries a part of the solar magnetic field throughout the solar system; the wind speed also exceeds the Alfven speed -a characteristic speed for magnetic disturbances in a plasma. Thus the solar wind is a supersonic, super-Alfvenic, strongly ionized flow that transports plasma, energy, angular momentum, and magnetic field past all the planets of the solar system. It is finally decelerated to subsonic speeds by its interaction with interstellar matter at a distance of a few hundred astronomical units.

Expanding hydromagnetic flows, like the solar wind, are common. Plasma streams out into space from the planets' ionospheres in miniature versions of the solar wind -polar winds. <u>Einstein</u> observations of stellar coronal x-rays indicate that nearly all stars that have convective outer layers like the sun have surface magnetic activity that generates stellar winds like the sun's. The solar wind has carried off much of the sun's primordial angular momentum over the sun's lifetime, and since these late stars are observed to rotate slowly like the sun, we infer they too have winds. More massive stars, with radiative outer layers, are observed to have much stronger stellar winds driven by radiation pressure. Much of the interstellar medium is filled with hot, low density plasma from the blended winds of early stars. Naturally, the interstellar plasma's parameters are similar to the solar wind's. The interstellar plasma may also expand out of our galaxy as a wind. MHD winds that are confined by surrounding gas pressure take the form of collimated bi-polar jets, which are observed to flow away from such diverse systems as stars in the early phases of formation, the exotic compact stellar system SS-433, and from radio galaxies and quasars. Super high energy, relativistic plasma winds appear to flow away from pulsars and active galactic nuclei.

The solar wind is the only astrophysical wind accessible to in situ measurement. Since the solar wind has been as completely diagnosed as any laboratory plasma, a detailed theoretical understanding of it is possible.

2.3b) Planetary and Astrophysical Magnetospheres

The planets' magnetospheres are cavities shielded from the solar wind by their intrinsic magnetic fields. Within each magnetosphere, the magnetic field organizes the behavior of charged particles, plasma waves, and electrical currents; it traps energetic particles to form radiation belts and confines ionospheric polar wind plasma escaping into space; finally it transmits hydromagnetic stresses between the magnetosphere and atmosphere, a process which leads to aurorae.

Each planetary body in the solar system has a distinctive magnetospheric interaction with the solar wind. The earth's magnetosphere was discovered in 1958. In the past 10 years, Pioneers 10 and 11 traversed the magnetosphere of Jupiter, and Mariner 10 discovered an unexpected magnetosphere at Mercury. Two Voyagers passed through the magnetospheres of Jupiter and Saturn, and Voyager 1 is now on its way to Uranus, which telescopic observations indicate has aurorae, and therefore a magnetosphere. The Pioneer Venus Orbiter subjected the plasma environment of Venus to especially detailed examination. In 1986, an international consortium of spacecraft will study the interactions of the solar wind with Comets Halley and Giacobini-Zinner.

In astrophysics, the concept of magnetosphere has been generalized to any magnetized plasma envelope of a compact central body. Our understanding of pulsars, stellar and galactic accreting x-ray sources, and tailed radio galaxies has benefitted from our awareness of analogous magnetospheric processes in the solar system. Although the parameters of space and astrophysical plasmas can differ so much that the day-to-day problems faced by researchers in these fields are quite different in detail, the fact that both types of system present similar problems of physics gives us confidence that there exists a deeper level at which the physics of planetary and astrophysical magnetospheres is unified.

2.3c) Reconnection

Suddenly the dark polar sky is pierced by a brilliant flash of light. Within minutes, dazzling array of auroral forms stretches from horizon-to-horizon, million ampere currents surge through the earth's atmosphere and out into space, and one hundred billion watts of power are dissipated in the earth's atmosphere -a magnetospheric substorm has begun. On the sun, a burst of x-rays near a dark sunspot signals the beginning of a catastrophic disruption of the solar corona -a solar flare. Relativistic flare electrons heat the chromosphere to x-ray temperatures. A strong shock wave moves through the corona and begins a journey into interplanetary space that will carry it beyond all the planets of the solar system. The optical and x-ray luminosity starts to build up in a distant quasar. Within a day, the quasar's luminosity will exceed the total power of a thousand galaxies. A sudden plasma loss occurs in a Tokamak fusion device. These diverse phenomena seem unrelated. Nonetheless, they may share a common origin -the explosive release of stored magnetic energy by the mixed MHD and plasma process of reconnection.

Violent reconnection can lead to spectacular events such as those above, but even in its more quiescent forms, reconnection is essential in determining the behavior of MHD systems. Consider the interaction between the magnetized solar wind and the earth's magnetosphere. Reconnection between solar wind and originally closed magnetospheric field lines opens some earth field lines to interplanetary space. Energetic particles that ordinarily would not hit the earth can be guided along open field lines into the earth's polar atmosphere. Thus, reconnection changes the topology of the earth's magnetic field. More importantly, reconnection enables the solar wind to do work on the magnetosphere, to set the plasma inside in motion. The basic energetics of the magnetosphere are in large part determined by the rate of reconnection. Or consider the magnetic fields in the solar corona. It is thought that a balance is set by the creation of magnetic field by turbulent convection below the solar surface and its destruction by reconnection, in "microflares," in the corona. Only when reconnection has been temporarily inhibited can the magnetic field increase enough to produce a spectacular flare by the reconnection that ultimately occurs.

In sum, the effects of reconnection must always be considered in MHD models of space and astrophysical objects. Not only can it influence their quiescent magnetic configurations, but it can cause sudden, dynamic reconfigurations of their structure.

2.3d) Particle Acceleration and Cosmic Rays

An astonishingly large fraction of the energy in space and astrophysical plasmas is in the form of energetic particles. For example, cosmic rays comprise about 1/3 the energy density of the interstellar medium. The energetic particles themselves lead to important diagnostics of astrophysical systems. The observed cosmic ray elemental and isotopic abundances are beginning to constrain models of nucleosynthesis and galactic evolution. We can infer the magnetic fields in regions containing relativistic electrons from the synchrotron radiation emitted by such electrons, a first step in constructing a global MHD model of the system.

Analogs of cosmic acceleration processes are observed in solar system plasmas. Energetic particles have been observed from explosive reconnection events in the earth's magnetic tail, and from solar flares. The ~ 10 KeV electrons responsible for the terrestrial aurora are accelerated at 1000-5000 km altitudes above the earth in regions that carry strong magnetic field aligned currents and, contrary to MHD reasoning, generate parallel electric fields. The auroral acceleration region generates strong radio emission. Aurorae and radio emissions have also been observed at Jupiter and Saturn, and aurorae, at Uranus. Finally, collisionless shock waves, some propagating in the solar wind, and some standing ahead of the planets, are observed to accelerate particles by the same processes now thought to generate cosmic rays.

Supernova shock waves are the primary energy input to the interstellar medium. In 1977, it was proposed that shock acceleration could account for the spectrum of galactic cosmic rays, but it was unclear whether it was efficient enough to account for the high cosmic ray energy density. The parameters of the interstellar shocks are similar to those studied in the solar system. Because the interstellar shocks are older and larger, they have more time to accelerate particles to the enormous energies observed. Although it is not possible to measure the plasma structure of interstellar shocks, theories of their structure can be tested by direct measurements of solar system shocks, and by measurements of galactic cosmic-ray energy spectra and composition.

2.4) The Unifying Thread

The unifying thread linking laboratory, space and astrophysical plasma research is the set of problems of magnetohydrodynamics and plasma physics of true intellectual significance that they share. The NAS Committee on Space Plasma Physics (1978), identified six of these:

1. Magnetic field reconnection
2. The interaction of turbulence with magnetic fields
3. The behavior of large scale plasma flows and their interactions with magnetic and gravitational fields
4. The acceleration of energetic particles
5. Particle confinement and transport
6. Collisionless shocks

To the above six we add two more:

7. Beam-plasma interactions, and the generation of

electromagnetic radiation
8. Collective interactions between neutral gases and plasmas

Problem 3 is concerned with large scale plasma systems, and the others relate to microscopic plasma processes occurring in such systems.

The fact that such problems emerge from a variety of contexts is proof of their general significance, and suggests that solutions to particular problems will find further applicability in contexts we cannot imagine today. The existence of such paradigm problems provides a basis upon which a network of common interests, personal interactions, and ultimately a common discipline, is being built.

3) TEN YEARS RESEARCH ON CRITICAL PROBLEMS OF SPACE AND ASTROPHYSICAL PLASMA PHYSICS.

To communicate succinctly the flavor of space and astrophysical plasma research, we highlight here the progress over the last decade on the eight central problems defined in 2.4. Because problem 3 subsumes all our investigations of large scale plasma systems, discussing it first puts all the other problems in context.

<u>Problem #3:</u> "The behavior of large scale plasma flows . . ."

<u>Planetary Magnetospheres.</u> Mariner 10 discovered a small highly active magnetoshpere at Mercury, that is energized by the intense solar wind near the sun. The Pioneer-Venus Mission has provided a large volume of information about the interaction of the solar wind with the Venusian ionosphere. The Pioneer 10 and 11 missions established the enormous scale and variability of Jupiter's magnetopshere, and Pioneer 11 made the first traversal of Saturn's magnetic field. The Voyager 1 and 2 missions established that Jupiter's rotation powers a radial outflow of heavy ion plasma injected by volcanic activity at the satellite Io. The Voyagers found that Saturn's magnetic dipole and spin axes are aligned, a fact which challenges current theories of planetary magnetic field generation. Saturn's magnetosphere has an important interaction with the dense atmosphere of the satellite Titan.

<u>Magnetohydrodynamic Properties of the Earth's Magnetic Tail.</u> The MHD flows in the earth's magnetic tail proved to be highly intermittent and variable, with intense flows often but not always correlated with substorm activity observed on the ground. Large scale convection cells were detected in the earth's magnetic tail. The field aligned current systems mapped near the earth and in the magnetic tail proved to be approximately consistent. Intense flows, and MHD and plasma turbulence, were associated with the field aligned currents in the tail; the field aligned currents corresponded to the auroral acceleration region near the earth.

Magnetohydrodynamic Structures near the Sun and in the solar wind.

Measurements made on Skylab revealed the important difference between open and closed magnetic field regions near the sun. Open magnetic regions —solar coronal holes —generate fast streams in solar wind. Closed magnetic regions —solar coronal loops —are the regions in which solar flares occur.

The European Helios mission extended measurements of the solar wind within the orbit of Mercury, and the Pioneer 11 mission extended these measurements past the orbits of Neptune and Pluto. The solar wind magnetic field proved to reverse direction across a time variable, warped neutral sheet, in accordance with simple stellar wind models. Strong, corotating shocks were found in the distant solar wind, changing our picture of how galactic cosmic rays are modulated by the solar wind.

Magnetospheres of Neutron Stars.

Our understanding of the two types of neutron star magnetospheres —those of pulsars and of accretion x-ray sources —was clarified. Phase resolved spectroscopy identified an x-ray line at the electron cyclotron frequency in an x-ray source, which proved that neutron stars can have superstrong magnetic fields, of order 10^{12} Gauss —a fundamental hypothesis of pulsar and x-ray source theories. This strengthened the picture that pulsars are rapidly rotating magnetized neutron stars. Detection of pulsed γ-ray lines from the Crab and Vela pulsars proved that they generate superhigh energy particles, and qualitatively supported the theoretical suggestion that electron-positron pair plasmas are created in pulsar magnetospheres. Simple MHD theories of radial transport in accretion disks proved inadequate, and it was suggested that processes similar to those in the solar corona may occur in accretion disk coronae.

Magnetohydrodynamic Jets.

Active galactic nuclei frequently produce pairs of high speed jets that propagate in anti-parallel directions through the surrounding galaxy and out into the intergalactic medium, where they create the two component radio emission characteristic of radio galaxies. VLBI measurements proved that the jets, which were theoretically anticipated, are accurately aligned on the light year scale of the nuclei and on the $\sim 10^{5-6}$ light year scale of the double radio components. Some jets appear to expand faster than the speed of light, a kinematic effect which indicates that the flow speed can be relativistic.

Recently, we have learned that similar jets are associated with galactic objects, such as the energetic system SS-433, or stars in the early stages of formation.

Black Hole Electrodynamics.

A new theory endows rotating black holes with electric and magnetic fields under appropriate circumstances, and suggests that the environment of black holes involves much more pulsar-like plasma physics than originally thought.

Problem #1:   Reconnection

Two achievements, one experimental and one theoretical, stood out in the past decade. Bursts of MeV particles were detected and associated with rapid plasma flows in the earth's magnetic tail. The electric-fields corresponding to the measured flow speeds could account for the observed particle energies. These results showed that inductive electromotive forces are important to magnetospheric dynamics, and suggested that tail reconnection is unsteady, and perhaps explosive. Rigorous analytical theory and numerical simulations established that a slow shock model proposed in 1964 is the correct description of reconnection in the MHD limit.

Theoretical understanding of the collisionless tearing limit of reconnection was consolidated, and an explosive tearing instability was proposed analytically and simulated numerically. A laboratory experiment diagnosed with high precision the turbulent transport processes occurring in a strong guide field plasma regime appropriate, perhaps, to solar flare conditions.

Problem #2:   Interaction of Turbulence with Magnetic Fields

The *Einstein* discovery that most stars exhibit solar-like surface activity, together with the fact that solar activity is determined by the interaction of solar magnetic fields with the ambient plasma, proved that plasma processes are central to the physics of stellar chromospheres and coronae. The observed correlation of stellar activity with stellar rotation will constrain theories of dynamo generation of stellar magnetic fields and the dissipation of these fields as they emerge through the stellar surface.

The solar surface magnetic field has proven to be concentrated in thin layers of strength $\sim 10^3$ G separated by larger regions of strength $< 10G$. There is essentially no evidence for a smooth uniform field at the solar surface. These remarkable observations challenge theories of turbulent magnetohydrodynamic convection.

A coherent program of active and passive radar experiments, chemical releases, rocket measurements, analytic theory, and numerical simulations led to clear understanding of the so-called "Spread-F Bubbles" in the equatorial ionosphere. This work is the most complete analysis of the nonlinear development of the Rayleigh-Taylor instability in plasma physics.

Problem #4:   Acceleration of Energetic Particles

Measurements of the Be^{10}/Be^9 ratio in cosmic rays showed that their age -the time since they had been accelerated -is about 10 million years. Abundance and isotope measurements showed that most cosmic rays are accelerated, not out of enriched supernova material, but out of an interstellar medium whose composition differs only slightly from solar, presumably because of chemical evolution since the birth of the sun. The discovery that much of the volume of the interstellar medium is in a low density plasma phase meant that supernova shocks could propagate much further than originally thought. This led to the suggestion that supernova shocks Fermi-accelerate the cosmic rays out of the interstellar medium. The Fermi-acceleration particle energy spectrum was shown to be consistent with cosmic ray observations. Detailed measurements of the energetic particle distributions and plasma turbulence associated with interplanetary shocks and planetary bow shocks began to be used to test self-consistent shock acceleration theories.

Measurements of field aligned currents, electron and ion densities, electrostatic and electromagnetic waves, and energetic ion and electron distribution functions, in the auroral acceleration region were systematically assembled and combined with measurements of auroral light, ionization, motions, and structures, thereby setting the stage for comprehensive understanding of auroral acceleration in the next decade.

It was found that processes within Jupiter's magnetosphere accelerate nearly all the ~ 10 MeV electrons found in the heliosphere.

Impulsive particle acceleration events, probably associated with reconnection, were found to accompany rapid flow reconfigurations of the earth's magnetic tail.

The discovery of an "anomalous" component of low energy cosmic rays that has unusual abundances of Oxygen, Nitrogen, Carbon, and Helium nuclei, and is modulated by the solar cycle like other cosmic rays, suggested the existence of a second source of cosmic rays outside, or in the outer reaches, of the heliosphere. Air shower observations suggested that 10^{14}-10^{16} eV cosmic rays are richer in heavier elements than those at lower energies, and that cosmic rays with 10^{17}-^{19} eV energies may be anisotropic, unlike cosmic rays of galactic origin.

Problem #5: Particle confinement and transport

Detailed models of the magnetic mirror confinement, radial diffusion, and turbulent pitch angle scattering of energetic ions and electrons were created and successfully tested by observations in the magnetospheres of earth and Jupiter.

A clear qualitative understanding of electron heat transport in the solar wind was achieved by means of systematic measurements of superthermal electrons, and analytic identification of the instabilities that can limit heat conduction in various solar wind conditions.

Quantitative studies of the conduction of heat by electrons between the solar corona and chromosphere, which promise to make interpretation of chomospheric line emissions more secure, were carried out.

Problem #6: Collisionless Shocks

The strong dependence of the earth's bowshock structure upon the parameters of the upstream solar wind was demonstrated by synopsis of individual detailed case studies, by the ISEE and other spacecraft.

The clear understanding of the discontinuous change in shock structure at the so-called first critical Mach number, achieved by a combination of spacecraft measurements, analytic theory, and numerical simulation, is the major single accomplishment in collisionless shock physics in the past ten years.

Quasi-parallel shocks, which propagate nearly parallel to the upstream magnetic field direction, proved to have extensive regions upstream which contain shock accelerated particles and large amplitude MHD turbulence, a property required by Fermi-acceleration theories.

Theories of the propagation, scattering, and energization of the upstream particles were created and tested by data from interplanetary shocks and planetary bow shocks. These results are providing a solid basis for theories of the acceleration of cosmic rays by supernova shocks.

Problem #7: Beam-plasma Interactions, and the Generation of radio emissions

The Jovian radio emissions were measured directly from space from the first time. They extended a factor 10 lower in frequency than was possible to measure from the ground, and their characteristic frequency-time structure illuminated how they were modulated by the interaction between Jupiter's magnetosphere and its satellite Io. Saturnian radio emissions were discovered, and shown by the Voyagers to have a modulation entirely different from the Jupiter-Io system's. Detailed studies of the intensity, frequency spectrum, and polarizations of the analogous Auroral Kilometric Radiation at earth, and their correlations with other auroral phenomena, motivated a generous development of theory. It now appears that the auroral electron beam may linearly excite the Auroral Kilometric Radiation directly, in which case the aurora would be a giant gyrotron in space.

The magnetospheres of Earth, Jupiter, and Saturn confine a diffuse continuum of radio noise, whose lower cutoff frequency is the most sensitive indicator of the plasma number density. Detection of Jovian continuum radiation proved that Jupiter's magnetic tail is 5 AU long and is the best vacuum so far encountered by man, with a density $\sim 10^{-5}$ particles -per cm^3. Detailed studies of the fine structure of continuum radiation illuminated how it is generated near the terrestrial plasmapause, and by the interaction of Saturn's magnetosphere with its satellites.

A nonlinear theory of Type III radio emissions in which streaming energetic electrons create electron plasma oscillations which then couple to radio waves in a background of plasma turbulence, was created and supported by observation in the solar wind.

Detailed studies of pulsar radio emissions provided the most complete diagnosis to date of the microstructure of a highly relativistic plasma. Some pulsars' radio emissions were shown to consist entirely of discrete bursts, suggesting the action of nonlinear plasma processes.

Problem #8: Interactions between plasmas and neutral gases

It was suggested in 1954 that a neutral gas would be rapidly ionized when its velocity through a plasma exceeds the so-called critical ionization velocity, given by equating an atom's kinetic energy to its ionization energy. This effect is now thought important to rotating plasma devices such as centrifuges, the formation of minor bodies in the early solar system, comets, satellite-magnetosphere interactions, neutral gas releases in space, and, to the interaction of the Space Shuttle with the plasma through which it moves. Recent laboratory experiments, together with theory, showed that the rapid ionization results from collective electron heating by lower hybrid plasma waves plus classical ionization.

An extensive series of chemical releases was used to diagnose the plasma flows in the magnetosphere and in the auroral acceleration region.

4) SPACE AND ASTROPHYSICAL PLASMA PHYSICS IN THE NEXT TEN YEARS

As it is more difficult to predict the future than to review the past, we restrict ourselves to a few general assessments of the state of our subject after the next ten years.

The exploration of Solar System Plasmas will have been nearly completed. The International Solar Polar Mission will study the solar wind, and its effects on cosmic rays, in three dimensions for the first time. A Pioneer or Voyager spacecraft might leave the heliosphere by the end of the decade, and thus detect interstellar matter and galactic cosmic rays directly. The first in situ measurements of comets and of Uranus' magnetosphere will occur. The Galileo mission will diagnose Jovian magnetospheric plasma as completely as any space plasma to date. In situ measurements of solar plasmas will be the primary unfinished task.

The plasma environment of the earth will have been subjected to controlled study, and, perhaps, to a measure of control, through the systematic use of active experiments, and by synoptic measurements provided by the OPEN spacecraft project.

High resolution optical and radio observations will have provided essential information defining quantitative models of solar surface magnetic fields and dynamics, solar flares, and coronal heating, thereby creating the basis for general understanding of stellar activity.

The growing ability to make series of detailed high resolution observations in many wavelength bands will render many astrophysical objects increasingly subject to theoretical models that explicitly take plasma processes into account.

Understanding of many space plasma processes will be sufficiently quantitative to make them reliable components of models of large scale space and astrophysical systems. In addition to radiation belt dynamics, the list of generally understood space plasma processes may include auroral acceleration, reconnection, collisionless shocks, and neutral-gas plasma interactions.

The first generation of large scale numerical models of space and astrophysical systems will have been completed. Foreseeable advances in computing technology will lead to their creation virtually simultaneously through space physics and astrophysics. Such models will probably have made plasma physics central to the observation of many astronomical observations, and have motivated new and different kinds of observations. One member of this class of models will have been tested by direct in situ measurements of the terrestrial magnetosphere's global dynamics provided by the OPEN spacecraft project.

5) MOTIVATIONS FOR RESEARCH ON SPACE AND ASTROPHYSICAL PLASMAS

Research in this century has revealed a chain of interactions, largely plasma physical, that connects activity at the surface of the sun to the solar wind, and then on to the magnetosphere and atmosphere. The most spectacular manifestation of this solar-terrestrial interaction-chain is the magnetic storm. The first evidence that a large solar flare might occur is the appearance of a complex sunspot group in the sun's photosphere. Prompt electromagnetic radiation arrives at earth a few minutes after the energy in the coronal magnetic fields associated with the sunspot group is explosively released. Energetic solar-flare protons are guided by the solar wind and magnetospheric magnetic field into the polar atmosphere soon afterward. The enhanced ionospheric plasma that they produce attenuates the radio noise received from cosmic radio sources. A day or so later, a shock wave passes over earth, enveloping it in dense, hot solar-flare plasma that compresses the magnetosphere. Substorms increase in frequency and strength, and inject hot plasma into the earth's inner magnetosphere to form a "ring current," which creates the geomagnetic field depression and activity that first motivated the name magnetic storm. The aurorae intensify and move to unusually low latitudes, creating a dense highly disturbed ionospheric plasma that interferes with radio communication and on occasion blacks it out altogether. Intense wind systems, sometimes of a world wide scale, are generated in the upper atmosphere.

The Committee on Solar-Terrestrial Research, in its report, <u>Solar-Terrestrial Research for the 1980's</u>, (NAS/1981) has identified four areas where research on the solar-terrestrial interaction chain can clarify important impacts on society and technology:

1. Predictions about the space environment
2. Effects on stratospheric ozone, which shields the life at the earth's surface from the harmful effects of solar ultraviolet radiation
3. Effects on ionospheric physics and radio propagation
4. Elucidating a potential connection between solar variability, weather, and climate

Although items (2) and (4), above, have significant long-term implications, item (1) is the most directly and immediately important. Many practical systems, both civilian and defense, and all our manned space endeavors, must operate in the highly variable and potentially hostile plasma environment of the earth and solar system. Plasma processes in this environment also influence and even disrupt important ground-based systems over local and regional scales. Ground based HF communication systems in the earth's polar regions can be blacked out by magnetic storms. Entire spacecraft have been electronically disabled by violent electrical discharges that occur when hot "ring-current" plasma envelops the spacecraft. The risk of such disasters can be reduced only by continuing attention to the effects of the plasma environment on space-craft systems. It is clear that to work in the space environment, we must understand it.

The Astronomy Survey Committee (NAS/1982) has characterized the motivations for astronomical research as follows:

> "Astronomy . . . is sustained by two of the most fundamental traits of human nature: the need to explore and the need to understand. Through the interplay of discovery, the aim of exploration, and analysis, the key to understanding, answers to questions about the universe have been sought since the earliest times, for astronomy is the oldest of the sciences. Yet it has never been, since its beginnings, more vigorous or exciting than it is today."

Our own branch of astronomy, astrophysical plasma physics, is driven by the need to understand the unusual plasmas surrounding some of the most exotic objects brought to light by recent astronomical research. Our imaginations are challenged, and we are forced to extend plasma physics to comprehend them.

Mature scientific disciplines are characterized by deep philosophical motivations, a unified body of powerful theoretical and experimental techniques, and a wide range of applications. It is our conviction that when space and astrophysical plasma physics are integrated with one another, and with laboratory and fusion research, plasma physics will have become mature. When a scientific discipline matures, technological innovation soon follows. Plasma physics is only beginning to have its impact.

REFERENCES

Colgate, S. A., et al., Space Plasma Physics: The Study of Solar System Plasmas, V. I, Space Science Board, National Academy of Sciences, Washington, D.C., 1978.

Field, G. B., et al., Astronomy and Astrophysics for the 1980's, V. I, Astronomy Survey Committee, National Academy of Sciences, Washington D.C., 1982

Friedman, H., D. S. Intriligator, et al., Solar Terrestrial Research for the 1980's, Committee on Solar-Terrestrial Research, Geophysics Research Board, National Academy of Sciences, Washington, D.C., 1981

Kennel, C. F., et al., Solar System Space Physics in the 1980's: a Research Strategy, Committee on Solar and Space Physics, Space Science Board, National Academy of Sciences, Washington, D.C., 1980

CONTRIBUTIONS FROM THE U.S.S.R.

SPATIO-TEMPORAL FEATURES OF THE DEVELOPMENT OF MICROWAVE EMISSION OF ACTIVE REGIONS AND FLARES

G.Ya.Smolkov
SibIZMIR, Irkutsk, U.S.S.R.

At SibIZMIR, a stepwise commissioning of the Siberian Solar Radio Telescope is being under way (Smolkov, 1982a; Smolkov et al., 1983a). This special-purpose instrument is designed to: 1) survey and monitor during daylight hours the state of solar activity (SA) with high two-dimensional resolution on a real time basis in a wave band permitting maximum possible detectability of active regions (AR's) and flares (F's); 2) study structures and development of AR's by day and the time of their observation on the solar disk; 3) study F's; 4) study the three-dimensional pattern of development of AR's and F's jointly with the Sayan Mountain and Baikal Astrophysical Observatories of SibIZMIR (Smolkov, 1982b), and 5) to make a synoptic study of SA during one or several solar rotations in the interests of solving challenging problems of Solar Physics and STP.

The SSRT operational wavelength is about 5.2 cm. The spatial resolving power has been chosen based on typical sizes of inhomogeneities of local sources (LS) of radio emission of AR's, as determined from eclipse observations (Gel'freikh, 1969; Smolkov et al., 1982), and to permit monitoring the state of SA, with simultaneous discrimination of salient features of AR's and F's. It is chosen to be 20"x20" at zenith during summer solstice.

The SSRT antenna system is a 256-element cross interferometer, consisting of two 128-element linear equidistant arrays pointing to the four cardinal points. Each array is of $1.2 \cdot 10^4 \lambda$ length. The mirrors are of 2.5 m diameter, spaced 94λ apart, as determined by the solar radio diameter. All the mirrors are in synchronism to track the Sun from sunrise to sunset, thus enabling the use of interference maxima of the multilobed beam of the cross, oriented within direction diagram (DD) of a single element.

Parallel synthesis of solar radio images is accomplished by using a 180-channel receiving system via a parallel-series composition of signals from the antennas, providing an equality of electrical lengths of waveguide lines and making it possible to use a frequency band, appropriate for solar coverage with a fun of lobes (vertical scanning), and owing to corotation of the DD with the Earth. During the daytime it is possible to record hundreds of distributions of circular polarization and intensity in AR's and hence to carefully investigate their development. It is also possible to study the development of flares, whose lifetimes exceed the time required for the Sun to transit the DD interference maxima. The dynamic range is 10^4 and the time constants are 1.4, 0.6 and 0.2 s.

Even one-dimensional radio brightness distributions obtained during phase adjustment of the operating model, an 8-element interferometer, and a stepwise commissioning of the W-beam (the resolution thus increasing from 4'.5 to 34") enabled us to refine some of our earlier understanding and to gain new insights into spatio-temporal features of development of microwave emission from AR's and F's (Smolkov et al., 1983b,c). Thus, the LS intensity may temporarily and substantially either decrease (burst in "absorption") or increase without or with reverting without any notable polarization changes, seemingly reflecting changes in physical conditions higher in AR's. Once a sunspot group is born in an AR and starts to develop, a polarized emission comes from it within half an hour, which corresponds to a thickness of the transition zone of 1000 km, with a constant rise velocity of the magnetic field with height. The LS polarization is reversed due to a change in conditions of either propagation or generation of the emission. In the former case the inversion proceeds stepwise in course of several hours with nearly the same LS intensity (even if there are gross changes in sunspot area) against fluctuations of the degree of polarization. The latter type of inversion applies to significant, simultaneous changes in LS intensity and area of related spots for the same time.

Spatial development of F's in AR's becomes traceable once LS features get separated above leading and following parts of a sunspot group. Flares, evolving above the interspot zone (at the tops of a loop-like configuration of the magnetic field), are manifest in the microwave emission either only above one of the parts or sequentially above different parts or, finally, above the entire spot group. The LS emission intensity first decreases above the part of a spot group, producing a flare thereafter. The degree of polarization temporarily intensifies, with a possible reversal of polarization.

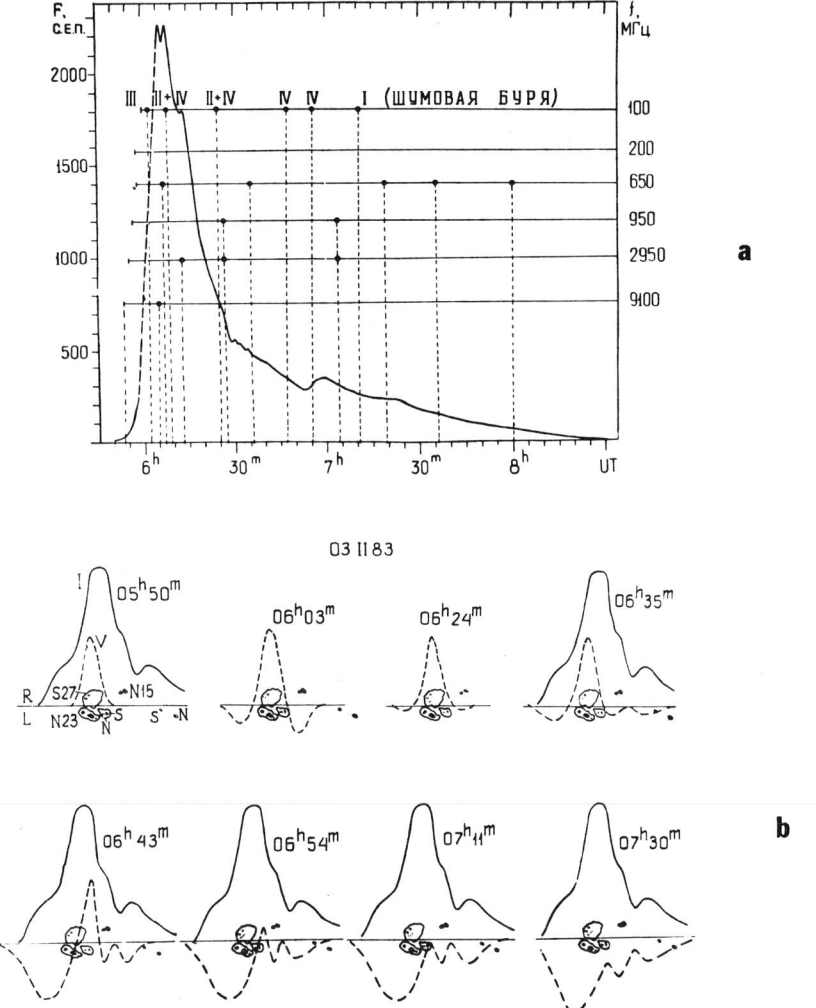

Fig. 1. Development of the burst on 3.02.83.

a) Development of the burst of an integrated solar flux at 5.2 cm wavelength and a schematic for its recordings at other wavelengths: the onset and maxima are labeled by vertical marks and full circles respectively and the bursts in metric wave band are denoted by their indices.

b) The variation of structure and sign of the LS circular polarization (34" resolution). A solid line shows the initial intensity distribution.

Prior to a flare, above a composite complex of activity at solar disk center, on 3.02.83, for example, a polarized emission from an LS was recorded above the greatest spots (Figure 1). Before the burst maximum, together with a slight increase of polarized flux of this spot component of the LS, oppositely-polarized features appeared on both sides of it. The west source is associated with individual spots of N-polarity but there is no photospheric cause of the east one. After some decrease from 06.35 the polarization was increasing intensively in these new features of the LS. The parent right hand polarized source is "devored" and the size as well as the degree of polarization of the east feature of the LS increase. The change of the structure and polarization sign of the LS is completed at 07.11. From that time onward there occurred three LS features, with a polarization sense opposite the original one. They corresponded to three anticipated loop-like configurations of the magnetic field of the activity complex. Restructuring and reversal of the polarization of the LS were provoked by the flare. A change of sign of the LS polarization due to central heliographic meridian passage by the spot group was observed from 4 to 5.02.83.

The burst activity correlation between the mutually separated LS's is most pronounced in a polarized flux. Both prior to and following the explosive phase of F's the intensity and degree of polarization of the bursts undergo simultaneous fluctuations with an increasing period.

References

Gel'freikh, G.B.: 1969, Vestnik AN SSSR, 4, p.46.
Smolkov, G.Ya.: 1982a, In: "Radio astronomical equipment, antennas and methods", Yerevan, p.230.
Smolkov, G.Ya.: 1982b, In: "Sun and Planetary Systems", W. Fricke and G.Teleki (eds.), D.Reidel Publ. Co., Dordrecht, p.123.
Smolkov, G.Ya., Dutov, A.S., Zandanov, V.G., et al.: 1977, In: "Issled. po geomagn. aeron. i fiz. Solntsa", Moscow, Nauka, 42, p.76; 1982, 62, p.181.
Smolkov, G.Ya., Treskov, T.A., Krissinel, G.B., and Potapov, N.N.: 1983a, In "Issled. po geomgn. aeron. i fiz. Solntsa", Moscow, Nauka, 64, p.130.
Smolkov, G.Ya., Treskov, T.A., and Nefedyev, V.P.: 1983b, In: Problems of solar activity, Alma-Ata, p.70.
Smolkov, G.Ya., Treskov, T.A., and Potapov, N.N.: 1983c, In: "Issled. po geomagn. aeron. i fiz. Solntsa, Moscow, Nauka, 65, p.204.

CONCERNING THE DYNAMICS OF ENERGETIC PROTONS IN CORONAL MAGNETIC LOOPS: DISPERSION EFFECTS OF ALFVEN WAVES

V.A. Mazur and A.V. Stepanov
SibIZMIR, Irkutsk, USSR

It is shown that the existence of plasma density inhomogeneities (ducts) elongated along the magnetic field in coronal loops, and of Alfven wave dispersion, associated with the taking into account of gyrotropy $U \equiv \omega/\omega_i \ll 1$ (Leonovich et al., 1983), leads to the possibility of a quasi-longitudinal $k_\perp < \sqrt{U} \, k_\parallel$ propagation (wave guiding) of Alfven waves. Here ω is the frequency of Alfven waves, ω_i is the proton gyrofrequency, and k is the wave number. It is found that with the parameter $\xi = \omega^2 R/\omega_i A > 1$, where R is the inhomogeneity scale of a loop across the magnetic field, and A is the Alfven wave velocity, refraction of Alfven waves does not lead, as contrasted to Wentzel's inference (1976), to the waves going out of the regime of quasi-longitudinal propagation. As the result, the amplification of Alfven waves in solar coronal loops can be important. A study is made of the cyclotron instability of Alfven waves under solar coronal conditions.

A typical value of the parameter ξ in flaring loops is $\xi \gtrsim 10^2$. Hence isotropization of the protons accelerated in loops, $\gtrsim 10$ MeV, on Alfven waves, causes the protons to escape into the loss cone and to decay in the solar chromosphere. This explains the deficit of energetic protons, observable in some flares, in the interplanetary medium, as compared to the number of particles expected to come from the gamma-ray emission. Pitch-angle diffusion of protons on Alfven waves determines the duration of a pulsation train (of several minutes). Tail ($\sim 1 \, R_0$) coronal magnetic loops without ducts are good candidates for storage of energetic protons for a long time ($\gtrsim 10$ hours).

References

Leonovich, A.S., Mazur, V.A., and Senatorov, V.N.: 1983, ZhETF, <u>85</u>, No. 7.
Wentzel, D.G.: 1976, Astroph. J., <u>208</u>, 595.

SUBJECT INDEX

Accretion disks   453
    coronae   458
    dynamos   456
    electrodynamics   88, 453
    magnetic reconnection in   453, 458, 477
    magnetic viscosity   457
    oscillations   477
    thick   88
    thin   454
    winds   459
Alfvén-ion cyclotron instability   203
Alfvén waves
    and cosmic ray propagation   341, 355
    and energetic protons in coronal loops   559
    heating by   97, 102, 365, 381, 391
    kinetic   336, 381, 383
    particle acceleration in jets   413
    shear   100, 102, 333, 334, 381, 391
    soliton collapse   355
    surface   99, 381, 384, 408
        and bending waves   493
Anomalous transport   315, 329
Anomalous viscosity   270
Auroral region   125, 332, 541, 543
"Ballerina effect"   494
Banded electrostatic ion waves   309
Beam cyclotron instability   320
Beam-plasma instabilities - dispersion surfaces   309
Bending waves   491
Birkeland currents   184
Black holes
    electrodynamics   88, 461
        equivalent circuit   464, 471
    resistivity   462, 468
Buneman instability   319
Cataclysmic variables   480, 481
Coalescence instability -- see Tearing instabilities
Coma cluster of galaxies   347, 353
Constant ψ approximation -- see Tearing instabilities
Coronal arcades   483
    stability   483, 486
Coronal heating   96
    accretion disk   458
    Alfvén waves   97
    coronal holes   111
    electric currents   97, 105
    electrodynamic coupling efficiency   140
    equivalent circuit   102, 109, 139
    flare stars   245

reconnection 106, 190, 297
shear Alfvén waves 102, 365
sound waves 96
surface waves 99
switch-on shocks 99
wave reflection and refraction 98
Coronal holes 497, 501
Coronal loops (see also Magnetic flux ropes)
flare stars 256, 258
heating 139, 142, 143
tearing instability 77, 79
Cosmic rays 542
acceleration 343, 471, 546
adiabatic deceleration 348
in clusters of galaxies 347, 353
isotropy 339, 346, 361
mirroring 349, 353
resonance gap 348, 353
self-confinement 341, 355
solar 350, 354
Cross field convection 475
Current-driven instabilities
laboratory experiments 47
threshold 58
Current sheets
anomalous transport in 315
bending waves in 491
geomagnetic tail 303
solar flares 233
de Hoffmann-Teller frame 15
de Laval nozzle 87
Diffusion region -- see Magnetic reconnection
Double Inverse Pinch Device (DIPD) 148
Double layers 113
Bohm criteria 114, 123
boundary conditions 115
Buneman instability 115
equivalent circuit 117
in solar flares 196
ion energization in stochastic DL 125
ion-acoustic 116, 124
laboratory experiments 51, 59, 159, 166
Langmuir condition 114, 123
multiple 120, 123
Dwarf novae 480
Electrode sheath drop 154
Electromagnetic coupling 329
shear Alfvén waves and 333, 334
Electron cyclotron drift instability 320
Fermi acceleration
first-order 343

    in jets 91, 407, 413
Filamentation instability 263
Finite Larmor radius effects 381, 385, 390
Flare stars 245
Flute instability 89, 441
Flux ropes -- see Magnetic flux ropes
Flux transfer 154
Flux transfer events 20, 29-33, 181
    Jupiter 32
    Mercury 32
Force-free fields 160, 221, 332, 448, 484
Frozen-in-field condition 315
Galactic disks - warps 492
Galactic magnetic field 361
Galactic nuclei 455, 471
    ion-pressure supported torus model 88
Galactic wind 345
Ideal magnetic instabilities 234
Impulsive flux transfer event (IFTE) 157, 165
International Sun-Earth Explorer (ISEE) 27
Ion acoustic instability 319
Jets 85, 398, 545
    cocoon 90, 434
    collimation, confinement 87, 94, 398, 411, 425, 433, 439, 442
                              497, 501
    Compton boosted radiation 131
    eletromagnetic tunneling 136
    electron-positron 468
    flow velocities 94
    formation 87, 459
    knot structures 90, 411, 443, 501
    Mach disks 90
    M87 439
    numerical simulations 401
    origin of magnetic field 94, 433
    particle acceleration 91, 413
    physical paramaters - extragalactic 396, 411
    reflection mode 89, 413
    shocks 91, 403
    stability 89, 402, 425, 441
    superluminal motion 86
    sweeping-pinch mechanism 287
    wind solutions 499
Jupiter 32, 217, 305, 333, 543
Kelvin-Helmholtz instability 375
    coronal heating 105, 111, 368
    jets 89, 402
    surface Alfvén waves 384
Kink instability 72, 204, 384, 447
    helical 43, 89, 403
    internal 73, 279, 448

Laboratory experiments 47, 147, 447
    wall effects 152
Line-tying 483, 486
Lower-hybrid drift instability 170, 209, 321, 328
Lundquist number 64, 237
Mach disks -- see Jets
Magnetic buoyancy 361, 455
Magnetic flux ropes 33, 43
    compound 36
    in jets 434, 437
    magnetopause 33
    stability 35, 40, 43
    sun 37, 295
    Venus 33, 43
Magnetic islands 69, 157, 274, 300, 303, 478
Magnetic merging -- see Magnetic reconnection
Megnetic reconnection 5, 25, 234, 544
    computer simulations 167, 200, 211, 266, 273, 278, 477
        numerical dissipation 183
            boundary conditions 183, 215
    definition 6, 22
    diffusion region 11, 18, 22, 315
    driven 175, 211, 235, 332
    electron inertial effects 18, 19
    explosive 21
    in accretion disks 453, 455, 471
    in jets 91, 405
    in planetary magnetospheres 167
    laboratory experiments 47, 147
    microinstabilities 321
    nonsteady 19
    observational evidence 185
    particle acceleration 13
    patchy 29
    Petschek model 11, 64, 236, 243
    reconnection rate 7, 8, 10, 12, 22, 58, 235, 244
    sequential 282
    supercritical 236
    Sweet-Parker model 8, 64, 199, 235
    viscosity 10, 12
Magnetic Reynolds number 234
Magnetic substorms 168, 242, 305, 332, 513
Magnetopause
    reconnection 12, 19, 26
Magnetosphere 26, 541, 543
    black hole 461, 471
    cross-tail current sheet 303
    fast rotating 217
    ring current 447
Marginal stability analysis 331
Mercury 32

SUBJECT INDEX

Modified two stream instability   209, 320, 505, 510, 517
Molecular clouds   361
Moreton wave   253
Nonlinear Schrodinger equation   355, 359
Ohm's law (generalized)   18
Panel on Space and Astrophysical Plasmas Report   537
Parker instability   361
Particle acceleration
    by lower-hybrid waves   451, 495, 513
    in black hole magnetospheres   463, 471
    in current sheets   13, 513
    in an electric field   191
    in jets   91, 413
    in shock waves   343
    ions, in stochastic double layers   125
    laboratory experiments   52, 451
Petschek model -- see Magnetic reconnection
Pinch instability   43, 89, 403, 433, 441
Plasma radiation
    from jets   134
    laboratory experiments   52, 58
Plasmoid   168, 171, 176
Pulsars   463
Radiative instability   263
Reconnection -- see Magnetic reconnection
Reflection mode -- see Jets
Resistive tearing instabilities -- see Tearing instabilities
Resonance broadening   353, 357
Resonance gap   348, 353
Return current   196, 519, 522
Rotational discontinuity   13
RS CVn binaries   281
Runaway electrons   191
Saturn   543
    rings   492
Scatter-free events   347
Separator   6
Separatrix   6
Shear Alfvén waves -- see Alfvén waves
Shock waves
    fast   241, 331
    in jets   91, 407, 499
    particle acceleration   343, 353, 407, 543
    Petschek   157, 165, 174, 240
    slow-mode   11, 14, 148, 157, 211, 236
    switch-on   97
Siberian Solar Radio Telescope   555
Solar active region magnetic fields - evolution   487, 555
Solar flares
    beam-return current systems   196, 519, 521, 525
    current sheets   233

    dissipative thermal model 509
    emerging flux model 238
    filament eruption 295, 297, 483
    Gold and Hoyle model 337, 340
    interplanetary particles 347, 350, 354, 560
    particle acceleration 186, 191, 197, 505, 510
    phases of energy release 293
    pre-flare energy build-up 487
    pulsations 201, 295, 297, 299
    reconnection in 21, 185, 197, 279
    two-ribbon 211, 239, 483
    Type II, III bursts 505, 508, 523
Solar wind 540
    high-speed streams 497
    ion heating 371
    sector structure 491
    velocity shear instabilities 371
Solitons 99, 355, 358, 522
Starspots 248, 281
Stellar flares 252
    pulsations 257, 260
    rise time 252, 260, 261
    RS CVn binaries 281
Summary of conference 529
Surface waves 99
    damping 101
    in jets 408
Sweeping-pinch mechanism 284, 287
Sweet-Parker model -- see Magnetic reconnection
Swith-on shocks 97
Tearing instabilities 61, 94, 299
    asymmetry of current sheet 80, 278
    coalescence instability 75, 81, 197, 242
    constant ψ approximation 70, 273, 278
    cylindrical equilibrium 71, 278
    double tearing mode 75, 81
    gravity 79
    kink-tearing mode 75
    linear 65, 265, 368
    line tying 81
    long wavelength 75, 80
    magnetic energy dissipation 73
    nonlinear 73, 273
    3-D 75
TS device 149
Uranus 543
Velocity shear instabilities 371
Venus 33, 43
X-line (X-point) 7, 234